Elementary Differential Equations

ELEMENTARY DIFFERENTIAL EQUATIONS

William E. Boyce

and

Richard C. DiPrima

Rensselaer Polytechnic Institute

Third Edition

JOHN WILEY & SONS
New York • Chichester • Brisbane • Toronto • Singapore

Library of Congress Cataloging in Publication Data:

Boyce, William E.
  Elementary differential equations.

  Includes index.
  1. Differential equations.   I.   DiPrima, Richard C.,
joint author.   II.   Title.
QA371.B77   1976   515'.35   75-35565
ISBN 0-471-09339-4

Printed in the United States of America

20 19 18 17 16 15 14 13

To My Mother, Ethel DiPrima,
and in Loving Memory
of My Father, Clyde DiPrima

Richard C. DiPrima

To My Mother, Marie S. Boyce,
and in Loving Memory
of My Father, Edward G. Boyce

William E. Boyce

Preface

A course in elementary differential equations is an excellent vehicle through which to convey to the student a feeling for the interrelation between pure mathematics and the physical sciences or engineering. Before the engineer or scientist can proceed confidently to use differential equations, he or she must have a mastery of the techniques of solution and at least a rudimentary knowledge of the underlying theory. Conversely, the student of pure mathematics often benefits greatly by a knowledge of some of the ways in which the necessity for solving specific problems has stimulated work of a more abstract nature.

We wrote this book from the intermediate viewpoint of the applied mathematician, whose interest in differential equations may be both highly theoretical and intensely practical. We have therefore sought to combine a sound and accurate (but not abstract) exposition of the elementary theory of differential equations with considerable material on methods of solution that have proved useful in a wide variety of applications. We have given principal attention to those methods that are capable of broad application and that can be extended to problems beyond the range of our book. We emphasize that these methods have a systematic and orderly structure, and are not merely a miscellaneous collection of unrelated mathematical tricks. At some point, however, the student should realize that many problems cannot be satisfactorily solved by elementary analytical techniques. Therefore, we have also discussed questions related to the qualitative behavior of solutions and ways of approximating solutions numerically.

When one is faced with a problem whose solution cannot be simply found, questions arise as to the existence and uniqueness of solutions. We answer these questions by stating the relevant theorems, although not necessarily in their most general form, and discussing their significance. In particular, the differences between linear and nonlinear equations, and between initial and boundary value problems are examined.

Although most students at this level can begin to grasp the importance and applications of the various theorems, the proofs in many cases involve concepts unknown to them, such as uniform convergence. Where this is true, we have omitted the proofs without apology. For example, we discuss some of the problems involved in proving the fundamental existence and uniqueness theorem for a first order initial value problem by the method of successive approximations; our approach, however, is rather intuitive, and the more difficult analytical details are avoided.

We have written this book primarily for students who have a knowledge of calculus gained from a normal two- or three-semester course; most of the book is well within the capabilities of such students. In Chapter 7, "Systems of First Order Linear Equations", experience with matrix algebra will be helpful, although we include two sections on matrices that summarize the information necessary to read the rest of the chapter. Sections marked by an asterisk probably require greater mathematical sophistication (although, strictly speaking, no more knowledge) than the rest of the book.

Since students frequently feel that courses and books on differential equations tend to be "cookbookish," we have made a special effort to combat this tendency. Whenever possible, the student's prior knowledge is drawn upon to suggest a method of attack for a new type of problem. Moreover, the discussions of applications are usually accompanied by a careful derivation of the relevant equations. In an elementary book, terseness is not a primary virtue, and we have generally striven for clarity rather than conciseness. We hope that whatever may have been lost in elegance has been regained in readability.

Although the scope of the book can better be judged from the table of contents, we will note here some of its principal features. Chapter 1 is a brief introduction to some of the terminology and history of differential equations. Chapter 2 deals with first order equations. The differences between linear and nonlinear equations are emphasized in the first three sections. This is followed by a discussion of the standard elementary integration methods and a consideration of applications in population dynamics, mixing, growth and decay, geometry, and in mechanics, including the straight line motion of a body with variable mass. At the end of the chapter the Picard existence and uniqueness theorem is discussed.

Chapters 3 to 7 deal with linear ordinary differential equations. In Chapter 3 we discuss second order linear equations. The concepts of fundamental sets of solutions, linear independence, and superposition are emphasized, together with methods of solution. Examples are drawn from the fields of mechanical vibrations and electrical networks. The ideas of this chapter are extended to higher order linear equations in Chapter 5. In Chapter 4, on power series solutions, we show why the classification of points as ordinary, regular singular, or irregular singular is necessary and natural, rather than arbitrary. We use the Euler equation as a model for handling more general equations having a regular singular point. The more difficult cases in which the roots of the indicial equation are equal, or differ by an integer, are discussed and illustrated by appropriate forms of the Bessel equation. In Chapter 6 we introduce the Laplace transform and emphasize its usefulness in solving initial value problems having piecewise continuous or impulsive forcing terms. Chapter 7 deals with systems of first order linear equations. The first two sections provide a brief introduction to systems and their solution by elimination methods, and can be read without any knowledge of matrices. In the remainder of the chapter we use vector-matrix notation. This clarifies the close relationship between the theory of single equations and that of systems. Numerous applications are discussed, with emphasis on the analysis of electrical networks.

In Chapters 8 and 9 we discuss problems, especially nonlinear ones, for which the previous methods fail. In Chapter 8 we carefully present discrete numerical techniques for solving initial value problems. Procedures ranging from the Euler tangent line method to the Runge-Kutta and Adams-Moulton predictor-corrector methods are discussed and compared. A great deal of emphasis is given to the underlying principles governing numerical procedures and their refinement, and to considering the kinds, sources, and control of errors. Chapter 9 introduces the qualitative theory of differential equations, with emphasis on stability questions for autonomous systems. The discussion in this chapter is motivated and illustrated by ecological applications to situations involving competing species and predator-prey relations. Critical points of linear systems are classified, and almost linear systems are studied with reference to a neighboring linear system. Liapounov's method is introduced from a physical point of view, and the chapter ends with a brief investigation of limit cycles.

Readers familiar with the previous editions will notice many minor changes and a few major ones in this revision. In general, changes have been made to improve the exposition, to provide further material on applications, and to add some new examples and problems. Many of the changes in this edition result from suggestions offered by interested students and faculty since the publication of the first edition. The most significant modifications are as follows:

1. We have included in the chapter on series solutions a more extensive review of properties of power series, especially manipulations involving shifting of indices.
2. We have reorganized the chapter on systems of equations and have added material on applications to network analysis.
3. We have rewritten the section on multistep numerical methods to emphasize the Adams-Moulton method.
4. We have extended the chapter on qualitative theory and stability to include a careful discussion of elementary ecological applications, competing species, and predator-prey equations.
5. We have added more than 150 new problems throughout the book, making a total of more than 1200 problems.

We have also made many minor changes in order to improve the clarity of the previous edition.

Based on our experience at Rensselaer we feel that this book contains slightly more material than can be covered in most three-hour one-semester courses. To take advantage of this fact, we have organized the book so as to permit a great deal of flexibility in its use. Beginning with Chapter 4, the chapters are substantially independent of each other. Futhermore, in most chapters the principal ideas are developed in the first few sections; the remaining ones are devoted to extensions and applications. Thus the instructor has maximum control over the selection and arrangement of course material, and over the depth to which the various topics are to be explored. For example, courses oriented toward systems analysis would probably emphasize Chapters 6, 7, and 9. Finally, we mention that if the students have been exposed to differential equations in their calculus course, then some of the introductory material in Chapters 2 and 3 can be dispensed with, and later chapters can be taken up more thoroughly.

Sections are numbered in decimal form.   Theorems and figures are numbered consecutively within each chapter.   Thus, Theorem 3.7 is the seventh theorem in Chapter 3, but is not necessarily in Section 3.7.   References are given at the end of each chapter.   Most sections are followed by problem sets for the student;   all answers are given at the end of the book.   Some of the more difficult problems are indicated by an asterisk.   Certain sections are also designated by an asterisk. These contain more advanced material, which can be omitted by a beginning student in a first reading.

Once again we thank the individuals, named in the prefaces to the first two editions, who have assisted in the preparation of those earlier editions.   We also appreciate the interest of many people, both faculty and students, who have indicated corrections or offered suggestions for this edition;   many of these have been adopted.

We express our special appreciation to Mrs. Helen D. Hayes, who expertly produced the final typescript.   Finally, we thank the editorial and production staff of Wiley for their assistance and cooperation.

William E. Boyce
Richard C. DiPrima
Troy, New York,
November, 1976

Contents

4. SERIES SOLUTIONS OF SECOND ORDER LINEAR EQUATIONS   145

5. HIGHER ORDER LINEAR EQUATIONS   207

6. THE LAPLACE TRANSFORM   226

7. SYSTEMS OF FIRST ORDER LINEAR EQUATIONS   265

8. NUMERICAL METHODS   336

9. NONLINEAR DIFFERENTIAL EQUATIONS AND STABILITY   378

CHAPTER 1

Introduction

Many important and significant problems in engineering, the physical sciences, and the social sciences, when formulated in mathematical terms, require the determination of a function satisfying an equation containing derivatives of the unknown function. Such equations are called *differential equations*. Perhaps the most familiar example is Newton's law

$$m\,\frac{d^2u(t)}{dt^2} = F\left[t, u(t), \frac{du(t)}{dt}\right] \tag{1}$$

for the position $u(t)$ of a particle acted on by a force F, which may be a function of time t, the position $u(t)$, and the velocity $du(t)/dt$. To determine the motion of a particle acted on by a given force F it is necessary to find a function u satisfying Eq. (1). If the force is that due to gravity, then

$$m\,\frac{d^2u(t)}{dt^2} = -mg. \tag{2}$$

On integrating Eq. (2) we have

$$\frac{du(t)}{dt} = -gt + c_1,$$

$$u(t) = -\tfrac{1}{2}gt^2 + c_1t + c_2, \tag{3}$$

where c_1 and c_2 are constants. To determine $u(t)$ completely it is necessary to specify two additional conditions, such as the position and velocity of the particle at some instant of time. These conditions can be used to determine the constants c_1 and c_2.

In developing the theory of differential equations in a systematic manner it is helpful to classify different types of equations. One of the more obvious classifications is based on whether the unknown function depends on a single independent variable or on several independent variables. In the first case only ordinary derivatives appear in the differential equation, and it is said to be an *ordinary differential equation*. In the second case the derivatives are partial derivatives, and the equation is called a *partial differential equation*.

Two examples of ordinary differential equations, in addition to Eq. (1), are

$$L \frac{d^2 Q(t)}{dt^2} + R \frac{dQ(t)}{dt} + \frac{1}{C} Q(t) = E(t), \tag{4}$$

for the charge $Q(t)$ on a condenser in a circuit with capacitance C, resistance R, inductance L, and impressed voltage $E(t)$; and the equation governing the decay with time of an amount $R(t)$ of a radioactive substance, such as radium,

$$\frac{dR(t)}{dt} = -kR(t), \tag{5}$$

where k is a known constant. Typical examples of partial differential equations are the potential equation

$$\frac{\partial^2 u(x, y)}{\partial x^2} + \frac{\partial^2 u(x, y)}{\partial y^2} = 0, \tag{6}$$

the diffusion or heat equation

$$\alpha^2 \frac{\partial^2 u(x, t)}{\partial x^2} = \frac{\partial u(x, t)}{\partial t}, \tag{7}$$

and the wave equation

$$a^2 \frac{\partial^2 u(x, t)}{\partial x^2} = \frac{\partial^2 u(x, t)}{\partial t^2}. \tag{8}$$

Here α^2 and a^2 are certain constants. The potential equation, the diffusion equation, and the wave equation arise in a variety of problems in the fields of electricity and magnetism, elasticity, and fluid mechanics. Each is typical of distinct physical phenomena (note the names), and each is representative of a large class of partial differential equations. In this book we will be concerned only with ordinary differential equations.

1.1   ORDINARY DIFFERENTIAL EQUATIONS

The *order* of an ordinary differential equation is the order of the highest derivative that appears in the equation. Thus Eqs. (1) and (4) of the previous section are second order ordinary differential equations, and Eq. (5) is a first order ordinary differential equation. More generally, the equation

$$F[x, u(x), u'(x), \ldots, u^{(n)}(x)] = 0 \tag{1}$$

is an ordinary differential equation of the nth order. Equation (1) represents a relation between the independent variable x and the values of the function u and its first n derivatives $u', u'', \ldots, u^{(n)}$. It is convenient and follows the usual notation in the theory of differential equations to write y for $u(x)$, with $y', y'', \ldots, y^{(n)}$ standing for $u'(x), u''(x), \ldots, u^{(n)}(x)$. Thus Eq. (1) is written as

$$F(x, y, y', \ldots, y^{(n)}) = 0. \tag{2}$$

For example,

$$y''' + 2e^x y'' + yy' = x^4 \tag{3}$$

is a third order differential equation for $y = u(x)$. Occasionally, other letters will be used instead of y; the meaning will be clear from the context.

We shall assume that it is always possible to solve a given ordinary differential equation for the highest derivative, obtaining

$$y^{(n)} = f(x, y, y', y'', \ldots, y^{(n-1)}). \tag{4}$$

We will only study equations of the form (4). This is mainly to avoid the ambiguity that may arise because a single equation of the form (2) may correspond to several equations of the form (4). For example, the equation

$$y'^2 + xy' + 4y = 0$$

leads to the two equations

$$y' = \frac{-x + \sqrt{x^2 - 16y}}{2} \quad \text{or} \quad y' = \frac{-x - \sqrt{x^2 - 16y}}{2}.$$

Solution. The fact that we have written Eq. (4) does not necessarily mean that there is a function $y = \phi(x)$ that satisfies it. Indeed, this is one of the questions we wish to investigate. By a *solution* of the ordinary differential equation (4) on the interval $\alpha < x < \beta$ we mean a function ϕ such that ϕ', ϕ'', $\ldots$, $\phi^{(n)}$ exist and satisfy

$$\phi^{(n)}(x) = f[x, \phi(x), \phi'(x), \ldots, \phi^{(n-1)}(x)] \tag{5}$$

for every x in $\alpha < x < \beta$. Unless stated otherwise, we shall assume that the function f of Eq. (4) is a real-valued function, and we will be interested in obtaining real-valued solutions $y = \phi(x)$.

It is easily verified by direct substitution that the first order equation

$$\frac{dR}{dt} = -kR \tag{6}$$

has the solution

$$R = \phi(t) = ce^{-kt}, \quad -\infty < t < \infty, \tag{7}$$

where c is an arbitrary constant. Similarly, the functions $y_1(x) = \cos x$ and $y_2(x) = \sin x$ are solutions of

$$y'' + y = 0 \tag{8}$$

for all x. A somewhat more complicated example is to verify that $\phi_1(x) = x^2 \ln x$ is a solution of

$$x^2 y'' - 3xy' + 4y = 0, \quad x > 0. \tag{9}$$

We have

$$\phi_1(x) = x^2 \ln x,$$

$$\phi_1'(x) = x^2 \frac{1}{x} + 2x \ln x = x + 2x \ln x,$$

$$\phi_1''(x) = 1 + 2x \frac{1}{x} + 2 \ln x = 3 + 2 \ln x.$$

On substituting in the differential equation (9) we obtain

$$x^2(3 + 2 \ln x) - 3x(x + 2x \ln x) + 4(x^2 \ln x)$$
$$= 3x^2 - 3x^2 + (2 - 6 + 4)x^2 \ln x = 0,$$

which verifies that $\phi_1(x) = x^2 \ln x$ is a solution of Eq. (9). It can also be shown that $\phi_2(x) = x^2$ is a solution of Eq. (9); this is left as an exercise.

One question that might come to mind is whether there are other solutions of Eqs. (6), (8), and (9) besides those just given. A question that might occur even earlier is the following: given an equation of the form (4), how do we know whether it even has a solution? This is the question of the *existence* of a solution. Not all differential equations have solutions, nor is the question of existence purely mathematical. If a meaningful physical problem is correctly formulated mathematically as a differential equation, then the mathematical problem should have a solution. In this sense an engineer or scientist has some check on the validity of his mathematical formulation.

Second, assuming a given equation has one solution, does it have other solutions? If so, what type of additional conditions must be specified in order to single out a particular solution? This is the question of *uniqueness*. Notice that there is an infinity of solutions of the first order equation (6) corresponding to the infinity of possible choices of the constant c in Eq. (7). If R is specified at some time t, this condition will determine a value for c; even so, however, we do not know yet that there may not be other solutions of Eq. (6) which also have the prescribed value of R at the prescribed time t. The questions of existence and uniqueness are difficult questions; they and related questions will be discussed as we proceed.

A third question, a more practical one, is: given a differential equation of the form (4), how do we actually determine a solution? We note that if we find a solution of the given equation we have at the same time answered the question of the existence of a solution. On the other hand, without knowledge of existence theory we might, for example, use a computer to find a numerical approximation to a "solution" that does not exist. Even though we may know that a solution exists, it may be that the solution is not expressible in terms of the usual elementary functions—polynomial, trigonometric, exponential, logarithmic, and hyperbolic functions. Unfortunately, this is the situation for most differential equations. However, before we can consider difficult problems it is first necessary to master some of the elementary theory of ordinary differential equations.

Linear and Nonlinear Equations.   A crucial classification of ordinary differential equations is according to whether they are linear or nonlinear. The differential equation

$$F(x, y, y', \ldots, y^{(n)}) = 0$$

is said to be *linear* if F is a linear function of the variables $y, y', \ldots, y^{(n)}$. Thus the general linear ordinary differential equation of order n is

$$a_0(x)y^{(n)} + a_1(x)y^{(n-1)} + \cdots + a_n(x)y = g(x). \tag{10}$$

Equations (2), (4), and (5) of the previous section and Eqs. (6), (8), and (9) of this section are linear equations. An equation that is not of the form (10) is a *nonlinear* equation. Equation (3) is nonlinear because of the term yy'. A simple physical problem that leads to a nonlinear differential equation is the oscillating pendulum. The angle θ that an oscillating pendulum of length l makes with the vertical direction (see Figure 1.1) satisfies the nonlinear equation

$$\frac{d^2\theta}{dt^2} + \frac{g}{l}\sin\theta = 0. \tag{11}$$

FIGURE 1.1

The mathematical theory and the techniques for solving linear equations are highly developed. In contrast, for nonlinear equations the situation is not as satisfactory. General techniques for solving nonlinear equations are largely lacking, and the theory associated with such equations is also more complicated than the theory for linear equations. In view of this it is fortunate that many significant problems lead to linear ordinary differential equations or, at least, in the first approximation to linear equations. For example, for the pendulum problem, if the angle θ is small, then $\sin\theta \simeq \theta$ and Eq. (11) can be replaced by the linear equation

$$\frac{d^2\theta}{dt^2} + \frac{g}{l}\theta = 0. \tag{12}$$

On the other hand, there are important physical phenomena, such as the current flow in an electron tube, in which it is not possible to approximate the governing nonlinear differential equation by a linear one—the nonlinearity is crucial.

In an elementary text it is natural to emphasize the discussion of linear equations. The greater part of this book is therefore devoted to linear equations and various methods for solving them. However, Chapters 8 and 9, as well as a large part of Chapter 2, are concerned with nonlinear equations. Throughout the text we attempt to show why nonlinear equations are, in general, more difficult, and why many of the techniques that are useful in solving linear equations cannot be applied to nonlinear equations.

1.2 HISTORICAL REMARKS

Without knowing something about differential equations and methods of solving them it is difficult to discuss the history and development of this important branch of mathematics. Further, the development of the theory of differential equations is intimately interwoven with the general development of mathematics and cannot be divorced from it. In these brief comments we will mention just a few, certainly not all, of the famous mathematicians of the seventeenth and eighteenth centuries who made important contributions to the theory and application of differential equations. We will follow the brief historical survey given by Ince* in Appendix A of his authoritative treatise on ordinary differential equations.

* References are listed at the end of each chapter.

The theory of differential equations dates back to the beginnings of the calculus with Isaac Newton (1642–1727) and Gottfried Wilhelm Leibniz (1646–1716) in the seventeenth century. Indeed Ince states,

> Yet our hazy knowledge of the birth and infancy of the science of differential equations condenses upon a remarkable date, the eleventh day of November, 1675, when Leibniz first set down on paper the equation
>
> $$\int y \, dy = \tfrac{1}{2}y^2,$$
>
> thereby not merely solving a simple differential equation, which was in itself a trivial matter, but what was an act of great moment, forging a powerful tool, the integral sign.

While Newton did relatively little work in the theory of differential equations per se, his development of the calculus and elucidation of the basic principles of mechanics provided a basis for the development of their applications in the eighteenth century, most notably by Euler. Newton classified first order equations according to the forms: $dy/dx = f(x)$, $dy/dx = f(y)$, and $dy/dx = f(x, y)$. For the latter equation he developed a method of solution using infinite series when $f(x, y)$ is a polynomial in x and y. He was very sensitive to criticism and as a consequence he was slow in publishing many of his discoveries.

Leibniz arrived at the fundamental results of the calculus independently, although a little later than Newton. Our modern notation for the derivative (dy/dx) and the integral sign are due to Leibniz. He discovered the method of separation of variables (Section 2.4), as well as procedures for solving first order homogeneous equations (Section 2.7) and linear first order equations (Section 2.1). He carried on a prolific correspondence with other mathematicians, especially the Bernoulli brothers. In the course of this correspondence many problems in differential equations were solved during the latter part of the seventeenth century.

Following Newton and Leibniz come the Bernoulli brothers, Jakob (1654–1705) and Johann (1667–1748), and Johann's son Daniel (1700–1782). These are just three of eight members of the Bernoulli family who were prominent scientists and mathematicians in their time. With the aid of the calculus, they formulated as differential equations and solved a number of problems in mechanics. One problem (1696–1697) to which both brothers contributed, and which led to much friction between them, was the *brachistochrone* problem—the determination of the curve of fastest descent. In Problem 13 of Section 2.10 we will see that this problem (Figure 1.2) leads to the nonlinear first order equation

$$y[1 + (y')^2] = c$$

where c is a constant. Newton also solved the problem early in 1697. It is said, perhaps apocryphally, that Newton learned of the problem (as a challenge) late

FIGURE 1.2

in the afternoon of a tiring day at the mint, and solved it that evening after dinner. He published the solution anonymously, but Johann Bernoulli on seeing it exclaimed, "Ah, I know the lion by his paw." It is Daniel whose name is associated with the famous Bernoulli equation in fluid mechanics. In 1690 Jakob Bernoulli published the solution of the differential equation, written in differential form $(b^2y^2 - a^3)^{1/2}\, dy = a^{3/2}\, dx$. Today this is a simple exercise but at that time to go from the equation $y' = [a^3/(b^2y^2 - a^3)]^{1/2}$ to the differential form, and then to assert that the integrals* of each side must be equal except for a constant was a major step. Indeed, for example, while Johann Bernoulli knew that $ax^p\, dx = d[ax^{p+1}/(p + 1)]$ was not meaningful for $p = -1$, he did not know that $dx/x = d(\ln x)$. Nevertheless, he was able to show that the equation $dy/dx = y/ax$, which we can solve by writing it as

$$a\frac{dy}{y} = \frac{dx}{x},$$

has the solution $y^a/x = c$ where c is a constant of integration (see Section 2.4).

By the end of the seventeenth century most of the elementary methods of solving first order ordinary differential equations (Chapter 2) were known, and attention was centered on higher order ordinary differential equations and partial differential equations. Jacopo Riccati (1676–1754), an Italian mathematician, considered equations of the form $f(y, y', y'') = 0$ (Problem 2 of Section 3.1). He also considered an important nonlinear equation known as the Riccati equation, $dy/dx = a_0(x) + a_1(x)y + a_2(x)y^2$, though not in such a general form (Problems 35 and 36 of Section 2.8).

Leonhard Euler (1707–1783), one of the greatest mathematicians of all time, also lived during the eighteenth century. He was a prolific mathematician; his collected works fill over seventy volumes. Even though he was blind during the last seventeen years of his life his work continued undiminished. Of particular interest here is his work on the formulation of problems in mechanics in mathematical language and his development of methods of solving these mathematical problems. Lagrange said of Euler's work in mechanics, "The first great work in which analysis is applied to the science of movement." Euler also considered such questions as the possibility of reducing second order equations to first order equations by a suitable change of variables. He introduced the concept of an integrating factor (Section 2.6), gave a general treatment of linear ordinary differential equations with constant coefficients (Chapters 3 and 5) in 1739, contributed to the method of power series solutions (Chapter 4), and gave a numerical procedure for solving differential equations (Section 8.2). He also made important contributions to the theory of Fourier series (Chapter 10) and gave the first systematic discussion of the calculus of variations.

Later in the eighteenth century the great French mathematicians Joseph-Louis Lagrange (1736–1813) and Pierre-Simon Laplace (1749–1827) made important contributions to the theory of ordinary differential equations and gave the first

---

* Jakob Bernoulli appears to be the first person to have used the word integral.

scientific treatment of partial differential equations. Indeed, possibly the most famous partial differential equation in mathematical physics, the potential equation $u_{xx} + u_{yy} = 0$, where subscripts denote partial derivatives, is also known as Laplace's equation. Lagrange's monumental work *Mècanique analytique* contains the general equations of motion of a dynamical system known today as *Lagrange's equations*. Laplace's five-volume work *Traité de mécanique céleste* earned him the title "the Newton of France." The latter volumes were published in the period 1789–1825 and included all of mechanics to that time. The attitudes of Lagrange and Laplace epitomize two philosophies of mathematics. For Laplace nature was essential, and mathematics was his tool in learning its secrets; for Lagrange mathematics was an art that justified its own being. However, both men made far-reaching advances in both the theory and application of mathematics.

The student who is interested in the history of mathematics (and differential equations) may wish to refer to one of the books dealing with the development of mathematics listed at the end of this chapter.

In more recent years part of the effort of mathematicians in the areas of ordinary and partial differential equations has been to develop a rigorous, systematic (but general) theory. The goal is not so much to construct solutions of particular differential equations, but rather to develop techniques suitable for treating classes of equations.

We conclude this short historical sketch with a remark that may bring some pleasure to the student who has noted with dismay how often phrases such as "It is clear ..." or "It can be easily shown ..." appear in mathematics texts. Nathaniel Bowditch (1773–1838), an American astronomer and mathematician, while translating Laplace's *Mécanique céleste* in the early 1800s, stated, "I never come across one of Laplace's, 'Thus it plainly appears' without feeling sure that I have hours of hard work before me to fill up the chasm and find out and show how it plainly appears".

PROBLEMS

1. For each of the following differential equations determine its order and whether or not the equation is linear.

(a) $x^2 \dfrac{d^2y}{dx^2} + x \dfrac{dy}{dx} + 2y = \sin x$ (b) $(1 + y^2) \dfrac{d^2y}{dx^2} + x \dfrac{dy}{dx} + y = e^x$

(c) $\dfrac{d^4y}{dx^4} + \dfrac{d^3y}{dx^3} + \dfrac{d^2y}{dx^2} + \dfrac{dy}{dx} + y = 1$ (d) $\dfrac{dy}{dx} + xy^2 = 0$

(e) $\dfrac{d^2y}{dx^2} + \sin(x + y) = \sin x$ (f) $\dfrac{d^3y}{dx^3} + x \dfrac{dy}{dx} + (\cos^2 x)y = x^3$

2. Verify for each of the following that the given function or functions is a solution of the differential equation.

(a) $y'' - y = 0$; $y_1(x) = e^x$, $y_2(x) = \cosh x$

(b) $y'' + 2y' - 3y = 0$; $y_1(x) = e^{-3x}$, $y_2(x) = e^x$

(c) $y'''' + 4y''' + 3y = x$; $y_1(x) = x/3$, $y_2(x) = e^{-x} + x/3$

(d) $2x^2y'' + 3xy' - y = 0$, $x > 0$; $y_1(x) = x^{\frac{1}{2}}$, $y_2(x) = x^{-1}$

(e) $x^2y'' + 5xy' + 4y = 0$, $x > 0$; $y_1(x) = x^{-2}$, $y_2(x) = x^{-2}\ln x$

(f) $y'' + y = \sec x$, $0 < x < \pi/2$; $y = \phi(x) = (\cos x)\ln \cos x + x \sin x$

(g) $y' - 2xy = 1$; $y = \phi(x) = e^{x^2}\int_0^x e^{-t^2}\,dt + e^{x^2}$

(h) $y'' + 4y' + 4y = 0$; $y_1(x) = e^{-2x}$, $y_2(x) = xe^{-2x}$

(i) $y' + y^2 \sin x = 0$, $-\pi/2 < x < \pi/2$; $y = \phi(x) = -\sec x$

(j) $y' = y - y^2$; $y = \phi(x) = \dfrac{1}{1 + e^{-x}}$

3. Determine for what values of r each of the following linear differential equations has solutions of the form $y = e^{rx}$.

(a) $y' + 2y = 0$

(b) $y'' - y = 0$

(c) $y'' + y' - 6y = 0$

(d) $y''' - 3y'' + 2y' = 0$

4. Determine for what values of r each of the following linear differential equations has solutions of the form $y = x^r$ for $x > 0$.

(a) $x^2y'' + 4xy' + 2y = 0$

(b) $x^2y'' - 4xy' + 4y = 0$

5. Consider the differential equation

$$y'^2 + x, \qquad x > 0. \tag{i}$$

Solving for y', we obtain

$$y' = \sqrt{x}, \qquad x > 0 \tag{ii}$$

or

$$y' = -\sqrt{x}, \qquad x > 0. \tag{iii}$$

Verify that both $y_1(x) = \frac{2}{3}x^{\frac{3}{2}}$ and $y_2(x) = -\frac{2}{3}x^{\frac{3}{2}}$ satisfy Eq. (i). Also verify that y_1 is a solution of Eq. (ii) but not of Eq. (iii), and that y_2 is a solution of Eq. (iii) but not of Eq. (ii).

REFERENCES

Bell, E. T., *Men of Mathematics*, Simon and Schuster, New York, 1937.

Boyer, C. B., *A History of Mathematics*, Wiley, New York, 1968.

Cajori, Florian, *A History of Mathematics*, Second Edition, Macmillan, New York, 1919.

Eves, H., *An Introduction to the History of Mathematics*, Third Edition, Holt, New York, 1964.

Ince, E. L., *Ordinary Differential Equations*, Longmans, Green, London, 1927; Dover, New York, 1956.

Kline, M., *Mathematical Thought from Ancient to Modern Times*, Oxford University Press, New York, 1972.

Struik, D. J., *A Concise History of Mathematics*, Third Edition, Dover, New York, 1967.

CHAPTER 2

First Order Differential Equations

2.1 LINEAR EQUATIONS

This chapter deals with differential equations of first order, that is, equations of the form

$$y' = f(x, y),\tag{1}$$

where f is a given function of two variables. Any function $y = \phi(x)$, which with its derivative y' identically satisfies Eq. (1), is called a solution, and our object is to determine whether such functions exist and, if so, how to find them.

The simplest type of first order differential equation occurs when f depends only on x. In this case

$$y' = f(x),\tag{2}$$

and we seek a function $y = \phi(x)$ whose derivative is the given function f. From elementary calculus we know that ϕ is an antiderivative of f, and we write

$$y = \phi(x) = \int^x f(t)\, dt + c,\tag{3}$$

where c is an arbitrary constant. For example, if

$$y' = \sin 2x,$$

then

$$y = \phi(x) = -\tfrac{1}{2}\cos 2x + c.$$

In Eq. (3) and elsewhere in this book we use the notation $\int^x f(t)\, dt$ to denote an antiderivative of the function f; that is, $F(x) = \int^x f(t)\, dt$ designates some particular representative of the class of functions whose derivatives are equal to f. All members of this class are included in the expression $F(x) + c$, where c is an arbitrary constant.

We will not discuss equations of the form (2) further except to note two features typical of first order differential equations generally. First, a single integration process is required to eliminate the derivative of y and obtain y itself. Second, this integration process leads to an expression for the solution that involves an arbitrary constant.

We now return to a consideration of Eq. (1). Unfortunately, there is no satisfactory general method for writing down solutions of this equation. Instead, we will present several methods, each of which is applicable to a certain class of equations of the form (1). First, we will consider the linear first order equation

$$y' + p(x)y = g(x), \tag{4}$$

where p and g are given continuous functions on some interval $\alpha < x < \beta$. In this section we will be concerned with methods for solving Eq. (4). More theoretical questions involving the existence and uniqueness of solutions in general will be discussed in Section 2.2.

Let us begin with the equation

$$y' + ay = 0, \tag{5}$$

where a is a real constant. This equation can be solved by inspection. We need a function y whose derivative y' is equal to $(-a)$ times y itself. Clearly, $y = e^{-ax}$ has this property, and therefore satisfies Eq. (5). Furthermore,

$$y = ce^{-ax}, \tag{6}$$

where c is an arbitrary constant, also does so. Since c is arbitrary, Eq. (6) represents infinitely many solutions of the differential equation (5). It is natural to ask whether Eq. (5) has any solutions other than those given by Eq. (6). We will show in the next section that there are no other solutions, but for the time being this question remains open.

Geometrically, Eq. (6) represents a one-parameter family of curves, called *integral curves* of Eq. (5). For $a = 1$ several members of this family are sketched in Figure 2.1. Each integral curve is the geometric representation of the corresponding solution of the differential equation. Specifying a particular solution is equivalent to picking out a particular integral curve from the one-parameter family. It is usually convenient to do this by prescribing a point (x_0, y_0) through which the integral curve must pass; that is, we seek a solution $y = \phi(x)$ such that

$$\phi(x_0) = y_0.$$

Such a condition is called an *initial condition*. Since y stands for $\phi(x)$ we could also write

$$y = y_0 \quad \text{at} \quad x = x_0.$$

However, it is common practice to express the initial condition in the form

$$y(x_0) = y_0, \tag{7}$$

and this is the notation we will usually use in this book. A first order differential equation together with an initial condition form an *initial value problem*.*

---

* This terminology is suggested by the fact that the independent variable often denotes time, the initial condition defines the situation at some fixed instant, and the solution of the initial value problem describes what happens later.

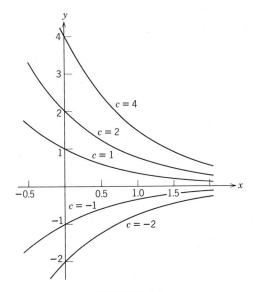

FIGURE 2.1

For example, the differential equation (5),

$$y' + ay = 0,$$

and the initial condition

$$y(0) = 2, \tag{8}$$

form an initial value problem. As noted above, all solutions of the differential equation (5) are given by Eq. (6). The solution satisfying the initial condition (8) is found by substituting $x = 0$ and $y = 2$ in Eq. (6). Then $c = 2$ and the desired solution is

$$y = \phi(x) = 2e^{-ax}. \tag{9}$$

This is the unique solution of the given initial value problem. More generally, it is possible to show that the initial value problem composed of the differential equation (4) and the initial condition (7) has a unique solution whenever the coefficients p and g are continuous functions. This is discussed in the next section.

Now let us consider the equation

$$y' + ay = g(x). \tag{10}$$

If $a = 0$, then the left-hand side of Eq. (10) is just the derivative of y, and the solution is given by Eq. (3) with f replaced by g. If $a \neq 0$, then the left-hand side of Eq. (10) is a combination of terms involving y and y'. We ask whether these terms can be identified as the derivative of some function. In other words, can we write

$$\frac{dy}{dx} + ay = \frac{d}{dx}(\quad)? \tag{11}$$

If so, then Eq. (10) has the form

$$\frac{d}{dx}(\ \ ) = g(x),$$

and we can immediately solve it by integrating both sides.

A clue as to how to proceed can be found by looking again at the solution (6) of Eq. (5). Rewriting $y = ce^{-ax}$ in the form

$$ye^{ax} = c \tag{12}$$

and then differentiating both sides of Eq. (12), we obtain

$$\frac{d}{dx}(ye^{ax}) = 0,$$

or

$$\frac{dy}{dx}e^{ax} + ae^{ax}y = e^{ax}\left(\frac{dy}{dx} + ay\right) = 0.$$

Thus, if we multiply Eq. (10) by e^{ax}, then we can write the left-hand side of the resulting equation as the derivative of the single function ye^{ax}:

$$\underbrace{e^{ax}y' + ae^{ax}y} = e^{ax}g(x)$$

$$\frac{d}{dx}(e^{ax}y) = e^{ax}g(x). \tag{13}$$

Equation (13) can be integrated, and we obtain

$$e^{ax}y = \int^{x} e^{at}g(t)\, dt + c,$$

where c is an arbitrary constant. Hence, a solution of Eq. (10) is

$$y = \phi(x) = e^{-ax}\int^{x} e^{at}g(t)\, dt + ce^{-ax}. \tag{14}$$

Thus, for a given function g the problem of determining a solution of Eq. (10) is reduced to that of evaluating the antiderivative in Eq. (14). The difficulty involved in this depends on g; nevertheless, Eq. (14) gives an explicit formula for the solution $y = \phi(x)$. The constant c can be determined if an initial condition is prescribed.

Example 1. Find the solution of the initial value problem

$$y' + 2y = e^{-x}, \tag{15}$$

$$y(0) = 3. \tag{16}$$

Comparing Eq. (15) with Eq. (10), we find that $a = 2$ in this problem. Thus, we multiply Eq. (15) by e^{2x}, thereby obtaining

$$e^{2x}y' + 2e^{2x}y = e^{2x}e^{-x}. \tag{17}$$

The left-hand side of Eq. (17) is the derivative of $e^{2x}y$, so this equation can be written in the form

$$(e^{2x}y)' = e^x. \tag{18}$$

Integrating Eq. (18) yields

$$ye^{2x} = e^x + c,$$

and hence

$$y = e^{-x} + ce^{-2x} \tag{19}$$

is a solution of Eq. (15) for any value of c. To satisfy the initial condition (16) we substitute $x = 0$ and $y = 3$ in Eq. (19). This gives $c = 2$, and thus the solution of the initial value problem (15), (16) is

$$y = e^{-x} + 2e^{-2x}. \tag{20}$$

Now let us return to the general first order linear equation (4)

$$y' + p(x)y = g(x).$$

By analogy with the procedure used above, we would like to choose a function μ so that if Eq. (4) is multiplied by $\mu(x)$, then the left-hand side of Eq. (4) can be written as the derivative* of the single function $\mu(x)y$. That is, we want to choose μ, if possible, so that

$$\mu(x)[y' + p(x)y] = [\mu(x)y]'$$
$$= \mu(x)y' + \mu'(x)y.$$

Thus $\mu(x)$ must satisfy

$$\mu(x)p(x)y = \mu'(x)y.$$

Assuming for the moment that $\mu(x) > 0$, we obtain

$$\frac{\mu'(x)}{\mu(x)} = p(x).$$

Since $\mu'(x)/\mu(x)$ is the derivative of $\ln \mu(x)$, we have

$$\ln \mu(x) = \int^x p(t)\, dt,$$

and finally

$$\mu(x) = \exp\left[\int^x p(t)\, dt\right]. \tag{21}$$

Note that $\mu(x)$, as defined by Eq. (21), is indeed positive. Further, since $\int^x p(t)\, dt$ is determined only up to an arbitrary additive constant, $\mu(x)$ is determined only up to an arbitrary multiplicative constant.

Returning to Eq. (4) and multiplying by $\mu(x)$, we obtain

$$[\mu(x)y]' = \mu(x)g(x).$$

---

* A function μ having this property is called an integrating factor. Integrating factors are discussed more fully in Section 2.6.

Therefore

$$\mu(x)y = \int^x \mu(s)g(s)\, ds + c,$$

or

$$y = \frac{1}{\mu(x)}\left[\int^x \mu(s)g(s)\, ds + c\right], \tag{22}$$

where $\mu(x)$ is given by Eq. (21). Equation (22) provides an explicit formula for the solution of the general first order linear equation (4), where p and g are given continuous functions. Two integrations are required, one to obtain $\mu(x)$ from Eq. (21) and the other to determine y from Eq. (22). The arbitrary constant c can be used to satisfy an initial condition.

Two additional things should be noted about the procedure we have developed for solving Eq. (4). First, before computing the integrating factor μ from Eq. (21), it is necessary to make sure that the differential equation is exactly in the form (4); specifically, the coefficient of y' must be one. Second, after finding $\mu(x)$ and multiplying Eq. (4) by it, one should make sure that the terms involving y are, in fact, the derivative of $\mu(x)y$, as they should be. This provides a check on the calculation of μ. Of course, once the solution y has been found, one can also check it by substituting it into the differential equation and initial condition.

Example 2. Find the solution of the initial value problem

$$y' - 2xy = x, \qquad y(0) = 1. \tag{23}$$

For this equation the function μ is given by

$$\mu(x) = \exp\left(-\int^x 2t\, dt\right) = e^{-x^2}.$$

Hence

$$e^{-x^2}y' - 2xe^{-x^2}y = xe^{-x^2},$$

so that

$$(e^{-x^2}y)' = xe^{-x^2}.$$

Therefore,

$$ye^{-x^2} = \int^x te^{-t^2}\, dt + c = -\tfrac{1}{2}e^{-x^2} + c,$$

and finally

$$y = -\tfrac{1}{2} + ce^{x^2}.$$

To satisfy the initial condition $y(0) = 1$ we must choose $c = \tfrac{3}{2}$. Hence

$$y = -\tfrac{1}{2} + \tfrac{3}{2}e^{x^2} \tag{24}$$

is the solution of the given initial value problem.

Example 3. Find the solution of the initial value problem

$$y' - 2xy = 1, \qquad y(0) = 1. \tag{25}$$

As in the previous example $\mu(x) = e^{-x^2}$, and we obtain

$$ye^{-x^2} = \int^x e^{-t^2}\, dt + c. \tag{26}$$

To evaluate c it is convenient to take the lower limit of integration* as the initial point $x = 0$. Then, multiplying Eq. (26) by e^{x^2} we obtain

$$y = e^{x^2}\int_0^x e^{-t^2}\, dt + ce^{x^2}.$$

The initial condition $y(0) = 1$ requires that $c = 1$, and hence

$$y = e^{x^2}\int_0^x e^{-t^2}\, dt + e^{x^2} \tag{27}$$

is the solution of the given problem.

Note that in the solution of Example 3 the integral of e^{-t^2} is not expressible as an elementary function. This illustrates the fact that it may be necessary to leave the solution of even a very simple problem in integral form. However, the function

$$erf(x) = \frac{2}{\sqrt{\pi}}\int_0^x e^{-t^2}\, dt, \tag{28}$$

known as the *error function*, has been extensively tabulated and can be regarded as a known function. From a computational point of view, therefore, Eq. (27) is a perfectly satisfactory expression for the solution of the initial value problem (25). It is just as satisfactory, in fact, as the solution $y = -\frac{1}{2} + \frac{3}{2}e^{x^2}$ of the initial value problem (23). In order to evaluate e^{x^2} for a given value of x we ordinarily consult a table (or use a small electronic calculator), and it is no more difficult to look up $erf(x)$ in a different table.

Alternatively, an integral expression such as Eq. (27) can provide a starting point for the numerical calculation of the solution of an initial value problem. For a fixed value of x the integral in Eq. (27) is a *definite integral*, and its value can be determined numerically by a method such as Simpson's rule. By repeating the numerical integration for different values of x one can construct a table of values of the solution y. There are also other numerical procedures that use the differential equation directly to compute approximate values of the solution for a given set of values of x; these methods are discussed in Chapter 8.

PROBLEMS

In each of Problems 1 through 6 solve the given differential equation.

1. $y' + 3y = x + e^{-2x}$ 2. $y' - 2y = x^2e^{2x}$

---

* The choice of the lower limit of integration is actually immaterial since the difference between $\int_a^x e^{-t^2} dt$ and $\int_b^x e^{-t^2} dt$ is merely a constant, which can be added to c.

3. $y' + y = xe^{-x} + 1$

4. $y' + (1/x)y = 3 \cos 2x, \qquad x > 0$

5. $y' - y = 2e^x$

6. $xy' + 2y = \sin x, \qquad x > 0$

In each of Problems 7 through 12 find the solution of the given initial value problem.

7. $y' - y = 2xe^{2x}, \qquad y(0) = 1$

8. $y' + 2y = xe^{-2x}, \qquad y(1) = 0$

9. $y' + y = \dfrac{1}{1 + x^2}, \qquad y(0) = 0$

10. $y' + \dfrac{2}{x} y = \dfrac{\cos x}{x^2}, \qquad y(\pi) = 0, \qquad x > 0$

11. $y' - 2y = e^{2x}, \qquad y(0) = 2$

12. $xy' + 2y = \sin x, \qquad y(\pi/2) = 1$

13. Find the solution of

$$\frac{dy}{dx} = \frac{1}{e^y - x}.$$

Hint: Consider x as the dependent variable instead of y.

14. (a) Show that $\phi(x) = e^{2x}$ is a solution of

$$y' - 2y = 0,$$

and that $y = c\phi(x)$ is also a solution of this equation for any value of the constant c.

(b) Show that $\phi(x) = 1/x$ is a solution of

$$y' + y^2 = 0,$$

for $x > 0$, but that $y = c\phi(x)$ is not a solution of this equation. Note that the equation of part (b) is nonlinear while that of part (a) is linear.

15. Show that if $y = \phi(x)$ is a solution of

$$y' + p(x)y = 0,$$

then $y = c\phi(x)$ is also a solution for any value of the constant c.

16. Let $y = y_1(x)$ be a solution of

$$y' + p(x)y = 0, \tag{i}$$

and let $y = y_2(x)$ be a solution of

$$y' + p(x)y = g(x). \tag{ii}$$

Show that $y = y_1(x) + y_2(x)$ is also a solution of Eq. (ii).

*17. Consider the following method of solving the general linear equation of first order:

$$y' + p(x)y = g(x). \tag{i}$$

(a) If $g(x)$ is identically zero, show that the solution is

$$y = A \exp\left[-\int^x p(t)\, dt\right], \tag{ii}$$

where A is constant.

(b) If $g(x)$ is not identically zero, assume that the solution is of the form

$$y = A(x) \exp\left[-\int^x p(t) \, dt \right], \tag{iii}$$

where A is now a function of x. By substituting for y in the given differential equation show that $A(x)$ must satisfy the condition

$$A'(x) = g(x) \exp\left[\int^x p(t) \, dt \right]. \tag{iv}$$

(c) Find $A(x)$ from Eq. (iv). Then substitute for $A(x)$ in Eq. (iii) and determine y. Verify that the solution obtained in this manner agrees with that of Eq. (22) in the text. This technique is known as the method of *variation of parameters;* it is discussed in detail in Section 3.6.2 in connection with second order linear equations.

*18. Use the method of Problem 17 to solve each of the following differential equations.

(a) $y' - 2y = x^2 e^{2x}$ 　　　　　　(b) $y' + (1/x)y = 3 \cos 2x, \qquad x > 0$

2.2 FURTHER DISCUSSION OF LINEAR EQUATIONS

In Section 2.1 we showed how to construct solutions of first order linear differential equations. Now we turn to a study of certain more theoretical questions. We will first state a fundamental theorem giving conditions under which an initial value problem for a first order linear equation will always have one and only one solution. The rest of the section will be devoted to a consideration of some of the implications of this theorem.

Theorem 2.1. *If the functions p and g are continuous on an open interval* $\alpha < x < \beta$ *containing the point* $x = x_0$, *then there exists a unique function* $y = \phi(x)$ *that satisfies the differential equation*

$$y' + p(x)y = g(x) \tag{1}$$

for $\alpha < x < \beta$, *and that also satisfies the initial condition*

$$y(x_0) = y_0, \tag{2}$$

where y_0 *is an arbitrary prescribed initial value.*

The proof of this theorem is essentially contained in the discussion in the last section leading to the formula

$$y = \frac{1}{\mu(x)}\left[\int^x \mu(s)g(s) \, ds + c \right], \tag{3}$$

where

$$\mu(x) = \exp\int^x p(t) \, dt. \tag{4}$$

Assuming that Eq. (1) has a solution, the derivation in Section 2.1 shows that it must be of the form (3). Note that since p is continuous for $\alpha < x < \beta$, it follows that

μ is defined in this interval, and is a nonzero differentiable function. This justifies the conversion of Eq. (1) into the form

$$[\mu(x)y]' = \mu(x)g(x). \tag{5}$$

The function μg has an antiderivative since μ and g are continuous, and Eq. (3) follows from Eq. (5). The initial assumption—that there is at least one solution of Eq. (1)—can now be verified by substituting the expression for y in Eq. (3) back into the differential equation. Finally, the initial condition (2) determines the constant c uniquely, thus completing the proof. Since Eq. (3) contains *all* solutions of Eq. (1), it is customary to call Eq. (3) the *general solution* of Eq. (1).

Several aspects of the above theorem should be noted. In the first place, it states that the given initial value problem *has* a solution and also that the problem has *only one* solution. In other words, the theorem asserts both the *existence* and *uniqueness* of the solution of the initial value problem (1) and (2). Further, the solution $y = \phi(x)$ is a differentiable function and, in fact, since $y' = -p(x)y + g(x)$, the derivative $y' = \phi'(x)$ is continuous. Finally, the solution satisfies the differential equation (1) throughout the interval $\alpha < x < \beta$ in which the coefficients p and g are continuous. This means that the solution will break down, if at all, only at points where either p or g is discontinuous. Thus a certain amount of qualitative information about the solution is obtained merely by identifying points of discontinuity of p and g.

As an example consider the initial value problem

$$xy' + 2y = 4x^2, \tag{6}$$

$$y(1) = 2. \tag{7}$$

Proceeding as in Section 2.1, we rewrite Eq. (6) as

$$y' + \frac{2}{x}y = 4x \tag{8}$$

and seek a solution in an interval containing $x = 1$. Since the coefficients in Eq. (8) are continuous except at $x = 0$, it follows from Theorem 2.1 that the solution of the given initial value problem is valid at least in the interval $0 < x < \infty$. To find this solution we first compute $\mu(x)$:

$$\mu(x) = \exp\left(\int^x \frac{2}{t}\,dt\right) = e^{2\ln x}$$

$$= x^2. \tag{9}$$

Multiplying Eq. (8) by $\mu(x) = x^2$ gives

$$x^2y' + 2xy = 4x^3,$$

or

$$(x^2y)' = 4x^3.$$

Hence $x^2y = x^4 + c$, and

$$y = x^2 + \frac{c}{x^2}, \tag{10}$$

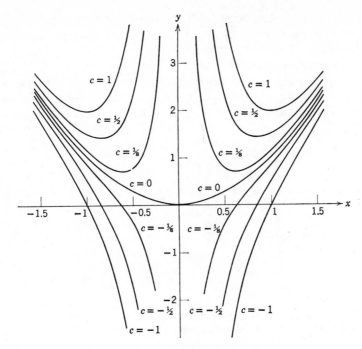

FIGURE 2.2

where c is arbitrary, is the general solution of Eq. (6). Integral curves of Eq. (6) for several values of c are sketched in Figure 2.2. In order to satisfy the initial condition (7) it is necessary that $c = 1$; thus,

$$y = x^2 + \frac{1}{x^2} \tag{11}$$

is the solution of the initial value problem (6), (7).

Note that the solution (11) becomes unbounded as $x \to 0$. This is not surprising since $x = 0$ is a point of discontinuity of the coefficient of y in the differential equation (8). However, if the initial condition (7) is changed to

$$y(1) = 1, \tag{12}$$

then it follows from Eq. (10) that $c = 0$. Hence the solution of the initial value problem (6), (12) is

$$y = x^2, \tag{13}$$

which is perfectly well behaved as $x \to 0$. This illustrates that Theorem 2.1 does *not* assert that the solution of an initial value problem must become singular whenever the functions p and g become discontinuous; rather, it asserts that the solution cannot become singular at other points.

The possible behavior of solutions of initial value problems for first order linear equations in the neighborhood of points where p and g are discontinuous is more varied than the previous discussion suggests. The possibilities are explored to some extent in Problems 9 through 12 and 15; a more detailed treatment appears in Chapter 4 in connection with second order linear equations.

Sometimes, equations that are not linear can be solved by first making a substitution converting the given equation into a linear equation. One type of equation for which this is possible is known as Bernoulli's equation (see Problems 16 and 17).

PROBLEMS

In each of Problems 1 through 4 find the general solution of the given differential equation.

1. $y' + (1/x)y = \sin x,$ $x > 0$
2. $x^2 y' + 3xy = (\sin x)/x,$ $x < 0$
3. $y' + (\tan x)y = x \sin 2x,$ $-\pi/2 < x < \pi/2$
4. $xy' + 2y = e^x,$ $x > 0$

In each of Problems 5 through 8 find the solution of the given initial value problem. State the interval in which the solution is valid.

5. $xy' + 2y = x^2 - x + 1,$ $y(1) = \frac{1}{2}$
6. $xy' + y = e^x,$ $y(1) = 1$
7. $y' + (\cot x)y = 2 \csc x,$ $y(\pi/2) = 1$
8. $xy' + 2y = \sin x,$ $y(\pi) = 1/\pi$

Each of the equations in Problems 9 through 12 has at least one discontinuous coefficient at $x = 0$. Solve each equation for $x > 0$ and describe the behavior of the solution as $x \to 0$ for various values of the constant of integration. Sketch several members of the family of integral curves.

9. $y' + (2/x)y = 1/x^2$ 10. $y' - (1/x)y = x$
11. $y' - (1/x)y = x^{1/2}$ 12. $y' + (1/x)y = (\cos x)/x$

13. (a) Show that the solution (3) of the general linear equation (1) can be written in the form

$$y = cy_1(x) + y_2(x),\qquad\qquad (i)$$

where c is an arbitrary constant. Identify the functions y_1 and y_2.

(b) Show that y_1 is a solution of the differential equation

$$y' + p(x)y = 0,\qquad\qquad (ii)$$

corresponding to $g(x) = 0$.

(c) Show that y_2 is a solution of the full linear equation (1). We will see later (for example, in Section 3.6) that solutions of higher order linear equations have a pattern similar to Eq. (i).

14. Show that if a and λ are positive constants, and b is any real number, then every solution of the equation

$$y' + ay = be^{-\lambda x}$$

has the property that $y \to 0$ as $x \to \infty$.

Hint: Consider the cases $a = \lambda$ and $a \neq \lambda$ separately.

15. **Discontinuous Coefficients.** Linear differential equations sometimes occur in which one or both of the functions p and g have jump discontinuities. If x_0 is such a point of discontinuity, then it is necessary to solve the equation separately for $x < x_0$ and for $x > x_0$. Afterwards the two solutions are matched so that y is continuous at x_0; this is accomplished by a proper choice of the arbitrary constants. The following two problems illustrate this situation. Note in each case that it is impossible also to make y' continuous at x_0.

(a) Solve the initial value problem

$$y' + 2y = g(x), \qquad y(0) = 0$$

where

$$g(x) = \begin{cases} 1, & 0 \leq x \leq 1, \\ 0, & x > 1, \end{cases}$$

(b) Solve the initial value problem

$$y' + p(x)y = 0, \qquad y(0) = 1$$

where

$$p(x) = \begin{cases} 2, & 0 \leq x \leq 1, \\ 1, & x > 1. \end{cases}$$

*Bernoulli Equations. Sometimes it is possible to solve a nonlinear equation by making a change of the dependent variable that converts it into a linear equation. The most important class of such equations is of the form

$$y' + p(x)y = q(x)y^n;$$

such equations are called Bernoulli equations after Jakob Bernoulli (1654–1705). Problems 16 and 17 deal with equations of this type.

16. (a) Solve Bernoulli's equation when $n = 0$; when $n = 1$.

(b) Show that if $n \neq 0, 1$, then the substitution $v = y^{1-n}$ reduces Bernoulli's equation to a linear equation. This method of solution was found by Leibniz in 1696.

17. Use the method of Problem 16(b) to solve each of the following differential equations.

(a) $x^2y' + 2xy - y^3 = 0$

(b) $y' = \epsilon y - \sigma y^2;$ $\qquad \epsilon > 0 \qquad$ and $\qquad \sigma > 0$

(c) $y' = \epsilon y - \sigma y^3;$ $\qquad \epsilon > 0 \qquad$ and $\qquad \sigma > 0$

(d) $y' = \epsilon y - f(x)y^3;$ $\quad \epsilon > 0$

Note: Equation (b) is important in the study of population dynamics. Equation (c) and its generalization (d) occur in hydrodynamic stability. Equations (b) and (c) can also be solved by the methods of Section 2.4. For a further discussion of Eq. (b), see Problem 19 of Section 2.4, Problem 11 of Section 2.9, and Section 9.1.

2.3 NONLINEAR EQUATIONS

We now turn to a study of differential equations of the form

$$y' = f(x, y), \tag{1}$$

subject to an initial condition

$$y(x_0) = y_0. \tag{2}$$

The differential equation (1) and the initial condition (2) together constitute an initial value problem. The basic questions to be considered are whether a solution of this initial value problem exists, whether such a solution is unique, over what interval a solution is defined, and how to construct a useful formula for the solution. All of these questions were answered with relative ease in Sections 2.1 and 2.2 for the case in which Eq. (1) is linear. However, if f is not a linear function of the dependent variable y, then our earlier treatment no longer applies. In this section we will discuss in a general way some features of nonlinear initial value problems. In particular, we will note several important differences between the nonlinear problem (1), (2) and the linear problem consisting of the differential equation

$$y' + p(x)y = g(x) \tag{3}$$

and the initial condition (2).

The reason why first order linear differential equations are relatively simple is that there is a formula giving the solution of such an equation in all cases. In contrast, there is no corresponding general method for solving first order nonlinear equations. In fact, the analytic determination of the solution $y = \phi(x)$ of a nonlinear equation is usually very difficult, and often impossible.

The lack of a general formula for the solution of a nonlinear equation has at least two important consequences. In the first place, methods which yield approximate, perhaps numerical, solutions and qualitative information about solutions assume greater significance for nonlinear equations than for linear ones. Chapters 8 and 9 contain an introduction to some of these methods. Second, questions dealing with the existence and uniqueness of solutions must now be dealt with by indirect methods, since a direct construction of the solution cannot be carried out in general.

Existence and Uniqueness. The following fundamental existence and uniqueness theorem is analogous to Theorem 2.1 for linear equations. However, its proof is a great deal more complicated, and it is postponed until Section 2.11.

Theorem 2.2. *Let the functions f and $\partial f/\partial y$ be continuous in some rectangle $\alpha < x < \beta, \gamma < y < \delta$ containing the point (x_0, y_0). Then, in some interval $x_0 - h < x < x_0 + h$ contained in $\alpha < x < \beta$, there is a unique solution $y = \phi(x)$ of the initial value problem (1), (2)*

$$y' = f(x, y), \qquad y(x_0) = y_0.$$

The conditions stated in Theorem 2.2 are sufficient to guarantee the existence of a unique solution of the initial value problem (1), (2) in some interval $x_0 - h < x < x_0 + h$. However, the determination of a value for h may be difficult. Further, even if f does not satisfy the hypotheses of the theorem, it is still possible that a unique solution may exist. Indeed, the conclusion of the theorem remains true if the hypothesis about the continuity of $\partial f/\partial y$ is replaced by certain weaker conditions. Further, the existence of a solution (but not its uniqueness) can be established on the basis of the continuity of f alone, without any additional hypotheses at all. The present form of the theorem, however, is satisfactory for most purposes.

It can be shown by means of examples that some conditions on f are essential in order to obtain the result stated in the theorem. For instance, the following example shows that the initial value problem (1), (2) may have more than one solution if the hypotheses of Theorem 2.2 are violated.

Example 1.   Find the solution of

$$y' = y^{1/3}, \qquad y(0) = 0 \tag{4}$$

for $x \geq 0$.

This problem is easily solved by the method of Section 2.4. For the present we can verify that the function

$$y = \phi(x) = (\tfrac{2}{3}x)^{3/2}, \quad x \geq 0$$

satisfies both of Eqs. (4). On the other hand, the function

$$y = \psi(x) = 0$$

is also a solution of the given initial value problem. Hence this problem does not have a unique solution. This fact does not contradict the existence and uniqueness theorem since

$$\frac{\partial}{\partial y} f(x, y) = \frac{\partial}{\partial y} (y^{1/3}) = \tfrac{1}{3} y^{-2/3},$$

and this function is not continuous, or even defined, at any point where $y = 0$. Hence the theorem does not apply in any region containing any part of the x axis. If (x_0, y_0) is any point not on the x axis, however, then there is a unique solution of the differential equation $y' = y^{1/3}$ passing through (x_0, y_0).

Interval of Definition.   For the linear problem (2), (3) the solution exists throughout any interval about $x = x_0$ in which the functions p and g are continuous. On the other hand, for the nonlinear initial value problem (1), (2), the interval in which a solution exists may be difficult to determine. The solution $y = \phi(x)$ exists as long as the point $[x, \phi(x)]$ remains within the region in which the hypotheses of the theorem are satisfied; however, since $\phi(x)$ is usually not known, it may be impossible to locate the point $[x, \phi(x)]$ with respect to this region. In any case, the interval in which a solution exists may have no simple relationship to the function f in Eq. (1). This is illustrated by the following example.

Example 2. Consider the initial value problem

$$y' = y^2, \qquad y(0) = 1. \tag{5}$$

It can be readily verified by direct substitution that

$$y = \frac{1}{1 - x} \tag{6}$$

is the solution of this initial value problem. Clearly the solution becomes unbounded as $x \to 1$, and therefore it is valid only for $-\infty < x < 1$. There is no indication from the differential equation itself, however, that the point $x = 1$ is in any way remarkable. Moreover, if the initial condition is replaced by

$$y(0) = 2, \tag{7}$$

it is again easy to verify that the solution of the differential equation (5) satisfying the initial condition (7) is

$$y = \frac{2}{1 - 2x}, \tag{8}$$

and that the solution now becomes unbounded as $x \to \frac{1}{2}$. This illustrates another disturbing feature of initial value problems for nonlinear equations, namely, the singularities of the solution may move about depending on the initial condition.

General Solution. Another way in which linear and nonlinear equations differ is in connection with the concept of a general solution. For a first order linear equation it is possible to obtain a solution containing one arbitrary constant, from which all possible solutions follow by specifying values for this constant. For nonlinear equations this may not be the case; even though a solution containing an arbitrary constant may be found, there may be other solutions that cannot be obtained by giving values to this constant. Specific examples are given in Problems 3 and 6. Thus we will use the term "general solution" only when discussing linear equations.

Implicit Solutions. We recall again that for a first order linear equation there is an explicit formula [Eq. (22) of Section 2.1] for the solution $y = \phi(x)$. As long as the necessary antiderivatives can be determined, the value of the solution at any point can be found merely by substituting the appropriate value of x into the formula. For a nonlinear equation it is rarely possible to find such an explicit solution. Often the best that we can do is to eliminate the derivative that appears in Eq. (1), thereby obtaining in place of the differential equation a derivative-free equation of the form

$$\psi[x, \phi(x)] = 0 \tag{9}$$

which is satisfied by some, perhaps all, solutions $y = \phi(x)$ of Eq. (1). Even this cannot be accomplished in most cases for nonlinear equations. However, if a relation such as Eq. (9) can be found, it is customary to say that we have an *implicit formula* for the solutions of Eq. (1). An equation of the form (9) is also called an *integral* (or first integral) of Eq. (1).

For example, consider the simple nonlinear equation

$$y' = -x/y. \tag{10}$$

Using the methods of the next section, it is not hard to show that all solutions of Eq. (10) also satisfy the algebraic equation

$$x^2 + y^2 = c^2, \tag{11}$$

where c is an arbitrary constant. To verify this statement we can differentiate Eq. (11) with respect to x, thereby obtaining $2x + 2yy' = 0$, or $y' = -x/y$, which is Eq. (10). Equation (11) is therefore an implicit formula for solutions of Eq. (10). For $x^2 \leq c^2$, there are many functions $y = \phi(x)$ that satisfy Eq. (11). Some of these are:

$$y = \phi_1(x) = \sqrt{c^2 - x^2}, \qquad -c \leq x \leq c \tag{12}$$

$$y = \phi_2(x) = -\sqrt{c^2 - x^2}, \qquad -c \leq x \leq c \tag{13}$$

$$y = \phi_3(x) = \begin{cases} \sqrt{c^2 - x^2}, & -c \leq x \leq 0 \\ -\sqrt{c^2 - x^2}, & 0 < x \leq c \end{cases} \tag{14}$$

$$y = \phi_4(x) = \begin{cases} -\sqrt{c^2 - x^2}, & -c \leq x \leq -c/2 \\ \sqrt{c^2 - x^2}, & -c/2 < x < c/2 \\ -\sqrt{c^2 - x^2}, & c/2 \leq x \leq c. \end{cases} \tag{15}$$

See Figure 2.3 for graphs of the functions given by Eqs. (12) to (15). However, only two of these functions satisfy the differential equation (10) over the *entire* interval $-c < x < c$, namely those given by Eqs. (12) and (13). If an initial condition is also given, then we want to select whichever of the functions ϕ_1 and ϕ_2 satisfies the specified condition, and also select the proper value of c. For example, if the initial condition is

$$y(0) = 3, \tag{16}$$

then we must discard ϕ_2, and keep ϕ_1 with c chosen to be 3. Hence

$$y = \sqrt{9 - x^2}, \qquad -3 < x < 3 \tag{17}$$

is a solution of the initial value problem (10) and (16). According to Theorem 2.2 there is no other solution of this problem on the interval $-3 < x < 3$.

In the foregoing example the explicit solutions given by Eqs. (12) and (13) were easily obtainable because the implicit relation (11) was quadratic in y. However, a little imagination will suggest that an implicit relation (9), assuming it can be found, will often be much more complicated than Eq. (11). If so, it will

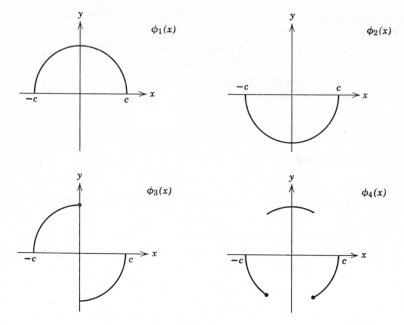

FIGURE 2.3

probably be impossible to solve it (analytically) for y; it may be quite difficult even to determine the intervals in which solutions exist. Furthermore, it must be kept in mind that there may be solutions of the implicit relation (9) that do not satisfy the differential equation; also, in some cases the differential equation may have other solutions that do not satisfy the implicit relation.

Graphical Construction of Integral Curves. Since there is no way, in general, to obtain exact analytic solutions of nonlinear differential equations, methods that yield approximate solutions or other qualitative information about solutions may be of great importance. One such method involves the graphical approximation of integral curves. At each point in the xy plane where $f(x, y)$ is defined, the differential equation (1)

$$y' = f(x, y)$$

provides a value for y', which can be thought of as the slope of a line segment through that point. The totality of all such line segments forms the *direction field* for the given differential equation. The integral curves are curves that, at every point, are tangent to the element of the direction field associated with that point.

The general shape of the integral curves can sometimes be visualized by drawing the elements of the direction field at a sufficiently large number of points. For example, consider

$$y' = 4y(1 - y), \tag{18}$$

a simple but important type of nonlinear equation. The construction of the direction field is greatly simplified by noting that the right-hand side of Eq. (18) is independent of x. Therefore, the elements of the direction field at all points on any line parallel to the x axis are themselves parallel. Further, two special solutions of Eq. (18) can be found by inspection, namely the constant solutions $y = \phi_1(x) = 0$ and $y = \phi_2(x) = 1$. These special solutions separate the xy plane into three zones, in each of which y' is of one sign. In fact,

$$\text{if} \qquad y < 0, \quad \text{then} \quad y' < 0;$$
$$\text{if} \qquad 0 < y < 1, \quad \text{then} \quad y' > 0;$$
$$\text{if} \qquad y > 1, \quad \text{then} \quad y' < 0.$$

Using this information it is relatively simple to construct Figure 2.4, which depicts the direction field for Eq. (18).

From this sketch certain qualitative properties of solutions can be easily deduced. In the first place, consider the solution corresponding to the initial condition $y(x_0) = y_0$ with $0 < y_0 < 1$. Such a solution tends to increase since $y' > 0$ in the vicinity of the initial point. However, this solution cannot increase beyond $y = 1$ because, if it reached $y = 1$, its slope would have become (and would remain) zero. In fact, the given solution cannot reach $y = 1$ for any finite value of x; if it did, then from that point as the initial point there would emanate two solutions [the solution in question and the solution $y = \phi_2(x) = 1$] in contradiction to Theorem 2.2. Thus a solution starting in the interval $0 < y < 1$ approaches $y = 1$ as $x \to \infty$. Similarly, a solution starting above $y = 1$ must decrease

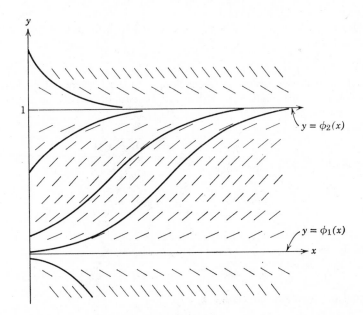

FIGURE 2.4

and approach $y = 1$ from above as $x \to \infty$. Finally, a solution that is originally negative tends to decrease further and eventually approaches $-\infty$.

It happens that Eq. (18) can be solved by the method of Section 2.4, or recognized as a Bernoulli equation (Problems 16 and 17 of Section 2.2). A few solutions are plotted in Figure 2.4. However, we emphasize that the qualitative behavior of the solutions of Eq. (18) was determined without solving the equation. The ideas touched on here are related to the concepts of stability and instability of solutions of differential equations, and are elaborated upon in Chapter 9.

PROBLEMS

1. For each of the following differential equations state the region in the xy plane where the existence of a unique solution through any specified point is guaranteed by the fundamental existence and uniqueness theorem.

(a) $y' = \dfrac{x - y}{2x + 5y}$ (b) $y' = (1 - x^2 - y^2)^{\frac{1}{2}}$

(c) $y' = 2xy/(1 + y^2)$ (d) $y' = 3(x + y)^{-2}$

(e) $y' = \dfrac{\ln |xy|}{1 - x^2 + y^2}$ (f) $y' = (x^2 + y^2)^{\frac{3}{2}}$

2. Show that $y = \phi(x) = (1 - x^2)^{-1}$ is a solution of the initial value problem

$$y' = 2xy^2, \qquad y(0) = 1.$$

In what interval is this solution valid?

3. Show that $y = \phi(x) = [2(x + c)]^{-\frac{1}{2}}$, where c is an arbitrary constant, satisfies the differential equation

$$y' + y^3 = 0.$$

Find the solution that satisfies the initial condition $y(1) = 2$. Where is this solution valid? Note that $y = 0$ is also a solution of the differential equation, but that this solution cannot be obtained by assigning a value to c in $y = \phi(x)$.

4. Verify that

$$y = \phi(x) = [1 + \tfrac{2}{3} \ln (1 + x^3)]^{\frac{1}{2}}$$

is a solution of the initial value problem

$$y' = \frac{x^2}{y(1 + x^3)}, \qquad y(0) = 1.$$

In what interval is this solution valid?

5. Verify that

$$y = (c^2 - 4x^2)^{\frac{1}{2}}, \qquad y = -(c^2 - 4x^2)^{\frac{1}{2}}$$

are solutions of the differential equation

$$y' = -4x/y.$$

For what parts of the xy plane are these solutions valid? Find the particular solution passing through the point $(0, 4)$; through the point $(1, -1)$.

6. (a) Verify that both $y_1(x) = 1 - x$ and $y_2(x) = -x^2/4$ are solutions of the initial value problem

$$y' = \frac{-x + (x^2 + 4y)^{1/2}}{2}, \qquad y(2) = -1.$$

Where are these solutions valid?

(b) Explain why the existence of two solutions of the given problem does not contradict the uniqueness part of Theorem 2.2.

(c) Show that $y = cx + c^2$, where c is an arbitrary constant, satisfies the differential equation in part (a) for $x \geq -2c$. If $c = -1$, the initial condition is also satisfied, and the solution $y = y_1(x)$ is obtained. Show that there is no choice of c that gives the second solution $y = y_2(x)$.

7. For each of the following differential equations, construct the direction field sufficiently well to be able to visualize the general behavior of the family of integral curves.

(a) $y' = x^2 + y^2$ (b) $y' = x^2 - xy + y^2 - 1$

(c) $y' = \dfrac{xy}{1 + x^2}$ (d) $y' = \dfrac{2x - 3y}{x + y}$

8. For each of the following differential equations determine the general pattern of the integral curves by a consideration of the direction field. Let $y = \phi(x; y_0)$ be the solution corresponding to the initial condition $y(0) = y_0$. For what values of y_0 is $\lim\limits_{x \to \infty} \phi(x; y_0)$ finite? Evaluate this limit in those cases.

(a) $y' = -y(1 + y^2)$ (b) $y' = y(1 - y^2)$

(c) $y' = (1 - y)(2 - y)$ (d) $y' = \epsilon y - \sigma y^2$; $\sigma > 0$ and $\epsilon > 0$

In each of Problems 9 through 12:

(a) Sketch the direction field.

(b) Sketch several members of the family of integral curves without solving the differential equation.

(c) Solve the given initial value problem.

(d) Sketch the solution obtained in (c) and compare with those obtained in (b).

9. $y' = -1$, $y(1) = 2$ 10. $y' = x + 1$, $y(0) = 1$

11. $y' + xy = 1$, $y(1) = 0$ 12. $y' - y = 1 - e^x$, $y(0) = 0$

2.4 SEPARABLE EQUATIONS

It is often convenient to write the equation

$$\frac{dy}{dx} = f(x, y) \tag{1}$$

in the form

$$M(x, y) + N(x, y)\frac{dy}{dx} = 0. \tag{2}$$

It is always possible to do this by setting $M(x, y) = -f(x, y)$ and $N(x, y) = 1$, but there may be other ways as well. In the event that M is a function of x only and N is a function of y only, then Eq. (2) takes the form

$$M(x) + N(y) \frac{dy}{dx} = 0. \tag{3}$$

For example, the equation

$$\frac{dy}{dx} = \frac{x^2}{1 + y^2}$$

can be written as

$$-x^2 + (1 + y^2) \frac{dy}{dx} = 0.$$

A differential equation that can be put in the form (3) is said to be *separable*; the reason for this name is clear if Eq. (3) is written in the differential form

$$M(x) \, dx = -N(y) \, dy. \tag{4}$$

Now suppose that H_1 and H_2 are any functions such that

$$H_1'(x) = M(x), \qquad H_2'(y) = N(y); \tag{5}$$

then Eq. (3) becomes

$$H_1'(x) + H_2'(y) \frac{dy}{dx} = 0. \tag{6}$$

If $y = \phi(x)$ is a differentiable function of x, then according to the chain rule

$$H_2'(y) \frac{dy}{dx} = \frac{d}{dx} H_2[\phi(x)]. \tag{7}$$

Consequently, if $y = \phi(x)$ is a solution of the differential equation (3), then Eq. (6) becomes

$$H_1'(x) + \frac{d}{dx} H_2[\phi(x)] = 0,$$

or

$$\frac{d}{dx} \{H_1(x) + H_2[\phi(x)]\} = 0. \tag{8}$$

Integrating Eq. (8) gives

$$H_1(x) + H_2[\phi(x)] = c, \tag{9}$$

or

$$H_1(x) + H_2(y) = c, \tag{10}$$

where c is an arbitrary constant. Thus the solution $y = \phi(x)$ of the differential equation (3) has been obtained in the implicit form (10). The functions H_1 and H_2 are any functions satisfying Eq. (5); that is, any antiderivatives of M and N, respectively. In practice the solution (10) is generally obtained from Eq. (4) by integrating the left side with respect to x and the right side with respect to y; the justification for this is the argument just given.

If, in addition to the differential equation, an initial condition

$$y(x_0) = y_0 \tag{11}$$

is prescribed, the solution of Eq. (3) satisfying this condition is obtained by setting $x = x_0$ and $y = y_0$ in Eq. (10). This gives

$$c = H_1(x_0) + H_2(y_0). \tag{12}$$

Substituting this value for c in Eq. (10), and noting that

$$H_1(x) - H_1(x_0) = \int_{x_0}^x M(t)\, dt, \qquad H_2(y) - H_2(y_0) = \int_{y_0}^y N(t)\, dt,$$

we obtain

$$\int_{x_0}^x M(t)\, dt + \int_{y_0}^y N(t)\, dt = 0. \tag{13}$$

Equation (13) is an implicit representation of the solution of the differential equation (3), which also satisfies the initial condition (11). The reader should bear in mind that the determination of the solution in explicit form, or even the determination of the precise interval in which the solution exists, generally requires that Eq. (13) be solved for y as a function of x. This may present formidable difficulties.

Example 1. Solve the initial value problem

$$\frac{dy}{dx} = \frac{3x^2 + 4x + 2}{2(y - 1)}, \qquad y(0) = -1. \tag{14}$$

The differential equation can be written as

$$2(y - 1)\, dy = (3x^2 + 4x + 2)\, dx.$$

Integrating the left-hand side with respect to y and the right-hand side with respect to x gives

$$y^2 - 2y = x^3 + 2x^2 + 2x + c, \tag{15}$$

where c is an arbitrary constant. To determine the solution satisfying the prescribed initial condition we substitute $x = 0$ and $y = -1$ into Eq. (15), obtaining $c = 3$. Hence the solution of the initial value problem (14) is given implicitly by

$$y^2 - 2y = x^3 + 2x^2 + 2x + 3. \tag{16}$$

To obtain the solution explicitly, we must solve Eq. (16) for y in terms of x. This is a simple matter in this case since Eq. (16) is quadratic in y, and we obtain

$$y = 1 \pm \sqrt{x^3 + 2x^2 + 2x + 4}. \tag{17}$$

Thus Eq. (17) gives two solutions of the differential equation, only one of which, however, satisfies the given initial condition. This is the solution corresponding to the minus sign in Eq. (17), so that we finally obtain

$$y = \phi(x) = 1 - \sqrt{x^3 + 2x^2 + 2x + 4} \tag{18}$$

as the solution of the initial value problem (14). Note that if the plus sign is chosen by mistake in Eq. (17), then we will obtain the solution of the same differential equation that satisfies the initial condition $y(0) = 3$. Finally, in order to determine the interval in which the solution (18) is valid, we must find the interval in which the quantity under the radical is positive. By plotting this expression as a function of x, the reader can show that the desired interval is $x > -2$.

Example 2. Find the solution of the initial value problem

$$\frac{dy}{dx} = \frac{y \cos x}{1 + 2y^2}, \qquad y(0) = 1. \tag{19}$$

First we write the differential equation in the form

$$\frac{1 + 2y^2}{y} dy = \cos x \, dx. \tag{20}$$

Integrating the left-hand side with respect to y and the right-hand side with respect to x, we obtain

$$\ln |y| + y^2 = \sin x + c. \tag{21}$$

To satisfy the initial condition we substitute $x = 0$ and $y = 1$ in Eq. (21); this gives $c = 1$. Hence the solution of the initial value problem (19) is given implicitly by

$$\ln |y| + y^2 = \sin x + 1. \tag{22}$$

Since Eq. (22) is not readily solved for y as a function of x, further analysis of this problem becomes more delicate. One fairly evident fact is that no solution crosses the x axis; that is, no point whose y coordinate is zero satisfies Eq. (22). To see this, observe that the left-hand side of Eq. (22) becomes infinite if $y = 0$; however, the right-hand side never becomes infinite, so no point on the x axis satisfies Eq. (22). Thus, for the solution of Eqs. (19) it follows that $y > 0$ always. Consequently, the absolute value bars in Eq. (22) can be dropped for this problem. It can also be shown that the interval of definition of the solution of the initial value problem (19) is the entire x axis, $-\infty < x < \infty$. Problem 20 indicates how to do this.

In Example 1 it was not difficult to solve explicitly for y as a function of x and to determine the exact interval in which the solution exists. However, this situation is exceptional, and often it will be necessary to leave the solution in implicit form, as in Example 2. Thus, in the problems below and in those following Sections 2.5 through 2.8, the terminology "Solve the following differential equation" means to find the solution explicitly if it is convenient to do so, but otherwise to find an implicit formula for the solution.

PROBLEMS

Solve each of the equations in Problems 1 through 8. State the regions of the xy plane in which the conditions of the fundamental existence and uniqueness theorem are satisfied.

1. $\dfrac{dy}{dx} = \dfrac{x^2}{y}$

2. $\dfrac{dy}{dx} = \dfrac{x^2}{y(1 + x^3)}$

3. $y' + y^2 \sin x = 0$

4. $y' = 1 + x + y^2 + xy^2$

5. $y' = (\cos^2 x)(\cos^2 2y)$

6. $xy' = (1 - y^2)^{1/2}$

7. $\dfrac{dy}{dx} = \dfrac{x - e^{-x}}{y + e^y}$

8. $\dfrac{dy}{dx} = \dfrac{x^2}{1 + y^2}$

For each of Problems 9 through 14 find the solution of the given initial value problem in explicit form, and determine (at least approximately) the interval in which it is defined.

9. $\sin 2x \, dx + \cos 3y \, dy = 0, \qquad y(\pi/2) = \pi/3$

10. $x \, dx + ye^{-x} \, dy = 0, \qquad y(0) = 1$

11. $\dfrac{dr}{d\theta} = r, \qquad r(0) = 2$

12. $\dfrac{dy}{dx} = \dfrac{2x}{y + x^2 y}, \qquad y(0) = -2$

13. $\dfrac{dy}{dx} = xy^3(1 + x^2)^{-1/2}, \qquad y(0) = 1$

14. $\dfrac{dy}{dx} = \dfrac{2x}{1 + 2y}, \qquad y(2) = 0$

15. Solve the equation
$$y^2(1 - x^2)^{1/2} \, dy = \sin^{-1} x \, dx$$
in the interval $-1 < x < 1$.

16. Solve the equation
$$\frac{dy}{dx} = \frac{ax + b}{cx + d},$$
where a, b, c, and d are constants.

17. Solve the equation
$$\frac{dy}{dx} = \frac{ay + b}{cy + d},$$
where a, b, c, and d are constants.

*18. Show that the equation
$$\frac{dy}{dx} = \frac{y - 4x}{x - y}$$
is not separable, but that if the variable y is replaced by a new variable v defined by $v = y/x$, then the equation is separable in x and v. Find the solution of the given equation by this technique.

19. (a) Solve the initial value problem $y' = \epsilon y$, $y(0) = y_0$, where ϵ is a positive constant. Show that the solution $y = \phi(x)$ has the following limiting behavior as $x \to \infty$:

$$\text{if } y_0 > 0, \qquad \lim_{x \to \infty} \phi(x) = \infty;$$

$$\text{if } y_0 = 0, \qquad \lim_{x \to \infty} \phi(x) = 0;$$

$$\text{if } y_0 < 0, \qquad \lim_{x \to \infty} \phi(x) = -\infty.$$

(b) Now suppose that the differential equation is changed by the addition of the term $-\sigma y^2$ where σ is a (small) positive number. Using a partial fraction expansion, solve the initial value problem

$$y' = \epsilon y - \sigma y^2, \qquad y(0) = y_0.$$

Show that $y = \phi(x)$ now has the following behavior as $x \to \infty$:

$$\text{if } y_0 > 0, \qquad \lim_{x \to \infty} \phi(x) = \frac{\epsilon}{\sigma}; \qquad\qquad \text{if } y_0 = 0, \qquad \lim_{x \to \infty} \phi(x) = 0.$$

Show that if $y_0 < 0$ then $\phi(x)$ becomes unbounded as x approaches a certain finite value depending on y_0. The nonlinear equation in part (b) is important in many applications (see Sections 2.9 and 9.1). Note also that this equation is a Bernoulli equation and can be solved by the method of Problem 16 in Section 2.2.

*20. Consider again the initial value problem (19) of the text. Denote the right-hand side of the differential equation by $f(x, y)$ and observe that

$$|f(x, y)| = \left| \frac{y}{1 + 2y^2} \right| |\cos x|.$$

(a) By finding the maximum and minimum values of $y/(1 + 2y^2)$, show that

$$\left| \frac{y}{1 + 2y^2} \right| \leq \frac{1}{2\sqrt{2}}$$

for all y, and hence that $|f(x, y)| \leq 1/2\sqrt{2}$ for all x and y.

(b) Use the result of part (a) to show that the interval of definition of the solution of the initial value problem (19) is $-\infty < x < \infty$.

2.5 EXACT EQUATIONS

Let us first consider the equation

$$\psi(x, y) = c \tag{1}$$

where c is a constant. Assuming that Eq. (1) defines y implicitly as a differentiable function of x, we can differentiate Eq. (1) with respect to x and obtain*

$$\psi_x(x, y) + \psi_y(x, y)y' = 0. \tag{2}$$

* Subscripts denote partial derivatives with respect to the variable indicated by the subscript.

Equation (2) is the differential equation whose solution is defined by Eq. (1).

Conversely, suppose that the differential equation

$$M(x, y) + N(x, y)y' = 0 \tag{3}$$

is given. If there exists a function ψ such that

$$\psi_x(x, y) = M(x, y), \qquad \psi_y(x, y) = N(x, y), \tag{4}$$

and such that $\psi(x, y) = c$ defines $y = \phi(x)$ implicitly as a differentiable function of x, then

$$M(x, y) + N(x, y)y' = \psi_x(x, y) + \psi_y(x, y)y'$$

$$= \frac{d}{dx}\,\psi[x, \phi(x)]. \tag{5}$$

As a result Eq. (3) becomes

$$\frac{d}{dx}\,\psi[x, \phi(x)] = 0. \tag{6}$$

In this case Eq. (3) is said to be an *exact differential equation*. The solution of Eq. (3), or the equivalent Eq. (6), is given implicitly by Eq. (1),

$$\psi(x, y) = c,$$

where c is an arbitrary constant.

In practice the differential equation (3) is often written in the symmetric differential form

$$M(x, y)\,dx + N(x, y)\,dy = 0. \tag{7}$$

Example 1.   Solve the differential equation

$$2xy^3 + 3x^2y^2\,\frac{dy}{dx} = 0.$$

By inspection it can be seen that the left-hand side is the derivative of x^2y^3. Thus the given equation can be rewritten as

$$\frac{d}{dx}\,(x^2y^3) = 0,$$

and the solution is given implicitly by $x^2y^3 = c$ and explicitly by

$$y = kx^{-2/3}.$$

In this simple example it was easy to see that the differential equation was exact and, in fact, easy to find its solution, by recognizing that the left-hand side was the derivative of x^2y^3. For more complicated equations it may not be possible to do this. A systematic procedure for determining whether a given differential equation is exact is provided by the following theorem.

Theorem 2.3. Let the functions M, N, M_y, and N_x be continuous in the rectangular* region R: $\alpha < x < \beta$, $\gamma < y < \delta$. Then Eq. (3),

$$M(x, y) + N(x, y)y' = 0,$$

is an exact differential equation in R if and only if

$$M_y(x, y) = N_x(x, y) \tag{8}$$

at each point of R. That is, there exists a function ψ satisfying Eqs. (4),

$$\psi_x(x, y) = M(x, y), \qquad \psi_y(x, y) = N(x, y),$$

if and only if M and N satisfy Eq. (8).

The proof of this theorem is in two parts. First we will show that if there is a function ψ such that Eq. (4) is true, then it follows that Eq. (8) is satisfied. Computing M_y and N_x from Eq. (4) yields

$$M_y(x, y) = \psi_{xy}(x, y), \qquad N_x(x, y) = \psi_{yx}(x, y). \tag{9}$$

Since M_y and N_x are continuous, it follows that ψ_{xy} and ψ_{yx} are also continuous. This guarantees their equality†, and Eq. (8) follows.

We will now show that if M and N satisfy Eq. (8), then Eq. (3) is exact. The proof involves the construction of a function ψ satisfying Eqs. (4),

$$\psi_x(x, y) = M(x, y), \qquad \psi_y(x, y) = N(x, y).$$

Integrating the first of Eqs. (4) with respect to x, holding y constant, gives

$$\psi(x, y) = \int^x M(t, y)\, dt + h(y). \tag{10}$$

The function h is an arbitrary function of y, playing the role of the arbitrary constant. Now we must show that it is always possible to choose $h(y)$ so that $\psi_y = N$. From Eq. (10)

$$\psi_y(x, y) = \frac{\partial}{\partial y} \int^x M(t, y)\, dt + h'(y)$$

$$= \int^x M_y(t, y)\, dt + h'(y).$$

Setting $\psi_y = N$ and solving for $h'(y)$ gives

$$h'(y) = N(x, y) - \int^x M_y(t, y)\, dt. \tag{11}$$

* It is not essential that the region R be rectangular, but only that it be simply connected; see Problem 16 for a discussion of the meaning of this term.

† The hypothesis of continuity is required, since otherwise ψ_{xy} and ψ_{yx} are not always equal. The exceptions are rather peculiar functions, however, and are seldom encountered.

In order to determine $h(y)$ from Eq. (11) it is essential that, despite its appearance, the right side of Eq. (11) be a function of y only. To establish this fact we can differentiate the quantity in question with respect to x, obtaining

$$N_x(x, y) - M_y(x, y),$$

which vanishes on account of Eq. (8). Hence, despite its apparent form, the right side of Eq. (11) does not in fact depend on x, and a single integration then gives $h(y)$. Substituting for $h(y)$ in Eq. (10), we obtain as the solution of Eqs. (4)

$$\psi(x, y) = \int^x M(t, y)\, dt + \int^y \left[N(x, s) - \int^x M_s(t, s)\, dt \right] ds. \qquad (12)$$

It should be noted that this proof contains a method for the computation of $\psi(x, y)$ and thus for solving the original differential equation (3). It is usually better to go through this process each time it is needed rather than to try to remember the result given in Eq. (12). Note also that the solution is obtained in implicit form. As in the previous section it may or may not be feasible to find the solution explicitly.

Example 2.   Solve

$$(y \cos x + 2xe^y) + (\sin x + x^2 e^y + 2)y' = 0. \qquad (13)$$

It is clear that

$$M_y(x, y) = \cos x + 2xe^y = N_x(x, y),$$

and the given equation is exact. Thus there is a $\psi(x, y)$ such that

$$\psi_x(x, y) = y \cos x + 2xe^y,$$

$$\psi_y(x, y) = \sin x + x^2 e^y + 2.$$

Integrating the first of these equations, we obtain

$$\psi(x, y) = y \sin x + x^2 e^y + h(y). \qquad (14)$$

Setting $\psi_y = N$ gives

$$\psi_y(x, y) = \sin x + x^2 e^y + h'(y) = \sin x + x^2 e^y + 2.$$

Thus

$$h'(y) = 2,$$

and

$$h(y) = 2y.$$

The constant of integration can be omitted since any solution of the previous differential equation will suffice; we do not require the most general one. Substituting for $h(y)$ in Eq. (14) gives

$$\psi(x, y) = y \sin x + x^2 e^y + 2y.$$

Hence the solution of the original equation is given implicitly by

$$y \sin x + x^2 e^y + 2y = c. \tag{15}$$

Example 3. Solve

$$(3x^2 + 2xy) + (x + y^2)y' = 0. \tag{16}$$

Here

$$M_y(x, y) = 2x, \qquad N_x(x, y) = 1;$$

since $M_y \neq N_x$, the given equation is not exact. To see that it cannot be solved by the procedure described above, let us seek a function ψ such that

$$\psi_x(x, y) = 3x^2 + 2xy, \qquad \psi_y(x, y) = x + y^2. \tag{17}$$

Integrating the first of Eqs. (17) gives

$$\psi(x, y) = x^3 + x^2 y + h(y), \tag{18}$$

where h is an arbitrary function of y only. To try to satisfy the second of Eqs. (17) we compute ψ_y from Eq. (18) and set it equal to N, obtaining

$$\psi_y(x, y) = x^2 + h'(y) = x + y^2$$

or

$$h'(y) = x + y^2 - x^2. \tag{19}$$

Since the right-hand side of Eq. (19) depends on x as well as y, it is impossible to solve Eq. (19) for $h(y)$. Thus there is no $\psi(x, y)$ satisfying both of Eqs. (17).

PROBLEMS

Determine whether or not each of the equations in Problems 1 through 12 is exact. If it is exact, find the solution.

1. $(2x + 3) + (2y - 2)y' = 0$ 2. $(2x + 4y) + (2x - 2y)y' = 0$

3. $(9x^2 + y - 1) - (4y - x)y' = 0$ 4. $(2xy^2 + 2y) + (2x^2 y + 2x)y' = 0$

5. $\dfrac{dy}{dx} = -\dfrac{ax + by}{bx + cy}$ 6. $\dfrac{dy}{dx} = -\dfrac{ax - by}{bx - cy}$

7. $(e^x \sin y - 2y \sin x)\, dx + (e^x \cos y + 2 \cos x)\, dy = 0$

8. $(e^x \sin y + 3y)\, dx - (3x - e^x \sin y)\, dy = 0$

9. $(ye^{xy} \cos 2x - 2e^{xy} \sin 2x + 2x)\, dx + (xe^{xy} \cos 2x - 3)\, dy = 0$

10. $\left(\dfrac{y}{x} + 6x\right) dx + (\ln x - 2)\, dy = 0, \qquad x > 0$

11. $(x \ln y + xy)\, dx + (y \ln x + xy)\, dy = 0; \qquad x > 0, \qquad y > 0$

12. $\dfrac{x\, dx}{(x^2 + y^2)^{3/2}} + \dfrac{y\, dy}{(x^2 + y^2)^{3/2}} = 0$

13. Find the value of b for which each of the following equations is exact, and then solve it, using that value of b.

(a) $(xy^2 + bx^2y) \, dx + (x + y)x^2 \, dy = 0$

(b) $(ye^{2xy} + x) \, dx + bxe^{2xy} \, dy = 0$

14. Consider the exact differential equation

$$M(x, y) \, dx + N(x, y) \, dy = 0.$$

Find an implicit formula $\psi(x, y) = c$ for the solution analogous to Eq. (12) by first integrating the equation $\psi_y = N$, rather than $\psi_x = M$, as in the text.

15. Show that any equation which is separable, that is, of the form

$$M(x) + N(y)y' = 0,$$

is also exact.

*16. In this problem we will point out the reason why it is necessary to have some restriction on the region R in Theorem 2.3. Let us consider the differential equation

$$\frac{x \, dy - y \, dx}{x^2 + y^2} = 0, \qquad (x, y) \neq (0, 0).$$

(a) Show that $M_y - N_x = 0$ at all points except the origin.

(b) Show that, except at $(0, 0)$, the given equation can be written as

$$d[\tan^{-1}(y/x)] = 0,$$

or, in polar coordinates

$$d\theta = 0.$$

(c) Now consider the solution of the given equation in the region R' (see Figure 2.5) interior to the square bounded by $|x| = 2$ and $|y| = 2$, but exterior to the circle $x^2 + y^2 = 1$. For this region the condition for exactness is satisfied at all points. From part (b) we obtain

$$\psi(x, y) = \tan^{-1}(y/x) = c$$

or

$$\theta = c,$$

where c is arbitrary. However, the expression $\psi(x, y)$ is not single-valued and hence does not define a function. For example, $\psi(1, 1) = (\pi/4) + 2n\pi$, where n is any integer. Thus there is no *function* ψ that satisfies Eqs. (4) in R', notwithstanding the fact that Eqs. (8) are satisfied throughout R'. The explanation is that the region R' is not the proper kind of region, and the hypothesis about the region R in the theorem is included precisely to exclude such regions as R'. Actually, R need not be rectangular, as long as it is bounded by a single continuous closed curve which does not cross itself. Such regions are called *simply connected*.

Simply connected regions can also be thought of as those into which a perfectly elastic circular rubber sheet can be deformed by stretching without tearing, or those into which a circle can be continuously deformed. In other words, simply connected

FIGURE 2.5

regions are regions without holes. Such matters are studied in the branch of mathematics known as topology.

2.6 INTEGRATING FACTORS

We will now show how the methods of Section 2.5 can be extended to a somewhat larger class of problems. Consider an equation of the form

$$M(x, y) \, dx + N(x, y) \, dy = 0. \tag{1}$$

If this equation is not already exact, we will try to choose a function μ, which may depend on both x and y, so that the equation

$$\mu(M \, dx + N \, dy) = 0 \tag{2}$$

is exact. The function μ is called an *integrating factor*. Equation (2) can then be solved by the methods of Section 2.5, and its solutions will also satisfy the original equation (1). This approach is an extension of the method developed in Section 2.1 for linear equations.

In order to investigate the possibility of carrying out this procedure, we recall from Section 2.5 that Eq. (2) is exact if and only if

$$(\mu M)_y = (\mu N)_x. \tag{3}$$

Since M and N are given functions, the integrating factor μ must satisfy the first order partial differential equation

$$M\mu_y - N\mu_x + (M_y - N_x)\mu = 0. \tag{4}$$

If a function μ satisfying Eq. (4) can be found, then Eq. (2) will be exact. The solution of Eq. (2) can then be obtained by the method of Section 2.5 and will be given implicitly in the form

$$\psi(x, y) = c.$$

This relation also defines the solution of Eq. (1), since the integrating factor μ can be canceled out of all terms in Eq. (2).

A partial differential equation of the form (4) may have more than one solution; if this is the case, any such solution may be used as an integrating factor of Eq. (1). This possible nonuniqueness of the integrating factor is illustrated in Example 3.

Unfortunately, Eq. (4), which determines the integrating factor, is ordinarily at least as difficult to solve as the original equation (1). Therefore, while in principle integrating factors are powerful tools for solving differential equations, in practice they can usually be found only in special cases. Some of these cases are indicated in the following examples and in Problems 10 and 11 at the end of the section.

Example 1. Verify that $\mu(x, y) = (xy^2)^{-1}$ is an integrating factor for the differential equation

$$(y^2 + xy) \, dx - x^2 \, dy = 0, \tag{5}$$

and then find its solution.

Since $M(x, y) = y^2 + xy$, and $N(x, y) = -x^2$, it follows that $M_y(x, y) = 2y + x$, and $N_x(x, y) = -2x$; thus the given equation is not exact. To show that $\mu(x, y) = (xy^2)^{-1}$ is an integrating factor for this equation, we multiply Eq. (5) by $\mu(x, y)$, thus obtaining

$$\left(\frac{1}{x} + \frac{1}{y}\right) dx - \frac{x}{y^2} dy = 0;$$

this latter equation is easily verified to be exact. By the method of Section 2.5 its solution is found to be given implicitly by

$$\ln |x| + \frac{x}{y} = c; \qquad x \neq 0, \qquad y \neq 0. \tag{6}$$

Note that $y = 0$ is also a solution of Eq. (5), although it is not given by Eq. (6).

Example 2. The two most important situations in which simple integrating factors can be found occur when μ is a function of only one of the variables x, y, instead of both. Let us determine necessary conditions on M and N so that $M\,dx + N\,dy = 0$ has an integrating factor μ, which depends on x only.

Assuming that μ is a function of x only, we have

$$(\mu M)_y = \mu M_y, \qquad (\mu N)_x = \mu N_x + N\frac{d\mu}{dx}.$$

Thus, if $(\mu M)_y$ is to equal $(\mu N)_x$, it is necessary that

$$\frac{d\mu}{dx} = \frac{M_y - N_x}{N}\mu. \tag{7}$$

If $(M_y - N_x)/N$ is a function of x only, then there is an integrating factor μ which also depends only on x; furthermore, $\mu(x)$ can be found by solving the first order *linear* equation (7).

A similar procedure can be used to determine a condition under which Eq. (1) has an integrating factor depending only on y (see Problem 10).

Example 3. Find an integrating factor for the equation

$$(3xy + y^2) + (x^2 + xy)\frac{dy}{dx} = 0, \tag{8}$$

and then solve the equation.

On computing the quantity $(M_y - N_x)/N$ we find that

$$\frac{M_y(x, y) - N_x(x, y)}{N(x, y)} = \frac{3x + 2y - (2x + y)}{x^2 + xy} = \frac{1}{x}.$$

Thus, according to Example 2, there is an integrating factor μ that is a function of x only and satisfies the differential equation

$$\frac{d\mu}{dx} = \frac{\mu}{x}.$$

Hence

$$\mu(x) = x. \tag{9}$$

Multiplying Eq. (8) by this integrating factor, we obtain

$$(3x^2y + xy^2) + (x^3 + x^2y)\frac{dy}{dx} = 0, \tag{10}$$

which is an exact equation. Its solution is easily found by the method of Section 2.5 to be given implicitly by

$$x^3y + \tfrac{1}{2}x^2y^2 = c. \tag{11}$$

Note that the quadratic equation (11) can be solved for y in terms of x. The reader may also verify, as in Example 1, that a second integrating factor of Eq. (8) is given by $\mu(x, y) = 1/xy(2x + y)$, and that the same solution is obtained, though with much greater difficulty, if this integrating factor is used (see Problem 12).

PROBLEMS

Show that the equations in Problems 1 through 3 are not exact, but become exact when multiplied by the given integrating factor. Then solve the equations.

1. $x^2y^3 + x(1 + y^2)y' = 0;$ $\mu(x, y) = 1/xy^3$

2. $\left(\dfrac{\sin y}{y} - 2e^{-x}\sin x\right) dx + \left(\dfrac{\cos y + 2e^{-x}\cos x}{y}\right) dy = 0;$ $\mu(x, y) = ye^x$

3. $y\,dx + (2x - ye^y)\,dy = 0;$ $\mu(x, y) = y$

In each of Problems 4 through 9 find an integrating factor, and solve the given equation.

4. $(3x^2y + 2xy + y^3)\,dx + (x^2 + y^2)\,dy = 0$

5. $y' = e^{2x} + y - 1$

6. $dx + \left(\dfrac{x}{y} - \sin y\right) dy = 0$

7. $y\,dx + (2xy - e^{-2y})\,dy = 0$

8. $e^x\,dx + (e^x \cot y + 2y \csc y)\,dy = 0$

9. $\left(3x + \dfrac{6}{y}\right) + \left(\dfrac{x^2}{y} + 3\dfrac{y}{x}\right)\dfrac{dy}{dx} = 0$

10. Show that if $(N_x - M_y)/M = Q$, where Q is a function of y only, then the differential equation

$$M + Ny' = 0$$

has an integrating factor of the form

$$\mu(y) = \exp \int^{y} Q(t) \, dt.$$

*11. Show that if $(N_x - M_y)/(xM - yN) = R$, where R depends on the quantity xy only, then the differential equation

$$M + Ny' = 0$$

has an integrating factor of the form $\mu(xy)$. Find a general formula for this integrating factor.

*12. Solve the differential equation

$$(3xy + y^2) + (x^2 + xy)y' = 0$$

using the integrating factor-

$$\mu(x, y) = [xy(2x + y)]^{-1}.$$

Verify that the solution is the same as that obtained in Example 3 with a different integrating factor.

2.7   HOMOGENEOUS EQUATIONS

The two main types of nonlinear first order equations that can be solved by direct processes of integration are those in which the variables separate and those which are exact. Section 2.6 dealt with the use of integrating factors to solve certain types of equations by reducing them to exact equations. This section will introduce another technique, that of changes of variables, which can sometimes be used to simplify a differential equation and thus make its solution possible (or more convenient). Generally speaking, the appropriate substitution will have to be suggested by the repeated appearance of some combination of the variables, or some other peculiarity in the structure of the equation.

The most important class of equations for which a definite rule can be laid down is the class of homogeneous* differential equations. An equation of the form

$$\frac{dy}{dx} = f(x, y)$$

is said to be homogeneous whenever the function f does not depend on x and y separately, but only on their ratio y/x or x/y. Thus homogeneous equations are of the form

$$\frac{dy}{dx} = F\left(\frac{y}{x}\right). \tag{1}$$

---

* The word homogeneous is used in more than one way in the study of differential equations. The reader must be alert to distinguish these as they occur.

Consider the following examples:

(a) $\dfrac{dy}{dx} = \dfrac{y^2 + 2xy}{x^2} = \left(\dfrac{y}{x}\right)^2 + 2\dfrac{y}{x}$;

(b) $\dfrac{dy}{dx} = \ln x - \ln y + \dfrac{x + y}{x - y} = \ln \dfrac{1}{y/x} + \dfrac{1 + (y/x)}{1 - (y/x)}$;

(c) $\dfrac{dy}{dx} = \dfrac{y^3 + 2xy}{x^2}$.

Thus (a) and (b) are homogeneous, since the right-hand side of each can be expressed as a function of y/x; since (c) cannot be so written, it is not homogeneous. Usually it is easy to tell by inspection whether or not a given equation is homogeneous; for complicated equations the criterion given in Problem 15 may be useful.

The form of a homogeneous equation suggests that it may be simplified by introducing a new variable,* which we will denote by v, to represent the ratio of y to x. Thus

$$y = xv, \tag{2}$$

and Eq. (1) becomes

$$\frac{dy}{dx} = F(v). \tag{3}$$

Looking on v as the new dependent variable (replacing y), we must consider v as a function of x, and replace dy/dx in Eq. (3) by a suitable expression in terms of v. Differentiating Eq. (2) gives

$$\frac{dy}{dx} = x\frac{dv}{dx} + v,$$

and hence Eq. (3) becomes

$$x\frac{dv}{dx} + v = F(v). \tag{4}$$

The most significant fact about Eq. (4) is that the variables x and v can *always* be separated, regardless of the form of the function F; in fact,

$$\frac{dx}{x} = \frac{dv}{F(v) - v}. \tag{5}$$

Solving Eq. (5) and then replacing v by y/x gives the solution of the original equation.

Thus any homogeneous equation can be transformed into one whose variables are separated by the substitution (2). As a practical matter, of course, it may or may not be possible to evaluate the integral required in solving Eq. (5) by elementary methods. Moreover, a homogeneous equation may also belong to one of the

* Leibniz in 1691 was the first to use this technique.

classes already discussed; it may well be exact, or even linear, for example. In such cases there is a choice of methods for finding its solution.

Example. Solve the differential equation

$$\frac{dy}{dx} = \frac{y^2 + 2xy}{x^2}.$$

(6)

Writing this equation as

$$\frac{dy}{dx} = \left(\frac{y}{x}\right)^2 + 2\frac{y}{x}$$

shows that it is homogeneous. The variables cannot be separated, nor is the equation exact, nor is an integrating factor obvious. Thus we are led to the substitution (2), which transforms the given equation into

$$x\frac{dv}{dx} + v = v^2 + 2v.$$

Hence

$$x\frac{dv}{dx} = v^2 + v$$

or, separating the variables,

$$\frac{dx}{x} = \frac{dv}{v(v+1)}.$$

Expanding the right-hand side by partial fractions, we obtain

$$\frac{dx}{x} = \left(\frac{1}{v} - \frac{1}{v+1}\right) dv.$$

Integrating both sides then yields

$$\ln|x| + \ln|c| = \ln|v| - \ln|v+1|,$$

where c is an arbitrary constant. Hence, combining the logarithms and taking the exponential of both sides, we obtain

$$cx = \frac{v}{v+1}.$$

Finally, substituting for v in terms of y gives the solution of Eq. (6) in the form

$$cx = \frac{y/x}{(y/x)+1} = \frac{y}{y+x}.$$

Solving for y we obtain

$$y = \frac{cx^2}{1 - cx}.$$

Sometimes it is helpful to transform both the independent and dependent variables. For example, in Problems 9, 10, and 11, a change of both variables is required to make the equation homogeneous, after which a further change of the dependent variable leads to a solution of the problem.

PROBLEMS

Show that the equations in Problems 1 through 8 are homogeneous, and find their solutions.

1. $\dfrac{dy}{dx} = \dfrac{x + y}{x}$ 2. $2y\,dx - x\,dy = 0$

3. $\dfrac{dy}{dx} = \dfrac{x^2 + xy + y^2}{x^2}$ 4. $\dfrac{dy}{dx} = \dfrac{x^2 + 3y^2}{2xy}$

5. $\dfrac{dy}{dx} = \dfrac{4y - 3x}{2x - y}$ 6. $\dfrac{dy}{dx} = -\dfrac{4x + 3y}{2x + y}$

7. $\dfrac{dy}{dx} = \dfrac{x + 3y}{x - y}$ 8. $(x^2 + 3xy + y^2)\,dx - x^2\,dy = 0$

9. (a) Find the solution of the equation

$$\frac{dy}{dx} = \frac{2y - x}{2x - y}.$$

(b) Find the solution of the equation

$$\frac{dy}{dx} = \frac{2y - x + 5}{2x - y - 4}.$$

Hint: In order to reduce the equation of part (b) to that of part (a), consider a preliminary substitution of the form $x = X - h$, $y = Y - k$. Choose the constants h and k so that the equation is homogeneous in the variables X and Y.

10. Solve $\dfrac{dy}{dx} = -\dfrac{4x + 3y + 15}{2x + y + 7}$.   See Hint, Problem 9(b).

11. Solve $\dfrac{dy}{dx} = \dfrac{x + 3y - 5}{x - y - 1}$.   See Hint, Problem 9(b).

12. Find the solution of the equation

$$(3xy + y^2)\,dx + (x^2 + xy)\,dy = 0$$

by the method of this section, and compare it with the solution obtained by other methods in Example 3 and Problem 12 of Section 2.6.

  *13. Show that if

$$M(x, y)\,dx + N(x, y)\,dy = 0$$

is a homogeneous equation, then it has

$$\mu(x, y) = \frac{1}{x M(x, y) + y N(x, y)}$$

for an integrating factor.

*14. Use the result of Problem 13 to solve each of the following equations:
  (a) The equation of Problem 2.   (b) The equation of Problem 4.

*15. Show that the equation $y' = f(x, y)$ is homogeneous if $f(x, y)$ is such that

$$f(x, tx) = f(1, t),$$

where t is a real parameter. Use this fact to determine whether each of the following equations is homogeneous.

(a) $y' = \dfrac{x^3 + xy + y^3}{x^2 y + xy^2}$

(b) $y' = \ln x - \ln y + \dfrac{x + y}{x - y}$

(c) $y' = \dfrac{(x^2 + 3xy + 4y^2)^{\frac{1}{2}}}{x + 2y}$

(d) $y' = \dfrac{\sin (xy)}{x^2 + y^2}$

2.8   MISCELLANEOUS PROBLEMS

This section consists of a list of problems. The first 32 problems can be solved by the methods of the previous sections. They are presented so that the reader may have some practice in identifying the method or methods applicable to a given equation. At the end of the list are a number of problems suggesting specialized techniques that are useful for certain types of equations. In particular, Problems 35 and 36 deal with Riccati (1676–1754) equations.

1. $\dfrac{dy}{dx} = \dfrac{x^3 - 2y}{x}$

2. $(x + y)\, dx - (x - y)\, dy = 0$

3. $\dfrac{dy}{dx} = \dfrac{2x + y}{3 + 3y^2 - x}, \quad y(0) = 0$

4. $(x + e^y)\, dy - dx = 0$

5. $\dfrac{dy}{dx} = -\dfrac{2xy + y^2 + 1}{x^2 + 2xy}$

6. $x\dfrac{dy}{dx} + xy = 1 - y, \quad y(1) = 0$

7. $\dfrac{dy}{dx} = \dfrac{x}{x^2 y + y^3}$ *Hint:* Let $u = x^2$.

8. $x\dfrac{dy}{dx} + 2y = \dfrac{\sin x}{x}, \quad y(2) = 1$

9. $\dfrac{dy}{dx} = -\dfrac{2xy + 1}{x^2 + 2y}$

10. $(3y^2 + 2xy)\, dx - (2xy + x^2)\, dy = 0$

11. $(x^2 + y)\, dx + (x + e^y)\, dy = 0$

12. $\dfrac{dy}{dx} + y = \dfrac{1}{1 + e^x}$

13. $x\, dy - y\, dx = (xy)^{\frac{1}{2}}\, dx$

14. $(x + y)\, dx + (x + 2y)\, dy = 0, \quad y(2) = 3$

15. $(e^x + 1)\dfrac{dy}{dx} = y - ye^x$

16. $\dfrac{dy}{dx} = \dfrac{x^2 + y^2}{x^2}$

17. $\dfrac{dy}{dx} = e^{2x} + 3y$

18. $(2y + 3x)\, dx = -x\, dy$

19. $x\, dy - y\, dx = 2x^2 y^2\, dy, \qquad y(1) = -2$

20. $y' = e^{x+y}$

21. $xy' = y + xe^{y/x}$

22. $\dfrac{dy}{dx} = \dfrac{x^2 - 1}{y^2 + 1}, \qquad y(-1) = 1$

23. $xy' + y - y^2 e^{2x} = 0$

× 24. $2 \sin y \cos x\, dx + \cos y \sin x\, dy = 0$

25. $\left(2\dfrac{x}{y} - \dfrac{y}{x^2 + y^2}\right) dx + \left(\dfrac{x}{x^2 + y^2} - \dfrac{x^2}{y^2}\right) dy = 0$

26. $(2y + 1)\, dx + \left(\dfrac{x^2 - y}{x}\right) dy = 0$

27. $(\cos 2y - \sin x)\, dx - 2 \tan x \sin 2y\, dy = 0$

28. $\dfrac{dy}{dx} = \dfrac{3x^2 - 2y - y^3}{2x + 3xy^2}$

29. $\dfrac{dy}{dx} = \dfrac{2y + \sqrt{x^2 - y^2}}{2x}$

30. $\dfrac{dy}{dx} = \dfrac{y^3}{1 - 2xy^2}, \qquad y(0) = 1$

31. $(x^2 y + xy - y)\, dx + (x^2 y - 2x^2)\, dy = 0$

32. $\dfrac{dy}{dx} = -\dfrac{3x^2 y + y^2}{2x^3 + 3xy}, \qquad y(1) = -2$

33. Show that an equation of the form

$$y = G(p),$$

where $p = dy/dx$, can be solved in the following manner.

(a) Differentiate with respect to x.

(b) Integrate the equation of part (a) to obtain x as a function of p. This equation and the original equation $y = G(p)$ form a parametric representation of the solution.

34. Solve the differential equation

$$y - \ln p = 0,$$

where $p = dy/dx$, by the method of Problem 33. Also solve this equation directly, and verify that the solutions are the same.

35. The equation

$$\frac{dy}{dx} = q_1(x) + q_2(x)y + q_3(x)y^2$$

is known as a *Riccati equation*. Suppose that some particular solution y_1 of this equation is known. A more general solution containing one arbitrary constant can be obtained through the substitution

$$y = y_1(x) + \frac{1}{v(x)}.$$

Show that $v(x)$ satisfies the first order *linear* equation

$$\frac{dv}{dx} = -(q_2 + 2q_3 y_1)v - q_3.$$

Note that $v(x)$ will contain a single arbitrary constant.

36. Using the method of Problem 35 and the given particular solution, solve each of the following Riccati equations.

(a) $y' = 1 + x^2 - 2xy + y^2;$ $\quad$ $y_1(x) = x$

(b) $y' = -\frac{1}{x^2} - \frac{y}{x} + y^2;$ $\quad$ $y_1(x) = 1/x$

(c) $\frac{dy}{dx} = \frac{2\cos^2 x - \sin^2 x + y^2}{2\cos x};$ $\quad$ $y_1(x) = \sin x$

2.9 APPLICATIONS OF FIRST ORDER EQUATIONS

Differential equations are of interest to nonmathematicians primarily because of the possibility of using them to investigate a wide variety of problems in the physical, biological, and social sciences. Three identifiable steps in this process are present regardless of the specific field of application.

In the first place, it is necessary to translate the physical situation into mathematical terms. This is generally done by making assumptions about what is happening that appear to be consistent with the observed phenomena. For example, it has been observed that radioactive materials decay at a rate proportional to the amount of the material present, that heat passes from a warmer to a cooler body at a rate proportional to the temperature difference, that objects move about in accordance with Newton's laws of motion, and that isolated insect populations grow at a rate proportional to the current population. Each of these statements involves a rate of change (derivative) and consequently, when expressed mathematically, takes the form of a differential equation.

It is important to realize that the mathematical equations are almost always only an approximate description of the actual process because they are based on observations that are themselves approximations. For example, bodies moving at speeds comparable to the speed of light are not governed by Newton's laws, insect populations do not grow indefinitely as stated because of eventual limitations on their food supply, and heat transfer is affected by factors other than the temperature difference. Moreover, the process of formulating a physical problem mathematically often involves the conceptual replacement of a discrete process by a continuous one. For instance, the number of members in an insect population changes by discrete amounts; however, if the population is large, it seems reasonable to consider it as a continuous variable and even to speak of its derivative. Alternatively, one can adopt the point of view that the mathematical equations exactly describe the operation of a simplified model, which has been constructed (or conceived of) so as to embody the most important features of the actual process.

In any case, once the problem has been formulated mathematically, one is often faced with the problem of solving one or more differential equations or, failing that, of finding out as much as possible about the properties of the solution. It may happen that this mathematical problem is quite difficult and, if so, further approximations may be indicated at this stage in order to make the problem mathematically tractable. For example, a nonlinear equation may be approximated by a linear one, or a slowly varying function may be replaced by its average value. Naturally, any such approximations must also be examined from the physical point of view to make sure that the simplified mathematical problem still reflects the essential features of the physical process under investigation. At the same time, an intimate knowledge of the physics of the problem may suggest reasonable mathematical approximations that will make the mathematical problem more amenable to analysis. This interplay of understanding of physical phenomena and knowledge of mathematical techniques and their limitations is characteristic of applied mathematics at its best, and is indispensable in successfully constructing mathematical models of intricate physical processes.

Finally, having obtained the solution (or at least some information about it), it must be interpreted in terms of the context in which the problem arose. In particular, one should always check as to whether the mathematical solution appears physically reasonable. This requires, at the very least, that the solution exist, that it be unique, and that it depend continuously on the data of the problem. This last consideration is important because the coefficients in the differential equation and initial conditions are often obtained as a result of measurements of some physical quantity and therefore are susceptible to small errors. If these small errors lead to large (or discontinuous) changes in the solution of the corresponding mathematical problem, which are not observed physically, then the relevance of the mathematical model to the physical problem must be reexamined. Of course, the fact that the mathematical solution appears to be reasonable does not guarantee that it is correct. However, if it is seriously inconsistent with careful observations of the physical system it purports to describe, this suggests either that errors have been made in solving the mathematical problem or that the mathematical model itself is too crude.

The examples in this section and in the next are typical of applications in which first order differential equations arise.

Example 1 (Radioactive Decay). The radioactive isotope thorium 234 disintegrates at a rate proportional to the amount present. If 100 milligrams of this material is reduced to 82.04 milligrams in one week, find an expression for the amount present at any time. Also, find the time interval that must elapse in order for the mass to decay to one half of its original value.

Let $Q(t)$ be the amount of thorium 234 present at any time t, where Q is measured in milligrams and t in days. The physical observation that thorium 234 disintegrates at a rate proportional to the amount present means that the time rate of change dQ/dt is proportional to Q. Thus Q satisfies the differential equation

$$\frac{dQ}{dt} = kQ, \tag{1}$$

where k is a negative constant of proportionality that must be determined. We seek the solution of Eq. (1) that also satisfies the initial condition

$$Q(0) = 100 \tag{2}$$

as well as the condition

$$Q(7) = 82.04. \tag{3}$$

Equation (1) can be solved either as a linear equation or by separating the variables, and its general solution is

$$Q(t) = ce^{kt}, \tag{4}$$

where c is an arbitrary constant. The initial condition (2) requires that $c = 100$, and therefore

$$Q(t) = 100e^{kt}. \tag{5}$$

In order to satisfy Eq. (3) we set $t = 7$ and $Q = 82.04$ in Eq. (5); this gives

$$82.04 = 100e^{7k},$$

and hence

$$k = \frac{\ln 0.8204}{7} \doteq -0.02828 \ (\text{days})^{-1}. \tag{6}$$

Thus the decay constant k has been determined. Using this value of k in Eq. (5), we obtain

$$Q(t) = 100e^{-0.02828t} \ (\text{milligrams}), \tag{7}$$

which gives the value of $Q(t)$ at any time.

The time period during which the mass is reduced to one half of its original value is known as the *half-life* of the material. Let τ be the time at which $Q(t)$ is equal to 50 milligrams. Then, from Eq. (5),

$$50 = 100e^{k\tau},$$

or

$$k\tau = -\ln 2. \tag{8}$$

The relation (8) between the decay constant and the half-life is valid not only for thorium 234 but for any material that obeys the differential equation (1); using the value of k given by Eq. (5), we find that for thorium 234

$$\tau = \frac{\ln 2}{0.02828} \cong 24.5 \ (\text{days}). \tag{9}$$

Example 2 (Compound Interest). Suppose that a sum of money S_0 is deposited in a bank that pays interest at a rate of 6 percent. The value $S(t)$ of the investment at the end of t years then depends on the frequency with which the interest is compounded. In this example we will investigate the effect of the frequency of compounding.

If interest is compounded once a year then

$$S(t) = S_0(1 + 0.06)^t.$$

If interest is compounded twice a year then

$$S(t) = S_0\left[\left(1 + \frac{0.06}{2}\right)\left(1 + \frac{0.06}{2}\right)\right]^t = S_0\left(1 + \frac{0.06}{2}\right)^{2t},$$

and in general if interest is compounded k times per year, then

$$S(t) = S_0\left(1 + \frac{0.06}{k}\right)^{kt}. \tag{10}$$

Let us approximate the actual situation by a mathematical model in which we assume that the interest is compounded *continuously*. With this assumption we can write the law of growth of the investment as a differential equation, namely

$$S'(t) = 0.06\, S(t).$$

The solution of this equation, subject to the initial condition $S(0) = S_0$, is

$$S(t) = S_0 e^{0.06t}. \tag{11}$$

Note that as $k \to \infty$ the right side of Eq. (10) approaches the right side of Eq. (11). In the second column in Table 2.1 we list the values of $S(t)/S_0$ as given by Eq. (11) for various values of t. The values of $S(t)/S_0$ in the third and fourth columns are found from Eq. (10), using $k = 4$ and $k = 365$, corresponding to quarterly and daily compounding respectively.

The results in the table show that the frequency with which interest is compounded is not particularly important in most cases. For example, over a ten-year period the difference between quarterly compounding and continuous compounding is less than one-half of one percent, that is, about \$8 per \$1000 invested, or about eighty cents a year. The difference would be greater for higher interest rates, and, of course, for a very large S_0 even a small percentage difference might be significant. In any case if the time interval is long (a few centuries or more), then the difference is sizeable.

From the table we also see that the effective annual interest rate for quarterly compounding is 6.14 percent and for daily or continuous compounding it is 6.18 percent. Many banks advertise an even higher effective annual yield than the 6.18 percent obtained by compounding continuously. This is accomplished by calculating a daily interest rate using a nominal year of 360 days, and then compounding this rate through the actual calendar year. Using this method for an interest rate of 6 percent and ignoring leap years, we find that

$$S(t) = S_0\left(1 + \frac{0.06}{360}\right)^{365t}. \tag{12}$$

TABLE 2.1

Years	$S(t)/S_0$ from Eq. (11)	$S(t)/S_0$ from Eq. (10) $k = 4$	$k = 365$	$S(t)/S_0$ from Eq. (12)
1	1.0618	1.0614	1.0618	1.0627
2	1.1275	1.1265	1.1275	1.1294
5	1.3499	1.3469	1.3498	1.3555
10	1.8221	1.8140	1.8220	1.8373
20	3.3201	3.2907	3.3198	3.3756
40	1.1023×10	1.0828×10	1.1021×10	1.1395×10
80	1.2151×10^2	1.1726×10^2	1.2146×10^2	1.2984×10^2
200	1.6275×10^5	1.4888×10^5	1.6260×10^5	1.9208×10^5
400	2.6489×10^{10}	2.2165×10^{10}	2.6439×10^{10}	3.6895×10^{10}

Results using Eq. (12) are given in the last column of Table 2.1. Using this procedure, the effective annual yield is 6.27 percent. Within the accuracy of the present calculations we can obtain the same result by using the effective rate, assuming continuous compounding, and then multiplying by the factor 365/360; thus $6.18 \times (365/360) = 6.27$. The effect is more dramatic for higher interest rates; for example, an annual interest rate of 7.50 percent yields an effective interest rate of 7.79 percent for continuous compounding and the higher rate of 7.90 percent using Eq. (12).

Example 3 (Mixing).   Consider a tank containing, at time $t = 0$, Q_0 lb of salt dissolved in 100 gal of water. Assume that water containing $\frac{1}{4}$ lb of salt per gallon is entering the tank at a rate of 3 gal/min, and that the well-stirred solution is leaving the tank at the same rate. Find an expression for the amount of salt $Q(t)$ in the tank at time t.

The rate of change of salt in the tank at time t, $Q'(t)$, must equal the rate at which salt enters the tank minus the rate at which it leaves. The rate at which salt enters is $\frac{1}{4}$ lb/gal times 3 gal/min. The rate at which salt leaves is $(Q(t)/100)$ lbs/gal times 3 gal/min. Thus

$$Q'(t) = \tfrac{3}{4} - \tfrac{3}{100}Q(t) \tag{13}$$

is the differential equation governing this process. Equation (13) is linear and its solution is

$$Q(t) = 25 + ce^{-0.03t}, \tag{14}$$

where c is arbitrary. The initial condition

$$Q(0) = Q_0 \tag{15}$$

requires that $c = Q_0 - 25$, and hence

$$Q(t) = 25(1 - e^{-0.03t}) + Q_0 e^{-0.03t}. \tag{16}$$

The second term on the right side of Eq. (16) represents the portion of the original salt remaining in the tank at time t. This term becomes very small with the passage of time as the original solution is drained from the tank. The first term on the right side of Eq. (16) gives the amount of salt in the tank at time t due to the action of the flow processes. As t becomes large this term approaches the constant value 25 (pounds). It is also clear physically that this must be the limiting value of Q as the original solution in the tank is more and more completely replaced by that entering with a concentration of $\frac{1}{4}$ lb/gal.

Example 4 (Epidemics). A mathematical theory of epidemics has been developed along the following lines. Assume that there is a community of n members containing p infected individuals and q uninfected but susceptible individuals, where $p + q = n$. Alternatively, let x be the proportion of sick members and y the proportion of well members. Then $x = p/n$, $y = q/n$, and $x + y = 1$. If n is large it is reasonable to think of x and y as continuous variables; then the rate at which the disease spreads is dx/dt. If we make the reasonable assumption that the disease spreads by contact between sick and well members of the community, then dx/dt is proportional to the number of contacts. If members of both groups move about freely among each other, the number of contacts in turn is proportional to the product of x and y. Thus we arrive at the differential equation

$$\frac{dx}{dt} = \beta xy$$

or

$$\frac{dx}{dt} = \beta x(1 - x), \tag{17}$$

where β is a positive proportionality factor. The initial condition is

$$x(0) = x_0, \tag{18}$$

where x_0 is the density of infected individuals at $t = 0$.

Equation (17) is separable, and may be written as

$$\frac{dx}{x(1 - x)} = \beta \, dt,$$

or, using a partial fraction expansion, in the form

$$\left(\frac{1}{x} + \frac{1}{1 - x}\right) dx = \beta \, dt.$$

Hence

$$\ln x - \ln (1 - x) = \beta t + C; \tag{19}$$

note that x and $1 - x$ are intrinsically positive so that absolute values are not needed. From Eq. (19) it follows that

$$\frac{x}{1 - x} = ce^{\beta t}, \tag{20}$$

where $c = e^C$. The initial condition requires that $c = x_0/(1 - x_0)$. Using this value of c in Eq. (20) and solving for x, we obtain

$$x = \frac{x_0 e^{\beta t}}{1 - x_0 + x_0 e^{\beta t}} = \frac{x_0}{x_0 + (1 - x_0)e^{-\beta t}}. \tag{21}$$

If $x_0 > 0$, then $x \to 1$ as $t \to \infty$ regardless of the actual value of x_0. Thus, no matter how small the initial number of infected individuals, the epidemic ultimately spreads throughout the entire population.

The mathematical model described above is clearly unrealistic in some respects. For one thing, if the disease is at all serious, a partial quarantine will occur naturally to the extent that sick individuals are restricted in their activities. An additional quarantine may be imposed by health authorities to restrict further the possibility of contact between sick and well individuals. Also, any diseases are contagious only for a limited period of time, whereas in the model used here sick individuals are assumed to remain so indefinitely. All of these factors would tend to slow the spread of the disease as compared to the rate predicted above.

Some diseases (such as malaria) are not spread by direct contact, but through the intervention of an animal or insect. A mathematical model for the spread of such a disease is considerably more complicated than the one in this example. Another type of situation, typical of typhoid fever, is described in Problem 19. Finally, it is often important to incorporate random effects in the study of epidemics, making use of appropriate probabilistic concepts.

PROBLEMS

1. Suppose that a sum S_0 is deposited in a bank that pays interest at an annual rate r compounded continuously.
(a) Find the time T required for the original sum to double in value as a function of the interest rate r.
(b) Determine T if $r = 5$ percent.
(c) Find the interest rate that must be paid if the initial investment is to double in ten years.

2. A young person with no initial capital invests k dollars per year at an annual interest rate r. Assume that investments are made continuously and that interest is compounded continuously.
(a) Determine the sum $S(t)$ accumulated at any time t.
(b) If $r = 5$ percent, determine k so that one million dollars will be available for retirement in forty years.

3. Determine the sum accumulated in twenty years by each of the following investment programs. Suppose that 5 percent interest is compounded, and deposits made, continuously.
(a) Nothing initially and $1500 per year for twenty years.
(b) $10,000 initially and $1000 per year for twenty years.
(c) $20,000 initially and $500 per year for twenty years.
(d) $30,000 initially and no additional deposits.
Note that in each case the total amount invested is $30,000.

4. A retired person has a sum $S(t)$ invested so as to draw interest at an annual rate r compounded continuously. Withdrawals for living expenses are made at a rate of k dollars per year; assume that the withdrawals are made continuously.

(a) If the initial value of the investment is S_0, determine $S(t)$ at any time.

(b) Assuming that S_0 and r are fixed, determine the withdrawal rate k_0 at which $S(t)$ will remain constant.

(c) If k exceeds the value k_0 found in part (b), then $S(t)$ will decrease and ultimately become zero. Find the time T at which $S(t) = 0$.

(d) Determine T if $r = 5$ percent and $k = 3k_0/2$.

(e) Suppose that a person retiring with capital S_0 wishes to withdraw funds at an annual rate k for not more than T years. Determine the maximum possible rate of withdrawal.

(f) How large an initial investment is required in order to permit an annual withdrawal of $5000 for twenty years, assuming an interest rate of 5 percent?

5. A certain man has a fortune that increases at a rate proportional to the square of his present wealth. If he had one million dollars a year ago, and has two million dollars today, how much will he be worth in six months? in two years?

6. The radioactive isotope plutonium 241 decays so as to satisfy the differential equation

$$\frac{dQ}{dt} = -0.0525Q,$$

where Q is measured in milligrams and t in years.

(a) Determine the half-life τ of plutonium 241.

(b) If 50 milligrams of plutonium are present today, how much will remain in ten years?

7. Einsteinium 253 decays at a rate proportional to the amount present. Determine the half-life τ if this material loses one third of its mass in 11.7 days.

8. Radium 226 has a half-life of 1620 years. Find the time period during which a body of this material is reduced to three-fourths of its original size.

9. **Radiocarbon Dating.** An important tool in archeological research is *radiocarbon dating*. This is a means of determining the age of certain wood and plant remains, and hence of animal or human bones or artifacts found buried at the same levels. The procedure was developed by the American chemist Willard Libby (1908–   ) in the early 1950s and resulted in his winning the Nobel prize for chemistry in 1960. Radiocarbon dating is based on the fact that some wood or plant remains contain residual amounts of carbon 14, a radioactive isotope of carbon. This isotope is accumulated during the lifetime of the plant and begins to decay at its death. Since the half-life of carbon 14 is long (approximately 5745 years), measurable amounts of carbon 14 remain after many thousands of years. If approximately 0.002 or more of the original amount of carbon 14 is still present, then Libby showed that by appropriate laboratory measurements the *proportion* of the original amount of carbon 14 that remains can be accurately determined. In other words, if $Q(t)$ is the amount of carbon 14 at time t and Q_0 is the original amount, then the ratio $Q(t)/Q_0$ can be determined, at least if this quantity is not too small.

(a) Assuming that Q satisfies the differential equation $Q' = kQ$, determine the decay constant k for carbon 14.

(b) Find an expression for $Q(t)$ at any time t, if $Q(0) = Q_0$.

(c) Suppose that certain remains are discovered in which the current residual amount of carbon 14 is 20 percent of the original amount. Determine the age of these remains.

10. Assume the population of the earth changes at a rate proportional to the current population. (A more accurate hypothesis concerning the law of population growth is considered in Problem 11.) Further, it is estimated that at time $t = 0$ (A.D. 1650) the earth's population was 600 million (6.0×10^8); at time $t = 300$ (A.D. 1950) its population was 2.8 billion (2.8×10^9). Find an expression giving the population of the earth at any time. Assuming that the greatest population the earth can support is 25 billion (2.5×10^{10}), when will this limit be reached?

11. In a certain isolated population $p(t)$ the rate of population growth dp/dt is a function of the population; thus

$$\frac{dp}{dt} = f(p).$$

(a) The simplest expression for $f(p)$ is $f(p) = \epsilon p$, where ϵ is a positive constant. Find $p(t)$ in this case if $p(0) = p_0$. Find the limit of $p(t)$ as $t \to \infty$.

(b) A better assumption in many situations is that $f(p) = \epsilon p - \sigma p^2$, where ϵ and σ are both positive constants. The term $-\sigma p^2$ may be needed, for example, if overcrowded conditions for large values of p result in an increase in the death rate and a decrease in the birth rate. If $p(0) = p_0$, find an expression for the population at any time t. Find the limiting population as $t \to \infty$. Compare these results with those of part (a). Note that, even if σ is very small, the nature of the solution is radically changed (for large p) as soon as the quadratic term is introduced into $f(p)$. This is due to the fact that dp/dt decreases as p increases and approaches zero as p approaches ϵ/σ from below. Thus p cannot increase beyond this limiting value.

12. It is known from experimental observations that, to an accuracy satisfactory in many circumstances, the surface temperature of an object changes at a rate proportional to the difference between the temperature of the object and that of the surroundings (the ambient temperature). This is sometimes known as Newton's law of cooling. Thus, if $u(t)$ is the temperature of the object at time t, and u_0 is the constant ambient temperature, we have the relation

$$\frac{du}{dt} = k(u - u_0),$$

where k is a constant proportionality factor.

(a) Find the solution satisfying the initial condition $u(0) = u_1$.

(b) Suppose the temperature of a cup of coffee is $200°$ F when freshly poured. One minute later it has cooled to $190°$ F in a room at $70°$ F. How long a period must elapse (assuming Newton's law of cooling to apply) before the coffee reaches a temperature of $150°$ F.

13. Assume that a spherical raindrop evaporates at a rate proportional to its surface area. If its radius originally is 3 mm, and 1 hour later has been reduced to 2 mm, find an expression for the radius of the raindrop at any time.

14. Consider a tank used in certain hydrodynamic experiments. After one experiment the tank contains 200 lit of a dye solution with a concentration of 1 gm/lit. To prepare for the next experiment the tank is to be rinsed with fresh water flowing in at a rate of 2 lit/min, the well-stirred solution flowing out at the same rate. Find the time which will elapse before the concentration of dye in the tank reaches 1 percent of its original value.

15. A tank originally contains 100 gal of fresh water. Then water containing $\frac{1}{2}$ lb of salt per gallon is poured into the tank at a rate of 2 gal/min, and the mixture is allowed to leave at the same rate. After 10 minutes the process is stopped, and fresh water is poured into the tank at a rate of 2 gal/min, with the mixture again leaving at the same rate. Find the amount of salt in the tank at the end of 20 minutes.

16. A tank with a capacity of 500 gal originally contains 200 gal of water with 100 lb of salt in solution. Water containing 1 lb of salt per gallon is entering at a rate of 3 gal/min, and the mixture is allowed to flow out of the tank at a rate of 2 gal/min. Find the amount of salt in the tank at any time prior to the instant when the solution begins to overflow. Find the concentration (in pounds per gallon) of salt in the tank when it is on the point of overflowing. Compare this concentration with the theoretical limiting concentration if the tank had infinite capacity.

17. A second order chemical reaction involves the interaction (collision) of one molecule of a substance P with one molecule of a substance Q in order to produce one molecule of a new substance X; $P + Q \to X$ is the usual notation. Suppose that p and q are the initial concentrations of substances P and Q, respectively, and let $x(t)$ be the concentration of X at time t. Then $p - x(t)$ and $q - x(t)$ are the concentrations of P and Q at time t, and the rate at which the reaction occurs is described by the equation

$$\frac{dx}{dt} = \alpha(p - x)(q - x),$$

where α is a positive constant.
   (a) If $x(0) = 0$, find $x(t)$.
   (b) If the substances P and Q are the same, then

$$\frac{dx}{dt} = \alpha(p - x)^2.$$

If $x(0) = 0$, find $x(t)$.

18. Suppose that in a certain chemical reaction a substance P is transformed into a substance X. Let p be the initial concentration of P and let $x(t)$ be the concentration of X at time t. Then $p - x(t)$ is the concentration of P at t. Suppose further that the reaction is autocatalytic, that is, the reaction is stimulated by the substance being produced. Hence dx/dt is proportional both to x and $p - x$, and

$$\frac{dx}{dt} = \alpha x(p - x),$$

where α is a positive constant. If $x(0) = x_0$, find $x(t)$.

19. Some diseases (such as typhoid) are spread largely by *carriers*, individuals who can transmit the disease but who exhibit no overt symptoms. Let x and y, respectively, denote the density of carriers and of well, but susceptible, people in the population at time t. Suppose that carriers are removed from the population at a rate β, so that

$$\frac{dx}{dt} = -\beta x. \tag{i}$$

Suppose also that the disease spreads at a rate proportional to the product of x and y. Thus

$$\frac{dy}{dt} = -\alpha x y. \tag{ii}$$

(a) Determine x at any time t by solving Eq. (i) subject to the initial condition $x(0) = x_0$.

(b) Use the result of part (a) to find y at any time t by solving Eq. (ii) subject to the initial condition $y(0) = y_0$.

(c) Find the proportion of the population that escapes the epidemic; that is, determine the limiting value of y as $t \to \infty$. Also determine how this quantity varies with α and β.

20. Suppose that a room containing 1200 cubic feet of air is originally free of carbon monoxide. Beginning at time $t = 0$ cigarette smoke, containing 4 percent carbon monoxide, is introduced into the room at a rate of 0.1 ft³/min, and the well-circulated mixture is allowed to leave the room at the same rate.

(a) Find an expression for the concentration $x(t)$ of carbon monoxide in the room at any time $t > 0$.

(b) Extended exposure to a carbon monoxide concentration as low as 0.00012 is harmful to the human body. Find the time τ at which this concentration is reached.

21. Consider a lake of constant volume V containing at time t an amount $Q(t)$ of pollutant, evenly distributed throughout the lake with a concentration $c(t)$, where $c(t) = Q(t)/V$. Assume that water containing a concentration k of pollutant enters the lake at a rate r, and that water leaves the lake at the same rate. Suppose that pollutants are also

TABLE 2.2

Lake	$V(\text{km}^3 \times 10^{-3})$	$r(\text{km}^3/\text{yr})$
Superior	12.2	65.2
Michigan	4.9	158
Erie	0.46	175
Ontario	1.6	209

added directly to the lake at a constant rate P. Note that the given assumptions neglect a number of factors that may, in some cases, be important; for example, the water added or lost by precipitation, absorption, and evaporation; the stratifying effect of temperature differences in a deep lake; the tendency of irregularities in the coastline to produce sheltered bays; and the fact that pollutants are not deposited evenly throughout the lake, but (usually) at isolated points around its periphery. The results below must be interpreted in the light of the neglect of such factors as these.

(a) If at time $t = 0$ the concentration of pollutant is c_0, find an expression for the concentration $c(t)$ at any time. What is the limiting concentration as $t \to \infty$?

(b) If the addition of pollutants to the lake is terminated ($k = 0$ and $P = 0$ for $t > 0$), determine the time interval T that must elapse before the concentration of pollutants is reduced to 50 percent of its original value; to 10 percent of its original value.

(c) Table 2.2 contains data* for several of the Great Lakes. Using these data determine from part (b) the time T necessary to reduce the contamination of each of these lakes to 10 percent of the original value.

* This problem is based on R. H. Rainey, "Natural Displacement of Pollution from the Great Lakes," *Science*, **155**, 1967, pp. 1242–1243; the information in the table was taken from that source.

22. **Orthogonal Trajectories.** A fairly common geometrical problem is to find the family of curves that intersects a given family of curves orthogonally at each point. Such families of curves are said to be orthogonal trajectories of each other.

(a) Consider the family of parabolas

$$y = kx^2, \qquad\qquad\qquad (i)$$

where k is a constant. Sketch the graph of Eq. (i) for several values of k. Find an expression for the slope of the parabola passing through a given point that involves the coordinates (x, y) of the point but not the parameter k.

Hint: Differentiate Eq. (i) and eliminate k.

(b) Making use of the fact that the slopes of orthogonal curves are negative reciprocals, write the differential equation for the orthogonal trajectories of Eq. (i).

(c) Solve the equation found in part (b) and determine the orthogonal trajectories. Sketch several numbers of this family of curves.

23. In each of the following cases use the method of Problem 22 to find the family of orthogonal trajectories of the given family of curves. Sketch both the given family and their orthogonal trajectories.

(a) The family of hyperbolas $xy = c$.

(b) The family of circles $(x - c)^2 + y^2 = c^2$.

(c) The family of ellipses $x^2 - xy + y^2 = c^2$.

(d) The family of parabolas $2cy + x^2 = c^2$, $c > 0$.

24. If two straight lines in the xy plane, having slopes m_1 and m_2 respectively, intersect at an angle θ, show that

$$(\tan \theta)(1 + m_1 m_2) = m_2 - m_1.$$

Using this fact, find the family of curves that intersects each of the following families at an angle of $45°$. In each case sketch both families of curves.

(a) $x - 2y = c$ $\qquad\qquad\qquad$ (b) $x^2 + y^2 = c^2$

25. Find all plane curves such that the tangent line at each point (x, y) passes through the fixed point (a, b).

26. The line normal to a given curve at each point (x, y) on the curve passes through the point $(2, 0)$. If the curve contains the point $(2, 3)$, find its equation.

27. Find all plane curves such that for each point on the curve the y axis bisects that part of the tangent line between the point of tangency and the x axis.

2.10 ELEMENTARY MECHANICS

Some of the most important applications of first order differential equations are found in the realm of elementary mechanics. In this section we will consider some problems involving the motion of a rigid body along a straight line. We will assume that such bodies obey Newton's (1642–1727) law of motion: *the product of the mass and the acceleration is equal to the external force.* In symbols,

$$F = ma, \qquad\qquad\qquad (1)$$

where F is the external force, m is the mass of the body, and a is its acceleration in the direction of F. If the mass of the body is constant, then Eq. (1) can be put into the form

$$F = \frac{d}{dt}(mv) = \frac{dP}{dt}, \qquad\qquad\qquad (2)$$

where v is the velocity, and $P = mv$ is the linear momentum.

Two systems of units are in common use. In the metric, or cgs system, the basic units are centimeters (length), grams (mass), and seconds (time). The unit of force, the dyne, is defined by Eq. (1) to be the force required to impart an acceleration of 1 cm/sec² to a mass of 1 gm. On the other hand, in the English system the basic units are feet (length), pounds (force), and seconds (time). The unit of mass, the slug, is defined by Eq. (1) to be that mass to which a force of 1 lb gives an acceleration of 1 ft/sec².

Now consider a body falling freely in a vacuum and close enough to the surface of the earth so that the only significant force acting on the body is its weight due to the earth's gravitational field. Equation (1) then takes the form

$$w = mg, \tag{3}$$

where w is the weight of the body and g denotes its acceleration due to gravity.

Even though the mass of the body remains constant, its weight and gravitational acceleration change with distance from the center of the earth's gravitational field. At sea level the value of g has been experimentally determined to be approximately 32 ft/sec² (English system), or 980 cm/sec² (metric system). It is customary to denote by g the gravitational acceleration at *sea level*; thus g is a constant. With this understanding Eq. (3) is exact only at sea level.

The general expression for the weight of a body of mass m is obtained from Newton's inverse-square law of gravitational attraction. If R is the radius of the earth, and x is the altitude above sea level, then

$$w(x) = \frac{K}{(x + R)^2},$$

where K is a constant. At $x = 0$ (sea level), $w = mg$; hence $K = mgR^2$, and

$$w(x) = \frac{mgR^2}{(x + R)^2}. \tag{4}$$

Expanding $(x + R)^{-2}$ in a Taylor series about $x = 0$ leads to

$$w(x) = mg\left(1 - 2\frac{x}{R} + \cdots\right). \tag{5}$$

It follows that if terms of the order of magnitude of x/R can be neglected in comparison with unity, then Eq. (4) can be replaced by the simpler approximation, Eq. (3). Thus, for motions in the vicinity of the earth's surface, or if extreme accuracy is not required, it is usually sufficient to use Eq. (3). In other cases, such as problems involving space flight, it may be necessary to use Eq. (4), or at least a better approximation to it than Eq. (3).

In many cases the external force F will include effects other than the weight of the body. For example, frictional forces due to the resistance offered by the air or other surrounding medium may need to be considered. Similarly, the effects of mass changes may be important, as in the case of a space vehicle which burns its fuel during flight.

In the problems we will consider here it is useful to describe the motion of the body with reference to some convenient coordinate system. We will always choose the x axis to be the line along which the motion takes place; the positive direction on the x axis can be chosen arbitrarily. Once the positive x direction is specified, a positive value of x indicates a displacement in this direction, a positive value of $v = dx/dt$ indicates the body is moving in the direction of the positive x axis, a positive value for a force F indicates it acts in the positive x direction, and so forth. The following examples are concerned with some typical problems.

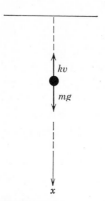

FIGURE 2.6 $v > 0$.

Example 1. An object of mass m is dropped from rest in a medium that offers resistance proportional to $|v|$, the magnitude of the instantaneous velocity of the object.* Assuming the gravitational force to be constant, find the position and velocity of the object at any time t.

In this case it is convenient to draw the x axis positive downward with the origin at the initial position of the object (see Figure 2.6). The weight mg of the object then acts in the downward (positive) direction, but the resistance $k\,|v|$, where k is a positive constant, acts to impede the motion. When $v > 0$ the resistance is in the upward (negative) direction, and is thus given by $-kv$. When $v < 0$ the resistance acts downward and is still given by $-kv$. Thus in all cases Newton's law can be written as

$$m\,\frac{dv}{dt} = mg - kv$$

or

$$\frac{dv}{dt} + \frac{k}{m}v = g. \tag{6}$$

Equation (6) is a linear first order equation and has the integrating factor $e^{kt/m}$. The solution of Eq. (6) is

$$v = \frac{mg}{k} + c_1 e^{-kt/m}. \tag{7}$$

The initial condition $v(0) = 0$ requires that $c_1 = -mg/k$; thus

$$v = \frac{mg}{k}(1 - e^{-kt/m}). \tag{8}$$

---

* In general, the resistance is a function of the speed, that is, the magnitude of the velocity. The assumption of a linear relation is reasonable only for comparatively low speeds. Where higher speeds are involved it may be necessary to assume that the resistance is proportional to some higher power of $|v|$, or even that it is given by some polynomial function of $|v|$.

To obtain the position x of the body we replace v by dx/dt in Eq. (8); integration and use of the second initial condition $x(0) = 0$ gives

$$x = \frac{mg}{k} t - \frac{m^2 g}{k^2} (1 - e^{-kt/m}). \tag{9}$$

The position and velocity of the body at any time t are given by Eqs. (9) and (8), respectively.

It is worth noting that as $t \to \infty$ in Eq. (7) the velocity approaches the limiting value

$$v_l = mg/k. \tag{10}$$

The limiting velocity in free fall v_l does not depend on the initial conditions, but only on the mass of the object and the coefficient of resistance of the medium. Given either v_l or k, Eq. (10) provides a formula for calculating the other. As k approaches zero (that is, as the resistance diminishes), the limiting velocity increases without bound.

Example 2. A body of constant mass m is projected upward from the earth's surface with an initial velocity v_0. Assuming there is no air resistance, but taking into consideration the variation of the earth's gravitational field with altitude, find the smallest initial velocity for which the body will not return to the earth. This is the so-called "escape velocity".

Take the x axis positive upward, with the origin on the surface of the earth (see Figure 2.7). The only force acting on the body is its weight, given by Eq. (4), which acts downward. Thus the equation of motion is

$$m \frac{dv}{dt} = -\frac{mgR^2}{(x+R)^2}, \tag{11}$$

and the initial condition is

$$v(0) = v_0. \tag{12}$$

Since the force appearing on the right side of Eq. (11) is a function of x only, it is convenient to think of x, rather than t, as the independent variable. This requires that we express dv/dt in terms of dv/dx by the chain rule for differentiation; hence

$$\frac{dv}{dt} = \frac{dv}{dx}\frac{dx}{dt} = v \frac{dv}{dx}, \tag{13}$$

and Eq. (11) is replaced by

$$v \frac{dv}{dx} = -\frac{gR^2}{(x+R)^2}. \tag{14}$$

FIGURE 2.7

Separating the variables and integrating yields

$$\tfrac{1}{2}v^2 = \frac{gR^2}{x + R} + c. \tag{15}$$

Since $x = 0$ when $t = 0$, the initial condition (12) at $t = 0$ can be replaced by the condition that $v = v_0$ when $x = 0$. This requires that $c = \tfrac{1}{2}v_0^2 - gR$; therefore

$$v^2 = v_0^2 - 2gR + \frac{2gR^2}{x + R}. \tag{16}$$

The escape velocity is found by requiring that the velocity v given by Eq. (16) remain positive for all (positive) values of x. This condition will be met provided that $v_0^2 \geq 2gR$. Thus the escape velocity v_e is given by

$$v_e = (2gR)^{1/2}; \tag{17}$$

the magnitude of v_e is approximately 6.9 miles/sec.

In calculating the escape velocity $v_e = (2gR)^{1/2}$, air resistance has been neglected. The actual escape velocity, including the effect of air resistance, is somewhat higher. On the other hand, the effective escape velocity can be significantly reduced if the body is transported a considerable distance above sea level before being launched. Both gravitational and frictional forces are thereby reduced; air resistance, in particular, diminishes quite rapidly with increasing altitude.

Bodies with Variable Mass.   For a space vehicle whose initial mass m_0 contains a substantial supply of fuel that is burned in flight it is necessary to consider the change in the mass of the vehicle. In order to apply Eq. (2) to this situation we must take account not only of the momentum of the rocket, but also of that of the exhaust gases produced by the combustion of the fuel. We will assume that the exhaust gases are ejected from the vehicle at a constant velocity u relative to the vehicle, and that the total mass of the vehicle-fuel system (including the exhaust gases) remains constant. If $m(t)$ denotes the mass of the vehicle and its cargo at time t, and if $m_e(t)$ denotes the mass of the exhaust gases at this instant, then

$$m(t) + m_e(t) = m_0. \tag{18}$$

Applying Eq. (2) to the vehicle-gas system gives*

$$-mg = m\frac{dv}{dt} - u\frac{dm}{dt}, \tag{19}$$

where v is the velocity of the rocket. The second term on the right side of Eq. (19) results from a consideration of the contribution of the exhaust gases to the momentum of the system.

* The details of the derivation are given in the Appendix at the end of this chapter.

Example 3. Consider a space vehicle of initial mass m_0 projected upward from the surface of the earth with an initial velocity v_0. Assume that a part of the mass $m_f < m_0$ consists of fuel that is consumed at a constant rate β during the interval $0 \leq t \leq t_1$. Assume that the exhaust speed $s = |u| = -u$ is also constant. Neglect air resistance and assume that the gravitational force is constant. Find an expression for the velocity of the vehicle as a function of time t for $0 \leq t \leq t_1$.

The stated conditions imply that the mass of the vehicle is given by

$$m(t) = m_0 - \beta t, \qquad t < t_1. \tag{20}$$

At the time t_1 the mass of the vehicle has been reduced to the value $m_1 = m_0 - \beta t_1 = m_0 - m_f$. For $t < t_1$ the equation of motion (19) takes the form

$$-(m_0 - \beta t)g = (m_0 - \beta t)\frac{dv}{dt} - s\beta,$$

or

$$\frac{dv}{dt} = -g + \frac{s\beta}{m_0 - \beta t}. \tag{21}$$

Integrating Eq. (21) gives

$$v = -gt - s \ln (m_0 - \beta t) + c.$$

The initial condition $v(0) = v_0$ requires that

$$c = v_0 + s \ln m_0,$$

and hence

$$v = v_0 - gt + s \ln \frac{m_0}{m_0 - \beta t}, \qquad 0 \leq t \leq t_1. \tag{22}$$

When the fuel is exhausted, at $t = t_1 = (m_0 - m_1)/\beta$, the vehicle has attained the velocity v_1,

$$v_1 = v_0 - \frac{gm_0}{\beta}\left(1 - \frac{m_1}{m_0}\right) + s \ln \frac{m_0}{m_1}. \tag{23}$$

If v_1 is at least as great as the escape velocity at the altitude attained at that instant, the vehicle will be able to leave the gravitational field of the earth. Otherwise, unless further stages are ignited, it will eventually return to the earth. It also follows from Eq. (22) that the velocity v increases with the exhaust speed s and the rate of burning β.

PROBLEMS

1. A body of constant mass m is projected vertically upward with an initial velocity v_0. Assuming the gravitational attraction of the earth to be constant, and neglecting all other forces acting on the body, find:
   (a) The maximum height attained by the body.
   (b) The time at which the maximum height is reached.
   (c) The time at which the body returns to its starting point.
2. A body is thrown vertically downward with an initial velocity v_0 in a medium offering resistance proportional to the magnitude of the velocity. Find the velocity v as

a function of time t. Find the limiting velocity v_l approached after a long time.

3. An object of mass m is dropped from rest in a medium offering resistance proportional to the magnitude of the velocity. Find the time interval that elapses before the velocity of the object reaches 90 percent of its limiting value.

4. A man and a motor boat together weigh 320 lb. If the thrust of the motor is equivalent to a constant force of 10 lb in the direction of motion, and if the resistance of the water to the motion is equal numerically to twice the speed in feet/second, and if the boat is initially at rest, find:

(a) The velocity of the boat at time t.

(b) The limiting velocity.

5. A body with mass m is projected vertically downward with initial velocity v_0 in a medium offering resistance proportional to the square root of the magnitude of the velocity. Find the relation between the velocity v and the time t. Find the limiting velocity.

6. A body of mass m falls from rest in a medium offering resistance proportional to the square of the velocity. Find the relation between the velocity v and the time t. Find the limiting velocity.

7. A body of mass m falls in a medium offering resistance proportional to $|v|^r$, where r is a positive constant. Assuming the gravitational attraction to be constant, find the limiting velocity of the body.

8. A body of constant mass m is projected vertically upward with an initial velocity v_0 in a medium offering a resistance $k|v|$, where k is a constant. Neglecting changes in the gravitational force, find:

(a) The maximum height x_m attained by the body.

(b) The time t_m at which this maximum height is reached. Show also that t_m and x_m can be expressed as

$$t_m = \frac{v_0}{g}\left[1 - \frac{1}{2}\frac{kv_0}{mg} + \frac{1}{3}\left(\frac{kv_0}{mg}\right)^2 - \cdots\right],$$

$$x_m = \frac{v_0^2}{2g}\left[1 - \frac{2}{3}\frac{kv_0}{mg} + \frac{1}{2}\left(\frac{kv_0}{mg}\right)^2 - \cdots\right],$$

both series converging when $kv_0/mg < 1$.

9. A body falling in a relatively dense fluid, oil for example, is acted on by three forces (see Figure 2.8): a resistive force R, a buoyant force B, and its weight w due to gravity. The buoyant force is equal to the weight of the fluid displaced by the object. For a slowly moving spherical body of radius a, the resistive force is given by Stokes'* law $R = 6\pi\mu a|v|$, where v is the velocity of the body, and μ is the coefficient of viscosity of the surrounding fluid.

FIGURE 2.8

---

* George Gabriel Stokes (1819–1903), professor at Cambridge, was one of the foremost applied mathematicians of the nineteenth century. The basic equations of fluid mechanics (the Navier-Stokes equations) are named partly in his honor, and one of the fundamental theorems of vector calculus bears his name. He was also one of the pioneers in the use of divergent (asymptotic) series, a subject of great interest and importance today.

(a) Find the limiting velocity of a solid sphere of radius a and density ρ falling freely in a medium of density ρ' and coefficient of viscosity μ.

(b) In 1910 the American physicist R. A. Millikan (1868–1953) determined the charge on an electron by studying the motion of tiny droplets of oil falling in an electric field. A field of strength E exerts a force Ee on a droplet with charge e. Assume that E has been adjusted so that the droplet is held stationary ($v = 0$), and that w and B are as given in part (a). Find a formula for e. Millikan was able to identify e as the charge on an electron and to determine that $e = 4.803 \times 10^{-10}$ esu.

10. (a) What is the weight of a body 1000 miles above the earth's surface if the body weighs 100 lb at sea level?

(b) The radius of the moon is approximately $\frac{27}{100}$ of the radius of the earth, and the average density of the moon is about $\frac{5}{8}$ of the average density of the earth. Find the apparent weight on the moon of an object weighing 100 lb on the surface of the earth.

11. A body of constant mass m is launched vertically upward from sea level with an initial velocity v_0 that does not exceed the escape velocity $v_e = (2gR)^{\frac{1}{2}}$. Neglecting air resistance, but considering the change of gravitational attraction with altitude, find the maximum altitude attained by the body (see Example 2).

12. Find the escape velocity for a body projected upward with an initial velocity v_0 from a point $x_0 = \xi R$ above the surface of the earth, where R is the radius of the earth and ξ is a constant. Neglect air resistance. Find the initial altitude from which the body must be launched in order to reduce the escape velocity to 85 percent of its value at the earth's surface.

13. **Brachistochrone Problem.**   One of the famous problems in the history of mathematics is the brachistochrone problem: to find the curve along which a particle will slide without friction in the minimum time from one given point P to another Q, the second point being lower than the first but not directly beneath it (see Figure 2.9). This problem was posed by Johann Bernoulli in 1696 as a challenge problem to the mathematicians of his day. Correct solutions were found by Johann Bernoulli himself, and also by his brother Jakob Bernoulli, Isaac Newton, Gottfried Leibniz, and Marquis de L'Hôpital. The brachistochrone problem is important in the development of mathematics as one of the forerunners of the calculus of variations.

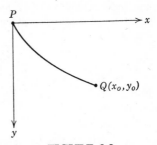

FIGURE 2.9

In solving this problem it is convenient to take the origin as the upper point P and to orient the axes as shown in Figure 2.9. The lower point Q has coordinates (x_0, y_0). It is then possible to show that the curve of minimum time is given by a function $y = \phi(x)$ that satisfies the differential equation

$$\left[1 + \left(\frac{dy}{dx}\right)^2 \right] y = k^2 \tag{i}$$

where k^2 is a certain positive constant to be determined later.

(a) Solve Eq. (i) for y'. Why is it necessary to choose the positive square root?

(b) Introduce the new variable t by the relation

$$y = k^2 \sin^2 t. \tag{ii}$$

Show that the equation found in part (a) then takes the form

$$2k^2 \sin^2 t \, dt = dx. \tag{iii}$$

(c) Letting $\theta = 2t$, show that the solution of Eq. (iii) for which $x = 0$ when $y = 0$ is given by

$$x = \frac{k^2}{2} (\theta - \sin \theta). \tag{iv}$$

From Eq. (ii) show that

$$y = \frac{k^2}{2} (1 - \cos \theta). \tag{v}$$

Equations (iv) and (v) are parametric equations of the solution of Eq. (i) that passes through $(0, 0)$. The graph of Eqs. (iv) and (v) is called a *cycloid*. If we make a proper choice of the constant k, then the cycloid also passes through the point (x_0, y_0), and is the solution of the brachistochrone problem. It is possible to eliminate θ and to obtain the solution in the form $y = \phi(x)$; however, it is easier to use the parametric equations.

14. Consider a rocket that, before launching, weighs 1200 lb, including 900 lb of fuel. Assume that the fuel burns evenly for a period of 60 sec, and that the exhaust speed of the rocket is 5000 ft/sec. If the rocket has no initial velocity, find its velocity at the end of 60 sec. Neglect air resistance and changes in the earth's gravitational field.

15. Consider a space vehicle of initial mass m_0 that starts from rest on the earth's surface. Assume that the rate of fuel consumption $\beta = -dm/dt$ is constant in the interval $0 < t < t_1$ and also that the exhaust speed s is constant. Assume further that the air resistance experienced by the vehicle is proportional to the magnitude of its velocity $v(t)$, but neglect the effect of air resistance on the exhaust gases, and also the change in the gravitational field with altitude. Show that the velocity of the vehicle satisfies the differential equation

$$m \frac{dv}{dt} + kv = -mg + s\beta, \qquad t < t_1$$

where $m(t) = m_0 - \beta t$. Solve this equation and obtain an expression for v at any time $t < t_1$.

*2.11 THE EXISTENCE AND UNIQUENESS THEOREM

In this section we will discuss the proof of Theorem 2.2, the fundamental existence and uniqueness theorem for first order initial value problems. This theorem states that under certain conditions on $f(x, y)$, the initial value problem

$$y' = f(x, y), \tag{1a}$$

$$y(x_0) = y_0 \tag{1b}$$

has a unique solution in some interval containing the point x_0.

In some cases (for example, if the differential equation is linear) the existence of a solution of the initial value problem (1) can be established directly by actually solving the problem and exhibiting a formula for the solution. However, in general, this approach is not feasible because there is no method of solving Eq. (1a) that applies in all cases. Therefore, for the general case it is necessary to adopt an indirect approach that establishes the existence of a solution of Eqs. (1) but usually does not provide a practical means of finding it. The heart of this method is the construction of a sequence of functions that converges to a limit function satisfying the initial value problem, although the members of the sequence individually do not. As a rule, it is impossible to compute explicitly more than a few members of the sequence; therefore, the limit function can be found explicitly only in rare cases. Nevertheless, under the restrictions on $f(x, y)$ stated in Theorem 2.2, it is possible to show that the sequence in question converges and that the limit function has the desired properties. The argument is fairly intricate and depends, in part, on techniques and results that are usually encountered for the first time in a course on advanced calculus. Consequently, we will not go into all of the details of the proof here; we will, however, indicate its main features and point out some of the difficulties involved.

First of all, we note that it is sufficient to consider the problem in which the initial point (x_0, y_0) is the origin; that is, the problem

$$y' = f(x, y), \tag{2a}$$

$$y(0) = 0. \tag{2b}$$

If some other initial point is given, then we can always make a preliminary change of variables, corresponding to a translation of the coordinate axes, that will take the given point (x_0, y_0) into the origin. Specifically, we introduce new dependent and independent variables w and s, respectively, defined by the equations

$$w = y - y_0, \qquad s = x - x_0. \tag{3}$$

Thinking of w as a function of s, we have by the chain rule

$$\frac{dw}{ds} = \frac{dw}{dx}\frac{dx}{ds} = \frac{d}{dx}(y - y_0)\frac{dx}{ds} = \frac{dy}{dx}.$$

Denoting $f(x, y) = f(s + x_0, w + y_0)$ by $F(s, w)$, the initial value problem (1) is converted into the form

$$w'(s) = F[s, w(s)], \tag{4a}$$

$$w(0) = 0. \tag{4b}$$

Except for the names of the variables, Eqs. (4) are the same as Eqs. (2). Hereafter, we will consider the slightly simpler problem (2) rather than the original problem (1).

The existence and uniqueness theorem can now be stated in the following way.

Theorem 2.4.   *If f and $\partial f/\partial y$ are continuous in a rectangle R: $|x| \leq a, |y| \leq b$, then there is some interval $|x| \leq h \leq a$ in which there exists a unique solution $y = \phi(x)$ of the initial value problem (2)*

$$y' = f(x, y), \qquad y(0) = 0.$$

In order to prove this theorem it is necessary to transform the initial value problem (2) into a more convenient form. If we suppose temporarily that there is a function $y = \phi(x)$ that satisfies the initial value problem, then $f[x, \phi(x)]$ is a continuous function of x only. Hence we can integrate Eq. (2a) from the initial point $x = 0$ to an arbitrary value of x, obtaining

$$\phi(x) = \int_0^x f[t, \phi(t)] \, dt, \tag{5}$$

where we have made use of the intial condition $\phi(0) = 0$.

Since Eq. (5) contains an integral of the unknown function ϕ, it is called an *integral equation*. This integral equation is not a formula for the solution of the initial value problem, but it does provide another relation satisfied by any solution of Eqs. (2). Conversely, suppose that there is a continuous function $y = \phi(x)$ that satisfies the integral equation (5); then this function also satisfies the initial value problem (2). To show this, we first substitute zero for x in Eq. (5), thus obtaining Eq. (2b). Further, since the integrand in Eq. (5) is continuous it follows from the fundamental theorem of calculus that $\phi'(x) = f[x, \phi(x)]$. Therefore the initial value problem and the integral equation are equivalent in the sense that any solution of one is also a solution of the other. It is more convenient to show that there is a unique solution of the integral equation in a certain interval $|x| \le h$. The same conclusion will then also hold for the initial value problem.

One method of showing that the integral equation (5) has a unique solution is known as the *method of successive approximations*, or Picard's (1856–1941) *iteration method*. In using this method we start by choosing an initial function ϕ_0, either arbitrarily or to approximate in some way the solution of the initial value problem. The simplest choice is

$$\phi_0(x) = 0; \tag{6}$$

then ϕ_0 at least satisfies the initial condition (2b) although presumably not the differential equation (2a). The next approximation ϕ_1 is obtained by substituting $\phi_0(t)$ for $\phi(t)$ in the right side of Eq. (5), and calling the result of this operation $\phi_1(x)$. Thus

$$\phi_1(x) = \int_0^x f[t, \phi_0(t)] \, dt. \tag{7}$$

Similarly, ϕ_2 is obtained from ϕ_1:

$$\phi_2(x) = \int_0^x f[t, \phi_1(t)] \, dt, \tag{8}$$

and in general,

$$\phi_{n+1}(x) = \int_0^x f[t, \phi_n(t)] \, dt. \tag{9}$$

In this manner we generate the sequence of functions $\{\phi_n\} = \phi_0, \phi_1, \ldots, \phi_n, \ldots$. Each member of the sequence satisfies the initial condition (2b), but in general none satisfies the differential equation. However, if at some stage, say for $n = k$, we find that $\phi_{k+1}(x) = \phi_k(x)$, then it follows that ϕ_k is a solution of the integral equation (5). Hence ϕ_k is also a solution of the initial value problem (2), and the sequence is terminated at this point. In general, this will not occur, and it is necessary to consider the entire infinite sequence.

In order to establish Theorem 2.4 four principal questions must be answered:

1. Do all members of the sequence $\{\phi_n\}$ exist, or may the process break down at some stage?

2. Does the sequence converge?

3. What are the properties of the limit function, in particular, does it satisfy the integral equation (5), and hence the initial value problem (2)?

4. Is this the only solution, or may there be others?

We will first show how these questions can be answered in a specific and relatively simple example, and then comment on some of the difficulties that may be encountered in the general case.

Example.   Consider the initial value problem

$$y' = 2x(1 + y), \qquad y(0) = 0. \tag{10}$$

To solve this problem by the method of successive approximations we note first that if $y = \phi(x)$, then the corresponding integral equation is

$$\phi(x) = \int_0^x 2t[1 + \phi(t)] \, dt. \tag{11}$$

If the initial approximation is $\phi_0(x) = 0$, it follows that

$$\phi_1(x) = \int_0^x 2t[1 + \phi_0(t)] \, dt = \int_0^x 2t \, dt = x^2. \tag{12}$$

Similarly

$$\phi_2(x) = \int_0^x 2t[1 + \phi_1(t)] \, dt = \int_0^x 2t[1 + t^2] \, dt = x^2 + \frac{x^4}{2}, \tag{13}$$

and

$$\phi_3(x) = \int_0^x 2t[1 + \phi_2(t)] \, dt = \int_0^x 2t\left[1 + t^2 + \frac{t^4}{2}\right] dt = x^2 + \frac{x^4}{2} + \frac{x^6}{2 \cdot 3}. \tag{14}$$

Equations (12), (13), and (14) suggest that

$$\phi_n(x) = x^2 + \frac{x^4}{2!} + \frac{x^6}{3!} + \cdots + \frac{x^{2n}}{n!} \tag{15}$$

for each $n \geq 1$, and this result can be established by mathematical induction. Clearly Eq. (15) is true for $n = 1$, and we must show that if it is true for $n = k$, then it also holds for $n = k + 1$. We have

$$\phi_{k+1}(x) = \int_0^x 2t[1 + \phi_k(t)] \, dt$$

$$= \int_0^x 2t\left(1 + t^2 + \frac{t^4}{2!} + \cdots + \frac{t^{2k}}{k!}\right) dt$$

$$= x^2 + \frac{x^4}{2!} + \frac{x^6}{3!} + \cdots + \frac{x^{2k+2}}{(k+1)!}, \tag{16}$$

and the inductive proof is complete.

It is clear from Eq. (15) that $\phi_n(x)$ is the nth partial sum of the infinite series

$$\sum_{k=1}^{\infty} x^{2k}/k!; \tag{17}$$

hence $\lim\limits_{n \to \infty} \phi_n(x)$ exists if and only if the series (17) converges. Applying the ratio test, we see that for each x

$$\left| \frac{x^{2k+2}}{(k+1)!} \frac{k!}{x^{2k}} \right| = \frac{x^2}{k+1} \to 0 \quad \text{as} \quad k \to \infty; \tag{18}$$

thus the series (17) converges for all x, and its sum $\phi(x)$ is the limit* of the sequence $\{\phi_n(x)\}$. Further, since the series (17) is a Taylor series, it can be differentiated or integrated term by term as long as x remains within the interval of convergence, which in this case is the whole x axis. Therefore we can verify by direct computation that $\phi(x) = \sum_{k=1}^{\infty} x^{2k}/k!$ is a solution of the integral equation (11). Alternatively, by substituting $\phi(x)$ for y in Eq. (10) we can verify that this function also satisfies the initial value problem.

Finally, to deal with the question of uniqueness, let us suppose that the initial value problem has another solution ψ, different from ϕ. We will show that this leads to a contradiction. Since ϕ and ψ both satisfy the integral equation (11), we have by subtraction that

$$\phi(x) - \psi(x) = \int_0^x 2t[\phi(t) - \psi(t)] \, dt.$$

Taking absolute values of both sides we have, if $x > 0$,

$$|\phi(x) - \psi(x)| \leq \int_0^x 2t \, |\phi(t) - \psi(t)| \, dt.$$

If we restrict x to lie in the interval $0 \leq x \leq A/2$, where A is arbitrary, then $2t \leq A$, and

$$|\phi(x) - \psi(x)| \leq A \int_0^x |\phi(t) - \psi(t)| \, dt. \tag{19}$$

---

* In this case it is possible to identify ϕ in terms of elementary functions, namely, $\phi(x) = e^{x^2} - 1$. However, this is irrelevant to the discussion of existence and uniqueness.

It is convenient at this point to introduce the function U defined by

$$U(x) = \int_0^x |\phi(t) - \psi(t)| \, dt. \tag{20}$$

Then it follows at once that

$$U(0) = 0, \tag{21}$$

$$U(x) \geq 0, \quad \text{for} \quad x \geq 0. \tag{22}$$

Further, U is differentiable, and $U'(x) = |\phi(x) - \psi(x)|$. Hence, by Eq. (19),

$$U'(x) - AU(x) \leq 0. \tag{23}$$

Multiplying Eq. (23) by the positive quantity e^{-Ax} gives

$$[e^{-Ax}U(x)]' \leq 0. \tag{24}$$

Then, upon integrating Eq. (24) from zero to x and using Eq. (21), we obtain

$$e^{-Ax}U(x) \leq 0 \quad \text{for} \quad x \geq 0.$$

Hence $U(x) \leq 0$ for $x \geq 0$, and in conjunction with Eq. (22), this requires that $U(x) = 0$ for each $x \geq 0$. Thus $U'(x) \equiv 0$, and therefore $\psi(x) \equiv \phi(x)$, which contradicts the original hypothesis. Consequently, there cannot be two different solutions of the initial value problem for $x \geq 0$. A slight modification of this argument leads to the same conclusion for $x \leq 0$.

Returning now to the general problem of solving the integral equation (5), let us consider briefly each of the questions raised earlier.

1. Do all members of the sequence $\{\phi_n\}$ exist? In the example f and $\partial f/\partial y$ were continuous in the whole xy plane, and each member of the sequence could be explicitly calculated. In contrast, in the general case f and $\partial f/\partial y$ are assumed to be continuous only in the rectangle R: $|x| \leq a$, $|y| \leq b$ (see Figure 2.10). Furthermore, the members of the sequence cannot as a rule be explicitly determined. The danger

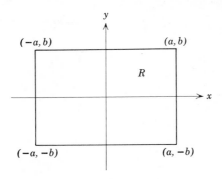

FIGURE 2.10

is that at some stage, say for $n = k$, the graph of $y = \phi_k(x)$ may contain points that lie outside of the rectangle R. Hence at the next stage—in the computation of $\phi_{k+1}(x)$—it would be necessary to evaluate $f(x, y)$ at points where it is not known to be continuous or even to exist. Thus the calculation of $\phi_{k+1}(x)$ might be impossible.

To avoid this danger it may be necessary to restrict x to a smaller interval than $|x| \leq a$. To find such an interval we make use of the fact that a continuous function on a closed region is bounded. Hence f is bounded on R; thus there exists a positive number M such that

$$|f(x, y)| \leq M, \qquad (x, y) \text{ in } R. \tag{25}$$

We have mentioned before that

$$\phi_n(0) = 0$$

for each n. Since $f[x, \phi_k(x)]$ is equal to $\phi'_{k+1}(x)$ the maximum absolute slope of the graph of the equation $y = \phi_{k+1}(x)$ is M. Since this graph contains the point $(0, 0)$, it must lie in the wedge-shaped shaded region in Figure 2.11. Hence the point $[x, \phi_{k+1}(x)]$ remains in R at least as long as R contains the wedge-shaped region, which is for $|x| \leq b/M$. We will hereafter consider only the rectangle D: $|x| \leq h$, $|y| \leq b$, where h is equal either to a or to b/M, whichever is smaller. With this restriction all members of the sequence $\{\phi_n(x)\}$ exist. Note that if $b/M < a$, then a larger value of h can be obtained by finding a better bound for $|f(x, y)|$, provided that M is not already equal to the maximum value of $|f(x, y)|$.

. 2. Does the sequence $\{\phi_n(x)\}$ converge? As in the example, we can identify $\phi_n(x) = \phi_1(x) + [\phi_2(x) - \phi_1(x)] + \cdots + [\phi_n(x) - \phi_{n-1}(x)]$ as the nth partial sum of the series

$$\phi_1(x) + \sum_{k=1}^{\infty} [\phi_{k+1}(x) - \phi_k(x)]. \tag{26}$$

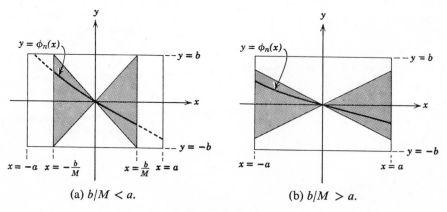

(a) $b/M < a$. (b) $b/M > a$.

FIGURE 2.11

The convergence of the sequence $\{\phi_n(x)\}$ is established by showing that the series (26) converges. To do this it is necessary to estimate the magnitude $|\phi_{k+1}(x) - \phi_k(x)|$ of the general term. The argument by which this is done is indicated in Problems 3 through 6 and will be omitted here. Assuming that the sequence converges, we will denote the limit function by ϕ, so that

$$\phi(x) = \lim_{n \to \infty} \phi_n(x). \tag{27}$$

3. What are the properties of the limit function ϕ? In the first place, we would like to know that ϕ is continuous. This is not, however, a necessary consequence of the convergence of the sequence $\{\phi_n(x)\}$, even though each member of the sequence is continuous itself. There are many cases in which a sequence of continuous functions converges to a limit function that is discontinuous. A simple example of this phenomenon is given in Problem 1. One way to show that ϕ is continuous is to show not only that the sequence $\{\phi_n\}$ converges but that it converges in a certain manner, known as uniform convergence. We will not take up this question here but will note only that the argument referred to in paragraph 2 is sufficient to establish the uniform convergence of the sequence $\{\phi_n\}$ and hence the continuity of the limit function ϕ in the interval $|x| \le h$.

Now let us return to Eq. (9),

$$\phi_{n+1}(x) = \int_0^x f[t, \phi_n(t)] \, dt.$$

Allowing n to approach ∞ on both sides, we obtain

$$\phi(x) = \lim_{n \to \infty} \int_0^x f[t, \phi_n(t)] \, dt. \tag{28}$$

We would like to interchange the operations of integration and taking the limit on the right side of Eq. (28), so as to obtain

$$\phi(x) = \int_0^x \lim_{n \to \infty} f[t, \phi_n(t)] \, dt. \tag{29}$$

In general, such an interchange is not permissible (see Problem 2, for example), but once again the fact that the sequence $\{\phi_n(x)\}$ not only converges but converges uniformly comes to the rescue and allows us to take the limiting operation inside the integral sign. Next we wish to take the limit inside the function f, which would give

$$\phi(x) = \int_0^x f\left[t, \lim_{n \to \infty} \phi_n(t)\right] dt \tag{30}$$

and hence

$$\phi(x) = \int_0^x f[t, \phi(t)] \, dt. \tag{31}$$

The statement that $\lim_{n \to \infty} f[t, \phi_n(t)] = f[t, \lim_{n \to \infty} \phi_n(t)]$ is equivalent to the statement that f is continuous in its second variable, which is known by hypothesis. Hence

Eq. (31) is valid and the function ϕ satisfies the integral equation (5). Therefore ϕ is also a solution of the initial value problem (2).

4. Are there other solutions of the integral equation (5) besides $y = \phi(x)$? To show the uniqueness of the solution $y = \phi(x)$ we can proceed much as in the example. First assume the contrary, that is, the existence of another solution $y = \psi(x)$. It is then possible to show (see Problem 7) that the difference $\phi(x) - \psi(x)$ satisfies the inequality

$$|\phi(x) - \psi(x)| \leq A \int_0^x |\phi(t) - \psi(t)|\, dt \tag{32}$$

for $0 \leq x \leq h$ and for a suitable positive number A. From this point the argument is identical to that given in the example, and we conclude that there is no solution of the initial value problem (2) other than that obtained by the method of successive approximations.

PROBLEMS

1. Let $\phi_n(x) = x^n$ for $0 \leq x \leq 1$, and show that

$$\lim_{n \to \infty} \phi_n(x) = \begin{cases} 0, & 0 \leq x < 1, \\ 1, & x = 1 \end{cases}$$

This example shows that a sequence of continuous functions may converge to a limit function that is discontinuous.

2. Consider the sequence $\phi_n(x) = 2nxe^{-nx^2}$, $0 \leq x \leq 1$.
(a) Show that $\lim_{n \to \infty} \phi_n(x) = 0$ for $0 \leq x \leq 1$ and hence that

$$\int_0^1 \lim_{n \to \infty} \phi_n(x)\, dx = 0.$$

(b) Show that $\int_0^1 2nxe^{-nx^2} dx = 1 - e^{-n}$ and hence that

$$\lim_{n \to \infty} \int_0^1 \phi_n(x)\, dx = 1.$$

This example shows that it is not necessarily true that

$$\lim_{n \to \infty} \int_a^b \phi_n(x)\, dx = \int_a^b \lim_{n \to \infty} \phi_n(x)\, dx$$

even though $\lim_{n \to \infty} \phi_n(x)$ exists and is continuous.

In Problems 3 through 6 we indicate how to prove that the sequence $\{\phi_n(x)\}$, defined by Eqs. (6) through (9), converges.

3. If $\partial f / \partial y$ is continuous in the rectangle D, show that there is a positive constant K such that

$$|f(x, y_1) - f(x, y_2)| \leq K|y_1 - y_2|$$

where (x, y_1) and (x, y_2) are any two points in D having the same x coordinate.
Hint: Hold x fixed and use the mean value theorem on f as a function of y only. Choose K to be the maximum value of $|\partial f/\partial y|$ in D.

4. If $\phi_{n-1}(x)$ and $\phi_n(x)$ are members of the sequence $\{\phi_n(x)\}$, use the result of Problem 3 to show that

$$|f[x, \phi_n(x)] - f[x, \phi_{n-1}(x)]| \leq K|\phi_n(x) - \phi_{n-1}(x)|.$$

5. (a) Show that if $|x| \leq h$, then

$$|\phi_1(x)| \leq M|x|$$

where M is chosen so that $|f(x, y)| \leq M$ for (x, y) in D.

(b) Use the results of Problem 4 and part (a) of Problem 5 to show that

$$|\phi_2(x) - \phi_1(x)| \leq \frac{MK\,|x|^2}{2}.$$

(c) Show, by mathematical induction, that

$$|\phi_n(x) - \phi_{n-1}(x)| \leq \frac{MK^{n-1}|x|^n}{n!} \leq \frac{MK^{n-1}h^n}{n!}.$$

6. Note that

$$\phi_n(x) = \phi_1(x) + [\phi_2(x) - \phi_1(x)] + \cdots + [\phi_n(x) - \phi_{n-1}(x)].$$

(a) Show that

$$|\phi_n(x)| \leq |\phi_1(x)| + |\phi_2(x) - \phi_1(x)| + \cdots + |\phi_n(x) - \phi_{n-1}(x)|.$$

(b) Use the results of Problem 5 to show that

$$|\phi_n(x)| \leq \frac{M}{K}\left[Kh + \frac{(Kh)^2}{2!} + \cdots + \frac{(Kh)^n}{n!}\right].$$

(c) Show that the sum in part (b) converges as $n \to \infty$, and hence show that the sum in part (a) also converges as $n \to \infty$. Conclude therefore that the sequence $\{\phi_n(x)\}$ converges, since it is the sequence of partial sums of a convergent infinite series.

7. In this problem we deal with the question of uniqueness of the solution of the integral equation (5),

$$\phi(x) = \int_0^x f[t, \phi(t)]\, dt.$$

(a) Suppose that ϕ and ψ are two solutions of Eq. (5). Show that

$$\phi(x) - \psi(x) = \int_0^x \{f[t, \phi(t)] - f[t, \psi(t)]\}\, dt.$$

(b) Show that

$$|\phi(x) - \psi(x)| \leq \int_0^x |f[t, \phi(t)] - f[t, \psi(t)]|\, dt.$$

(c) Use the result of Problem 3 to show that

$$|\phi(x) - \psi(x)| \leq K\int_0^x |\phi(t) - \psi(t)|\, dt,$$

where K is an upper bound for $|\partial f/\partial y|$ in D. This is the same as Eq. (32), and the rest of the proof may be constructed as indicated in the text.

APPENDIX

Derivation of Equation of Motion of a Body with Variable Mass.   To derive Eq. (19) of Section 2.10 we can proceed in the following way. Assuming that the total mass of the vehicle-fuel-exhaust system remains at the constant value m_0, we have

$$m(t) + m_e(t) = m_0, \tag{1}$$

where $m(t)$ and $m_e(t)$ are the mass of the vehicle (including fuel) and the mass of the exhaust gases, respectively. Applying Newton's law to the total system gives

$$F = \frac{dP}{dt} = \frac{dP_r}{dt} + \frac{dP_e}{dt}, \tag{2}$$

where $P_r(t)$, $P_e(t)$, and $P(t)$ refer to the linear momenta of the vehicle, the exhaust gases, and the entire system, respectively. In the present discussion we will assume that the only external force acting on the entire system is due to gravity. Neglecting resistive forces and variations in the gravitational field, it follows that

$$F = -m_0 g. \tag{3}$$

If $v(t)$ denotes the velocity of the vehicle at time t, then $P_r(t) = m(t)v(t)$, and

$$\frac{dP_r}{dt} = \frac{dm}{dt} v + m \frac{dv}{dt}. \tag{4}$$

On the other hand, the calculation of the momentum P_e of the exhaust gases is rendered more complicated by the fact that different portions of the exhaust gases have different velocities. Suppose that we think of the gas as being ejected continuously at a velocity u relative to the vehicle, and let $\Delta m_e(\tau)$ be the mass of gas ejected during the short time interval $(\tau, \tau + d\tau)$. The initial velocity of this mass element is approximately $v(\tau) + u$. Assuming that since its ejection it has been subject to the influence of gravity only, the equation of motion of the mass element $\Delta m_e(\tau)$ is

$$\Delta m_e \frac{dv_e}{dt} = -g \, \Delta m_e,$$

where $v_e(t, \tau)$ is the velocity at time t of the gas element created at time τ. Hence $v_e(t, \tau) = -gt + c(\tau)$. The quantity $c(\tau)$ is determined from the condition that at time $t = \tau$ the velocity v_e is given by $v(\tau) + u$. Hence $c(\tau) = v(\tau) + u + g\tau$, and

$$v_e(t, \tau) = -(t - \tau)g + v(\tau) + u.$$

The momentum of the mass element Δm_e is thus equal to $v_e(t, \tau) \Delta m_e$; the total momentum $P_e(t)$ of all of the ejected gas is therefore given by the integral

$$P_e(t) = \int v_e(t, \tau) \, dm_e$$

taken over the entire gas. If the propulsion process originated at time $t = 0$, and has continued until the present time t, then $P_e(t)$ can also be represented by a time integral over the interval $(0, t)$. Thus

$$P_e(t) = \int_0^t v_e(t, \tau) \frac{dm_e}{d\tau} (\tau) \, d\tau$$

$$= \int_0^t [-(t - \tau)g + v(\tau) + u] \frac{dm_e}{d\tau} (\tau) \, d\tau,$$

or, since $dm_e/dt = -dm/dt$ from Eq. (1),

$$P_e(t) = \int_0^t [(t - \tau)g - v(\tau) - u] \frac{dm}{d\tau} (\tau) \, d\tau. \tag{5}$$

To compute dP_e/dt it is necessary to differentiate the integral in Eq. (5). Contributions to dP_e/dt result from the presence of the parameter t under the integral sign, as well as from its presence in the upper limit. We obtain

$$\frac{dP_e}{dt} (t) = g \int_0^t \frac{dm}{d\tau} (\tau) \, d\tau - [v(t) + u] \frac{dm}{dt} (t)$$

$$= g[m(t) - m(0)] - [v(t) + u] \frac{dm}{dt} (t)$$

$$= g[m(t) - m_0] - [v(t) + u] \frac{dm}{dt} (t), \tag{6}$$

where the fact that initially the mass of the vehicle comprises the total mass of the system has been used in the last step.

The equation of motion of the space vehicle of varying mass is obtained by substituting for F, dP_r/dt, and dP_e/dt in Eq. (2) the expressions found in Eqs. (3), (4), and (6), respectively. The result, after some obvious cancelations, is

$$-mg = m \frac{dv}{dt} - u \frac{dm}{dt}. \tag{7}$$

REFERENCES

A fuller discussion of the proof of the fundamental existence and uniqueness theorem can be found in many more advanced books on differential equations. Two that are reasonably accessible to elementary readers are:

Coddington, E. A., *An Introduction to Ordinary Differential Equations*, Prentice-Hall, Englewood Cliffs, N.J., 1961.

Brauer, F. and Nohel, J., *Ordinary Differential Equations*, 2nd ed., Benjamin, New York, 1973.

A useful catalog of differential equations and their solutions is contained in the following book:

Kamke, E., *Differentialgleichungen Lösungsmethoden und Lösungen*, Chelsea, New York, 1948.

Although the text is in German, very little knowledge of German is required to consult the list of solved problems.

CHAPTER 3

Second Order Linear Equations

3.1 INTRODUCTION

The general second order differential equation is an equation of the form

$$F(x, y, y', y'') = 0. \tag{1}$$

The theory associated with such an equation is quite complicated. We will therefore begin by restricting our attention to equations which can be solved for y'', that is, equations which can be written in the form

$$y'' = f(x, y, y'). \tag{2}$$

For the first order differential equation $y' = f(x, y)$ we found that there is a solution containing one arbitrary constant. Since a second order equation involves a second derivative and hence, roughly speaking, two integrations are required to find a solution, it is natural to expect to find solutions of Eq. (2) containing two arbitrary constants. For example, the solution of

$$y'' = g(x) \tag{3}$$

is

$$y = \phi(x) = c_1 + c_2 x + \int^x \left[\int^t g(s) \, ds \right] dt, \tag{4}$$

where c_1 and c_2 are arbitrary constants. For a first order equation it was sufficient to specify the value of the solution at one point to determine a unique integral curve. Since we expect the solution of a second order equation to contain two arbitrary constants, in order to obtain a unique solution it will be necessary to specify two conditions, for example, the value of the solution y_0 and its derivative y_0' at a point x_0. These conditions are referred to as *initial conditions*. Thus to determine uniquely an integral curve of a second order equation it is necessary to specify not only a point through which it passes, but also the slope of the curve at the point.

To insure the existence of a solution of Eq. (2) satisfying prescribed initial conditions it is necessary to postulate certain properties for the function f. The situation is governed by the following existence and uniqueness theorem.

Theorem 3.1.   *If the functions f, f_y, and f_v are continuous in an open region R of the three-dimensional xyv space, and if the point (x_0, y_0, y_0') is in R, then in some interval about x_0 there exists a unique solution $y = \phi(x)$ of the differential equation (2),*

$$y'' = f(x, y, y'),$$

that satisfies the prescribed initial conditions

$$y(x_0) = y_0, \qquad y'(x_0) = y_0'. \tag{5}$$

The proof of this theorem is similar to that for the first order equation given in Section 2.11. Usually in carrying out the proof it is convenient to replace the second order equation by an equivalent system of two first order equations (see Chapter 7). Proofs can be found in many more advanced books, for example, Coddington (Chapter 6) or Ince (Chapter 3).

Even though the existence of a solution of Eq. (2) is guaranteed under the conditions of Theorem 3.1, it may not be possible to determine a convenient analytical expression for the solution unless f is a sufficiently simple function. Just as in the case of first order equations, we distinguish between linear and nonlinear second order equations. The *general second order linear equation* has the form

$$P(x) \frac{d^2y}{dx^2} + Q(x) \frac{dy}{dx} + R(x)y = G(x), \tag{6}$$

where P, Q, R, and G are given functions.

A simple but important example of a second order linear differential equation is the equation governing the motion of a mass on a spring

$$m \frac{d^2u}{dt^2} + c \frac{du}{dt} + ku = F(t), \tag{7}$$

where m, c, and k are constants and F is a prescribed function. This equation is derived in Section 3.7. Other examples are Legendre's* equation

$$(1 - x^2)y'' - 2xy' + \alpha(\alpha + 1)y = 0, \tag{8}$$

of order α, and Bessel's† equation of order ν,

$$x^2y'' + xy' + (x^2 - \nu^2)y = 0, \tag{9}$$

where α and ν are constants, often integers. Bessel's equation arises in many different physical situations, particularly in problems involving circular geometry, such as the determination of the temperature distribution in a circular plate.

---

* A. M. Legendre (1752–1833) was a distinguished French mathematician whose primary work was in the fields of elliptic functions and theory of numbers. The Legendre functions, which are solutions of the differential equation bearing his name, arose from his studies of the attraction of spheroids.

† F. W. Bessel (1784–1846) was a German mathematician who made fundamental contributions in the fields of astronomy, geodesy, and celestial mechanics. He was the first person to make a systematic analysis of the solutions (known as Bessel functions) of the equation that bears his name.

Legendre's equation often occurs in physical situations involving spherical geometry.

If Eq. (2) is not of the form (6) it is said to be *nonlinear*. Although the theory of nonlinear second order differential equations is fairly difficult, there are two special cases where it is possible to simplify the general second order nonlinear equation (2). These occur when either the variable x or the variable y is missing in $f(x, y, y')$; that is, when Eq. (2) is of the form

$$y'' = f(x, y') \tag{10}$$

or

$$y'' = f(y, y'). \tag{11}$$

In these cases it is always possible to reduce Eq. (10) or Eq. (11) to a first order equation for $v = y'$. Provided that the first order equation is of a type discussed in Chapter 2, we can solve that equation for v, and one more integration yields the solution of the original differential equation. This is discussed in Problems 1, 2, and 3 at the end of this section.

The rest of this chapter and the next chapter are devoted to methods of solving second order linear differential equations. Although there is no specific formula for the solution of Eq. (6), as was the case for the first order linear equation, there is an extensive mathematical theory for second order linear equations. In the following discussion we will assume, unless otherwise stated, that the functions P, Q, R, and G in Eq. (6) are continuous on some interval $\alpha < x < \beta$ (in some problems the interval may be unbounded, that is, α may be $-\infty$ and/or β may be $+\infty$), and that, further, the function P is nowhere zero in the interval. In this case we can divide Eq. (6) by $P(x)$ and obtain an equation of the form

$$\frac{d^2y}{dx^2} + p(x)\frac{dy}{dx} + q(x)y = g(x). \tag{12}$$

Notice that the above assumptions are fulfilled for Eq. (7) on $-\infty < x < \infty$; for Legendre's equation on the intervals, $-1 < x < 1$, or $x > 1$, or $x < -1$; and for Bessel's equation on any interval not including the origin.

If we write Eq. (12) in the form of Eq. (2), the function f is given by

$$f(x, y, y') = -p(x)y' - q(x)y + g(x). \tag{13}$$

Since $\partial f(x, y, y')/\partial y = -q(x)$ and $\partial f(x, y, y')/\partial y' = -p(x)$ are continuous, the existence and uniqueness of a solution of Eq. (12) satisfying the initial conditions $y(x_0) = y_0$, $y'(x_0) = y'_0$, $\alpha < x_0 < \beta$, in some interval about x_0 follows from Theorem 3.1. However, just as for first order equations, the existence and uniqueness theorem for second order equations can be stated in a stronger form for linear equations than for nonlinear ones.

Theorem 3.2. *If the functions p, q, and g are continuous on the open interval $\alpha < x < \beta$, then there exists one and only one function $y = \phi(x)$ satisfying the differential equation (12),*

$$y'' + p(x)y' + q(x)y = g(x),$$

on the entire interval $\alpha < x < \beta$, *and the prescribed initial conditions* (5),

$$y(x_0) = y_0, \qquad y'(x_0) = y_0',$$

at a particular point x_0 *in the interval.*

To illustrate one use of this theorem consider the following simple examples.

Example 1.   Find the solution of the differential equation

$$y'' + y = 0,$$

satisfying the initial conditions $y(0) = 0$, $y'(0) = 1$.

It is easily verified that $\sin x$ and $\cos x$ are solutions of the differential equation. Further $y = \sin x$ satisfies $y(0) = 0$, $y'(0) = 1$; hence, according to Theorem 3.2, $y = \sin x$ is the unique solution of the problem.

Example 2.   What is the only solution of

$$y'' + p(x)y' + q(x)y = 0, \qquad \alpha < x < \beta,$$

satisfying the initial conditions $y(x_0) = 0$, $y'(x_0) = 0$, where x_0 is a point in the interval $\alpha < x < \beta$?

Since $y = 0$ satisfies both the differential equation and the initial conditions, it is the unique solution.

In Theorems 3.1 and 3.2 it is important to note that the initial conditions that determine a unique solution of Eq. (2) or Eq. (12) are conditions on the value of the solution and its first derivative at a fixed point in the interval. In contrast, the question of finding a solution of, say, Eq. (12) on the interval $x_0 < x < x_1$, and satisfying $y(x_0) = A, y(x_1) = B$ is not covered by this theorem. Indeed, the latter problem may not always have a solution. Such problems are known as boundary value problems.

In order to solve the second order differential equation (12),

$$y'' + p(x)y' + q(x)y = g(x),$$

we will find it is necessary to solve only the *homogeneous,** or *reduced*, or *complementary* equation

$$y'' + p(x)y' + q(x)y = 0 \qquad (14)$$

obtained from Eq. (12) by setting $g(x) = 0$. Once the solution of the homogeneous equation (14) is known we can, by a general method, solve the nonhomogeneous equation (12). The next three sections contain some general results about the homogeneous equation (14). Then in Section 3.5 we shall show how Eq. (14) can be solved in the particular case that the functions p and q are constants. Even this case is of considerable practical importance; for example, the equation for the

---

* Note that the use of the word homogeneous here is not related to its use in the discussion of first order homogeneous differential equations in Section 2.7.

motion of a mass on a spring has constant coefficients. In Section 3.6 we shall turn our attention to the nonhomogeneous equation (12). The rest of the chapter is devoted to applications in the areas of mechanical vibrations and electric networks.

PROBLEMS

1. For a second order differential equation of the form $y'' = f(x, y')$, the substitution $v = y'$, $v' = y''$ leads to a first order equation of the form $v' = f(x, v)$. Provided that this equation can be solved for v, y can be obtained by integrating $dy/dx = v(x)$. Notice that one arbitrary constant is obtained in solving the first order equation for v, and a second is obtained in the integration for y. Solve the following differential equations.

(a) $x^2 y'' + 2xy' - 1 = 0$, $\quad x > 0$ $\qquad$ (b) $xy'' + y' = 1$, $\quad x > 0$

(c) $y'' + x(y')^2 = 0$ $\qquad\qquad\qquad$ ✓(d) $2x^2 y'' + (y')^3 = 2xy'$, $\quad x > 0$

2. Consider a second order differential equation of the form $y'' = f(y, y')$. If we let v equal y', we obtain $v' = f(y, v)$. This equation contains the variables v, x, and y and hence is not of the form of the first order equations discussed in Chapter 2. It is possible to eliminate the variable x by thinking of y as the independent variable; then by the chain rule

$$\frac{dv}{dx} = \frac{dv}{dy}\frac{dy}{dx} = v\frac{dv}{dy},$$

and hence the original differential equation can be written as

$$v\frac{dv}{dy} = f(y, v).$$

Provided that this first order equation can be solved, we obtain v as a function of y. A relation between y and x results from solving $dy/dx = v(y)$. Again, there will be two arbitrary constants in the final result. Solve the following differential equations.

(a) $yy'' + (y')^2 = 0$ $\qquad\qquad\qquad$ (b) $y'' + y = 0$

(c) $y'' + y(y')^3 = 0$ $\qquad\qquad\qquad$ (d) $2y^2 y'' + 2y(y')^2 = 1$

3. Using the techniques of Problems 1 and 2, solve the following differential equations. If initial conditions are prescribed, find the solution satisfying the stated conditions.

(a) $y'y'' = 2$, $\qquad y(0) = 1$, $\qquad y'(0) = 2$

(b) $y'' - 3y^2 = 0$, $\qquad y(0) = 2$, $\qquad y'(0) = 4$

(c) $(1 + x^2)y'' + 2xy' + 3x^{-2} = 0$, $\qquad x > 0$

(d) $y'y'' - x = 0$, $\qquad y(1) = 2$, $\qquad y'(1) = 1$

4. Determine the intervals for which a unique solution of each of the following linear differential equations satisfying the initial conditions $y(x_0) = y_0$, $y'(x_0) = y_0'$, where x_0 is any point in the interval, is certain to exist.

(a) $xy'' + 3y = x$ $\qquad\qquad\qquad$ (b) $y'' + 6y' + 7y = 2\sin x$

(c) $x(x - 1)y'' + 3xy' + 4y = 2$ (d) $y'' + (\cos x)y' + 3(\ln |x|)y = 0$

(e) $(1 + x^2)y'' + 4y' = e^x$ (f) $e^x y'' + x^2 y' + y = \tan x$

5. Assuming that p and q are continuous on an open interval including the origin and that $y = \phi(x)$ is a solution of

$$y'' + p(x)y' + q(x)y = 0, \qquad y(0) = a_0, \qquad y'(0) = a_1,$$

determine $\phi''(0)$. If p and q are polynomials it can be shown that the solution $y = \phi(x)$ of the above differential equation can be differentiated infinitely many times. Assuming that p and q are polynomials, determine $\phi'''(0)$ in terms of a_0, a_1, $p(0)$, $q(0)$, $p'(0)$, and $q'(0)$. Can this process be continued indefinitely?

6. The solution of a second order equation of the form $y'' = f(x, y, y')$ will generally involve two arbitrary constants. Conversely, a given family of functions involving two arbitrary constants can be shown to be the solution of some second order differential equation. By eliminating the constants c_1 and c_2 among y, y', and y'', find the differential equation satisfied by each of the following families of functions.

(a) $y = c_1 e^x + c_2 e^{-x}$ (b) $y = c_1 \cos x + c_2 \sin x$

(c) $y = c_1 x + c_2 \sin x$ (d) $y = (c_1 + c_2 x)e^x$

(e) $y = c_1 x + c_2 x^2$ (f) $y = c_1 \cosh x + c_2 \sinh x$

3.2 FUNDAMENTAL SOLUTIONS OF THE HOMOGENEOUS EQUATION

In developing the theory of linear differential equations, it is helpful to introduce a differential operator notation. Let p and q be continuous functions on an open interval $\alpha < x < \beta$. Then for any twice differentiable function ϕ on $\alpha < x < \beta$, we define the *differential operator* L by the equation

$$L[\phi] = \phi'' + p\phi' + q\phi. \tag{1}$$

Note that $L[\phi]$ is a function on $\alpha < x < \beta$; in fact L itself can be considered as a function whose domain and range are themselves functions of a single real variable. The operator L is often written as $L = D^2 + pD + q$, where D is the derivative operator. The value of the function $L[\phi]$ at the point x is

$$L[\phi](x) = \phi''(x) + p(x)\phi'(x) + q(x)\phi(x).$$

For example, if $p(x) = x^2$, $q(x) = 1 + x$, and $\phi(x) = \sin 3x$ then

$$L[\phi](x) = (\sin 3x)'' + x^2(\sin 3x)' + (1 + x) \sin 3x$$
$$= -9 \sin 3x + 3x^2 \cos 3x + (1 + x) \sin 3x.$$

In this section we will study the second order linear homogeneous equation

$$L[\phi](x) = \phi''(x) + p(x)\phi'(x) + q(x)\phi(x) = 0, \tag{2}$$

where it is understood that the functions p and q are continuous on an open interval $\alpha < x < \beta$. Recall that it is customary to use the symbol y to denote $\phi(x)$. Thus we will often write in place of Eq. (2)

$$L[y] = y'' + p(x)y' + q(x)y = 0. \tag{3}$$

We emphasize that $L[\phi]$ as given in Eq. (1) is a function, while $L[y]$ as given in Eq. (3) is the value of the function $L[\phi]$ at the point x. Thus the symbol y is treated in a special way.

Using the fact that if u_1 and u_2 are differentiable functions, then

$$[c_1 u_1(x) + c_2 u_2(x)]' = c_1 u_1'(x) + c_2 u_2'(x),$$

where c_1 and c_2 are arbitrary constants, we can derive the following important theorem.

Theorem 3.3. *If $y = y_1(x)$ and $y = y_2(x)$ are solutions of the differential equation* (3),

$$L[y] = y'' + p(x)y' + q(x)y = 0,$$

then the linear combination $y = c_1 y_1(x) + c_2 y_2(x)$, where c_1 and c_2 are arbitrary constants, is also a solution of Eq. (3).

We must show that if $L[y_1] = y_1'' + py_1' + qy_1 = 0,$* and $L[y_2] = y_2'' + py_2' + qy_2 = 0$, then $L[c_1 y_1 + c_2 y_2] = 0$. But

$$
\begin{aligned}
L[c_1 y_1 + c_2 y_2] &= (c_1 y_1 + c_2 y_2)'' + p(c_1 y_1 + c_2 y_2)' + q(c_1 y_1 + c_2 y_2) \\
&= c_1(y_1'' + py_1' + qy_1) + c_2(y_2'' + py_2' + qy_2) \\
&= c_1 L[y_1] + c_2 L[y_2] \\
&= 0,
\end{aligned}
$$

which proves the theorem. If we set $c_2 = 0$ in the above proof we obtain the result that if the function y_1 is a solution of Eq. (3), then any constant multiple of y_1 is also a solution of Eq. (3).

In the process of proving Theorem 3.3, we have shown that for any two functions u_1 and u_2 possessing continuous second derivatives, and for any two arbitrary constants c_1 and c_2,

$$L[c_1 u_1 + c_2 u_2] = c_1 L[u_1] + c_2 L[u_2].$$

An operator with this property is known as a *linear operator* and, in particular, the differential operator L is a second order linear differential operator. Another example of a linear operator is given in Problem 10.

The fact that a linear combination of solutions of a *linear, homogeneous* equation is also a solution of the equation is of fundamental importance, and is often referred to as the *superposition principle*. The theory of linear homogeneous equations, including higher order ordinary differential equations and partial

---

* Notice that since $L[y_1]$ is a function, the 0 on the right-hand side of the statement $L[y_1] = 0$ actually stands for the function that is identically zero on $\alpha < x < \beta$.

differential equations, depends strongly on the superposition principle. This principle is illustrated by the following simple examples.

Example 1. Verify by direct calculation that $\phi(x) = c_1 \cos x + c_2 \sin x$ is a solution of the differential equation $y'' + y = 0$.

Substituting $\phi(x)$ for y, we have

$$\phi''(x) + \phi(x) = (c_1 \cos x + c_2 \sin x)'' + (c_1 \cos x + c_2 \sin x)$$
$$= c_1[(\cos x)'' + \cos x] + c_2[(\sin x)'' + \sin x]$$
$$= c_1[-\cos x + \cos x] + c_2[-\sin x + \sin x] = 0.$$

Example 2. Verify that $\phi(x) = x + 1$ is a solution of the differential equation $y'' + 3y' + y = x + 4$, but that $\psi(x) = 2\phi(x)$ is not a solution.

Since $\phi'(x) = 1$, $\phi''(x) = 0$ we have

$$\phi''(x) + 3\phi'(x) + \phi(x) = 0 + 3(1) + (x + 1) = x + 4.$$

However,

$$\psi''(x) + 3\psi'(x) + \psi(x) = 0 + 3(2) + 2(x + 1) \neq x + 4.$$

This is not a contradiction of Theorem 3.3, since the differential equation is not homogeneous.

Example 3. Show that if the functions y_1 and y_2 are solutions of the equation $L[y] = y'' + y^2 = 0$, it does *not* necessarily follow that the linear combination $c_1 y_1 + c_2 y_2$ is a solution.

Substituting, we have

$$L[c_1 y_1 + c_2 y_2] = (c_1 y_1 + c_2 y_2)'' + (c_1 y_1 + c_2 y_2)^2$$
$$= c_1 y_1'' + c_2 y_2'' + c_1^2 y_1^2 + 2c_1 c_2 y_1 y_2 + c_2^2 y_2^2$$
$$\neq c_1 L[y_1] + c_2 L[y_2].$$

The superposition principle fails because the differential equation is nonlinear.

We have seen that if the functions y_1 and y_2 are solutions of Eq. (3), then the linear combination $c_1 y_1 + c_2 y_2$ is a solution of Eq. (3) and, corresponding to the infinity of values we can assign to c_1 and c_2, we can construct an infinity of solutions of Eq. (3). But does this infinity of solutions include all possible solutions of Eq. (3)?

Two solutions y_1 and y_2 of Eq. (3) are said to form a *fundamental set* of solutions if every solution of Eq. (3) can be expressed as a linear combination of y_1 and y_2. To prove that two solutions, y_1 and y_2, form a fundamental set of solutions, we must show that for every solution $y = \phi(x)$ of Eq. (3) we can find constants c_1 and c_2 such that $\phi(x) = c_1 y_1(x) + c_2 y_2(x)$. We proceed as follows.

Let $y = \phi(x)$ be any solution of Eq. (3). Choose a point x_0 in the interval $\alpha < x < \beta$; then $\phi(x_0)$ and $\phi'(x_0)$, respectively, are the values of the chosen solution and its derivative at this point. We now seek constants c_1 and c_2 so that the function $c_1 y_1 + c_2 y_2$ and its derivative take on the same values $\phi(x_0)$ and $\phi'(x_0)$ at $x = x_0$. Thus c_1 and c_2 must satisfy the following system of linear algebraic equations:

$$\begin{aligned} c_1 y_1(x_0) + c_2 y_2(x_0) &= \phi(x_0), \\ c_1 y_1'(x_0) + c_2 y_2'(x_0) &= \phi'(x_0). \end{aligned} \tag{4}$$

The system of equations (4) has a unique solution for c_1 and c_2 provided that the determinant of coefficients is not zero; that is,

$$\begin{vmatrix} y_1(x_0) & y_2(x_0) \\ y_1'(x_0) & y_2'(x_0) \end{vmatrix} = y_1(x_0)y_2'(x_0) - y_1'(x_0)y_2(x_0) \neq 0. \tag{5}$$

Then with c_1 and c_2 determined by Eqs. (4), the functions $c_1 y_1 + c_2 y_2$ and ϕ satisfy the same initial conditions at $x = x_0$ as well as the same differential equation (3). Since Theorem 3.2 asserts that there is only one solution of this initial value problem, we conclude that $\phi(x) = c_1 y_1(x) + c_2 y_2(x)$ for each x in $\alpha < x < \beta$.

We will always be able to choose an x_0 such that Eq. (5) holds if we know that it holds for all x_0 in $\alpha < x < \beta$. In this case Eqs. (4) can always be solved regardless of the values x_0, $\phi(x_0)$, and $\phi'(x_0)$. Thus we have proved the following theorem.

Theorem 3.4. *If the functions p and q are continuous on the open interval $\alpha < x < \beta$ and if y_1 and y_2 are solutions of the differential equation (3),*

$$L[y] = y'' + p(x)y' + q(x)y = 0,$$

satisfying the condition

$$y_1(x)y_2'(x) - y_1'(x)y_2(x) \neq 0 \tag{6}$$

at every point in $\alpha < x < \beta$, then any solution of Eq. (3) on the interval $\alpha < x < \beta$ can be expressed as a linear combination of y_1 and y_2.

It is customary to call the linear combination $c_1 y_1 + c_2 y_2$, with c_1 and c_2 arbitrary, the *general solution* of Eq. (3).

Example 4. Find a fundamental set of solutions of the equation

$$y'' + y = 0, \qquad -\infty < x < \infty. \tag{7}$$

By inspection we observe that $y_1(x) = \cos x$ and $y_2(x) = \sin x$ are solutions of Eq. (7). Further,

$$y_1(x)y_2'(x) - y_1'(x)y_2(x) = \cos x \cos x - (-\sin x)\sin x = 1.$$

Hence if $y = \phi(x)$ is a solution of Eq. (7), then there exist constants c_1 and c_2 such that $\phi(x) = c_1 \cos x + c_2 \sin x$.

Given two differentiable functions y_1 and y_2 on some open interval, the function $y_1 y_2' - y_1' y_2$ is referred to as the Wronskian* of y_1 and y_2, and is usually written as

$$W(y_1, y_2) = \begin{vmatrix} y_1 & y_2 \\ y_1' & y_2' \end{vmatrix} = y_1 y_2' - y_1' y_2. \tag{8}$$

The value of the Wronskian of y_1 and y_2 at the point x will be denoted by $W(y_1, y_2)(x)$ or simply $W(x)$ if it is clear what functions are being considered.

Now suppose we have found two solutions, y_1 and y_2, of Eq. (3). How do we decide whether the functions y_1 and y_2 form a fundamental set; that is, how do we decide whether $W(y_1, y_2)$ is ever zero in the interval $\alpha < x < \beta$? In Example 4, $W(\cos x, \sin x) = 1$, and hence it is clear that $\cos x$ and $\sin x$ form a fundamental set for the differential equation $y'' + y = 0$ on any interval. However, in general, $W(y_1, y_2)$ will not be constant, and it may be difficult to determine whether this function vanishes at some point in an interval $\alpha < x < \beta$. Fortunately this question is answered by the following remarkable theorem.

Theorem 3.5. *If the functions p and q are continuous on the open interval $\alpha < x < \beta$, and if the functions y_1 and y_2 are solutions of the differential equation (3),*

$$L[y] = y'' + p(x)y' + q(x)y = 0,$$

on $\alpha < x < \beta$, then $W(y_1, y_2)$ either is identically zero, or else is never zero in $\alpha < x < \beta$.

The proof of this theorem can be based on an argument similar to that used in proving Theorem 3.4; however, we will present an alternate proof in which we also will derive a formula for $W(y_1, y_2)$. We make use of the fact that the functions y_1 and y_2 satisfy

$$\begin{aligned} y_1'' + py_1' + qy_1 &= 0, \\ y_2'' + py_2' + qy_2 &= 0. \end{aligned} \tag{9}$$

Multiplying the first equation by $-y_2$, the second by y_1, and adding gives

$$(y_1 y_2'' - y_2 y_1'') + p(y_1 y_2' - y_1' y_2) = 0. \tag{10}$$

Letting $W_{12}(x) = W(y_1, y_2)(x)$, and noting that

$$W_{12}' = y_1 y_2'' - y_1'' y_2, \tag{11}$$

allows us to write Eq. (10) in the form

$$W_{12}' + pW_{12} = 0. \tag{12}$$

This is a separable equation (Section 2.4) and also a first order linear equation (Section 2.1), and it can be integrated immediately to give

$$W_{12}(x) = c \exp\left[-\int^x p(t)\, dt\right], \tag{13}$$

* In honor of the Polish mathematician Wronski (1778–1853).

where c is a constant.* Since the exponential function never vanishes, $W_{12}(x)$ will vanish only if $c = 0$, and if $c = 0$ then $W_{12}(x)$ is identically zero, thus proving the theorem. In addition, Eq. (13) gives a formula for determining the Wronskian of a fundamental set of solutions of Eq. (3) up to a multiplicative constant without solving the equation. Also note that the Wronskians of any two fundamental sets of solutions can differ only by a multiplicative constant.

Combining the results of Theorems 3.4 and 3.5 allows us to state the following theorem.

Theorem 3.6. *If the functions p and q are continuous on the open interval* $\alpha < x < \beta$ *and if the functions* y_1 *and* y_2 *are solutions of the differential equation* (3),

$$L[y] = y'' + p(x)y' + q(x)y = 0,$$

on the interval $\alpha < x < \beta$, *and if there is at least one point in* $\alpha < x < \beta$ *where* $W(y_1, y_2)$ *is not zero, then every solution* $y = \phi(x)$ *of Eq.* (3) *can be expressed in the form*

$$\phi(x) = c_1 y_1(x) + c_2 y_2(x).$$

Finally we must show that there actually does exist a fundamental set of solutions of Eq. (3); that is, we must prove the following theorem.

Theorem 3.7. *If the functions p and q are continuous on the open interval* $\alpha < x < \beta$, *then there exists a fundamental set of solutions of the differential equation* (3),

$$L[y] = y'' + p(x)y' + q(x)y = 0,$$

on the interval $\alpha < x < \beta$.

Let c be a point in $\alpha < x < \beta$. It follows from Theorem 3.2 that there exist unique solutions y_1 and y_2 of the initial value problems

$$y'' + p(x)y' + q(x)y = 0; \qquad y(c) = 1, \qquad y'(c) = 0$$
$$y'' + p(x)y' + q(x)y = 0; \qquad y(c) = 0, \qquad y'(c) = 1$$

on the interval $\alpha < x < \beta$. It is readily seen that $W(y_1, y_2)(c) = 1 \neq 0$. Hence it follows from Theorem 3.6 that y_1 and y_2 form a fundamental set of solutions of Eq. (3).

We will conclude this section by giving an interesting alternative proof of Theorem 3.6, which does not require the use of the existence and uniqueness theorem (note the proof of Theorem 3.4). Suppose that the functions y_1, y_2, and

---

* The result given in Eq. (13) was derived by the Norwegian mathematician N. H. Abel (1802–1829) in 1827, and is known as Abel's identity. He also showed that there is no general formula for solving a quintic equation in terms of explicit algebraic operations on the coefficients in the polynomial equation. His greatest contribution was in analysis in the area known as elliptic functions, although his memoir was not noticed until after his death. The distinguished French mathematician Legendre described it as "a monument more lasting than bronze."

y_3 are solutions of Eq. (3), and that $W(y_1, y_2)(x)$ is never zero in $\alpha < x < \beta$. To show that $y_3 = c_1 y_1 + c_2 y_2$ consider

$$y_1'' + py_1' + qy_1 = 0,$$
$$y_2'' + py_2' + qy_2 = 0,$$
$$y_3'' + py_3' + qy_3 = 0.$$

Multiplying the first equation by $-y_2$, the second by y_1, adding, and using Eq. (11) gives

$$W_{12}' + pW_{12} = 0, \tag{14}$$

where $W_{12}(x) = W(y_1, y_2)(x)$. Similar computations using the second and third equations, and the first and third equations, give

$$W_{23}' + pW_{23} = 0, \tag{15}$$

$$W_{13}' + pW_{13} = 0, \tag{16}$$

where the notation is clear. Equations (14), (15), and (16) are first order linear equations. Their solutions are

$$W_{12}(x) = y_1(x)y_2'(x) - y_1'(x)y_2(x) = k_{12} \exp\left[-\int^x p(t)\,dt\right], \tag{17}$$

$$W_{23}(x) = y_2(x)y_3'(x) - y_2'(x)y_3(x) = k_{23} \exp\left[-\int^x p(t)\,dt\right], \tag{18}$$

$$W_{13}(x) = y_1(x)y_3'(x) - y_1'(x)y_3(x) = k_{13} \exp\left[-\int^x p(t)\,dt\right], \tag{19}$$

where k_{12}, k_{23}, and k_{13} are constants; and in particular $k_{12} \neq 0$ since W_{12} is never zero by hypothesis. Multiplying Eq. (18) by $-y_1(x)$, Eq. (19) by $y_2(x)$, and adding gives

$$[y_1(x)y_2'(x) - y_1'(x)y_2(x)]y_3(x) = [-k_{23}y_1(x) + k_{13}y_2(x)] \exp\left[-\int^x p(t)\,dt\right]. \tag{20}$$

Finally, substituting for $y_1(x)y_2'(x) - y_1'(x)y_2(x)$ from Eq. (17) and canceling the exponentials gives

$$y_3(x) = -\frac{k_{23}}{k_{12}}y_1(x) + \frac{k_{13}}{k_{12}}y_2(x).$$

Hence y_3 can be expressed as a linear combination of y_1 and y_2, as was to be proved.

PROBLEMS

1. Verify that e^x and e^{-2x} and the linear combination $c_1 e^x + c_2 e^{-2x}$, where c_1 and c_2 are arbitrary constants, are solutions of the differential equation

$$y'' + y' - 2y = 0.$$

2. In Problem 1 find the unique solution of the differential equation that satisfies the initial conditions $y(0) = 1$, $y'(0) = 0$. What is the unique solution of this problem if the initial conditions are $y(1) = 0$, $y'(1) = 0$?

3. Verify that e^x and e^{-x} are solutions of $y'' - y = 0$. Hence show that $\sinh x = (e^x - e^{-x})/2$ and $\cosh x = (e^x + e^{-x})/2$ are also solutions of this differential equation.

4. Verify that $\sin x$ and $\cos x$ are solutions of $y'' + y = 0$. Accepting for the moment that $(cf)' = cf'$ where f is a real-valued function and c is a complex number, show that the linear combination $(1 + i) \sin x + (2 - i) \cos x$ is also a solution of $y'' + y = 0$. Here i is the imaginary unit, $i^2 = -1$.

5. Verify that x^2 and x^{-1} and the linear combination $c_1 x^2 + c_2 x^{-1}$, where c_1 and c_2 are arbitrary constants, are solutions of the differential equation $x^2 y'' - 2y = 0$, $x > 0$.

6. Verify that 1 and $x^{\frac{1}{2}}$ are solutions of the differential equation $yy'' + (y')^2 = 0$, $x > 0$; but that the linear combination $c_1 + c_2 x^{\frac{1}{2}}$ is not, in general, a solution. Why?

7. Show that if $y = \phi(x)$ is a solution of the differential equation $y'' + p(x)y' + q(x)y = g(x)$, $g(x) \neq 0$, then $y = c\phi(x)$, where c is any constant other than one, is not a solution. Why?

8. If $L[y] = ay'' + by' + cy$, where a, b, and c are constants, compute

(a) $L[x]$ (b) $L[\sin x]$

(c) $L[e^{rx}]$, r a constant   (d) $L[x^r]$, r a constant

9. If $L[y] = ax^2 y'' + bxy' + cy$, $x > 0$, where a, b, and c are constants, compute

(a) $L[x^2]$ (b) $L[e^{rx}]$, r a constant

(c) $L[x^r]$, r a constant   (d) $L[\ln x]$

10. Show that the operator M defined by

$$M[u](x) = \int_\alpha^\beta K(x - t)u(t)\, dt, \qquad \alpha < x < \beta,$$

where the function K is continuous on the interval $\alpha - \beta \leq s \leq \beta - \alpha$, is a linear operator; that is, show that

$$M[c_1 u_1 + c_2 u_2] = c_1 M[u_1] + c_2 M[u_2]$$

where c_1 and c_2 are constants.

11. Compute the Wronskians of the following pairs of functions:

(a) e^{mx}, e^{nx}, where m and n are integers, and $m \neq n$

(b) $\sinh x$, $\cosh x$ (c) x, xe^x

(d) $e^x \sin x$, $e^x \cos x$ (e) $\cos^2 x$, $1 + \cos 2x$

12. In the following problems verify that the functions y_1 and y_2 are solutions of the given differential equation and determine in what intervals they form a fundamental set of solutions by computing $W(y_1, y_2)$.

(a) $y'' + \lambda^2 y = 0$; $y_1(x) = \sin \lambda x$, $y_2(x) = \cos \lambda x$, where λ is a real number

(b) $y'' - y' - 2y = 0$; $y_1(x) = e^{-x}$, $y_2(x) = e^{2x}$

(c) $y'' - 2y' + y = 0$; $y_1(x) = e^x$, $y_2(x) = xe^x$

(d) $x^2 y'' - x(x + 2)y' + (x + 2)y = 0$; $y_1(x) = x$, $y_2(x) = xe^x$

Notice that if the equation of part (d) is put in the standard form $y'' + p(x)y' + q(x)y = 0$, the coefficients $p(x) = -(x + 2)/x$ and $q(x) = (x + 2)/x^2$ become unbounded as $x \to 0$, but the solutions x and xe^x are perfectly well behaved as $x \to 0$. Thus it does not necessarily follow that at a point where the coefficients are discontinuous the solution will be discontinuous—but this is often the case!

13. Verify that if ϕ, ϕ_1, and ϕ_2 are differentiable functions, then $W(\phi\phi_1, \phi\phi_2) = \phi^2 W(\phi_1, \phi_2)$.

14. In the following problems verify that the functions y_1 and y_2 form a fundamental set of solutions of the given differential equations and determine the solution satisfying the prescribed initial conditions.

(a) $y'' - y = 0$, $y(0) = 0$, $y'(0) = 1$; $y_1(x) = e^x$, $y_2(x) = e^{-x}$

(b) $y'' - y = 0$, $y(0) = 0$, $y'(0) = 1$; $y_1(x) = \sinh x$, $y_2(x) = \cosh x$
Compare with the result for part (a).

(c) $y'' + 5y' + 6y = 0$, $y(0) = 1$, $y'(0) = 1$; $y_1(x) = e^{-2x}$, $y_2(x) = e^{-3x}$

(d) $y'' + y' = 0$, $y(1) = 0$, $y'(1) = 1$; $y_1(x) = 2$, $y_2(x) = e^{-x}$

In Problems 15, 16, and 17 assume that p and q are continuous, and that the functions y_1 and y_2 are solutions of the differential equation $y'' + p(x)y' + q(x)y = 0$ on the interval $\alpha < x < \beta$.

15. Prove that if y_1 and y_2 vanish at the same point in $\alpha < x < \beta$, then they cannot form a fundamental set of solutions on that interval.

16. Prove that if y_1 and y_2 have maxima or minima at the same point in $\alpha < x < \beta$, then they cannot form a fundamental set of solutions on that interval.

*17. Prove that if y_1 and y_2 are a fundamental set of solutions, then they cannot have a common point of inflection in $\alpha < x < \beta$ unless p and q vanish simultaneously there.

*18. The concept of exactness which was discussed for first order differential equations can be extended to second order linear equations. The equation $P(x)y'' + Q(x)y' + R(x)y = 0$ is said to be *exact* if it can be written in the form $[P(x)y']' + [f(x)y]' = 0$, where $f(x)$ is to be determined in terms of $P(x)$, $Q(x)$, and $R(x)$. The latter equation can be integrated once immediately to give a first order linear equation for y which can be solved by the method of Section 2.1. Show by equating the coefficients of the above equations, and then eliminating $f(x)$, that a necessary condition for exactness is $P''(x) - Q'(x) + R(x) = 0$. It can also be shown that this is a sufficient condition for exactness. Determine whether each of the following equations is exact; if so, find its solution.

(a) $y'' + xy' + y = 0$ (b) $y'' + 3x^2y' + xy = 0$

(c) $xy'' - (\cos x)y' + (\sin x)y = 0$, $x > 0$

(d) $x^2y'' + xy' - y = 0$, $x > 0$

*19. If a second order linear homogeneous equation is not exact, it can be made exact by multiplying by an appropriate integrating factor $\mu(x)$. Thus we require that $\mu(x)$ be such that $\mu(x)P(x)y'' + \mu(x)Q(x)y' + \mu(x)R(x)y = 0$ can be written in the form $[\mu(x)P(x)y']' + [f(x)y]' = 0$. Show by equating coefficients in these two equations and then eliminating $f(x)$ that the function μ must satisfy

$$P\mu'' + (2P' - Q)\mu' + (P'' - Q' + R)\mu = 0.$$

This equation is known as the *adjoint* equation. It plays a very important role in the advanced theory of differential equations. In general the problem of solving the adjoint differential equation is as difficult as that of solving the original equation. Determine the adjoint equation for each of the following differential equations.

(a) The Bessel equation of order v:

$$x^2y'' + xy' + (x^2 - v^2)y = 0$$

(b) The Legendre equation of order α:

$$(1 - x^2)y'' - 2xy' + \alpha(\alpha + 1)y = 0$$

(c) The Airy equation:
$$y'' - xy = 0$$

(d) The Whittaker equation:

$$x^2y'' + \left(-\frac{x^2}{4} + kx + \frac{1}{4} - m^2\right)y = 0,$$

where k and m are integers.

*20. Show for the second order linear equation $P(x)y'' + Q(x)y' + R(x)y = 0$ that the adjoint of the adjoint is the original equation.

*21. A second order linear equation $P(x)y'' + Q(x)y' + R(x)y = 0$ is said to be *self-adjoint* if its adjoint is identical with the original equation. Show that a necessary condition for this differential equation to be self-adjoint is $P'(x) = Q(x)$. Determine whether the equations of Problem 19 are self-adjoint.

*22. In developing certain aspects of the theory of the equation $y'' + p(x)y' + q(x)y = 0$ it is often convenient to transform the dependent variable so that the equation takes the form $u'' + f(x)u = 0$. This is known as the *normal form* of a second order linear homogeneous equation. In order to transform the original equation into normal form set $y = u(x)v(x)$ and choose $v(x)$ so that the coefficient of u' is zero. Determine $v(x)$ and $f(x)$.

3.3 LINEAR INDEPENDENCE

The concept of the general solution of a second order linear differential equation as a linear combination of two solutions whose Wronskian does not vanish is intimately related to the concept of linear independence of two functions. This is a very important concept and has significance far beyond the present problem; we will briefly discuss it in this section.

Two functions f and g are said to be *linearly dependent* on an interval $\alpha < x < \beta$ if there exist two constants c_1 and c_2, not both zero, such that

$$c_1f(x) + c_2g(x) = 0 \tag{1}$$

for all x in $\alpha < x < \beta$. Two functions f and g are said to be *linearly independent* on an interval $\alpha < x < \beta$ if they are not linearly dependent; that is, if Eq. (1) holds for all x only if $c_1 = c_2 = 0$. The same definitions apply on any interval, open or not.

Example 1.   The functions $\sin x$ and $\cos (x + \pi/2)$ are linearly dependent on any interval since

$$c_1 \sin x + c_2 \cos \left(x + \frac{\pi}{2} \right) = 0$$

for all x if we choose $c_1 = 1$, $c_2 = 1$.

Example 2.   Show that the functions e^x and e^{2x} are linearly independent on any interval.

We will show that the functions are linearly independent by showing that it is impossible to find constants c_1 and c_2, not both zero, such that

$$c_1 e^x + c_2 e^{2x} = 0 \tag{2}$$

for all x in the given interval. To establish this result, we assume the contrary: that Eq. (2) does hold at all points in $\alpha < x < \beta$. Then choose two points x_0 and $x_1 \neq x_0$ in the interval. We have

$$c_1 e^{x_0} + c_2 e^{2x_0} = 0,$$
$$c_1 e^{x_1} + c_2 e^{2x_1} = 0.$$

Since the determinant of the coefficients is $e^{2x_1 + x_0} - e^{2x_0 + x_1} \neq 0$, it follows that the only solution of these equations is $c_1 = c_2 = 0$. Hence e^x and e^{2x} are linearly independent.

We can now restate Theorem 3.6 using the concept of linear independence.

Theorem 3.8.   *If the functions p and q are continuous on the open interval $\alpha < x < \beta$, and if the functions y_1 and y_2 are linearly independent solutions of the differential equation*

$$L[y] = y'' + p(x)y' + q(x)y = 0, \tag{3}$$

then $W(y_1, y_2)$ is nonzero on $\alpha < x < \beta$, and hence any solution of Eq. (3) can be expressed as a linear combination of the solutions y_1 and y_2.

To prove this theorem we must show that if y_1 and y_2 are solutions of Eq. (3) and are linearly independent on $\alpha < x < \beta$, then $W(y_1, y_2)$ is nonzero on $\alpha < x < \beta$. Let us assume that there exists a point x_0, $\alpha < x_0 < \beta$, such that $W(y_1, y_2)(x_0) = 0$; we will now show that this leads to a contradiction. If $W(y_1, y_2)(x_0) = 0$, then the system of equations

$$c_1 y_1(x_0) + c_2 y_2(x_0) = 0$$
$$c_1 y_1'(x_0) + c_2 y_2'(x_0) = 0 \tag{4}$$

for c_1 and c_2 has a nontrivial solution. Using these values of c_1 and c_2, let $\phi(x) = c_1 y_1(x) + c_2 y_2(x)$. Then ϕ is a solution of Eq. (3) and, further, by Eqs. (4) satisfies the initial conditions $\phi(x_0) = \phi'(x_0) = 0$. Therefore by the existence and uniqueness theorem, $\phi(x) = 0$ for all x in $\alpha < x < \beta$ (see Example 2 of Section 3.1).

Then $c_1y_1(x) + c_2y_2(x) = 0$ for all x in $\alpha < x < \beta$, which implies that y_1 and y_2 are linearly dependent, which is a contradiction.

The converse of Theorem 3.8 is also true; namely, if $L[y_1] = 0$, $L[y_2] = 0$, and $W(y_1, y_2) \neq 0$ on $\alpha < x < \beta$, then the functions y_1 and y_2 are linearly independent on $\alpha < x < \beta$. To establish this fact, again assume the contrary: that y_1 and y_2 are linearly dependent on $\alpha < x < \beta$. Then there exist c_1 and c_2, not both zero, such that

$$c_1y_1(x) + c_2y_2(x) = 0 \tag{5}$$

for each x in $\alpha < x < \beta$. It follows that

$$c_1y_1'(x) + c_2y_2'(x) = 0 \tag{6}$$

on $\alpha < x < \beta$ as well. In order for nonzero values of c_1 and/or c_2 to satisfy Eqs. (5) and (6), it is necessary (and sufficient) for $W(y_1, y_2)$ to vanish for each x, which is a contradiction. Hence two solutions of Eq (3) are linearly independent if and only if their Wronskian is nonzero at each point in the interval.

We emphasize that this relationship between the Wronskian and linear independence no longer holds if the functions are not solutions of Eq. (3). An example of two linearly independent functions whose Wronskian vanishes identically is given in Problem 10.

It is interesting to note the similarity of this result to several results in two dimensional vector algebra. Two vectors **a** and **b** are said to be linearly dependent if there are two scalars c_1 and c_2, not both zero, such that $c_1\mathbf{a} + c_2\mathbf{b} = \mathbf{0}$; otherwise they are said to be linearly independent. Let **i** and **j** be unit vectors directed along the positive x and y axes, respectively. Since there are no scalars c_1 and c_2, not both zero, such that $c_1\mathbf{i} + c_2\mathbf{j} = \mathbf{0}$, the vectors **i** and **j** are linearly independent. Further we know that any vector with components a_1 and a_2 can be written as $a_1\mathbf{i} + a_2\mathbf{j}$; that is, as a linear combination of the two linearly independent vectors **i** and **j**. It is not difficult to show that any vector in two dimensions can be represented as a linear combination of any two linearly independent two-dimensional vectors (see Problem 7). Such a pair of linearly independent vectors is said to form a basis for, or to span, the vector space of two-dimensional vectors.

The term vector space is also applied to other collections of mathematical objects that satisfy the same laws of addition and multiplication by scalars that geometric vectors do. For example, it can be shown that the set of functions that are twice differentiable on $\alpha < x < \beta$ forms a vector space. Similarly the set V of functions satisfying Eq. (3) also forms a vector space.

Since every member of V can be expressed as a linear combination of two linearly independent members y_1 and y_2, we say that such a pair spans V or forms a basis for V. This leads to the conclusion that V is two-dimensional—Eq. (3) is second order—and analogous in many respects to the space of geometric vectors in a plane. Later we will find that the set of solutions of an nth order linear homogeneous differential equation forms a vector space of dimension n, which is spanned by any set of n linearly independent solutions of the differential equation.

PROBLEMS

In Problems 1 through 4 prove that the functions y_1 and y_2 are linearly independent solutions of the given differential equation.

1. $y'' - y = 0;$ $y_1(x) = e^x,$ $y_2(x) = e^{-x}$

2. $y'' - y = 0;$ $y_1(x) = \cosh x,$ $y_2(x) = \sinh x$

3. $y'' - y' - 6y = 0;$ $y_1(x) = e^{-2x},$ $y_2(x) = e^{3x}$

4. $x^2 y'' + xy' - 4y = 0,$ $x > 0;$ $y_1(x) = x^2,$ $y_2(x) = x^{-2}$

5. Prove that if the functions y_1 and y_2 are linearly independent solutions of $y'' + p(x)y' + q(x)y = 0$, then $c_1 y_1$ and $c_2 y_2$ are linearly independent solutions provided neither c_1 nor c_2 equals 0.

6. Prove that if the functions y_1 and y_2 are linearly independent solutions of the differential equation $y'' + p(x)y' + q(x)y = 0$, then $y_3 = y_1 + y_2$, and $y_4 = y_1 - y_2$ also form a fundamental set of solutions.

7. (a) Prove that any two-dimensional vector can be written as a linear combination of the vectors $\mathbf{i} + \mathbf{j}$ and $\mathbf{i} - \mathbf{j}$.
Hint: Any vector $\mathbf{a}$ can be written as $a_1 \mathbf{i} + a_2 \mathbf{j}$. Show that it is possible to determine c_1 and c_2 such that $\mathbf{a} = c_1(\mathbf{i} + \mathbf{j}) + c_2(\mathbf{i} - \mathbf{j})$ by equating the two expressions for $\mathbf{a}$ and solving for c_1 and c_2 in terms of a_1 and a_2.

(b) Prove that if the vectors $\mathbf{x} = x_1 \mathbf{i} + x_2 \mathbf{j}$ and $\mathbf{y} = y_1 \mathbf{i} + y_2 \mathbf{j}$ are linearly independent, then any vector $\mathbf{z} = z_1 \mathbf{i} + z_2 \mathbf{j}$ can be expressed as a linear combination of $\mathbf{x}$ and $\mathbf{y}$. Note that if $\mathbf{x}$ and $\mathbf{y}$ are linearly independent then $x_1 y_2 - y_1 x_2 \neq 0$. Why?

*8. Verify that x and x^2 are linearly independent on $-1 < x < 1$, but that $W(x, x^2)$ vanishes at $x = 0$. From this what can you conclude about the possibility of x and x^2 being solutions of Eq. (3) of the text? Show that x and x^2 are solutions of $x^2 y'' - 2xy' + 2y = 0$. Does this contradict your conclusion? Does it contradict Theorem 3.5 of Section 3.2?

*9. If the functions y_1 and y_2 are linearly independent solutions of the differential equation $y'' + p(x)y' + q(x)y = 0$, show that between consecutive zeros of y_1 there is one and only one zero of y_2.
Hint: Use Rolle's (1652–1719) theorem to prove the conclusion by contradiction. This result is often referred to as Sturm's (1803–1855) theorem. As a particular example, note that $\sin x$ and $\cos x$ are linearly independent solutions of $y'' + y = 0$, and that the zeros $(0, \pm\pi, \pm 2\pi, \cdots)$ of $\sin x$ and the zeros $(\pm\pi/2, \pm 3\pi/2, \cdots)$ of $\cos x$ are interlaced.

10. Show that the functions $y_1(x) = x|x|$ and $y_2(x) = x^2$ are linearly dependent on $0 < x < 1$ and on $-1 < x < 0$ but are linearly independent on $-1 < x < 1$. Note that while y_1 and y_2 are linearly independent, $W(y_1, y_2)$ is identically zero on $-1 < x < 1$; hence y_1 and y_2 cannot be solutions of Eq. (3) of the text.

3.4 REDUCTION OF ORDER

A very important and useful fact is the following: if one solution of a second order linear homogeneous differential equation is known, a second linearly independent solution (and hence a fundamental set of solutions) can be determined.

The procedure, which is due to D'Alembert,* is usually referred to as the method of reduction of order.

Suppose we know one solution y_1, not identically zero, of

$$y'' + p(x)y' + q(x)y = 0. \tag{1}$$

Then cy_1, where c is any constant, is also a solution of Eq. (1). This suggests the following question: can we determine a function v, not a constant, such that $y = v(x)y_1(x)$ is a solution of Eq. (1)? The answer is yes, and further, we shall see that v can be determined in a straightforward manner. If we set

$$y = v(x)y_1(x), \tag{2}$$

then

$$y' = v(x)y_1'(x) + v'(x)y_1(x),$$
$$y'' = v(x)y_1''(x) + 2v'(x)y_1'(x) + v''(x)y_1(x).$$

Substituting for y, y', and y'' in Eq. (1) and collecting terms gives

$$v(y_1'' + py_1' + qy_1) + v'(2y_1' + py_1) + v''y_1 = 0. \tag{3}$$

Since y_1 is a solution of Eq. (1) the quantity in the first parentheses is zero. In any interval in which y_1 does not vanish, we can divide by y_1 obtaining

$$v'' + \left(p + 2\frac{y_1'}{y_1}\right)v' = 0. \tag{4}$$

Equation (4) is a *first order linear equation* for v' and can be solved immediately (Section 2.1). The solution is

$$v'(x) = c \exp\left[-\int^x \left(p(s) + \frac{2y_1'(s)}{y_1(s)}\right) ds\right] = cu(x), \tag{5}$$

where c is an arbitrary constant, and

$$u(x) = \frac{1}{[y_1(x)]^2} \exp\left[-\int^x p(s)\, ds\right]. \tag{6}$$

Then

$$v(x) = c\int^x u(t)\, dt + k, \tag{7}$$

where k is also an arbitrary constant. However, we can omit the constant k since

$$y = y_1(x)v(x)$$
$$= cy_1(x)\int^x u(t)\, dt + ky_1(x), \tag{8}$$

* Jean D'Alembert (1717–1783), a French mathematician, is particularly famous for his work in mechanics and hydrodynamics.

and hence it only adds a multiple of $y_1(x)$ to the second solution. Thus two solutions of Eq. (1) are

$$y = y_1(x) \qquad \text{and} \qquad y = y_1(x) \int^x u(t)\, dt. \qquad (9)$$

Since the antiderivative of the function u cannot be a constant, the solutions are linearly independent. In using the method of reduction of order it is *not* important to try to memorize Eqs. (9) and (6); what is important to remember is that if one solution y_1 is known, a second solution of the form $y = v(x)y_1(x)$ can be found by the above procedure.

While the method of reduction of order does not tell us how to find the first solution of Eq. (1), it is encouraging to know that we have *reduced* the problem of solving Eq. (1) to that of finding just one solution.

It is also possible to derive the second linearly independent solution given in Eq. (9) by using Abel's formula for the Wronskian of two linearly independent solutions of Eq. (1). This is discussed in Problems 12 and 13.

The method of reduction of order is also useful for nonhomogeneous equations; this is discussed in Problem 15 of Section 3.6.2.

Example. Show that $y = x$ is a solution of the Legendre equation of order one

$$(1 - x^2)y'' - 2xy' + 2y = 0, \qquad -1 < x < 1, \qquad (10)$$

and find a second linearly independent solution.

First if $y = x$, then $y' = 1$ and $y'' = 0$. Substituting for y, y', and y'' in Eq. (10) gives

$$(1 - x^2) \cdot 0 - 2x + 2x = 0,$$

so indeed $y = x$ is a solution. To find a second solution let $y = xv(x)$; then

$$y' = xv' + v, \qquad y'' = xv'' + 2v'.$$

Substituting for y, y', and y'' in Eq. (10) gives

$$(1 - x^2)(xv'' + 2v') - 2x(xv' + v) + 2xv = 0.$$

Collecting terms and dividing by $x(1 - x^2)$ we obtain

$$v'' + \left(\frac{2}{x} - \frac{2x}{1 - x^2}\right)v' = 0. \qquad (11)$$

Note that the coefficient of v' is not defined at $x = 0$, the zero of $y_1(x)$. However, this will not cause any difficulty in the final solution. First we will obtain a solution of Eq. (11) in the intervals $-1 < x < 0$ and $0 < x < 1$.

Equation (11) is a first order linear equation for v', and the integrating factor $x^2(1 - x^2)$; hence

$$[x^2(1 - x^2)v']' = 0,$$
$$x^2(1 - x^2)v' = c,$$

so

$$v(x) = c \int^x \frac{dt}{t^2(1 - t^2)} = c \int^x \left(\frac{1}{t^2} + \frac{1}{1 - t^2} \right) dt$$

$$= c \left(-\frac{1}{x} + \frac{1}{2} \ln \frac{1 + x}{1 - x} \right).$$

Consequently, a second solution (suppressing a constant multiplier without loss of generality) of Eq. (10) is

$$y_2(x) = xv(x) = 1 - \frac{x}{2} \ln \frac{1 + x}{1 - x}. \tag{12}$$

Although the function v is undefined at $x = 0$, clearly the limit of $y_2(x) = xv(x)$ as $x \to 0$ exists. The function y_2 defined by Eq. (12) on $-1 < x < 1$ actually satisfies the differential equation (10) on $-1 < x < 1$, not just on $-1 < x < 0$ and $0 < x < 1$. On the other hand, $y_2(x)$ becomes unbounded as $x \to \pm 1$; this is closely related to the fact that in Eq. (10) the coefficient of y'' vanishes at $x = \pm 1$, while the coefficients of y' and y are nonzero. This will be discussed further in Chapter 4.

PROBLEMS

In Problems 1 through 7 find a second solution of the given differential equation by the method of reduction of order.

1. $y'' - 4y' - 12y = 0$, $y_1(x) = e^{6x}$
2. $y'' + 2y' + y = 0$, $y_1(x) = e^{-x}$
3. $x^2y'' + 2xy' = 0$, $y_1(x) = 1$. For what range of x would you expect the solution to be valid?
4. $x^2y'' + 2xy' - 2y = 0$, $y_1(x) = x$. For what range of x would you expect the solution to be valid?
5. $x^2y'' + 3xy' + y = 0$, $x > 0$; $y_1(x) = x^{-1}$
6. $x^2y'' - x(x + 2)y' + (x + 2)y = 0$, $x > 0$; $y_1(x) = x$
7. $(1 - x \cot x)y'' - xy' + y = 0$, $y_1(x) = x$. Consider the interval $0 < x < \pi$.

Hint: $$\int \frac{x \, dx}{1 - x \cot x} = \ln |x \cos x - \sin x|.$$

8. Verify that $y_1(x) = x^{-1/2} \sin x$ is one solution of Bessel's equation of order one-half,

$$x^2y'' + xy' + (x^2 - \tfrac{1}{4})y = 0,$$

and determine a second solution. Consider the interval $0 < x < \infty$.

9. The differential equation

$$xy'' - (x + N)y' + Ny = 0,$$

where N is a nonnegative integer has been discussed by several authors.* One reason it is of interest is because it has an exponential solution and a polynomial solution.

(a) Verify that one solution is $y_1(x) = e^x$.

(b) Show that a second solution has the form $y_2(x) = c e^x \int x^N e^{-x} \, dx$. Calculate $y_2(x)$ for $N = 1$ and $N = 2$ and convince yourself that with $c = -1/N!$

$$y_2(x) = 1 + \frac{x}{1!} + \frac{x^2}{2!} + \cdots + \frac{x^N}{N!}.$$

It is interesting to note that $y_2(x)$ is exactly the first $N + 1$ terms of the Taylor series for e^x.

*10. Verify that $y_1(x) = 3x^2 - 1$ is one solution of Legendre's equation of order two,

$$(1 - x^2)y'' - 2xy' + 6y = 0,$$

and determine a second solution. Consider the intervals $-1 < x < -1/\sqrt{3}$, $-1/\sqrt{3} < x < 1/\sqrt{3}$, and $1/\sqrt{3} < x < 1$ separately, but observe that the final result for y_2 is valid on $-1 < x < 1$.

Hint: Using the method of partial fractions it can be shown that

$$\frac{1}{(3x^2 - 1)^2(1 - x^2)} = A\left[\frac{1}{(\sqrt{3}x - 1)^2} + \frac{1}{(\sqrt{3}x + 1)^2}\right] + B\left[\frac{1}{1 - x} + \frac{1}{1 + x}\right].$$

Determine the constants A and B.

11. Consider the two-dimensional flow of a uniform stream around a fixed circular cylinder whose axis is perpendicular to the stream. In the analysis of turbulent flow in the wake of the cylinder, the differential equation

$$f'' + \delta(\xi f' + f) = 0$$

arises. Here ξ is a dimensionless space variable perpendicular both to the axis of the cylinder and to the stream, $f(\xi)$ measures the reduction of the stream velocity in the wake, a prime denotes differentiation with respect to ξ, and δ is a physical parameter that is inversely proportional to the turbulent diffusivity. Show that $f_1(\xi) = e^{-\delta\xi^2/2}$ is one solution and find an integral form for the general solution.

12. Suppose that y_1 is a nonvanishing solution of $y'' + p(x)y' + q(x)y = 0$ and that we wish to find a second linearly independent solution y_2. Show that $(y_2/y_1)' = W(y_1, y_2)/y_1^2$ and then use Abel's formula for $W(y_1, y_2)$ to obtain y_2.

13. Verify that $y_1(x) = x$ is a solution of $x^2 y'' + 2xy' - 2y = 0$, $x > 0$, and then, using the result of Problem 12, determine a second linearly independent solution.

3.5   HOMOGENEOUS EQUATIONS WITH CONSTANT COEFFICIENTS

We will now turn from the general theory of second order linear homogeneous

* T. A. Newton, "On Using a Differential Equation to Generate Polynomials", *American Mathematical Monthly*, **81**, 1974, pp. 592–601. Also see the references given there.

equations to methods of actually solving such equations. In this section we will consider the problem of finding the general solution of a second order linear homogeneous differential equation with constant coefficients.* The corresponding problem with variable coefficients, which is much more difficult, will be considered in Chapter 4.†

Consider the equation

$$L[y] = ay'' + by' + cy$$
$$= (aD^2 + bD + c)y = 0, \tag{1}$$

where $a \neq 0$, b, and c are real numbers. Since a, b, and c are constants it follows immediately from the existence and uniqueness Theorem 3.2 that the solutions of Eq. (1) are valid on the interval $-\infty < x < \infty$.

A clue to a method of solving Eq. (1) can be found simply by reading the equation; that is, what function $y = \phi(x)$ satisfies the relationship that a times its second derivative plus b times its first derivative plus c times the function itself adds up to zero for all x? With only constant coefficients it is natural first to consider functions $y = \phi(x)$ such that y, y', y'' differ only by constant multiplicative factors.

One function that has been studied extensively in the elementary calculus has just the property we wish—the exponential function e^{rx}. Hence we will try to find solutions of Eq. (1) of the form e^{rx} for suitably chosen values of r. Substituting $y = e^{rx}$ in Eq. (1) leads to the equation

$$L[e^{rx}] = a(e^{rx})'' + b(e^{rx})' + ce^{rx} = 0 \tag{2}$$

or

$$e^{rx}(ar^2 + br + c) = 0. \tag{3}$$

Since e^{rx} is not zero, we must have

$$ar^2 + br + c = 0. \tag{4}$$

If r is a root of this quadratic equation, often called the *characteristic* or *auxiliary equation*, then e^{rx} is a solution of Eq. (1). Notice that the coefficients in the characteristic equation (4) are the same as those in the differential equation (1). The roots r_1 and r_2 of Eq. (4) are given by

$$r_1 = \frac{-b + (b^2 - 4ac)^{1/2}}{2a}, \qquad r_2 = \frac{-b - (b^2 - 4ac)^{1/2}}{2a}. \tag{5}$$

The nature of the solutions of Eq. (1) clearly depends on the values of r_1 and r_2, which in turn depend on the constant coefficients in the differential equation through the relations (5). Just as in elementary algebra we must examine separately the cases $b^2 - 4ac$ positive, zero, and negative.

---

* Both Daniel Bernoulli (1700–1782) and Euler (1707–1783) knew how to solve such equations before 1740. The first published account was given by Euler in 1743.

† In some cases it is possible to reduce a differential equation with variable coefficients to one with constant coefficients by a suitable change of variable (see Problems 16, 17, and 18 of Section 3.5.1).

The first two cases are considered in this section; the case $b^2 - 4ac < 0$ is considered in the next section.

Real and Unequal Roots. For $b^2 - 4ac > 0$, Eqs. (5) give two real, unequal values for r_1 and r_2. Hence $e^{r_1 x}$ and $e^{r_2 x}$ are solutions of Eq. (1). Further, it is easy to verify that $W(e^{r_1 x}, e^{r_2 x})$ is nowhere zero (see Problem 15). Hence the general solution of Eq. (1) is

$$y = c_1 e^{r_1 x} + c_2 e^{r_2 x}. \tag{6}$$

Example 1. Find the solution of the differential equation $y'' + 5y' + 6y = 0$ satisfying the initial conditions $y(0) = 0$, $y'(0) = 1$.

Substituting $y = e^{r x}$ leads to

$$r^2 + 5r + 6 = 0,$$
$$(r + 3)(r + 2) = 0;$$

hence $r = -2, -3$. The general solution is

$$y = c_1 e^{-2x} + c_2 e^{-3x}.$$

To satisfy the initial conditions at $x = 0$ we must have

$$c_1 + c_2 = 0, \qquad -2c_1 - 3c_2 = 1.$$

Solving these equations gives

$$c_1 = 1, \qquad c_2 = -1.$$

Hence the solution of the differential equation satisfying the prescribed conditions is

$$y = e^{-2x} - e^{-3x}.$$

Real and Equal Roots. When $b^2 - 4ac = 0$, it follows from Eqs. (5) that $r_1 = r_2 = -b/2a$, and we have only the one solution, $e^{-(b/2a)x}$. However, we can use the method of reduction of order (see Section 3.4) to reduce the order of the equation and find a second solution. Let

$$y = v(x)e^{-(b/2a)x};$$

then

$$y' = v'(x)e^{-(b/2a)x} - \frac{b}{2a}v(x)e^{-(b/2a)x},$$

$$y'' = \left[v''(x) - \frac{b}{a}v'(x) + \frac{b^2}{4a^2}v(x) \right]e^{-(b/2a)x}.$$

Substituting for y, y', and y'' in Eq. (1) and dividing out the common factor $e^{-(b/2a)x}$ gives the following equation for v:

$$a\left(v'' - \frac{b}{a}v' + \frac{b^2}{4a^2}v \right) + b\left(v' - \frac{b}{2a}v \right) + cv = 0.$$

After collecting terms we find that

$$av'' - \left(\frac{b^2}{4a} - c\right)v = 0.$$

Since $b^2 - 4ac = 0$ the last term drops out, and we obtain

$$v'' = 0.$$

Hence

$$v(x) = c_1 x + c_2,$$

where c_1 and c_2 are arbitrary constants. Consequently, a second solution of Eq. (1) is $(c_1 x + c_2)e^{-(b/2a)x}$. In particular, corresponding to $c_2 = 0$, $c_1 = 1$ we obtain $xe^{-(b/2a)x}$. We know from our earlier discussion of the method of reduction of order that $e^{-(b/2a)x}$ and $xe^{-(b/2a)x}$ are linearly independent solutions of Eq. (1). This can also be demonstrated by calculating their Wronskian (see Problem 16). Thus in the case $b^2 - 4ac = 0$ the general solution of Eq. (1) is

$$y = c_1 e^{r_1 x} + c_2 x e^{r_1 x}, \qquad r_1 = -b/2a. \tag{7}$$

Example 2. Find the general solution of

$$y'' + 4y' + 4y = 0.$$

Substituting $y = e^{rx}$ leads to

$$r^2 + 4r + 4 = 0$$
$$(r + 2)(r + 2) = 0;$$

hence $r = -2$ is a repeated root of the characteristic equation. One solution is e^{-2x}, and a second linearly independent solution, according to the above theory, is xe^{-2x}. Thus the general solution is

$$y = c_1 e^{-2x} + c_2 x e^{-2x}.$$

There is an interesting alternative way of determining the second solution when the roots of the characteristic equation are equal. Since r_1 is a repeated root of $ar^2 + br + c = 0$, it follows that $ar^2 + br + c = a(r - r_1)^2$. Then, for all r

$$L[e^{rx}] = a(e^{rx})'' + b(e^{rx})' + ce^{rx} = ae^{rx}(r - r_1)^2. \tag{8}$$

The right-hand side of Eq. (8) vanishes when $r = r_1$, which shows that $e^{r_1 x}$ is a solution of Eq. (1), as we already know. Differentiating both sides of Eq. (8) with respect to r and interchanging differentiation with respect to r and x gives

$$L[xe^{rx}] = axe^{rx}(r - r_1)^2 + 2ae^{rx}(r - r_1). \tag{9}$$

When $r = r_1$ the right side of Eq. (9) vanishes and hence $L[xe^{r_1 x}] = 0$; thus $xe^{r_1 x}$ is also a solution of Eq. (1).

PROBLEMS

In each of Problems 1 through 13 determine the general solution of the given differential equation. If initial conditions are given, find the solution satisfying the stated conditions.

1. $y'' + 2y' - 3y = 0$

2. $4y'' + 4y' + y = 0$

3. $6y'' - y' - y = 0$

4. $2y'' - 3y' + y = 0$

5. $y'' - y = 0$

6. $y'' - 2y' + y = 0$

7. $y'' + 5y' = 0$

8. $y'' - 9y' + 9y = 0$

9. $y'' - 2y' - 2y = 0$

10. $y'' + 2y' + y = 0$

11. $y'' + y' - 2y = 0$,　　$y(0) = 1$,　　$y'(0) = 1$

12. $y'' - 6y' + 9y = 0$,　　$y(0) = 0$,　　$y'(0) = 2$

13. $y'' + 8y' - 9y = 0$,　　$y(1) = 1$,　　$y'(1) = 0$

14. Show that the general solution of $y'' - 4y = 0$ is

$$y = c_1 \sinh 2x + c_2 \cosh 2x.$$

15. Show that if $r_1 \neq r_2$ then the functions $e^{r_1 x}$ and $e^{r_2 x}$ are linearly independent on $-\infty < x < \infty$. This completes the proof of the result given in Eq. (6) of the text.

16. Show that $W(e^{rx}, xe^{rx}) \neq 0$ for any value of r. Hence conclude that if $e^{r_1 x}$ and $xe^{r_1 x}$ are solutions of $ay'' + by' + cy = 0$, then they are a fundamental set of solutions.

17. In this problem we present another argument that suggests the form of the second solution of $ay'' + by' + cy = 0$ when the roots r_1 and r_2 of the characteristic equation $ar^2 + br + c = 0$ are equal. First, suppose that $r_2 \neq r_1$ so that $e^{r_1 x}$ and $e^{r_2 x}$ are solutions of the differential equation. Verify that $\phi(x; r_1, r_2) = (e^{r_2 x} - e^{r_1 x})/(r_2 - r_1)$ is also a solution of the differential equation for $r_2 \neq r_1$. Now think of r_1 as fixed and evaluate the limit of $\phi(x; r_1, r_2)$ as $r_2 \to r_1$ by using L'Hôpital's rule.

3.5.1 COMPLEX ROOTS

To complete our discussion of the equation

$$ay'' + by' + cy = 0, \tag{1}$$

we consider the case where the roots of the characteristic equation

$$ar^2 + br + c = 0 \tag{2}$$

are complex numbers of the form $\lambda + i\mu$ where λ and μ are real. This immediately raises the question of what we mean by an expression of the form $e^{(\lambda + i\mu)x}$.

First recall that in elementary calculus it was shown that the Taylor (1685–1731) series for e^x about $x = 0$ is

$$e^x = \sum_{n=0}^{\infty} \frac{x^n}{n!}, \qquad -\infty < x < \infty. \tag{3}$$

If we substitute ix for x in Eq. (3) and take the result as the definition of e^{ix}, we find that

$$e^{ix} = \sum_{n=0}^{\infty} \frac{(ix)^n}{n!}$$

$$= \sum_{n=0}^{\infty} \frac{(-1)^n x^{2n}}{(2n)!} + i \sum_{n=1}^{\infty} \frac{(-1)^{n-1} x^{2n-1}}{(2n-1)!} , \qquad (4)$$

where we have made use of the fact that $i^2 = -1$, $i^3 = -i$, $i^4 = 1$, etc. The first series in Eq. (4) is precisely the Taylor series for $\cos x$, and the second is the Taylor series for $\sin x$. Hence we *define*

$$e^{ix} = \cos x + i \sin x. \qquad (5)$$

This relationship is referred to as Euler's formula. Substituting $-x$ for x in Eq. (5) and making use of the relations $\cos (-x) = \cos x$, $\sin (-x) = - \sin x$ gives

$$e^{-ix} = \cos x - i \sin x. \qquad (6)$$

The functions $\cos x$ and $\sin x$ can be expressed in terms of e^{ix} and e^{-ix} by adding and subtracting Eqs. (5) and (6), respectively. We obtain

$$\cos x = \frac{e^{ix} + e^{-ix}}{2} , \qquad \sin x = \frac{e^{ix} - e^{-ix}}{2i} .$$

Next we *define*

$$e^{(\lambda + i\mu)x} = e^{\lambda x} e^{i\mu x} = e^{\lambda x}(\cos \mu x + i \sin \mu x). \qquad (7)$$

With this definition it can be shown that the usual laws of algebra hold. For example,

$$\frac{e^{(\lambda + i\mu)x}}{e^{(\alpha + i\beta)x}} = e^{(\lambda + i\mu)x - (\alpha + i\beta)x} = e^{(\lambda - \alpha)x + i(\mu - \beta)x},$$

and

$$[e^{(\lambda + i\mu)x}]^n = e^{n(\lambda + i\mu)x} = e^{n\lambda x + in\mu x},$$

where n is a positive or negative integer. Actually, the last result is also valid for nonintegral values of n.

Finally, we must consider the question of differentiating $e^{(\lambda + i\mu)x}$ with respect to x. While we know that

$$\frac{d}{dx} (e^{rx}) = re^{rx} \qquad (8)$$

for r a real number, we do not know that this is true if r is complex. However, it can be shown by direct computation using the definition given by Eq. (7) that Eq. (8) is true for r complex. Indeed, the usual laws of the elementary calculus are valid for complex exponentials with the understanding that e^{rx} is defined by Eq. (7) for r complex. More generally for a complex-valued function $f(x) = u(x) + iv(x)$ where u and v are real-valued functions, and for a complex constant c, we have $f'(x) = u'(x) + iv'(x)$ and $[cf(x)]' = cf'(x)$.

Now we can return to the problem of solving Eq. (1) when the roots of the characteristic equation are complex. Since a, b, and c are real, the roots occur in conjugate pairs $r_1 = \lambda + i\mu$ and $r_2 = \lambda - i\mu$; hence the general solution of Eq. (1) is

$$y = c_1 e^{(\lambda+i\mu)x} + c_2 e^{(\lambda-i\mu)x}. \tag{9}$$

The general solution (9) has the disadvantage that the functions $e^{(\lambda+i\mu)x}$ and $e^{(\lambda-i\mu)x}$ are complex-valued. Since the differential equation is real it would seem desirable, if possible, to express the general solution of Eq. (1) as a linear combination of real-valued solutions. This can be done as follows. Since $e^{(\lambda+i\mu)x}$ and $e^{(\lambda-i\mu)x}$ are solutions of Eq. (1), their sums and differences are also solutions; thus

$$e^{(\lambda+i\mu)x} + e^{(\lambda-i\mu)x} = e^{\lambda x}(\cos \mu x + i \sin \mu x) + e^{\lambda x}(\cos \mu x - i \sin \mu x)$$
$$= 2e^{\lambda x} \cos \mu x$$

and

$$e^{(\lambda+i\mu)x} - e^{(\lambda-i\mu)x} = e^{\lambda x}(\cos \mu x + i \sin \mu x) - e^{\lambda x}(\cos \mu x - i \sin \mu x)$$
$$= 2ie^{\lambda x} \sin \mu x$$

are solutions of Eq. (1). Hence, neglecting the constant multipliers 2 and $2i$, respectively, the functions

$$e^{\lambda x} \cos \mu x, \qquad e^{\lambda x} \sin \mu x \tag{10}$$

are real-valued solutions of Eq. (1). It can easily be shown that

$$W(e^{\lambda x} \cos \mu x, e^{\lambda x} \sin \mu x) = \mu e^{2\lambda x},$$

and since this never vanishes these two functions form a fundamental set of real-valued solutions. Hence we can express the general solution of Eq. (1) in the form

$$y = c_1 e^{\lambda x} \cos \mu x + c_2 e^{\lambda x} \sin \mu x,$$

where c_1 and c_2 are arbitrary constants.

Example. Find the general solution of

$$y'' + y' + y = 0. \tag{11}$$

The characteristic equation is

$$r^2 + r + 1 = 0,$$

and the roots are

$$r = \frac{-1 \pm (1 - 4)^{1/2}}{2} = -\frac{1}{2} \pm i \frac{\sqrt{3}}{2}.$$

Hence the general solution of Eq. (11) is

$$y = e^{-x/2} \left(c_1 \cos \frac{\sqrt{3}}{2} x + c_2 \sin \frac{\sqrt{3}}{2} x \right).$$

Let us consider again the real-valued solutions $e^{\lambda x} \cos \mu x$ and $e^{\lambda x} \sin \mu x$ of Eq. (1). Notice that these functions are the real and imaginary parts, respectively, of the complex-valued solution $e^{(\lambda + i\mu)x}$. This is just a special case of a general result that depends only on the fact that the coefficients in the differential equation are real-valued, whether or not they are constants. We state this result as a theorem.

Theorem 3.9. *Let the real-valued functions p and q be continuous on the open interval $\alpha < x < \beta$. Let $y = \phi(x) = u(x) + iv(x)$ be a complex-valued solution of the differential equation*

$$L[y] = y'' + p(x)y' + q(x)y = 0, \tag{12}$$

where u and v are real-valued functions. Then u and v are also solutions of the differential equation (12).

To prove this theorem, observe that

$$\begin{aligned} L[u + iv] &= (u + iv)'' + p(u + iv)' + q(u + iv) \\ &= (u'' + pu' + qu) + i(v'' + pv' + qv) \\ &= L[u] + iL[v] = 0. \end{aligned}$$

But a complex number is zero if, and only if, both its real and imaginary parts are zero; thus $L[u] = 0$ and $L[v] = 0$.

If the constants a, b, and c in Eq. (1) are complex numbers, it is still possible to find solutions of the form e^{rx} where r must satisfy

$$ar^2 + br + c = 0.$$

However, in general the roots of the characteristic equation will be complex numbers but not complex conjugates, and the corresponding solutions of Eq. (1) will be complex-valued. This is discussed briefly in Problems 13, 14, and 15.

PROBLEMS

In each of Problems 1 through 8 determine the general solution of the given differential equation. If initial conditions are given, find the solution satisfying the stated conditions.

1. $y'' - 2y' + 2y = 0$ 2. $y'' - 2y' + 6y = 0$

3. $y'' + 2y' - 8y = 0$ 4. $y'' + 2y' + 2y = 0$

5. $9y'' - 6y' + y = 0$ 6. $y'' + 6y' + 13y = 0$

7. $y'' + 4y = 0, \quad y(0) = 0, \quad y'(0) = 1$

8. $y'' + 4y' + 5y = 0, \quad y(0) = 1, \quad y'(0) = 0$

9. Verify that $W(e^{\lambda x} \cos \mu x, e^{\lambda x} \sin \mu x) = \mu e^{2\lambda x}$.

10. In this problem a prime is used to denote differentiation with respect to time t.

(a) Show that if a, b, and c are positive constants then all solutions of $ay'' + by' + cy = 0$ approach zero as $t \to \infty$. Does this conclusion hold if either $b = 0$ or $c = 0$?

(b) Show that $y = \sin t$ is a solution of

$$y'' + (k \sin^2 t)y' + (1 - k \cos t \sin t)y = 0$$

for any value of the constant k. Show that if $0 < k < 2$ then $(1 - k \cos t \sin t) > 0$ and $(k \sin^2 t) \geq 0$. Thus observe that even though the coefficients of this variable coefficient differential equation are nonnegative (and the coefficient of y' is zero only at the points $t = 0, \pi, 2\pi, \cdots$), it still has a solution that does not approach zero as $t \to \infty$. Thus we observe a not unusual situation in the theory of differential equations: apparently very similar equations can have quite different properties.

11. Show that the general solution of

$$y'' + y = 0$$

can be expressed as $y = A_1 \cos (x + \delta_1)$ or as $y = A_2 \sin (x + \delta_2)$, where A_1, δ_1, and A_2, δ_2 are constants.

12. For r a real number show that

$$(\cos x + i \sin x)^r = \cos rx + i \sin rx.$$

For r a positive integer this is known as De Moivre's (1667–1754) formula. Use this result to obtain the double angle formulas $\sin 2x = 2 \sin x \cos x$ and $\cos 2x = \cos^2 x - \sin^2 x$.

*13. Show that any complex number $a + ib$ can be written in the form $Ae^{i\theta}$, where A and θ are positive real numbers, and $0 \leq \theta < 2\pi$.
Hint: Expand $Ae^{i\theta}$ using Euler's formula and solve for A and θ in terms of a and b.

*14. Defining the two possible square roots of a complex number $Ae^{i\theta}$ as $\pm A^{1/2}e^{i\theta/2}$, compute the square roots of $1 + i$, $1 - i$, i.

*15. Solve

(a) $y'' + iy' + 2y = 0$ (b) $y'' + 2y' + iy = 0$

Note that the roots of the characteristic equations are not complex conjugates. Does the real or imaginary part of either of the solutions satisfy the differential equation?

*16. Now that we know how to solve second order equations with constant coefficients, let us determine the conditions under which the equation $y'' + p(x)y' + q(x)y = 0$ can be transformed into one having constant coefficients by a suitable change of independent variable. Let z be the new independent variable, $z = u(x)$, where for the moment the relationship between z and x is unspecified. Show that

(a) $$\frac{dy}{dx} = \frac{dz}{dx}\frac{dy}{dz}$$

$$\frac{d^2y}{dx^2} = \left(\frac{dz}{dx}\right)^2 \frac{d^2y}{dz^2} + \frac{d^2z}{dx^2}\frac{dy}{dz}$$

(b) The differential equation becomes

$$\left(\frac{dz}{dx}\right)^2 \frac{d^2y}{dz^2} + \left(\frac{d^2z}{dx^2} + p(x)\frac{dz}{dx}\right)\frac{dy}{dz} + q(x)y = 0$$

(c) The equation of part (b) will have constant coefficients if we choose

$$z = u(x) = \int^x [q(t)]^{1/2}\, dt,$$

provided that

$$\frac{z'' + p(x)z'}{q(x)} = \frac{q'(x) + 2p(x)q(x)}{2[q(x)]^{3/2}}$$

is a constant. Notice that for the transformation $z = u(x)$ to be real-valued it is necessary that $q(x) \geq 0$. This means that in the original differential equation $q(x)$ must be of one sign; if $q(x) < 0$ multiply the differential equation by -1.

Hence conclude that a necessary and sufficient condition for transforming $y'' + p(x)y' + q(x)y = 0$ into an equation with constant coefficients by a change of the independent variable is that the function $(q' + 2pq)/q^{3/2}$ be a constant.

*17. Using the result of Problem 16, determine whether each of the following equations can be transformed into an equation with constant coefficients by a change of the independent variable. If so, find the general solution of the given equation.

(a) $y'' + xy' + e^{-x^2}y = 0$, $-\infty < x < \infty$

(b) $y'' + 3xy' + x^2y = 0$, $-\infty < x < \infty$

(c) $xy'' + (x^2 - 1)y' + x^3y = 0$, $0 < x < \infty$

*18. Show, using the result of Problem 16, that it is always possible to transform an equation of the form

$$x^2y'' + \alpha xy' + \beta y = 0, \qquad x > 0,$$

where α and β are real constants, into an equation with constant coefficients by letting $z = \ln x$. An equation of this type is called an Euler equation; it is discussed in Section 4.4. Find the general solution of $x^2y'' + xy' + y = 0$, $x > 0$.

19. By suitable changes of variables we can establish a relation between a second order linear homogeneous equation $y'' + p(x)y' + q(x)y = 0$ and a first order equation.

(a) Set $y'/y = -u$ and determine y in terms of u. Then show that u satisfies

$$du/dx = q(x) - p(x)u + u^2.$$

This equation is known as a Riccati equation; it is discussed in Problem 35 of Section 2.8. Note that the price for going from a second order to a first order equation is that of going from a linear to a nonlinear equation.

(b) Now consider the general Riccati equation

$$dw/dx = q_0(x) + q_1(x)w + q_2(x)w^2.$$

Show that the transformation $w = -y'/yq_2$ leads to the second order linear homogeneous equation

$$q_2(x)y'' - [q_2'(x) + q_1(x)q_2(x)]y' + q_2^2(x)q_0(x)y = 0.$$

3.6 THE NONHOMOGENEOUS PROBLEM

In the previous sections of this chapter we discussed the general theory of second order linear homogeneous differential equations, and in the case of constant coefficients have shown how to construct solutions. We will now turn our attention to solving the nonhomogeneous differential equation

$$L[y] = y'' + p(x)y' + q(x)y = g(x). \tag{1}$$

Throughout the discussion we will assume that the functions p, q, and g are continuous on the interval of interest.

In modern engineering the problem of solving Eq. (1) is often referred to as an input-output problem. Supposing for the moment that the differential equation (1) represents a certain mechanical or electrical system, it is natural to refer to the solution $y = \phi(x)$ as the output of the system. The coefficients p and q are determined by the physical mechanism and the nonhomogeneous term g is the input function. The problem of solving Eq. (1) corresponds to determining the output of the system for different input functions.

Before considering particular cases of Eq. (1), we will prove several simple but very useful general results that will simplify our succeeding work.

Theorem 3.10.   *The difference of any two solutions of the differential equation* (1),

$$L[y] = y'' + p(x)y' + q(x)y = g(x),$$

is a solution of the corresponding homogeneous differential equation

$$L[y] = y'' + p(x)y' + q(x)y = 0. \tag{2}$$

To prove this theorem suppose that the functions u_1 and u_2 are solutions of Eq. (1).   Then

$$L[u_1] = g,$$

and

$$L[u_2] = g.$$

Subtracting the second equation from the first equation gives

$$L[u_1] - L[u_2] = 0,$$

or, since L is a linear operator,

$$L[u_1 - u_2] = 0,$$

which is the desired result.   With the aid of this theorem we can prove the following important theorem.

Theorem 3.11.   *Given one solution y_p of the nonhomogeneous linear differential equation* (1)

$$L[y] = y'' + p(x)y' + q(x)y = g(x),$$

then any solution $y = \phi(x)$ of this equation can be expressed as

$$\phi(x) = y_p(x) + c_1 y_1(x) + c_2 y_2(x), \tag{3}$$

where y_1 and y_2 are linearly independent solutions of the corresponding homogeneous equation.

To prove this theorem, observe that according to Theorem 3.10 $\phi - y_p$ is a solution of the homogeneous equation. Thus, according to Theorem 3.4 of Section 3.2, $\phi - y_p$ can be expressed as a linear combination of y_1 and y_2, which establishes the theorem.

It is customary to refer to the linear combination (3) as the *general solution* of Eq. (1).

Consequently, to find the general solution of Eq. (1) we must find the general solution of the homogeneous equation (2) and then find *any* solution of the non-homogeneous equation. As we could have anticipated (see Section 3.1), the general solution of Eq. (1) involves two arbitrary constants, and hence to specify a unique solution of Eq. (1) it is necessary to specify two additional conditions, namely, the initial conditions $y(x_0) = y_0$ and $y'(x_0) = y_0'$.

The general solution of the homogeneous equation (2) is often referred to as the *complementary solution* and denoted by y_c. Thus $y_c(x) = c_1 y_1(x) + c_2 y_2(x)$. A solution of the nonhomogeneous equation y_p is usually called a *particular solution*.* Notice that y_p is not uniquely determined since if Y is a solution of the nonhomogeneous equation then Y plus any multiple of y_1 or y_2 is still a solution of the nonhomogeneous equation. It follows from Theorem 3.11 that the general solution of the nonhomogeneous equation (1) is

$$y = y_c(x) + y_p(x). \tag{4}$$

In many problems the nonhomogeneous term g may be very complicated; however, if g can be expressed as the sum of a finite number of functions we can make use of the linearity of the differential equation to replace the original problem by several simpler ones. For example, suppose that it is possible to write $g(x)$ as $g_1(x) + g_2(x) + \cdots + g_m(x)$; then Eq. (1) becomes

$$L[y] = y'' + p(x)y' + q(x)y = g_1(x) + g_2(x) + \cdots + g_m(x). \tag{5}$$

If we can find particular solutions y_{p_i} of the differential equations

$$L[y] = g_i(x), \qquad i = 1, 2, \ldots, m, \tag{6}$$

then it follows by direct substitution that

$$y_p(x) = y_{p_1}(x) + y_{p_2}(x) + \cdots + y_{p_m}(x) \tag{7}$$

is a particular solution of Eq. (5). Hence the general solution of Eq. (5) is of the form

$$y = y_c(x) + y_{p_1}(x) + \cdots + y_{p_m}(x). \tag{8}$$

In general, it is easier to find solutions of Eqs. (6) and to add the results than to solve Eq. (5) as it stands. This method of constructing the solution of a complicated problem by adding solutions of simpler problems is known as the method of superposition.

Example. Find the general solution of

$$y'' + 4y = 1 + x + \sin x. \tag{9}$$

---

* This is rather an unfortunate usage, since the term particular solution may also refer to a solution satisfying prescribed initial conditions. The meaning is usually clear from the context.

First consider the corresponding homogeneous equation, $y'' + 4y = 0$. Substituting $y = e^{rx}$ yields the characteristic equation $r^2 + 4 = 0$. Thus the complementary solution is

$$y_c(x) = c_1 \cos 2x + c_2 \sin 2x.$$

To determine a particular solution of Eq. (9) we superimpose particular solutions of $y'' + 4y = 1$, $y'' + 4y = x$, and $y'' + 4y = \sin x$. It can be readily verified that $\frac{1}{4}$, $\frac{1}{4}x$, and $\frac{1}{3} \sin x$ are particular solutions of these equations, respectively. Thus the general solution of Eq. (9) is

$$y = c_1 \cos 2x + c_2 \sin 2x + \tfrac{1}{4} + \tfrac{1}{4}x + \tfrac{1}{3} \sin x.$$

3.6.1 THE METHOD OF UNDETERMINED COEFFICIENTS

A number of methods can be used to obtain particular solutions of second order nonhomogeneous differential equations. When applicable, the method of undetermined coefficients is one of the simplest. Basically, the method consists of making an intelligent guess as to the form of the particular solution and then substituting this function, which in general will involve one or more unknown coefficients, into the differential equation. For this method to be successful we must be able to determine the unknown coefficients so that the function actually does satisfy the differential equation.

Clearly such a method depends to a considerable extent on the ability to discover in advance the general form of a particular solution. For a completely arbitrary differential equation very little can be said. However, if the equation is of the form

$$a \frac{d^2y}{dx^2} + b \frac{dy}{dx} + cy = g(x), \tag{1}$$

where a, b, and c are real constants, and the nonhomogeneous term $g(x)$ is an exponential function $(e^{\alpha x})$, or a polynomial $(a_0 x^n + \cdots + a_n)$, or sinusoidal in character ($\sin \beta x$ or $\cos \beta x$), then definite rules can be given for the determination of a particular solution by the method of undetermined coefficients. These rules also cover the more general case in which $g(x)$ is a product of terms of the above types, such as

$$g(x) = e^{\alpha x}(a_0 x^n + a_1 x^{n-1} + \cdots + a_n) \begin{cases} \cos \beta x \\ \sin \beta x. \end{cases} \tag{2}$$

Notice that any other product of exponentials, polynomials, and sines and cosines is equivalent to a sum of terms of the type (2). In particular, the product of two or more sinusoidal terms can always be reduced to a sum of individual sinusoidal terms by the use of trigonometric identities. We can treat the case in which $g(x)$ is a sum of terms of type (2) by using the method of superposition which was discussed in the last section. For more general nonhomogeneous terms than (2) or for equations with variable coefficients the method of undetermined coefficients is rarely useful.

Before discussing the general procedure let us consider a few simple examples which will illustrate the method.

Example 1. Find a particular solution of the differential equation

$$y'' - 3y' - 4y = 2 \sin x. \tag{3}$$

We want to find a function y_p such that the sum of its second derivative minus three times its first derivative minus four times the function itself adds up to $2 \sin x$. There is little hope in trying such functions as $\ln x$, e^x, or x^2 for $y_p(x)$ since no matter how we combine such functions and their derivatives it is impossible to obtain $\sin x$. The obvious functions to consider for $y_p(x)$ are $\sin x$ and $\cos x$. So we *assume* that $y_p(x)$ is of the form

$$y_p(x) = A \cos x + B \sin x,$$

where A and B are, for the moment, unknown constants. Then

$$y_p'(x) = -A \sin x + B \cos x,$$
$$y_p''(x) = -A \cos x - B \sin x.$$

Substituting for y, y', and y'' in Eq. (3) and collecting terms gives

$$(-A - 3B - 4A) \cos x + (-B + 3A - 4B) \sin x = 2 \sin x.$$

This equation will be satisfied identically if and only if

$$-5A - 3B = 0, \qquad 3A - 5B = 2.$$

Hence $A = \frac{3}{17}$, $B = -\frac{5}{17}$, and a particular solution of Eq. (3) is

$$y_p(x) = \tfrac{1}{17}(3 \cos x - 5 \sin x).$$

Example 2. Find a particular solution of the differential equation

$$y'' - 3y' - 4y = 4x^2. \tag{4}$$

It is natural to try $y_p(x) = Ax^2$ where A is a constant to be determined. Then $y_p'(x) = 2Ax$, $y_p''(x) = 2A$, and substituting in Eq. (4) gives

$$2A - 6Ax - 4Ax^2 = 4x^2.$$

If this equation is to be satisfied for all x the coefficients of like powers of x on each side of the equation must be identical. This leads to three equations and there is no choice of A that will satisfy all three. Hence it is impossible to find a particular solution of Eq. (4) of the form Ax^2.

However, if we think of the nonhomogeneous term $4x^2$ in Eq. (4) as the polynomial $4x^2 + 0x + 0$ it now appears reasonable to assume that the form of $y_p(x)$ is $Ax^2 + Bx + C$, where A, B, and C are constants to be determined. Substituting for y, y', and y'' in Eq. (4) and equating like powers of x on both sides of the equation gives three simultaneous linear algebraic nonhomogeneous equations for the three unknowns A, B, and C. Their solution is $A = -1$, $B = \frac{3}{2}$, and $C = -\frac{13}{8}$; hence a particular solution of Eq. (4) is

$$y_p(x) = -x^2 + \tfrac{3}{2}x - \tfrac{13}{8}.$$

The student may wonder what would happen if a higher degree polynomial, such as $Ax^4 + Bx^3 + Cx^2 + Dx + E$, were assumed for $y_p(x)$. The answer is that all coefficients beyond the quadratic term would turn out to be zero. Thus, with certain exceptions to be noted later, it is unnecessary to assume for $y_p(x)$ a polynomial of higher degree than the degree of the polynomial in the nonhomogeneous term.

Example 3.   Find a particular solution of the equation

$$y'' - 3y' - 4y = e^{-x}. \tag{5}$$

Following the reasoning used before we assume that $y_p(x) = Ae^{-x}$. Substituting in Eq. (5) gives

$$(A + 3A - 4A)e^{-x} = e^{-x},$$

or

$$0 \cdot Ae^{-x} = e^{-x},$$

which does not permit the determination of A. Hence the particular solution is not of the form Ae^{-x}. The difficulty in this case stems from the fact that e^{-x} is a solution of the homogeneous differential equation. Consequently, the operations of the left-hand side of Eq. (5), when applied to any constant multiple of e^{-x}, yield zero. This raises the general question of whether the method of undetermined coefficients can still be used when the nonhomogeneous term is a solution of the homogeneous equation. We shall soon see that the answer to this question is yes.

Let us now turn to the general cases in which $g(x)$ is of the form

$$g(x) = \begin{cases} P_n(x) = a_0 x^n + a_1 x^{n-1} + \cdots + a_n \\ e^{\alpha x} P_n(x) \\ e^{\alpha x} P_n(x) \sin \beta x \\ e^{\alpha x} P_n(x) \cos \beta x. \end{cases}$$

$g(x) = P_n(x)$.   In this case Eq. (1) becomes

$$ay'' + by' + cy = a_0 x^n + a_1 x^{n-1} + \cdots + a_n. \tag{6}$$

To obtain a particular solution we assume

$$y_p(x) = A_0 x^n + A_1 x^{n-1} + \cdots + A_{n-2} x^2 + A_{n-1} x + A_n. \tag{7}$$

Substituting in Eq. (6) we obtain

$$a[n(n-1)A_0 x^{n-2} + \cdots + 2A_{n-2}] + b(nA_0 x^{n-1} + \cdots + A_{n-1})$$
$$+ c(A_0 x^n + A_1 x^{n-1} + \cdots + A_n) = a_0 x^n + \cdots + a_n. \tag{8}$$

Equating the coefficients of like powers of x gives

$$cA_0 = a_0,$$
$$cA_1 + nbA_0 = a_1,$$

$$.$$
$$.$$
$$.$$

$$cA_n + bA_{n-1} + 2aA_{n-2} = a_n.$$

Provided $c \neq 0$ the solution of the first equation is $A_0 = a_0/c$, and the remaining equations determine $A_1, A_2, \ldots, A_n$ successively. If $c = 0$, but $b \neq 0$, the polynomial on the left-hand side of Eq. (8) is of degree $n - 1$, and we cannot satisfy Eq. (8). In order to insure that $ay_p''(x) + by_p'(x)$ will be a polynomial of degree n we must choose $y_p(x)$ to be a polynomial of degree $n + 1$. Hence we assume

$$y_p(x) = x(A_0x^n + \cdots + A_n).$$

There is no constant term in this expression for $y_p(x)$, but there is no need to include such a term since when $c = 0$ a constant is a solution of the homogeneous differential equation. Since $b \neq 0$ we have $A_0 = a_0/b(n + 1)$, and the coefficients $A_1, \cdots, A_n$ can be determined similarly. If both c and b vanish, we would assume

$$y_p(x) = x^2(A_0x^n + \cdots + A_n).$$

The term $ay_p''(x)$ gives rise to a term of degree n, and we can proceed as before.* Again the constant and linear terms in $y_p(x)$ are omitted since in this case they are both solutions of the homogeneous equation.

$g(x) = e^{\alpha x}P_n(x)$.   The problem of determining a particular solution of

$$ay'' + by' + cy = e^{\alpha x}P_n(x) \tag{9}$$

can be reduced to the one just solved.† Let

$$y_p(x) = e^{\alpha x}u(x);$$
then
$$y_p'(x) = e^{\alpha x}[u'(x) + \alpha u(x)]$$
and
$$y_p''(x) = e^{\alpha x}[u''(x) + 2\alpha u'(x) + \alpha^2 u(x)].$$

Substituting for y, y', and y'' in Eq. (9), canceling the factor $e^{\alpha x}$, and collecting terms give

$$au''(x) + (2a\alpha + b)u'(x) + (a\alpha^2 + b\alpha + c)u(x) = P_n(x). \tag{10}$$

---

* The student should note that in the case $c = 0$, $b \neq 0$, Eq. (6) can be integrated once, giving a first order linear equation whose nonhomogeneous term is a polynomial of degree $n + 1$. A particular solution of this equation could then be found by the method of undetermined coefficients with $y_p(x)$ a polynomial of degree $n + 1$. If both c and b vanish, then Eq. (6) can be integrated immediately, the solution being a polynomial of degree $n + 2$.

† One of a mathematician's favorite devices is to reduce a new problem to one that has already been solved.

The determination of a particular solution of Eq. (10) is precisely the problem that we just solved. If $a\alpha^2 + b\alpha + c$ does not vanish we assume $u(x) = (A_0x^n + \cdots + A_n)$, and hence a particular solution of Eq. (9) is of the form

$$y_p(x) = e^{\alpha x}(A_0x^n + A_1x^{n-1} + \cdots + A_n). \tag{11}$$

On the other hand, if $a\alpha^2 + b\alpha + c$ vanishes but $(2a\alpha + b)$ does not, we must take $u(x)$ to be of the form $x(A_0x^n + \cdots + A_n)$. The corresponding form for $y_p(x)$ is x times the expression in the right member of Eq. (11). It should be noted that the vanishing of $a\alpha^2 + b\alpha + c$ implies that $e^{\alpha x}$ is a solution of the homogeneous equation. If both $a\alpha^2 + b\alpha + c$ and $2a\alpha + b$ vanish (and this implies that both $e^{\alpha x}$ and $xe^{\alpha x}$ are solutions of the homogeneous equation) then the correct form for $u(x)$ is $x^2(A_0x^n + \cdots + A_n)$; and hence $y_p(x)$ is x^2 times the expression on the right-hand side of Eq. (11).

$g(x) = e^{\alpha x}P_nx \cos \beta x$ or $e^{\alpha x}P_n(x) \sin \beta x$. The two cases are similar; consider the latter case. We can reduce this problem to the previous one by recalling that $\sin \beta x = (e^{i\beta x} - e^{-i\beta x})/2i$. Hence $g(x)$ is of the form

$$g(x) = P_n(x)\frac{e^{(\alpha+i\beta)x} - e^{(\alpha-i\beta)x}}{2i}$$

and we should choose

$$y_p(x) = e^{(\alpha+i\beta)x}(A_0x^n + \cdots + A_n) + e^{(\alpha-i\beta)x}(B_0x^n + \cdots + B_n),$$

or equivalently

$$y_p(x) = e^{\alpha x}(A_0x^n + \cdots + A_n)\cos \beta x + e^{\alpha x}(B_0x^n + \cdots + B_n)\sin \beta x.$$

Usually the latter form is preferred. If $\alpha \pm i\beta$ satisfy the characteristic equation corresponding to the homogeneous equation we must, of course, multiply each of the polynomials by x to increase their degree by one.

If the nonhomogeneous term involves expressions such as $e^{\alpha x}\cos \beta x$ and $e^{\alpha x}\sin \beta x$ it is convenient to treat these together, since each one individually gives rise to the same form for a particular solution. For example, if $g(x) = x\sin x + 2\cos x$, the form for $y_p(x)$ would be

$$(A_0x + A_1)\sin x + (B_0x + B_1)\cos x$$

provided that $\sin x$ and $\cos x$ were not solutions of the homogeneous equation.

We can summarize the preceding results in Table 3.1.

TABLE 3.1   The Particular Solution of $ay'' + by' + cy = g(x)$

$g(x)$	$y_p(x)$
$P_n(x) = a_0x^n + a_1x^{n-1} + \cdots + a_n$	$x^s(A_0x^n + A_1x^{n-1} + \cdots + A_n)$
$P_n(x)e^{\alpha x}$	$x^s(A_0x^n + A_1x^{n-1} + \cdots + A_n)e^{\alpha x}$
$P_n(x)e^{\alpha x}\begin{cases}\sin \beta x\\ \cos \beta x\end{cases}$	$x^s[(A_0x^n + A_1x^{n-1} + \cdots + A_n)e^{\alpha x}\cos \beta x$ $+ (B_0x^n + B_1x^{n-1} + \cdots + B_n)e^{\alpha x}\sin \beta x]$

Here s is the smallest nonnegative integer ($s = 0, 1,$ or 2) which will insure that no term in $y_p(x)$ is a solution of the corresponding homogeneous equation.

These results, coupled with the fact that the principle of superposition can be used to find the corresponding particular solution when the nonhomogeneous term consists of the sum of several functions, allow us to solve fairly rapidly a wide class of linear constant coefficient nonhomogeneous differential equations.

If the differential equation has variable coefficients or the nonhomogeneous term is more complicated than Eq. (2), it is usually better to use the method of variation of parameters, which is discussed in the next section.

Example 4. Using the method of undetermined coefficients, determine the correct form for $y_p(x)$ for the differential equation

$$y'' + 4y = xe^x + x \sin 2x. \tag{12}$$

First we solve the corresponding homogeneous equation. We readily find that

$$y_c(x) = c_1 \cos 2x + c_2 \sin 2x.$$

In order to construct a particular solution y_p of Eq. (12) we use the superposition principle, and consider the problems

$$y'' + 4y = xe^x \tag{13}$$

and

$$y'' + 4y = x \sin 2x \tag{14}$$

separately. For the first problem we assume $y_p(x) = (A_0 x + A_1)e^x$. Since e^x is not a solution of the homogeneous equation it is not necessary to revise our initial guess. For the second problem we assume $y_p(x) = (B_0 x + B_1) \cos 2x + (C_0 x + C_1) \sin 2x$, but since $\cos 2x$ and $\sin 2x$ are solutions of the homogeneous equation it is necessary to multiply their polynomial coefficients by x. Hence the correct form for $y_p(x)$ for Eq. (12) is

$$y_p(x) = (A_0 x + A_1)e^x + (B_0 x^2 + B_1 x) \cos 2x + (C_0 x^2 + C_1 x) \sin 2x.$$

In determining the coefficients in this expression it is easier to compute the particular solution corresponding to xe^x, Eq. (13), and $x \sin 2x$, Eq. (14), separately than to try to do all the algebra at one time.

PROBLEMS

Using the method of undetermined coefficients to find a particular solution of the nonhomogeneous equation, find the general solution of the following differential equations. Where specified find the solution satisfying the given initial conditions.

1. $y'' + y' - 2y = 2x$, $y(0) = 0$, $y'(0) = 1$
2. $2y'' - 4y' - 6y = 3e^{2x}$
3. $y'' + 4y = x^2 + 3e^x$, $y(0) = 0$, $y'(0) = 2$
4. $y'' + 2y' = 3 + 4 \sin 2x$ 5. $y'' + 9y = x^2 e^{3x} + 6$
6. $y'' - 2y' + y = xe^x + 4$, $y(0) = 1$, $y'(0) = 1$

7. $2y'' + 3y' + y = x^2 + 3 \sin x$

8. $y'' + y = 3 \sin 2x + x \cos 2x$ 9. $y'' + 2y' + y = e^x \cos x$

10. $u'' + \omega_0^2 u = \cos \omega t,$ $\omega^2 \neq \omega_0^2$ 11. $u'' + \omega_0^2 u = \cos \omega_0 t$

12. $u'' + \mu u' + \omega_0^2 u = \cos \omega t$ $\mu^2 - 4\omega_0^2 < 0$

13. $y'' + y' + y = \sin^2 x$

14. $y'' + y' + 4y = 2 \sinh x$ *Hint:* $\sinh x = (e^x - e^{-x})/2$

15. $y'' - y' - 2y = \cosh 2x$ *Hint:* $\cosh x = (e^x + e^{-x})/2$

In Problems 16 through 22 determine a suitable form for $y_p(x)$ if the method of un-determined coefficients is to be used. Do not evaluate the constants.

16. $y'' + 3y' = 2x^4 + x^2 e^{-3x} + \sin 3x$ 17. $y'' + y = x(1 + \sin x)$

18. $y'' - 5y' + 6y = e^x \cos 2x + e^{2x}(3x + 4) \sin x$

19. $y'' + 2y' + 2y = 3e^{-x} + 2e^{-x} \cos x + 4e^{-x}x^2 \sin x$

20. $y'' - 4y' + 4y = 2x^2 + 4xe^{2x} + x \sin 2x$

21. $y'' + 4y = x^2 \sin 2x + (6x + 7) \cos 2x$

22. $y'' + 3y' + 2y = e^x(x^2 + 1) \sin 2x + 3e^{-x} \cos x + 4e^x$

23. Determine the general solution of

$$y'' + \lambda^2 y = \sum_{m=1}^{N} a_m \sin m\pi x$$

where $\lambda > 0$ and $\lambda \neq m\pi$, $m = 1, 2, \ldots, N$.

24. If a, b, and c are positive and Y_1 and Y_2 are solutions of

$$ay'' + by' + cy = g(t),$$

show that $Y_1(t) - Y_2(t) \to 0$ as $t \to \infty$. Is this result valid if $b = 0$?

25. In many physical problems the input function (that is, the nonhomogeneous term) may be specified by different formulas in different time periods. As a simple example of such a problem determine the solution $y = \phi(t)$ of

$$y'' + y = \begin{cases} t, & 0 \leq t \leq \pi, \\ \pi e^{\pi - t}, & t > \pi, \end{cases}$$

satisfying the initial conditions $y(0) = 0$ and $y'(0) = 1$, and the requirement that y and y' be continuous for all t. Plot the nonhomogeneous term and the solution $y = \phi(t)$ as a function of time.
Hint: First solve the initial value problem for $t \leq \pi$, then solve for $t > \pi$ determining the arbitrary constants so that y and y' are continuous at $t = \pi$.

3.6.2   THE METHOD OF VARIATION OF PARAMETERS

In Section 3.6.1 we discussed a simple method of determining particular solutions of nonhomogeneous differential equations with constant coefficients provided the nonhomogeneous term is of a suitable form. In this section we will consider a *general method* of determining a particular solution of the equation

$$y'' + p(x)y' + q(x)y = g(x) \tag{1}$$

where the functions p, q, and g are continuous on the interval of interest. In order to use this method, known as the method of *variation of parameters*,* it is necessary to know a fundamental set of solutions of the corresponding homogeneous equation

$$y'' + p(x)y' + q(x)y = 0. \tag{2}$$

Suppose y_1 and y_2 are linearly independent solutions of the homogeneous equation (2). Then the general solution of Eq. (2) is

$$y_c(x) = c_1y_1(x) + c_2y_2(x).$$

The method of variation of parameters involves the replacement of the constants c_1 and c_2 by functions u_1 and u_2. We then seek to determine the two functions u_1 and u_2 so that

$$y_p(x) = u_1(x)y_1(x) + u_2(x)y_2(x) \tag{3}$$

satisfies the nonhomogeneous differential equation (1). The importance of this method is due to the fact that it is possible to determine the functions u_1 and u_2 in a simple manner. In order to determine u_1 and u_2 two conditions are required. One condition on u_1 and u_2 stems from the fact that y_p must satisfy Eq. (1). A second condition can be imposed arbitrarily, and will be selected so as to facilitate the calculations. Differentiating Eq. (3) gives

$$y_p' = (u_1'y_1 + u_2'y_2) + (u_1y_1' + u_2y_2'). \tag{4}$$

As just indicated we will simplify this by requiring that u_1 and u_2 satisfy

$$u_1'y_1 + u_2'y_2 = 0. \tag{5}$$

With this condition on u_1' and u_2', y_p'' is given by

$$y_p'' = u_1'y_1' + u_2'y_2' + u_1y_1'' + u_2y_2''. \tag{6}$$

Notice that as a result of imposing condition (5), only *first derivatives* of u_1 and u_2 appear in the expression for y_p''. Substituting for y_p' and y_p'' in Eq. (1) gives

$$u_1(y_1'' + py_1' + qy_1) + u_2(y_2'' + py_2' + qy_2) + u_1'y_1' + u_2'y_2' = g.$$

The terms in parentheses vanish since y_1 and y_2 are solutions of the homogeneous equation (2); hence the condition that y_p satisfy Eq. (1) leads to the requirement

$$u_1'y_1' + u_2'y_2' = g. \tag{7}$$

Rewriting Eqs. (5) and (7), we have the following system of two equations

$$u_1'y_1 + u_2'y_2 = 0$$
$$u_1'y_1' + u_2'y_2' = g$$

* The method is also referred to a Lagrange's method in honor of the French mathematician, J. L. Lagrange (1736–1813), who used it in 1774. Lagrange is famous for his work in celestial mechanics, analytical mechanics, and mathematical analysis. By the age of twenty-five he was recognized as one of the greatest living mathematicians.

for the two unknown functions u_1' and u_2'. Solving this system of equations gives

$$u_1' = \frac{-y_2 g}{W(y_1, y_2)}, \qquad u_2' = \frac{y_1 g}{W(y_1, y_2)}, \tag{8}$$

where $W(y_1, y_2) = y_1 y_2' - y_1' y_2$. Division by $W(y_1, y_2)$ is permissible since y_1 and y_2 are linearly independent solutions of the homogeneous equation and hence $W(y_1, y_2)$ cannot vanish in the interval. Integrating Eqs. (8) and substituting in Eq. (3) gives a particular solution of the nonhomogeneous equation (1). We state this result in a theorem.

Theorem 3.12.   *If the functions p, q, and g are continuous on $\alpha < x < \beta$, and if the functions y_1 and y_2 are linearly independent solutions of the homogeneous equation associated with the differential equation (1),*

$$y'' + p(x)y' + q(x)y = g(x),$$

then a particular solution of Eq. (1) is given by

$$y_p(x) = -y_1(x) \int^x \frac{y_2(t)g(t)}{W(y_1, y_2)(t)}\, dt + y_2(x) \int^x \frac{y_1(t)g(t)}{W(y_1, y_2)(t)}\, dt. \tag{9}$$

Equation (9) can also be written as

$$y_p(x) = \int^x \frac{y_1(t)y_2(x) - y_1(x)y_2(t)}{y_1(t)y_2'(t) - y_1'(t)y_2(t)}\, g(t)\, dt. \tag{10}$$

Note that in Eq. (10) t is a dummy variable of integration and x is the independent variable.

In using the method of variation of parameters for a particular problem it is usually safer to substitute $y_p(x) = u_1(x)y_1(x) + u_2(x)y_2(x)$ and to proceed as above rather than to try to remember either formula (9) or (10). If one of these formulas is used care should be taken that the differential equation is written in the standard form (1). It should also be mentioned that while Eqs. (9) and (10) provide formulas for computing $y_p(x)$, it may not always be easy or even possible to evaluate these integrals in closed form. Nevertheless, even in these cases the formulas for $y_p(x)$ provide a starting point for the numerical evaluation of $y_p(x)$, and that may be the best that can be done. From a numerical point of view it is usually advantageous to have an integral form of the solution, since numerical evaluation of an integral is usually, but not always, considerably easier than direct numerical integration of a differential equation.

Again we emphasize that in computing a particular solution by the method of variation of parameters it is not necessary that the coefficients in the differential equation be constants; all that is required is that we know two linearly independent solutions of the homogeneous equation.

Example.   Determine the general solution of the differential equation

$$y'' + y = \sec x, \qquad 0 < x < \pi/2. \tag{11}$$

Two linearly independent solutions of the homogeneous equation are $y_1(x) = \cos x$, $y_2(x) = \sin x$. Even though Eq. (11) has constant coefficients and we know $y_c(x)$, we cannot use the method of undetermined coefficients because the right-hand side of Eq. (11) is not of the form $e^{\alpha x}P_n(x) \cos \beta x$ or $e^{\alpha x}P_n(x) \sin \beta x$. Instead we use the method of variation of parameters. We write

$$y_p(x) = u_1(x) \cos x + u_2(x) \sin x.$$

Then

$$y_p'(x) = [-u_1(x) \sin x + u_2(x) \cos x] + [u_1'(x) \cos x + u_2'(x) \sin x].$$

Setting the second term in brackets equal to zero, differentiating again, and substituting in Eq. (11) gives

$$u_1'(x) \cos x + u_2'(x) \sin x = 0,$$
$$- u_1'(x) \sin x + u_2'(x) \cos x = \sec x.$$

Solving we obtain

$$u_1'(x) = -\tan x, \qquad u_2'(x) = 1,$$
$$u_1(x) = \ln \cos x, \qquad u_2(x) = x.$$

Hence a particular solution of Eq. (11) is

$$y_p(x) = x \sin x + (\cos x) \ln \cos x,$$

and the general solution is

$$y = c_1 \cos x + c_2 \sin x + x \sin x + (\cos x) \ln \cos x.$$

We can summarize the results of the preceding sections of this chapter, which are concerned with methods of solving the differential equation (1)

$$y'' + p(x)y' + q(x)y = g(x),$$

where the functions p, q, and g are continuous on $\alpha < x < \beta$, as follows.

1. If p and q are constants it is always possible to find two linearly independent solutions of the homogeneous equation. A particular solution can then be obtained by the method of variation of parameters, and hence the complete solution is known. If g is of the appropriate form it will probably be easier to use the method of undetermined coefficients rather than the method of variation of parameters to determine the particular solution.

2. If p and q are not constants, there is no general way to solve Eq. (1) in terms of a finite number of elementary functions.* However, if one solution of the homogeneous equation can be found, a second solution can be obtained, in principle, by reducing the order of the equation. Then a particular solution can be obtained by the method of variation of parameters, and hence the complete solution is known. Actually both calculations can be done together. If we know one solution of the homogeneous equation, we can solve the nonhomogeneous equation by the method

---

* Methods involving infinite series, which are useful for a large class of problems involving variable coefficients, are discussed in Chapter 4. Numerical methods are discussed in Chapter 8.

of reduction of order and obtain both a particular solution and a second linearly independent solution of the homogeneous equation. This is illustrated in Problems 15 and 16. Thus, in general, the problem of solving Eq. (1) rests on the possibility of finding *one* solution of the corresponding homogeneous equation.

PROBLEMS

In each of Problems 1 through 7 determine a particular solution using the method of variation of parameters.

1. $y'' - 5y' + 6y = 2e^x$ 2. $y'' - y' - 2y = 2e^{-x}$
3. $y'' + 2y' + y = 3e^{-x}$ 4. $y'' + y = \tan x,$ $0 < x < \pi/2$
5. $y'' + 9y = 9\sec^2 3x,$ $0 < x < \pi/6$
6. $y'' + 4y' + 4y = e^{-2x}/x^2,$ $x > 0$
7. $y'' + 4y = 3\csc 2x,$ $0 < x < \pi/2$

8. Verify that x and xe^x are solutions of the homogeneous equation corresponding to

$$x^2 y'' - x(x+2)y' + (x+2)y = 2x^3, \qquad x > 0,$$

and find the general solution.

9. Verify that $(1+x)$ and e^x are solutions of the homogeneous equation corresponding to

$$xy'' - (1+x)y' + y = x^2 e^{2x}, \qquad x > 0,$$

and find the general solution.

10. Two linearly independent solutions of Bessel's equation of order one-half,

$$x^2 y'' + xy' + (x^2 - \tfrac{1}{4})y = 0, \qquad x > 0,$$

are $x^{-\frac{1}{2}} \sin x$ and $x^{-\frac{1}{2}} \cos x$. Find the general solution of

$$x^2 y'' + xy' + (x^2 - \tfrac{1}{4})y = 3x^{\frac{3}{2}} \sin x, \qquad x > 0.$$

11. Verify that e^x and x are solutions of the homogeneous equation corresponding to

$$(1-x)y'' + xy' - y = 2(x-1)^2 e^{-x}, \qquad 0 < x < 1,$$

and find the general solution.

12. Find a formula for a particular solution of the differential equation

$$y'' - 5y' + 6y = g(x).$$

13. Find a formula for a particular solution of the differential equation

$$x^2 y'' + xy' + (x^2 - \tfrac{1}{4})y = g(x), \qquad x > 0.$$

(See Problem 10.)

*14. Consider the problem of solving the initial value problem

$$y'' + y = g(t), \qquad y(0) = 0, \qquad y'(0) = 0,$$

where a prime denotes differentiation with respect to time t.
(a) Show that the general solution of $y'' + y = g(t)$ is

$$y = \phi(t) = \left(c_1 - \int_\alpha^t g(s) \sin s \; ds\right) \cos t + \left(c_2 + \int_\beta^t g(s) \cos s \; ds\right) \sin t$$

where c_1 and c_2 are arbitrary constants and α and β are any conveniently chosen points
(b) Using (a) show that $y(0) = 0$ and $y'(0) = 0$ if

$$c_1 = \int_\alpha^0 g(s) \sin s \, ds, \qquad c_2 = -\int_\beta^0 g(s) \cos s \, ds;$$

then show that the solution of the above initial value problem for arbitrary $g(t)$ can be written as

$$y = \phi(t) = \int_0^t g(s) \sin (t - s) \, ds.$$

Notice that this equation gives a formula for computing the solution of the original initial value problem for any given nonhomogeneous term $g(t)$. The function ϕ not only satisfies the differential equation but will also *automatically* satisfy the initial conditions. The formula also shows the relation between the input $g(t)$ and the output $\phi(t)$. Further, we see that the output at time t_0 depends only on the behavior of the input from the initial time 0 to the time of interest t_0. This integral is often referred to as the *convolution* of $\sin t$ and $g(t)$.

(c) Now that we have the solution of the linear nonhomogeneous differential equation satisfying homogeneous initial conditions we can solve the same problem with nonhomogeneous initial conditions by superimposing a solution of the homogeneous equation satisfying the nonhomogeneous initial conditions. Show that the solution of $y'' + y = g(t)$ with $y(0) = y_0$, $y'(0) = y'_0$ is

$$y = \phi(t) = \int_0^t g(s) \sin (t - s) \, ds + y_0 \cos t + y'_0 \sin t.$$

15. Provided that one solution of the homogeneous equation is known, the non-homogeneous equation

$$y'' + p(x)y' + q(x)y = g(x) \qquad \text{(i)}$$

can also be solved by the method of reduction of order (see Section 3.4). Suppose that y_1 is a known solution of the homogeneous equation.
(a) Show that $y = y_1(x)v(x)$ is a solution of Eq. (i) provided that v satisfies

$$y_1 v'' + (2y_1' + py_1)v' = g. \qquad \text{(ii)}$$

(b) Equation (ii) is a first order linear equation for v'. Show that the solution is

$$y_1^2(x)h(x)v'(x) = \int^x y_1(s)h(s)g(s) \, ds + k,$$

where $h(x) = \exp\left[\int^x p(s) \, ds\right]$ and k is a constant.
(c) Using part (b) show that the general solution of Eq. (i) is

$$y = c_1 y_1(x) + c_2 y_1(x) \int^x \frac{ds}{y_1^2(s)h(s)} + y_1(x) \int^x \frac{1}{y_1^2(s)h(s)} \left[\int^s y_1(t)h(t)g(t) \, dt\right] ds.$$

16. Using the procedure developed in Problem 15 solve the following differential equations.

(a) $x^2y'' - 2xy' + 2y = 4x^2$, $x > 0$; $y_1(x) = x$

(b) $x^2y'' + 7xy' + 5y = x$, $x > 0$; $y_1(x) = x^{-1}$

3.7 MECHANICAL VIBRATIONS

In Sections 3.1 through 3.6.2 of this chapter methods of solving second order linear ordinary differential equations have been developed. We found that when the differential equation has constant coefficients it can be solved in a straightforward manner. Two important areas of application of this theory are in the fields of mechanical and electrical oscillations where a large number of phenomena are governed by equations of the form

$$m\frac{d^2u}{dt^2} + c\frac{du}{dt} + ku = F(t). \tag{1}$$

Here m, c, and k are constants, F is a given function, and $u = \phi(t)$ represents the response of a certain physical system as a function of the time t. In particular, this equation governs the motion of a vibrating mass at the end of a vertical spring. The derivation of Eq. (1) for this problem is worth considering in detail since the principles are common to many problems.

First consider the case of the static elongation of a spring of natural length l due to the addition of a mass m. Let Δl denote the elongation (see Figure 3.1). The forces acting on the mass are the force of gravity acting downward, $mg = w$, where g is the acceleration due to gravity and w is the weight of the mass, and the spring force acting upward. Since the mass is in equilibrium in this position these forces are numerically equal. If the displacement Δl is small compared to the length l, the spring force, according to Hooke's law,* is proportional to Δl and hence has

FIGURE 3.1

---

* Hooke (1635–1703) first published his law in 1676 as an anagram: *ceiiinosssttuv*; and in 1678 gave the solution *ut tensio sic vis* which means, roughly, "as the force so is the displacement."

magnitude $k\Delta l$. The constant k depends on the spring and is called the spring constant. For a known weight w it can be computed by measuring Δl and using

$$k\Delta l = mg. \tag{2}$$

Notice that k has the dimensions of $mg/\Delta l$—that is, force/length.

In the corresponding dynamic problem we are interested in studying the motion of the mass when it is acted on by an external force or is initially displaced. Let u, measured positive downward, denote the displacement of the mass from the equilibrium position. The displacement u depends on t and is related to the forces acting on the mass through Newton's law, which states that md^2u/dt^2 must have the same magnitude and be in the same direction as the net resultant force acting on the mass m. In deriving the equation governing the motion of the mass, care must be exercised to determine the correct signs of the various forces that act on the mass. In the derivation below we adopt the convention that a force in the downward direction is positive, while a force in the upward direction is negative. Four forces act on the mass:

1. Its weight $w = mg$, which always acts downward.

2. The force due to the spring F_s, which is proportional to the elongation $\Delta l + u$ and always acts to restore the spring to its natural position. If $\Delta l + u > 0$, the spring is extended and the force, which is directed upward, is given by $F_s(t) = -k(\Delta l + u)$. Note that the magnitude of the force on the mass due to the spring is $k(\Delta l + u)$; the minus sign tells us that this force is directed upward in this particular case. If $u + \Delta l < 0$ (that is, $-u > \Delta l$), the spring is compressed a distance* $(-u) - (\Delta l)$ and the force, which is directed downward, is $k[(-u) - (\Delta l)]$. Hence for all u

$$F_s(t) = -k(\Delta l + u). \tag{3}$$

We emphasize again that Eq. (3) gives not only the magnitude but also the direction of the force exerted on the mass by the spring at any position u.

3. The damping or resisting force $F_d(t)$. This force may be due to the viscous properties of the fluid in which the mass is moving (air resistance, for instance); or the moving mass may be attached to an oil or dashpot mechanism (see Figure 3.2.) In either case the damping force will act in a direction opposite to the direction of motion of the mass. We know from experimental evidence that as long as the speed of the mass is not too large the resistance may be taken to be proportional to the speed $|du/dt|$. If the mass is moving downward, u is increasing and hence du/dt is positive, so $F_d(t)$, which is directed upward, is given by $-c(du/dt)$. The positive constant c is referred to as the damping constant; it has the units of force/velocity. If the mass is moving up-

FIGURE 3.2   ward, u is decreasing, du/dt is negative, and $F_d(t)$, which is now

* Remember that if the mass is above the equilibrium position its *distance* above the equilibrium position is $-u$, not u.

directed downward, is again given by $-c(du/dt)$. Hence for all u

$$F_d(t) = -c \frac{du}{dt}. \tag{4}$$

4. An applied force $F(t)$, which is directed downward or upward as $F(t)$ is positive or negative. This could be a force due to the motion of the mount to which the spring is attached, or it could be a force applied directly to the mass.

According to Newton's law

$$mg + F_s(t) + F_d(t) + F(t) = m \frac{d^2u}{dt^2}.$$

Substituting for $F_s(t)$ and $F_d(t)$ gives

$$mg - k(\Delta l + u) - c \frac{du}{dt} + F(t) = m \frac{d^2u}{dt^2}. \tag{5}$$

Since $k\Delta l = mg$, Eq. (5) reduces to the differential equation (1),

$$m\ddot{u} + c\dot{u} + ku = F(t).$$

In writing this equation we have followed a common convention of indicating time derivatives by dots.

It is important to remember that in Eq. (1), the mass m, the damping constant c, and the spring constant k are all *positive* constants. It is also important to recognize that Eq. (1) represents only an approximate equation for the determination of u. In reality the relationship between $F_s(t)$ and $\Delta l + u$ is not linear; however, for most of the usual materials, such as steel, and for $(\Delta l + u)/l \ll 1$, the relationship is, to a high degree of accuracy, approximately linear. Similarly, the expression for the damping force is an approximate one. Also we have neglected the mass of the spring compared to the mass of the attached body in deriving Eq. (1). Fortunately, for a wide variety of applications, Eq. (1) is a satisfactory mathematical model of the physical system.

The complete formulation of the vibration problem requires the specification of two initial conditions—the initial displacement u_0, and the initial velocity $\dot{u}_0$. It may not be obvious from physical reasoning that we can only specify two initial conditions, or that these are the two that should be specified. However, it does follow from the existence and uniqueness Theorem 3.2 that these conditions give a mathematical problem that has a unique solution. The solution of Eq. (1) satisfying these initial conditions gives the position of the mass as a function of time.

Example. Derive the initial value problem for the following spring-dashpot-mass system. It is known that a 5-lb weight stretches the spring 1 in. The dashpot mechanism exerts a force of 0.02 lb for a velocity of 2 in./sec. A 2-lb weight is attached to the spring and released from a position 2 in. below equilibrium.

We have

$$m = \frac{w}{g} = \frac{2 \text{ lb}}{32 \text{ ft/sec}^2} = \frac{1}{16} \frac{\text{lb-sec}^2}{\text{ft}},$$

$$c = \frac{0.02 \text{ lb}}{2 \text{ in./sec}} = \frac{0.02 \text{ lb}}{(1/6) \text{ ft/sec}} = 0.12 \frac{\text{lb-sec}}{\text{ft}},$$

$$k = \frac{5 \text{ lb}}{1 \text{ in.}} = \frac{5 \text{ lb}}{(1/12) \text{ ft}} = 60 \frac{\text{lb}}{\text{ft}};$$

and hence

$$\tfrac{1}{16}\ddot{u} + 0.12\dot{u} + 60u = 0,$$

or

$$\ddot{u} + 1.92\dot{u} + 960u = 0,$$

where u is measured in feet and t in seconds. The initial conditions are $u(0) = \tfrac{1}{6}$, $\dot{u}(0) = 0$.

3.7.1 FREE VIBRATIONS

Undamped Free Vibrations. If there is no external force and no damping, then Eq. (1) from the previous section reduces to

$$m\ddot{u} + ku = 0. \tag{1}$$

The solution of this equation is

$$u = A \cos \omega_0 t + B \sin \omega_0 t, \tag{2}$$

where

$$\omega_0^2 = k/m, \tag{3}$$

and ω_0, called the circular frequency, clearly has the units of 1/time. For a particular problem, the constants A and B are determined by the prescribed initial conditions.

In discussing the solution of Eq. (1) it is convenient to write Eq. (2) in the form

$$u = R \cos (\omega_0 t - \delta). \tag{4}$$

Regardless of the values of A and B this can always be done by taking $R = (A^2 + B^2)^{1/4}$ and solving the equations $A = R \cos \delta$ and $B = R \sin \delta$ for δ.

Because of the periodic character of the cosine function, Eq. (4) represents a periodic motion, or simple harmonic motion, of period

$$T = \frac{2\pi}{\omega_0} = 2\pi \left(\frac{m}{k}\right)^{1/2}. \tag{5}$$

To distinguish between problems of free vibrations without damping and problems of damped vibrations or forced vibrations, the circular frequency ω_0 and period $2\pi/\omega_0$ of the undamped, unforced motion are usually referred to as the natural circular frequency and natural period of the vibration.

Since $|\cos(\omega_0 t - \delta)| \le 1$, u always lies between the lines $u = \pm R$. The maximum displacement occurs at the times $\omega_0 t - \delta = 0, \pm\pi, \pm 2\pi, \ldots$. The constants R and δ are referred to as the amplitude and phase angle, respectively, of the motion. A sketch of the motion represented by Eq. (4) is shown in Figure 3.3. Notice that this periodic motion does not die out with increasing time. This is a consequence of neglecting the damping force, which would allow a means of dissipating the energy supplied to the system by the original displacement and velocity.

FIGURE 3.3 Simple harmonic motion.

Example. Find the natural period of a spring-mass system if the mass weighs 10 lb and stretches a steel spring 2 in. If the spring is stretched an additional 2 in. and then released, determine the subsequent motion.

The spring constant is $k = 10\,\text{lb}/2\,\text{in.} = 60\,\text{lb/ft}$. The mass $m = w/g = \frac{10}{32}$ lb/(ft/sec²). Hence

$$T = 2\pi \left(\frac{m}{k}\right)^{1/2} = 2\pi \left[\frac{10}{(32)60}\right]^{1/2} = \frac{\pi\sqrt{3}}{12}\ \text{sec}.$$

The equation of motion is

$$\tfrac{10}{32}\ddot{u} + 60u = 0;$$

hence

$$u = A \cos(8\sqrt{3}t) + B \sin(8\sqrt{3}t).$$

The solution satisfying the initial conditions $u(0) = \tfrac{1}{6}$ ft and $\dot{u}(0) = 0$ ft/sec is

$$u = \tfrac{1}{6}\cos(8\sqrt{3}t),$$

where t is measured in seconds and u is measured in feet.

Damped Free Vibrations. If we include the effect of damping, the differential equation governing the motion of the mass is

$$m\ddot{u} + c\dot{u} + ku = 0. \tag{6}$$

The roots of the corresponding characteristic equation are

$$r_1, r_2 = \frac{-c \pm \sqrt{c^2 - 4km}}{2m} = -\frac{c}{2m} \pm \left(\frac{c^2}{4m^2} - \frac{k}{m}\right)^{1/2}. \tag{7}$$

Since c, k, and m are positive, $c^2 - 4km$ is always less than c^2; hence if $c^2 - 4km \geq 0$ the values of r_1 and r_2 given by Eq. (7) are *negative*. If $c^2 - 4km < 0$ the values of r_1 and r_2 given by Eq. (7) are complex, but with *negative* real part. In tabular form the solution of Eq. (6) is

$$c^2 - 4km > 0, \qquad u = Ae^{r_1 t} + Be^{r_2 t}, \qquad r_1 \text{ and } r_2 < 0; \tag{8}$$
$$c^2 - 4km = 0, \qquad u = (A + Bt)e^{-(c/2m)t}; \tag{9}$$
$$c^2 - 4km < 0, \qquad u = e^{-(c/2m)t}(A \cos \mu t + B \sin \mu t),$$
$$\mu = (4km - c^2)^{1/2}/2m > 0. \tag{10}$$

In all three cases, regardless of the initial conditions, that is, regardless of the values of A and B, $u \to 0$ as $t \to \infty$, and hence the motion dies out with increasing time. Thus the solutions of Eqs. (1) and (6) confirm what we would intuitively expect—without damping the motion always continues unabated and with damping the motion must decay to zero with increasing time.

The first two cases, Eqs. (8) and (9), which are referred to as *overdamped* and *critically damped* respectively, represent motions in which the originally displaced mass "creeps" back to its equilibrium position. Depending on the initial conditions, it may be possible to overshoot the equilibrium position once, but this is not a "vibratory motion." Two typical examples of such a motion are sketched in Figure 3.4. The theory is discussed further in Problems 15 and 16.

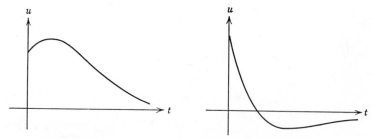

FIGURE 3.4 Overdamped or critically damped motions.

The third case, which is referred to as an *underdamped motion*, often occurs in mechanical systems and represents a "damped vibration." To see this, we let $A = R \cos \delta$ and $B = R \sin \delta$ in Eq. (10), and we obtain

$$u = Re^{-(c/2m)t} \cos (\mu t - \delta). \tag{11}$$

The displacement u must lie between the curves $u = \pm Re^{-(c/2m)t}$, and hence

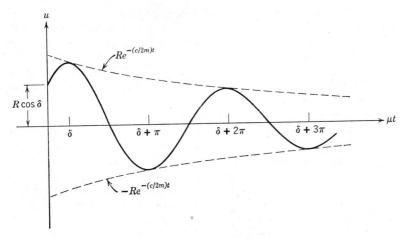

FIGURE 3.5 Damped vibration.

resembles a cosine curve with decreasing amplitude. A typical example is sketched in Figure 3.5.

While the motion is not truly periodic, we can define a *quasi-period* $T_d = 2\pi/\mu$ as the time between successive maxima of the displacement. It is interesting to note the relation between T_d and T. We have

$$T_d = \frac{2\pi}{\mu} = 2\pi\left(\frac{k}{m} - \frac{c^2}{4m^2}\right)^{-\frac{1}{2}} = 2\pi\left(\frac{k}{m}\right)^{-\frac{1}{2}}\left(1 - \frac{c^2}{4km}\right)^{-\frac{1}{2}}$$

$$= T\left(1 - \frac{c^2}{4km}\right)^{-\frac{1}{2}}.$$

Hence if $c^2/4km$ is small,

$$T_d \cong T\left(1 + \frac{c^2}{8km}\right).$$

Consequently, when the damping is very small (that is, when the dimensionless quantity $c^2/4km \ll 1$), we can neglect damping in computing the quasi-period of vibration. In many physical problems this is precisely the situation. On the other hand, if we want to study the detailed motion of the mass for all time, we can *never* neglect the damping force, no matter how small. The importance of the damping force will be seen even more emphatically in the next section where we discuss forced vibrations.

PROBLEMS

1. A 2-lb weight stretches a 2-ft spring 6 in. If the weight is pulled down an additional 3 in. and released, determine the subsequent motion neglecting air resistance. What are the amplitude, circular frequency, and period of the motion?

2. A mass of 100 gm is attached to a steel spring of natural length 50 cm. The spring is stretched 5 cm by the addition of this mass. If the mass is started in motion with a velocity of 10 cm/sec in the downward direction, determine the subsequent motion. Neglect air resistance.

3. A 3-lb weight stretches a spiral spring 3 in. If the weight is pushed upward, contracting the spring a distance 1 in., and then released with a velocity in the downward direction of 2 ft/sec, determine the subsequent motion. Neglect air resistance. What are the amplitude, circular frequency, and period of the motion?

4. Show that the period of motion of an undamped vibration of a mass hanging from a vertical spring is $2\pi\sqrt{\Delta l/g}$ where Δl is the elongation due to the weight w.

5. In Problem 3 suppose air resistance exerts a damping force which is given by $0.3\dot{u}$ lb. Compare the quasi-period of the damped motion with the natural period of the undamped motion. What is the percentage change based on the natural period?

6. In Problem 5, how large would the damping coefficient have to be in order to cause a 50 percent change in the period of the motion?

7. Show that $A\cos\omega_0 t + B\sin\omega_0 t$ can be written in the form $r\sin(\omega_0 t - \theta)$. Determine r and θ in terms of A and B. What is the relationship between r, R, θ, and δ if $R\cos(\omega_0 t - \delta) = r\sin(\omega_0 t - \theta)$?

8. Determine the solution of the differential equation $m\ddot{u} + ku = 0$ satisfying the following initial conditions. Express your answer in the form $R\cos(\omega_0 t - \delta)$.

(a) $u(0) = u_0$, $\dot{u}(0) = 0$;   (b) $u(0) = 0$, $\dot{u}(0) = \dot{u}_0$;   (c) $u(0) = u_0$, $\dot{u}(0) = \dot{u}_0$.

9. Determine the natural period of oscillation of the pendulum of length l shown in Figure 3.6. Neglect air resistance and the weight of the string compared to the weight of the mass. Assume that θ is small enough that $\sin\theta \simeq \theta$.
Hint: Write Newton's law for the tangential component of the motion.

10. A cubic block of side l and mass density ρ per unit volume is floating in a fluid of mass density $\hat{\rho}$ per unit volume. If the block is slightly depressed and then released it oscillates in the vertical direction. Assuming that the viscous damping of the fluid and air can be neglected, derive the differential equation of motion and determine the period of the motion.

FIGURE 3.6

11. A mass of 20 gm stretches a spiral spring a distance of 5 cm. If the spring is attached to an oil dashpot damper with a damping constant of 400 dyne sec/cm, determine the subsequent motion if the mass is pulled down an additional 2 cm and released.

12. A 16-lb weight stretches a spring 3 in. The weight is attached to a dashpot mechanism with a damping constant of 2 lb sec/ft. Determine the subsequent motion if the weight is released from its equilibrium position with a velocity of 3 in./sec in the downward direction.

13. A weight of 8 lb is attached to a steel spring of length 15 in. The weight extends the spring a distance of $\frac{1}{8}$ ft. If the weight is also attached to an oil dashpot damper with a damping constant c, determine the values of c for which the motion will be overdamped, critically damped, and underdamped. Be sure to give the units for c.

14. Determine a solution of the differential equation $m\ddot{u} + c\dot{u} + ku = 0$ in the case $c^2 - 4km < 0$ satisfying the following initial conditions. Express your answer in the form

$Re^{-(c/2m)t} \cos{(\mu t - \delta)}$.

(a) $u(0) = u_0$, $\dot{u}(0) = 0$;   (b) $u(0) = 0$, $\dot{u}(0) = \dot{u}_0$;   (c) $u(0) = u_0$, $\dot{u}(0) = \dot{u}_0$.

15. Show that if the damping constant c in the equation $m\ddot{u} + c\dot{u} + ku = 0$ is such that the motion is overdamped or critically damped, then the mass can pass through its zero displacement position at most once, regardless of the initial conditions.
Hint: Determine all possible values of t such that $u = 0$.

16. For the case of critical damping, find the general solution of Eq. (6) satisfying the initial conditions $u(0) = u_0$, $\dot{u}(0) = \dot{u}_0$. If $\dot{u}_0 = 0$, show that $u \to 0$ as $t \to \infty$, but $u \neq 0$ for any finite value of t. Assuming u_0 is positive, determine a condition on $\dot{u}_0$ that will insure that the mass passes through the zero displacement position after it is released.

*17. For the damped oscillation given by Eq. (11) the time between successive maxima is $T_d = 2\pi/\mu$. The ratio of the displacement at two successive maxima (time t and time $t + T_d$) is given by $e^{(c/2m)T_d}$. Thus successive displacements at time intervals T_d form a geometric progression with ratio $e^{(c/2m)T_d}$. The natural logarithm of this ratio, denoted by $\Delta = (c/2m)T_d = (\pi c/m\mu)$, is referred to as the *logarithmic decrement*. Since m, μ, and Δ are quantities which can easily be measured for a mechanical system, this equation provides a convenient and *practical* method of computing the damping constant of the system. In particular, for the motion of a vibrating mass in a viscous fluid the damping constant depends on the viscosity of the fluid; for simple geometric shapes the form of this dependence is known, and the above relation allows the determination of the viscosity experimentally. This is one of the most accurate ways of determining the viscosity of a gas at high pressure.

(a) In Problem 12, what is the logarithmic decrement?

(b) In Problem 13 if the measured logarithmic decrement is 3, and $T_d = 0.3$ sec, determine the damping constant c.

3.7.2 FORCED VIBRATIONS

Consider now the case in which a periodic external force, say $F_0 \cos{\omega t}$, is applied to a spring-mass system. In this case the equation of motion is

$$m\ddot{u} + c\dot{u} + ku = F_0 \cos{\omega t}. \tag{1}$$

First suppose there is no damping; then Eq. (1) reduces to

$$m\ddot{u} + ku = F_0 \cos{\omega t}. \tag{2}$$

Provided $\omega_0 = \sqrt{k/m} \neq \omega$, the general solution of Eq. (2) is

$$u = c_1 \cos{\omega_0 t} + c_2 \sin{\omega_0 t} + \frac{F_0}{m(\omega_0^2 - \omega^2)} \cos{\omega t}. \tag{3}$$

The constants c_1 and c_2 are determined by the initial conditions. The resultant motion is, in general, the sum of two periodic motions of different frequencies (ω_0 and ω) and amplitudes. We consider two examples.

Beats. Suppose the mass is initially at rest, that is, $u(0) = 0$, $\dot{u}(0) = 0$. Then it turns out that the constants c_1 and c_2 in Eq. (3) are given by

$$c_1 = \frac{-F_0}{m(\omega_0^2 - \omega^2)}, \qquad c_2 = 0, \tag{4}$$

and the solution of Eq. (2) is

$$u = \frac{F_0}{m(\omega_0{}^2 - \omega^2)} (\cos \omega t - \cos \omega_0 t).$$
(5)

This is the sum of two periodic functions of different periods but the same amplitude. Making use of the trigonometric identities $\cos (A \pm B) = \cos A \cos B \mp \sin A \sin B$ with $A = (\omega_0 + \omega)t/2$ and $B = (\omega_0 - \omega)t/2$, we can write Eq. (5) in the form

$$u = \frac{2F_0}{m(\omega_0{}^2 - \omega^2)} \sin \frac{(\omega_0 - \omega)t}{2} \sin \frac{(\omega_0 + \omega)t}{2}.$$
(6)

If $|\omega_0 - \omega|$ is small, then $\omega_0 + \omega \gg |\omega_0 - \omega|$, and consequently $\sin (\omega_0 + \omega)t/2$ is a rapidly oscillating function compared to $\sin (\omega_0 - \omega)t/2$. Then the motion is a rapid oscillation with circular frequency $(\omega_0 + \omega)/2$, but with a slowly varying sinusoidal amplitude. This type of motion, possessing a periodic variation of amplitude, exhibits what is called a *beat*. Such a phenomenon occurs in acoustics when two tuning forks of nearly equal frequency are sounded simultaneously. In this case the periodic variation of amplitude is quite apparent to the unaided ear. In electronics the variation of the amplitude with time is called *amplitude modulation*. The function $u = \phi(t)$ given in Eq. (6) is sketched in Figure 3.7.

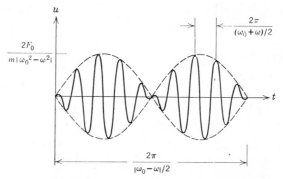

FIGURE 3.7   The phenomenon of beats.

Resonance.   As a second example, consider the case $\omega = \omega_0$; that is, the period of the forcing function is the same as the natural period of the system. Then the nonhomogeneous term $F_0 \cos \omega t$ is a solution of the homogeneous equation. In this case, the solution of Eq. (2) is

$$u = c_1 \cos \omega_0 t + c_2 \sin \omega_0 t + \frac{F_0}{2m\omega_0} t \sin \omega_0 t.$$
(7)

Because of the presence of the term $t \sin \omega_0 t$ in Eq. (7) it is clear, regardless of the

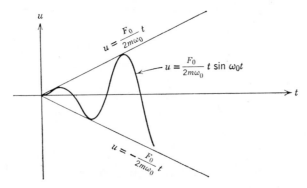

FIGURE 3.8   The phenomenon of resonance.

values of c_1 and c_2, that the motion will become unbounded as $t \to \infty$ (see Figure 3.8). This is the phenomenon known as *resonance*. In actual practice, however, the spring would probably break. Of course, as soon as u becomes large, our theory is no longer valid since we have assumed that u is small when we used a linear relation to determine the spring constant. If damping is included in the system the motion will remain bounded; however, there may still be a large response to the input function $F_0 \cos \omega t$ if the damping is small and ω is close to ω_0.

Damped Forced Vibrations.   The motion of the spring-mass system with damping and the forcing function $F_0 \cos \omega t$ can be determined in a straightforward manner. Although the computations are rather lengthy they are not difficult. The solution of Eq. (1) is

$$u = c_1 e^{r_1 t} + c_2 e^{r_2 t} + \frac{F_0}{\sqrt{m^2(\omega_0{}^2 - \omega^2)^2 + c^2 \omega^2}} \cos(\omega t - \delta), \qquad (8)$$

where δ is given by $\cos \delta = m(\omega_0{}^2 - \omega^2)/\Delta$ and $\sin \delta = c\omega/\Delta$ with $\Delta = \sqrt{m^2(\omega_0{}^2 - \omega^2)^2 + c^2\omega^2}$. Here r_1 and r_2 are the roots of the characteristic equation associated with Eq. (1). As was shown in the last section both $e^{r_1 t}$ and $e^{r_2 t}$ approach zero as $t \to \infty$. Hence as $t \to \infty$,

$$u \to u_p(t) = \frac{F_0}{\sqrt{m^2(\omega_0{}^2 - \omega^2)^2 + c^2 \omega^2}} \cos(\omega t - \delta). \qquad (9)$$

For this reason $u_c(t) = c_1 e^{r_1 t} + c_2 e^{r_2 t}$ is often called the *transient solution* and $u_p(t)$ is called the *steady state solution*. To speak roughly, the transient solution allows us to satisfy the imposed initial conditions; with increasing time the energy put into the system by the initial displacement and velocity is dissipated through the damping force, and the motion then represents the response of the system to the external force $F(t)$. Without damping (which, of course, is impossible in any physical system) the effect of the initial conditions would persist for all time.

Notice that $m^2(\omega_0{}^2 - \omega^2)^2 + c^2\omega^2$ is never zero, even for $\omega = \omega_0$; hence with damping the motion is always bounded. For fixed values of m, c, and k the amplitude of the steady state solution will be a maximum when $m^2(\omega_0{}^2 - \omega^2)^2 + c^2\omega^2$ is a minimum; this corresponds to choosing

$$\omega^2 = \omega_0{}^2 - \frac{1}{2}\left(\frac{c}{m}\right)^2. \tag{10}$$

In designing a spring-mass system* to detect periodic forces in a narrow frequency range around ω it is clear that we would want to choose k, c, and m in such a way that Eq. (10) is satisfied or nearly satisfied. In this way we obtain the maximum response of the system to such forces thus making their detection easier. A similar situation occurs in problems involving the detection of electrical signals in electrical networks.

PROBLEMS

1. Express $\cos 9t - \cos 7t$ in the form $A \sin \alpha t \sin \beta t$.
2. Express $\sin 7t - \sin 6t$ in the form $A \sin \alpha t \cos \beta t$.
3. A spiral spring is stretched $\frac{1}{8}$ ft by a weight of 4 lb. The spring is acted on by an external force $2 \cos 3t$ lb. If the weight is displaced a distance 2 in. from its position of equilibrium and then released, determine the subsequent motion. Neglect damping. If the external force is $4 \sin \omega t$ lb, for what value of ω will resonance occur?
4. If an undamped spring-mass system with weight 6 lb and spring constant 1 lb/in. is suddenly set in motion at $t = 0$ by an external force of $4 \cos 7t$ lb, determine the subsequent motion and draw a graph of the displacement versus t.
5. Determine the solution of the differential equation $m\ddot{u} + ku = F_0 \cos \omega t$, $\omega \neq \sqrt{k/m}$, satisfying the following initial conditions:

(a) $u(0) = u_0$, $\dot{u}(0) = 0$;   (b) $u(0) = 0$, $\dot{u}(0) = \dot{u}_0$;   (c) $u(0) = u_0$, $\dot{u}(0) = \dot{u}_0$.

6. A generalization of the phenomenon of beats occurs when the motion is the sum of two periodic functions of different amplitudes $A \cos \omega t$ and $B \cos \omega_0 t$. Show that such a motion can be expressed as

$$R(t) \cos\left[\frac{\omega_0 + \omega}{2} t - \delta(t)\right].$$

If $|A - B| \ll 1$ and $|\omega - \omega_0| \ll 1$, show that R is a slowly varying function. The variation of the phase angle δ with t is referred to as *frequency modulation*.

*7. Determine the motion $u = \phi(t)$ of a mass m hanging on a spring with spring constant k if the mass is acted on by a force

$$F(t) = F_0 \begin{cases} t, & 0 \le t \le \pi \\ 2\pi - t, & \pi < t \le 2\pi \\ 0, & 2\pi < t. \end{cases}$$

* Such instruments are usually referred to as seismic instruments since they are similar in principle to the seismograph used to detect motions of the earth's surface. The periodic motion of the mount to which the spring is attached is equivalent to a problem in which a periodic force acts on the mass.

For convenience assume $\omega_0^2 = k/m = 1$. Neglect air resistance and assume that the mass is initially at rest.

Hint: Treat each time interval separately, and match the solutions in the different intervals by requiring that ϕ and $\dot\phi$ be continuous functions of t.

8. A spiral spring is stretched 6 in. by a weight of 8 lb. The mass is attached to a dashpot mechanism which has a damping constant of 0.25 lb sec/ft. If the mass is subjected to an external force of the form $4 \cos 2t$ lb, determine the steady state response of the system. What is the optimum choice of the mass m in order to maximize the response of the system to this external force?

9. Determine the solution of the differential equation $m\ddot u + c\dot u + ku = F_0 \sin \omega t$ satisfying the following initial conditions. Assume $c^2 - 4km < 0$.

(a) $u(0) = u_0$, $\dot u(0) = 0$; (b) $u(0) = 0$, $\dot u(0) = \dot u_0$; (c) $u(0) = u_0$, $\dot u(0) = \dot u_0$.

3.8   ELECTRICAL NETWORKS

As a second example of the application of the theory of linear second order differential equations with constant coefficients, we consider the flow of electric current in a simple series circuit. The circuit is depicted in Figure 3.9. The current $I = \phi(t)$ (measured in amperes) is a function of the time t. The resistance R

FIGURE 3.9

(ohms), the capacitance C (farads), and the inductance L (henrys) are all positive, and in general may depend on the time t and the current I. For a wide variety of applications this dependence can be neglected, and we will assume that R, C, and L are known constants. The impressed voltage E (volts) is a given function of time—often of the form $E_0 \cos \omega t$. Another physical quantity that will appear in our discussion is the total charge $Q = \psi(t)$ (coulombs) on the capacitor at the time t. The charge Q is related to the current I by $I = dQ/dt$.

The flow of current in the circuit under consideration is governed by Kirchhoff's* second law: *In a closed circuit the impressed voltage is equal to the sum of the voltage drops in the rest of the circuit.*

---

* Gustav Kirchhoff (1824–1887), a German physicist, is famous for his contributions in electricity (the laws that bear his name) and for his work in spectroscopy.

According to the elementary laws of electricity we know that

The voltage drop across the resistance $= IR$,

The voltage drop across the capacitor $= \dfrac{1}{C} Q$,

The voltage drop across the inductance $= L \dfrac{dI}{dt}$.

Hence

$$L \frac{dI}{dt} + RI + \frac{1}{C} Q = E(t). \tag{1}$$

The units have been chosen so that 1 volt $=$ 1 ohm $\cdot$ 1 ampere $=$ 1 coulomb/ 1 farad $=$ 1 henry $\cdot$ 1 ampere/1 second.

Since $I = dQ/dt$ we can substitute for I in Eq. (1), obtaining the following second order linear nonhomogeneous equation for Q,

$$L\ddot{Q} + R\dot{Q} + \frac{1}{C} Q = E(t), \tag{2}$$

where dots denote differentiation with respect to t. The initial conditions that must be specified, say at time $t = 0$, are

$$Q(0) = Q_0, \qquad \dot{Q}(0) = I(0) = I_0. \tag{3}$$

Thus we must know the initial charge on the capacitor and the initial current in the circuit.

Alternatively, we can obtain a second order differential equation for the current I by differentiating Eq. (1) with respect to t, and then substituting for dQ/dt. This gives

$$L\ddot{I} + R\dot{I} + \frac{1}{C} I = \dot{E}(t). \tag{4}$$

The initial conditions that must be specified are

$$I(0) = I_0, \qquad \dot{I}(0) = \dot{I}_0. \tag{5}$$

Notice from Eq. (1) that

$$\dot{I}_0 = \frac{1}{L} \left[E(0) - RI_0 - \frac{1}{C} Q_0 \right]. \tag{6}$$

Hence $\dot{I}_0$ is known if we specify $I(0)$ and $Q(0)$. The initial charge and current are physically measureable quantities and a knowledge of these quantities allows us to determine the initial conditions (3) corresponding to Eq. (2) or the initial conditions (5) corresponding to Eq. (4).

It is interesting to note the analogy between the mechanical vibration problem discussed earlier and the present electric circuit problem. This is clearly shown in Table 3.2. This analogy between the two systems allows us to solve problems in mechanical vibrations by building the corresponding electrical circuit and then measuring the charge Q. This is one of the basic ideas in the field of analog computers.*

TABLE 3.2

Mechanical System	Electric Circuit
$m\ddot{u} + c\dot{u} + ku = F(t)$	$L\ddot{Q} + R\dot{Q} + \dfrac{1}{C}Q = E(t)$
Displacement u	Charge Q
Velocity $\dot{u}$	Current $I = \dot{Q}$
Mass m	Inductance L
Damping c	Resistance R
Spring constant k	Elastance $1/C$
Impressed force $F(t)$	Impressed voltage $E(t)$

It is apparent that the general initial value problem

$$a\ddot{y} + b\dot{y} + cy = g(t); \qquad y(0) = y_0, \qquad \dot{y}(0) = \dot{y}_0 \qquad (7)$$

for a linear second order equation with constant coefficients corresponds to at least two physical problems: the motion of a mass on a vibrating spring, and the flow of electric current in a simple series circuit. Actually, other physical problems, such as the torsional oscillation of a shaft with a flywheel, also lead to the initial value problem (7). This situation illustrates a fundamental relationship between mathematics and physics: *many physical problems, when formulated mathematically, are identical.* Thus the solution of several physical problems can be obtained from the solution of a single mathematical problem by an appropriate interpretation of the symbols.

Let us now consider Eq. (2) when the impressed voltage is a periodic function, $E_0 \cos \omega t$. Just as for the spring-mass system discussed in Sections 3.7.1 and 3.7.2 we can develop the concepts of a natural frequency, critical damping, resonance, beats, transient solution, and steady-state solution for the series circuit. This is done in the following example and in the problems.

Example. Determine the behavior of the solution of Eq. (2) as $t \to \infty$ if $E(t) = E_0 \cos \omega t$ and $R \neq 0$.

First, since L, R, and C are positive it follows that all solutions of the homogeneous equation corresponding to Eq. (2) approach zero as $t \to \infty$ (see Section 3.7.1). To find a particular solution of

$$L\ddot{Q} + R\dot{Q} + \dfrac{1}{C}Q = E_0 \cos \omega t \qquad (8)$$

* For a particularly simple introduction to the theory of analog computers, see T. D. Truitt and A. E. Rogers.

we use the method of undetermined coefficients. Substituting

$$Q_p(t) = A \cos \omega t + B \sin \omega t \tag{9}$$

we find that A and B must satisfy

$$\left(\frac{1}{C} - L\omega^2\right)A + \omega R B = E_0$$

$$-\omega R A + \left(\frac{1}{C} - L\omega^2\right)B = 0.$$

Solving for A and B and substituting for them in Eq. (9) gives

$$Q_p(t) = \frac{E_0(1/C - L\omega^2) \cos \omega t + \omega R E_0 \sin \omega t}{(1/C - L\omega^2)^2 + \omega^2 R^2}. \tag{10}$$

Equation (10) can also be written as

$$Q_p(t) = \frac{E_0 \sin(\omega t + \delta)}{\sqrt{(1/C - L\omega^2)^2 + \omega^2 R^2}}, \tag{11}$$

where δ is given by $\sin \delta = (1/C - L\omega^2)/\Delta$, $\cos \delta = \omega R/\Delta$ and $\Delta = \sqrt{(1/C - L\omega^2)^2 + \omega^2 R^2}$. Since $Q = Q_c(t) + Q_p(t)$ and $Q_c(t) \to 0$ as $t \to \infty$, it follows that $Q \to Q_p(t)$ as $t \to \infty$. For this reason $Q_p(t)$ is often referred to as the steady-state solution of Eq. (8).

Generally, we wish to know the steady-state current in the circuit, and this can be found by differentiating $Q_p(t)$. We obtain

$$I_p(t) = \frac{E_0 \cos(\omega t + \delta)}{\sqrt{R^2 + [\omega L - 1/(\omega C)]^2}}. \tag{12}$$

Notice that the steady-state current has the same frequency as the impressed voltage. The term $\omega L - 1/(\omega C)$ that appears in Eq. (12) is known as the *reactance* of the circuit, and the expression $\sqrt{R^2 + [\omega L - 1/(\omega C)]^2}$ is called the *impedance* of the circuit.

PROBLEMS

1. If there is no resistance present in a simple series circuit, show that in the absence of an impressed voltage the charge Q on the capacitor is periodic in time with frequency $\omega_0 = \sqrt{1/LC}$. The quantity $\sqrt{1/LC}$ is referred to as the natural frequency of the circuit.

2. Show that if there is no resistance in the circuit and the impressed voltage is of the form $E_0 \cos \omega t$, then the charge on the capacitor will become unbounded as $t \to \infty$ if $\omega = \sqrt{1/LC}$. This is the phenomenon of resonance. Show that the charge will always be bounded, no matter what the choice of ω, provided that there is some resistance, no matter how small, in the circuit.

3. Show that the solution of $L\ddot{I} + R\dot{I} + (1/C)I = 0$ is of the form

$$I = Ae^{r_1t} + Be^{r_2t}, \qquad \text{for } R^2 - 4L/C > 0;$$
$$= (A + Bt)e^{\lambda t}, \qquad \text{for } R^2 - 4L/C = 0;$$
$$= e^{\lambda t}(A \cos \mu t + B \sin \mu t) \quad \text{for } R^2 - 4L/C < 0.$$

The circuit is said to be overdamped, critically damped, or underdamped corresponding to these three cases, respectively.

4. Suppose that a series circuit consisting of an inductor, a resistor, and a capacitor is open, and there is an initial charge $Q_0 = 10^{-6}$ on the capacitor. Determine the variation of the charge and the current after the switch is closed for the following cases:

(a) $L = 0.2,$ $C = 10^{-5},$ $R = 3 \times 10^2$

(b) $L = 1,$ $C = 4 \times 10^{-6},$ $R = 10^3$

(c) $L = 2,$ $C = 10^{-5},$ $R = 4 \times 10^2$

5. If $L = 0.2$ henry and $C = 0.8 \times 10^{-6}$ farads, determine the resistance R so that the circuit is critically damped.

6. A series has a capacitor of 0.25×10^{-6} farad, a resistor of 5×10^3 ohms, and an inductor of 1 henry. The initial charge on the capacitor is zero. If a 12-volt battery is connected to the circuit and the circuit is closed at $t = 0$, determine the charge on the capacitor at $t = 0.001$ sec, at $t = 0.01$ sec, and the steady-state charge.

7. Determine the steady-state current in a series circuit if the impressed voltage is $E(t) = 110 \cos 120\pi t$ volts, and $L = 10$ henrys, $R = 3 \times 10^3$ ohms, and $C = 0.25 \times 10^{-5}$ farad.

8. For a series circuit with given values of L, R, and C and an impressed voltage $E_0 \cos \omega t$, for what value of ω will the steady-state current be a maximum?

*9. In electrical engineering practice it is often convenient to think of $E_0 \cos \omega t$ as the real part of $E_0 e^{j\omega t}(j = \sqrt{-1}$, which is standard notation in electrical engineering). Then corresponding to Eq. (8) of the text we would consider

$$L\ddot{Q} + R\dot{Q} + \frac{1}{C}Q = E_0 e^{jwt}, \tag{i}$$

where it is understood that when we finish our computations we must take the real part of the solution.* Assume that $R \neq 0$.

(a) Show that the steady-state solution of Eq. (i) is

$$Q = \frac{E_0}{j\omega R - \omega^2 L + 1/C} e^{jwt}.$$

Hint: Substitute $Q = Ae^{j\omega t}$.

* The use of complex notation for electrical circuit problems was pioneered by the mathematician and inventor Charles P. Steinmetz (1865–1923). The use of his symbolic notation in studying alternating-current phenomena was largely responsible for the rapid progress made in the commercial introduction of alternating current apparatus. He also discovered the law of hysteresis and made important contributions to the study of lightning phenomena. As a young man he was a socialist agitator who left Germany in 1888 just ahead of the police, and without getting his doctoral degree. He came to the United States in 1889 and was employed by what is now the General Electric Company.

(b) Hence show that the steady-state current is

$$I = \frac{E_0}{R + j(\omega L - 1/\omega C)} e^{jwt}.$$

(c) Using the fact that $\alpha + i\beta = \sqrt{\alpha^2 + \beta^2} e^{j\gamma}$ where $\gamma = \tan^{-1}(\beta/\alpha)$ show that

$$I = \frac{E_0}{\sqrt{R^2 + (\omega L - 1/\omega C)^2}} e^{j(wt-\delta)},$$

where $\delta = \tan^{-1}[(\omega L - 1/\omega C)/R]$. Finally, show that the real part of this expression is identical with that derived in the example of this section.

(d) What is the steady-state current if the impressed voltage is $E(t) = E_0 \sin \omega t$?

(e) The complex parameter $Z = R + j(\omega L - 1/\omega C)$ is known as the *complex impedance*. The reciprocal of Z is called the *admittance*, and the real and imaginary parts of $1/Z$ are called the *conductance* and *susceptance*. Determine the admittance, conductance, and susceptance.

REFERENCES

Coddington, E. A., *An Introduction to Ordinary Differential Equations*, Prentice-Hall, Englewood Cliffs, N.J., 1961.

Ince, E. L., *Ordinary Differential Equations*, Longmans, Green, London, 1927; Dover, New York, 1956.

Two books on analog computers are:

Truitt, T. D. and Rogers, A. E., *Basics of Analog Computers*, Rider, New York, 1960.

Jenners, Roger R., *Analog Computation and Simulation: Laboratory Approach*, Allyn and Bacon, Boston, 1965.

There are many books on mechanical vibrations and electric circuits. A classic book on mechanical vibrations is:

Den Hartog, J. P., *Mechanical Vibrations*, McGraw-Hill, New York, 1947.

More recent intermediate level books are:

Thomson, W. T., *Theory of Vibrations with Applications*, Prentice-Hall, Englewood Cliffs, N.J., 1972.

Vierck, R. K., *Vibration Analysis*, International, Scranton, Pa., 1967.

An elementary book with material on electrical circuits is:

Smith, J. R. J., *Circuits, Devices, and Systems*, Second Edition, Wiley, New York, 1976.

CHAPTER 4

Series Solutions of Second
Order Linear Equations

4.1 INTRODUCTION. REVIEW OF POWER SERIES

In Chapter 3 we described methods of solving second order linear differential equations with constant coefficients. We will now consider methods of solving second order linear equations when the coefficients are functions of the independent variable. It is sufficient to consider the homogeneous equation

$$P(x)\frac{d^2y}{dx^2} + Q(x)\frac{dy}{dx} + R(x)y = 0, \tag{1}$$

since the procedure for the corresponding nonhomogeneous equation is similar.

A wide class of problems in mathematical physics leads to equations of the form (1) having polynomial coefficients; for example, the Bessel equation

$$x^2y'' + xy' + (x^2 - \nu^2)y = 0,$$

where ν is a constant, and the Legendre equation

$$(1 - x^2)y'' - 2xy' + \alpha(\alpha + 1)y = 0,$$

where α is a constant. For this reason, as well as to simplify the algebraic computations, we will primarily consider the case in which the functions P, Q, and R are polynomials. However, as we will see, the method of solution is applicable for a class of functions more general than polynomials.

Although we will consider generalizations later, suppose for the present that P, Q, and R are polynomials. Suppose also that we are interested in solving Eq. (1) in the neighborhood of a point x_0, which is often, but not always, a point where initial conditions are specified. If the functions P, Q, and R have $x - x_0$ as a common factor we will assume that this common factor has been canceled from Eq. (1). Hence if $P(x_0) = 0$, then at least one of $Q(x_0)$ and $R(x_0)$ is not equal to zero. The solution of Eq. (1) in an interval containing x_0 is closely associated with the behavior of P in that interval. In particular, if P does not vanish in the interval, we can divide Eq. (1) by $P(x)$ to obtain $y'' + p(x)y' + q(x)y = 0$ where

145

$p(x) = Q(x)/P(x)$ and $q(x) = R(x)/P(x)$ are continuous functions. Hence the existence of a solution is guaranteed by Theorem 3.2. In Sections 4.2 and 4.2.1 we will consider the solution of Eq. (1) in the neighborhood of a point x_0 where $P(x_0) \neq 0$. Such a point is called an *ordinary point* of Eq. (1). On the other hand, if $P(x_0) = 0$, Theorem 3.2 does not apply since either $p(x)$ and/or $q(x)$ become unbounded as $x \to x_0$. Such a point is called a *singular point* of Eq. (1). Sections 4.3 through 4.7 deal with the determination of solutions of Eq. (1) in the neighborhood of a singular point.

In either case the method of solution is based on the idea of expressing y as an infinite series in powers of $x - x_0$, where x_0 is some specified point. While at first sight it may appear discouraging to think of finding solutions of differential equations in the form of infinite series, from a computational point of view this may be the most convenient way of solving the differential equation. Indeed, tables of $\sin x$ and $\cos x$ (the solutions of $y'' + y = 0$) can be computed from their Taylor series about $x = 0$:

$$\sin x = x - \frac{x^3}{3!} + \frac{x^5}{5!} + \cdots + (-1)^n \frac{x^{2n+1}}{(2n+1)!} + \cdots = \sum_{n=0}^{\infty} \frac{(-1)^n x^{2n+1}}{(2n+1)!}, \qquad (2)$$

$$\cos x = 1 - \frac{x^2}{2!} + \frac{x^4}{4!} + \cdots + (-1)^n \frac{x^{2n}}{(2n)!} + \cdots = \sum_{n=0}^{\infty} \frac{(-1)^n x^{2n}}{(2n)!}, \qquad (0! = 1) \quad (3)$$

or similar series.

Since we will be using infinite series a great deal in this chapter, it is worthwhile to summarize without proof some of the pertinent results concerning infinite series, and in particular, power series.

1. A power series $\sum_{n=0}^{\infty} a_n(x - x_0)^n$ is said to converge at a point x if

$$\lim_{m \to \infty} \sum_{n=0}^{m} a_n(x - x_0)^n$$

exists. It is clear that the series converges for $x = x_0$; it may converge for all x, or it may converge for some values of x and not for others.

2. The series $\sum_{n=0}^{\infty} a_n(x - x_0)^n$ is said to converge absolutely at a point x if the series

$$\sum_{n=0}^{\infty} |a_n(x - x_0)^n|$$

converges. It can be shown that if the series converges absolutely, then the series also converges; however, the converse is not necessarily true.

3. One of the most useful tests for the absolute convergence of a power series is the ratio test. If for a fixed value of x

$$\lim_{n \to \infty} \left| \frac{a_{n+1}(x - x_0)^{n+1}}{a_n(x - x_0)^n} \right| = |x - x_0| \lim_{n \to \infty} \left| \frac{a_{n+1}}{a_n} \right| = l,$$

the power series converges absolutely at that value of x if $l < 1$, and diverges if $l > 1$. If $l = 1$ the test is inconclusive.

Example 1.   For which values of x does the power series

$$\sum_{n=1}^{\infty} (-1)^{n+1} n (x - 2)^n$$

converge?

To test for convergence we use the ratio test. We have

$$\lim_{n \to \infty} \left| \frac{(-1)^{n+2}(n + 1)(x - 2)^{n+1}}{(-1)^{n+1} n (x - 2)^n} \right| = |x - 2| \lim_{n \to \infty} \frac{n + 1}{n} = |x - 2|.$$

According to Statement 3 the series converges absolutely for $|x - 2| < 1$, or $1 < x < 3$, and diverges for $|x - 2| > 1$. The values of x corresponding to $|x - 2| = 1$ are $x = 1$ and $x = 3$. The series diverges for each of these values of x since the nth term of the series does not approach zero as $n \to \infty$.

---

4. If the power series $\sum_{n=0}^{\infty} a_n (x - x_0)^n$ converges at $x = x_1$, it converges absolutely for $|x - x_0| < |x_1 - x_0|$; and if it diverges at $x = x_1$, it diverges for $|x - x_0| > |x_1 - x_0|$.

5. There is a number $\rho \geq 0$, called the *radius of convergence*, such that $\sum_{n=0}^{\infty} a_n (x - x_0)^n$ converges absolutely for $|x - x_0| < \rho$ and diverges for $|x - x_0| > \rho$. For a series that converges nowhere except at x_0, we define ρ to be zero; for a series that converges for all x, we say that ρ is infinite. The interval of convergence is depicted by the hatched lines in Figure 4.1.

$$x_0 - \rho \qquad\qquad x_0 \qquad\qquad x_0 + \rho$$

FIGURE 4.1

Example 2.   Determine the radius of convergence of the power series

$$\sum_{n=1}^{\infty} \frac{(x + 1)^n}{n 2^n}.$$

We apply the ratio test:

$$\lim_{n \to \infty} \left| \frac{(x + 1)^{n+1}}{(n + 1) 2^{n+1}} \frac{n 2^n}{(x + 1)^n} \right| = |x + 1| \lim_{n \to \infty} \frac{n}{2(n + 1)} = \frac{|x + 1|}{2}.$$

Thus the series converges absolutely for $|x + 1| < 2$, or $-3 < x < 1$, and diverges for $|x + 1| > 2$. The radius of convergence of the power series is $\rho = 2$. Finally, we check the end points of the interval of convergence. At $x = -3$ we have

$$\sum_{n=1}^{\infty} \frac{(-3 + 1)^n}{n2^n} = \sum_{n=1}^{\infty} \frac{(-1)^n}{n}$$

which converges, but does not converge absolutely. The series is said to converge conditionally at $x = -3$. At $x = 1$ the series becomes

$$\sum_{n=1}^{\infty} \frac{1}{n},$$

which diverges. To summarize, the series converges for $-3 \leq x < 1$, converges absolutely for $-3 < x < 1$, and has radius of convergence 2.

If $\sum_{n=0}^{\infty} a_n(x - x_0)^n$ and $\sum_{n=0}^{\infty} b_n(x - x_0)^n$ converge to $f(x)$ and $g(x)$, respectively, for $|x - x_0| < \rho$, $\rho > 0$, then the following are true for $|x - x_0| < \rho$.

6. The series can be added or subtracted termwise, and

$$f(x) \pm g(x) = \sum_{n=0}^{\infty} (a_n \pm b_n)(x - x_0)^n.$$

7. The series can be formally multiplied, and

$$f(x)g(x) = \left[\sum_{n=0}^{\infty} a_n(x - x_0)^n \right] \left[\sum_{n=0}^{\infty} b_n(x - x_0)^n \right] = \sum_{n=0}^{\infty} c_n(x - x_0)^n$$

where $c_n = a_0 b_n + a_1 b_{n-1} + \ldots + a_n b_0$. Further, if $g(x_0) \neq 0$, the series can be formally divided and

$$\frac{f(x)}{g(x)} = \sum_{n=0}^{\infty} d_n(x - x_0)^n,$$

although the calculation of the d_n is somewhat complicated. Also, in the case of division the radius of convergence of the resulting power series may be less than ρ.

8. The function f is continuous and has derivatives of all orders for $|x - x_0| < \rho$. Further, f', f'', ... can be computed by differentiating the series termwise; that is,

$$f'(x) = a_1 + 2a_2(x - x_0) + \cdots + na_n(x - x_0)^{n-1} + \cdots$$
$$= \sum_{n=1}^{\infty} na_n(x - x_0)^{n-1},$$

etc., and each of the series converges absolutely for $|x - x_0| < \rho$.

9. The value of a_n is given by

$$a_n = \frac{f^{(n)}(x_0)}{n!}.$$

The series is called the Taylor (1685–1731) series for the function f about $x = x_0$.

10. If $\sum\limits_{n=0}^{\infty} a_n(x - x_0)^n = \sum\limits_{n=0}^{\infty} b_n(x - x_0)^n$ for each x, then $a_n = b_n$, $n = 0$,

1, 2, ... In particular, if $\sum\limits_{n=0}^{\infty} a_n(x - x_0)^n = 0$ for each x, then $a_0 = a_1 = \ldots =$

$a_n = \ldots = 0$.

A function f that has a Taylor series expansion about $x = x_0$,

$$f(x) = \sum_{n=0}^{\infty} \frac{f^{(n)}(x_0)}{n!}(x - x_0)^n,$$

with a radius of convergence $\rho > 0$ is said to be *analytic* at $x = x_0$. According to the above statements, if f and g are analytic at x_0 then $f \pm g$, $f \cdot g$, and f/g (provided $g(x_0) \neq 0$) are analytic at $x = x_0$. The result that will be used most often in the following sections is that a polynomial is analytic at every point; thus sums, differences, products, and quotients (except at zeros of the denominator) of polynomials are analytic at every point.

Shift of Index of Summation.   The index of summation in an infinite series is a dummy parameter just as the integration variable in a definite integral is a dummy variable. For example,

$$\sum_{n=0}^{\infty} \frac{2^n x^n}{n!} = \sum_{j=0}^{\infty} \frac{2^j x^j}{j!}.$$

Just as we make changes of the variable of integration in a definite integral, we will find it convenient to make changes of summation indices in calculating series solutions of differential equations. We illustrate how to shift the summation index by several examples.

1. $\sum\limits_{n=2}^{\infty} a_n x^n = a_2 x^2 + a_3 x^3 + a_4 x^4 + \cdots + a_n x^n + \cdots$

$\qquad = a_{0+2} x^{0+2} + a_{1+2} x^{1+2} + \cdots + a_{m+2} x^{m+2} + \cdots$

$\qquad = \sum\limits_{m=0}^{\infty} a_{m+2} x^{m+2} = \sum\limits_{n=0}^{\infty} a_{n+2} x^{n+2}.$

We have shifted the counting index by $+2$ $(n \to n + 2)$ and have started counting 2 lower.

2. $\sum\limits_{n=2}^{\infty} (n + 2)(n + 1)a_n(x - a)^{n-2}.$

It may be desirable to write this series as a series with generic term $(x - a)^n$ rather than $(x - a)^{n-2}$. To accomplish this we shift the index by $+2$ $(n \to n + 2)$ and start counting 2 lower as in the previous example. We obtain

$$\sum_{n=0}^{\infty} (n + 4)(n + 3)a_{n+2}(x - a)^n.$$

It can be readily verified that the terms in the two series are exactly the same.

3. $x^2 \sum_{n=0}^{\infty} (n + r)a_n x^{n+r-1}$.

First we take the x^2 inside the summation obtaining

$$\sum_{n=0}^{\infty} (n + r)a_n x^{n+r+1}.$$

To write this series as a new series with generic term x^{n+r} we shift the index down by 1 $(n \rightarrow n - 1)$ and start counting 1 higher. Thus

$$\sum_{n=0}^{\infty} (n + r)a_n x^{n+r+1} = \sum_{n=1}^{\infty} (n + r - 1)a_{n-1} x^{n+r}.$$

Again, it can be readily verified that the terms in the two series are exactly the same.

4. As a final example we consider the implication of the equation

$$\sum_{n=1}^{\infty} na_n x^{n-1} = \sum_{n=0}^{\infty} a_n x^n.$$

We wish to use Statement 10 to determine the a_n. In order to equate coefficients of like powers of x, it is helpful to have the generic term in each series involve x^n. To accomplish this we shift the index of summation on the left-hand side by the rule $n \rightarrow n + 1$ and start counting one lower. We obtain

$$\sum_{n=0}^{\infty} (n + 1)a_{n+1} x^n = \sum_{n=0}^{\infty} a_n x^n.$$

According to Statement 10 we conclude that

$$(n + 1)a_{n+1} = a_n, \qquad n = 0, 1, 2, 3, \ldots,$$

or

$$a_{n+1} = \frac{1}{n + 1} a_n.$$

Hence

$$a_1 = \frac{1}{1} a_0, \qquad a_2 = \frac{1}{2} a_1 = \frac{1}{2 \cdot 1} a_0 = \frac{a_0}{2!}$$

$$a_3 = \frac{1}{3} a_2 = \frac{1}{3} \frac{a_0}{2!} = \frac{a_0}{3!}, \ldots,$$

and in general

$$a_n = \frac{1}{n!} a_0, \qquad n = 1, 2, \ldots.$$

Thus the original relation determines all of the coefficients in terms of a_0. Moreover,

$$\sum_{n=0}^{\infty} a_n x^n = \sum_{n=0}^{\infty} a_0 \frac{x^n}{n!} = a_0 \sum_{n=0}^{\infty} \frac{x^n}{n!},$$

and this is precisely $a_0 e^x$. In writing the above series we have used the common convention that $0! = 1$.

PROBLEMS

1. Determine the radius of convergence of each of the following power series.

(a) $\sum_{n=0}^{\infty} (x - 3)^n$

(b) $\sum_{n=0}^{\infty} \dfrac{n}{2^n} x^n$

(c) $\sum_{n=0}^{\infty} \dfrac{x^{2n}}{n!}$

(d) $\sum_{n=0}^{\infty} 2^n x^n$

(e) $\sum_{n=1}^{\infty} \dfrac{(2x + 1)^n}{n^2}$

(f) $\sum_{n=1}^{\infty} \dfrac{(x - x_0)^n}{n}$

2. Determine the Taylor series about the point x_0 for each of the following functions. Also determine the radius of convergence of the series.

(a) $\sin x$, $x_0 = 0$

(b) e^x, $x_0 = 0$

(c) x, $x_0 = 1$

(d) x^2, $x_0 = -1$

(e) $\ln x$, $x_0 = 1$

(f) $\dfrac{1}{1 + x}$, $x_0 = 0$

(g) $\dfrac{1}{1 - x}$, $x_0 = 0$

(h) $\dfrac{1}{1 - x}$, $x_0 = 2$

3. Given that $y = \sum_{n=0}^{\infty} nx^n$, compute y' and y'' and write out the first four terms of each series as well as the coefficient of x^n in the general term.

4. Given that $y = \sum_{n=0}^{\infty} a_n x^n$, compute y' and y'' and write out the first four terms of each series as well as the coefficient of x^n in the general term. Show that if $y'' = y$, then the coefficients a_0 and a_1 are arbitrary, and determine a_2 and a_3 in terms of a_0 and a_1. Show that $a_{n+2} = a_n/(n + 2)(n + 1)$, $n = 0, 1, 2, 3, \ldots$.

5. Verify the following:

(a) $\sum_{n=0}^{\infty} a_n(x - 1)^{n+1} = \sum_{n=1}^{\infty} a_{n-1}(x - 1)^n$

(b) $\sum_{n=2}^{\infty} n(n - 1)a_n x^{n-2} = \sum_{n=0}^{\infty} (n + 2)(n + 1)a_{n+2} x^n$

(c) $\sum_{n=0}^{\infty} a_n x^{n+2} = \sum_{n=2}^{\infty} a_{n-2} x^n$

(d) $\sum_{n=k}^{\infty} a_{n+m} x^{n+p} = \sum_{n=0}^{\infty} a_{n+m+k} x^{n+p+k}$, $x > 0$,

where k and m are given integers and p is a constant.

(e) $x^2 \sum_{n=0}^{\infty} (n+r)(n+r-1)a_n x^{n+r-2} + \alpha x \sum_{n=0}^{\infty} (n+r)a_n x^{n+r-1} + \beta x^2 \sum_{n=0}^{\infty} a_n x^{n+r}$

$= [r(r-1) + \alpha r]a_0 x^r + [(r+1)r + \alpha(r+1)]a_1 x^{r+1}$

$+ \sum_{n=2}^{\infty} [(n+r)(n+r-1)a_n + \alpha(n+r)a_n + \beta a_{n-2}]x^{n+r}, \qquad x > 0,$

where r is a constant.

6. Determine the a_n so that the equation

$$\sum_{n=1}^{\infty} na_n x^{n-1} + 2 \sum_{n=0}^{\infty} a_n x^n = 0$$

is satisfied. Try to identify the function represented by the series $\sum_{n=0}^{\infty} a_n x^n$.

4.2 SERIES SOLUTIONS NEAR AN ORDINARY POINT, PART I

In this section we will illustrate the method of solving

$$P(x)y'' + Q(x)y' + R(x)y = 0, \tag{1}$$

when the functions P, Q, and R are polynomials, by considering several specific examples. We are interested in solving Eq. (1) in the neighborhood of a point x_0 at which $P(x_0) \neq 0$; that is, x_0 is an ordinary point. Since $P(x_0) \neq 0$ and P is continuous, it follows that there is some interval about the point x_0 in which P is never zero;* in this interval Q/P and R/P are continuous functions. Hence according to the existence and uniqueness Theorem 3.2 there exists a unique solution of Eq. (1) satisfying the initial conditions $y(x_0) = y_0$, $y'(x_0) = y_0'$ for an arbitrary choice of y_0 and y_0'.

We will look for solutions of Eq. (1) of the form

$$y = a_0 + a_1(x - x_0) + \cdots + a_n(x - x_0)^n + \cdots = \sum_{n=0}^{\infty} a_n(x - x_0)^n. \tag{2}$$

We must consider two questions. The first is whether we can formally determine† the a_n so that y as given by Eq. (2) satisfies Eq. (1). The second is whether the series thus determined actually converges, and if so, for what values of $x - x_0$. If we can show that the series does converge for $|x - x_0| < \rho$, $\rho > 0$, then all the formal procedures such as termwise differentiation can be justified, and we will have constructed a solution of Eq. (1) that is valid for $|x - x_0| < \rho$.

We will postpone until Section 4.2.1 all consideration of more theoretical matters such as the radius of convergence of the series (2). For the present we

---

* Geometrically this is plausible, and it can be proven rigorously.
† By "formally determine" we mean "carry out all the algebraic computations necessary for the determination of a solution without justifying at each step that it is permissible to perform the computation."

will simply assume that such a solution exists, and show how to determine the a_n. The most practical way to do this is simply to substitute the series (2) and its derivatives for y, y', and y'' in Eq. (1); the a_n are then determined so that the differential equation is formally satisfied. The following examples will illustrate the procedure. The differential equations are also of considerable importance in their own right.

Example 1. We start with the problem of finding a series solution of the equation

$$y'' + y = 0, \qquad -\infty < x < \infty. \tag{3}$$

As we well know, two linearly independent solutions of this equation are $\sin x$ and $\cos x$. To obtain practice in finding a series solution we will illustrate how the $\sin x$ and $\cos x$ solutions can be obtained using power series. For Eq. (3), $P(x) = 1$, $Q(x) = 0$, and $R(x) = 1$; hence, $x = 0$ is an ordinary point.

We look for a solution of the form

$$y = a_0 + a_1 x + a_2 x^2 + \cdots + a_n x^n + a_{n+1}x^{n+1} + a_{n+2}x^{n+2} + \cdots = \sum_{n=0}^{\infty} a_n x^n. \tag{4}$$

Differentiating term by term yields

$$y' = a_1 + 2a_2 x + \cdots + na_n x^{n-1} + (n + 1)a_{n+1}x^n$$
$$+ (n + 2)a_{n+2}x^{n+1} + \cdots = \sum_{n=1}^{\infty} na_n x^{n-1}, \tag{5}$$

$$y'' = 2a_2 + \cdots + n(n - 1)a_n x^{n-2} + (n + 1)na_{n+1}x^{n-1}$$
$$+ (n + 2)(n + 1)a_{n+2}x^n + \cdots = \sum_{n=2}^{\infty} n(n - 1)a_n x^{n-2}. \tag{6}$$

Substituting the series (4) and (6) for y and y'' in Eq. (3) gives

$$\sum_{n=2}^{\infty} n(n - 1)a_n x^{n-2} + \sum_{n=0}^{\infty} a_n x^n = 0.$$

Next, in the first sum, we shift the index of summation by replacing n by $n + 2$ and starting the sum at 0 rather than 2. We obtain

$$\sum_{n=0}^{\infty} (n + 2)(n + 1)a_{n+2}x^n + \sum_{n=0}^{\infty} a_n x^n = 0$$

or

$$\sum_{n=0}^{\infty} [(n + 2)(n + 1)a_{n+2} + a_n]x^n = 0.$$

For this equation to be satisfied for all x it is necessary that the coefficient of each power of x be zero; hence, we conclude that

$$(n + 2)(n + 1)a_{n+2} + a_n = 0, \qquad n = 0, 1, 2, 3, \ldots. \tag{7}$$

Equation (7) is referred to as a *recurrence relation*. According to this relation the even a's ($a_0, a_2, a_4, \ldots$) and the odd a's ($a_1, a_3, a_5, \ldots$) are determined separately. We have

$$a_2 = -\frac{a_0}{2 \cdot 1} = -\frac{a_0}{2!}, \qquad a_4 = -\frac{a_2}{4 \cdot 3} = +\frac{a_0}{4!},$$

$$a_6 = -\frac{a_4}{6 \cdot 5} = -\frac{a_0}{6!}, \qquad \ldots,$$

and in general, if $n = 2k$, then

$$a_n = a_{2k} = \frac{(-1)^k}{(2k)!}\, a_0, \qquad k = 1, 2, 3, \ldots. \tag{8}$$

Similarly,

$$a_3 = -\frac{a_1}{2 \cdot 3} = -\frac{a_1}{3!}, \qquad a_5 = -\frac{a_3}{5 \cdot 4} = +\frac{a_1}{5!},$$

$$a_7 = -\frac{a_5}{7 \cdot 6} = -\frac{a_1}{7!}, \qquad \ldots,$$

and in general, if $n = 2k + 1$, then

$$a_n = a_{2k+1} = \frac{(-1)^k}{(2k+1)!}\, a_1, \qquad k = 1, 2, 3, \ldots. \tag{9}$$

Thus the series (4) takes the form

$$y = a_0 + a_1 x - \frac{a_0}{2!} x^2 - \frac{a_1}{3!} x^3 + \frac{a_0}{4!} x^4 + \frac{a_1}{5!} x^5$$

$$+ \cdots + \frac{(-1)^n a_0}{(2n)!} x^{2n} + \frac{(-1)^n a_1 x^{2n+1}}{(2n+1)!} + \cdots$$

$$= a_0 \left[1 - \frac{x^2}{2!} + \frac{x^4}{4!} + \cdots + \frac{(-1)^n}{(2n)!} x^{2n} + \cdots \right]$$

$$+ a_1 \left[x - \frac{x^3}{3!} + \frac{x^5}{5!} + \cdots + \frac{(-1)^n}{(2n+1)!} x^{2n+1} + \cdots \right]$$

$$= a_0 \sum_{n=0}^{\infty} \frac{(-1)^n}{(2n)!} x^{2n} + a_1 \sum_{n=0}^{\infty} \frac{(-1)^n}{(2n+1)!} x^{2n+1}. \tag{10}$$

The first series is exactly the Taylor series for $\cos x$; the second series is the Taylor series for $\sin x$. Thus, as we expected at the outset, we obtain the solution $y = a_0 \cos x + a_1 \sin x$. Notice that no conditions are imposed on a_0 and a_1; hence they are arbitrary. This is true in general for a series solution at an ordinary point; upon reflection this is not surprising. From Eq. (2) we see that y and y' evaluated at x_0 are a_0 and a_1, respectively. Since the initial conditions $y(x_0)$ and

$y'(x_0)$ can be chosen arbitrarily, it follows that a_0 and a_1 should be arbitrary until specific initial conditions are stated.

We conclude this example with an aside. We started out knowing that the solutions of Eq. (3) should be sin x and cos x. But suppose we had not known that and had simply solved Eq. (3) using series methods and had obtained the solution (10). In recognition of the fact that the differential equation (3) often occurs in applications we might decide to give the two solutions of Eq. (10) special names; perhaps,

$$\cos x \quad \text{for} \quad \sum_{n=0}^{\infty} \frac{(-1)^n}{(2n)!} x^{2n}, \quad \sin x \quad \text{for} \quad \sum_{n=0}^{\infty} \frac{(-1)^n}{(2n+1)!} x^{2n+1}.$$

Then we might ask what properties these functions have. First, it is reasonably obvious that $\sin 0 = 0$, $\cos 0 = 1$, $\cos(-x) = \cos x$, and $\sin(-x) = -\sin x$. Another formula that is easily derived is

$$\frac{d}{dx} \sin x = \frac{d}{dx} \sum_{n=0}^{\infty} \frac{(-1)^n}{(2n+1)!} x^{2n+1} = \sum_{n=0}^{\infty} \frac{(-1)^n(2n+1)}{(2n+1)!} x^{2n} = \sum_{n=0}^{\infty} \frac{(-1)^n}{(2n)!} x^{2n} = \cos x.$$

Similarly, $d(\cos x)/dx = -\sin x$. Moreover, calculating with the infinite series, we can derive all of the usual analytical and algebraic properties of the sine and cosine functions such as $\sin^2 x + \cos^2 x = 1$.

It is interesting that these very important functions in mathematical physics can be defined as solutions of a simple, but frequently occurring, second order linear differential equation, and their properties derived* without any recourse to the area in which we probably first learned about them—plane trigonometry. Precisely, the function sin x can be defined as the solution of the initial value problem $y'' + y = 0$, $y(0) = 0$, $y'(0) = 1$; similarly, the function cos x can be defined as the solution of the initial value problem $y'' + y = 0$, $y(0) = 1$, $y'(0) = 0$. Other initial value problems can also be used to define the same functions. For example, the nonlinear initial value problem $(y')^2 = 1 - y^2$, $y(0) = 0$, $y'(0) > 0$ defines the function sin x (see Problem 5).

Example 2. Find a series solution in powers of x of Airy's† equation

$$y'' = xy, \quad -\infty < x < \infty. \tag{11}$$

---

* Such an analysis is given in Section 24 of K. Knopp, *Theory and Applications of Infinite Series*, Hafner, New York, 1951.

† Sir George Airy (1801–1892), an English astronomer and mathematician, was made the Lucasian Professor of Mathematics at Cambridge at the age of 25 and was director of the Greenwich Observatory from 1835 to 1881. One reason Airy's equation is of interest is that for x negative the solutions are oscillatory, similar to trigonometric functions, and for x positive one solution grows exponentially and the other decays exponentially. Can you explain why it is reasonable to expect such behavior?

For this equation $P(x) = 1$, $Q(x) = 0$, and $R(x) = -x$; hence, $x = 0$ is an ordinary point. We assume that

$$y = \sum_{n=0}^{\infty} a_n x^n. \tag{12}$$

The series for y'' is given by Eq. (6); but, as explained in the previous example, we can rewrite it as

$$y'' = \sum_{n=0}^{\infty} (n + 2)(n + 1)a_{n+2} x^n. \tag{13}$$

Substituting the series (12) and (13) for y and y'' in Eq. (11), we obtain

$$\sum_{n=0}^{\infty} (n + 2)(n + 1)a_{n+2} x^n = x \sum_{n=0}^{\infty} a_n x^n = \sum_{n=0}^{\infty} a_n x^{n+1}. \tag{14}$$

Next, we shift the index of summation in the series on the right-hand side of the above equation by replacing n by $n - 1$ and starting the summation at 1 rather than zero. Thus, we have

$$2 \cdot 1 a_2 + \sum_{n=1}^{\infty} (n + 2)(n + 1)a_{n+2} x^n = \sum_{n=1}^{\infty} a_{n-1} x^n.$$

Again, for this equation to be satisfied for all x it is necessary that the coefficients of like powers of x be equal; hence $a_2 = 0$, and we obtain the recurrence relation

$$(n + 2)(n + 1)a_{n+2} = a_{n-1} \quad \text{for} \quad n = 1, 2, 3, \dots. \tag{15}$$

Since a_{n+2} is given in terms of a_{n-1}, it is clear that the a's will be determined in steps of three. Thus a_0 determines a_3, which in turn determines $a_6, \dots$; a_1 determines a_4, which in turn determines $a_7, \dots$; and a_2 determines a_5, which in turn determines $a_8, \dots$. Since $a_2 = 0$, we immediately conclude that $a_5 = a_8 = a_{11} = \dots = 0$.

For the sequence $a_0, a_3, a_6, a_9, \dots$ we set $n = 1, 4, 7, 10, \dots$ in the recurrence relation:

$$a_3 = \frac{a_0}{3 \cdot 2}, \quad a_6 = \frac{a_3}{6 \cdot 5} = \frac{a_0}{(6 \cdot 5)(3 \cdot 2)}, \quad a_9 = \frac{a_6}{(9 \cdot 8)} = \frac{a_0}{(9 \cdot 8)(6 \cdot 5)(3 \cdot 2)}, \dots$$

For this sequence of coefficients it is convenient to write a formula for a_{3n}, $n = 1, 2, 3, \dots$. The above results suggest the general formula

$$a_{3n} = \frac{1}{[(3n)(3n - 1)][(3n - 3)(3n - 4)] \cdots [6 \cdot 5][3 \cdot 2]} a_0, \quad n = 1, 2, \dots.$$

This expression can also be obtained by rewriting the recurrence relation (15) as follows. First replace n by $n - 2$ to obtain $n(n - 1)a_n = a_{n-3}$; next, to generate the coefficients corresponding to a_0, replace n by $3n$, which gives $(3n)(3n - 1)a_{3n} = a_{3n-3}$. Repeated application of this formula gives the above formula for a_{3n}.

For the sequence $a_1, a_4, a_7, a_{10}, \ldots$ we set $n = 2, 5, 8, 11, \ldots$ in the recurrence relation:

$$a_4 = \frac{a_1}{4 \cdot 3}, \qquad a_7 = \frac{a_4}{7 \cdot 6} = \frac{a_1}{(7 \cdot 6)(4 \cdot 3)}, \qquad a_{10} = \frac{a_7}{10 \cdot 9} = \frac{a_1}{(10 \cdot 9)(7 \cdot 6)(4 \cdot 3)}, \ldots$$

In the same way as before we find that

$$a_{3n+1} = \frac{1}{[(3n+1)(3n)][(3n-2)(3n-3)] \cdots [7 \cdot 6][4 \cdot 3]} a_1, \qquad n = 1, 2, 3, \ldots.$$

The solution of Airy's equation is

$$y = a_0 \left[1 + \frac{x^3}{3 \cdot 2} + \frac{x^6}{6 \cdot 5 \cdot 3 \cdot 2} + \cdots + \frac{x^{3n}}{(3n)(3n-1) \cdots 3 \cdot 2} + \cdots \right]$$

$$+ a_1 \left[x + \frac{x^4}{4 \cdot 3} + \frac{x^7}{7 \cdot 6 \cdot 4 \cdot 3} + \cdots + \frac{x^{3n+1}}{(3n+1)(3n)(3n-2)(3n-3) \cdots 4 \cdot 3} + \cdots \right]$$

$$= a_0 \left[1 + \sum_{n=1}^{\infty} \frac{x^{3n}}{(3n)(3n-1)(3n-3)(3n-4) \cdots 3 \cdot 2} \right]$$

$$+ a_1 \left[x + \sum_{n=1}^{\infty} \frac{x^{3n+1}}{(3n+1)(3n)(3n-2)(3n-3) \cdots 4 \cdot 3} \right]. \tag{16}$$

Because of the rapid growth of the denominators of the terms in the series (16), we might expect these series to have a large radius of convergence. Indeed, as we shall see in the next section, both of these series converge for all x. Also see Problem 4.

Assume for the moment that the series do converge for all x, and let y_1 and y_2 denote the functions defined by the first and second sets of brackets, respectively, in Eq. (16). Then, by first choosing $a_0 = 1$, $a_1 = 0$ and then $a_0 = 0$, $a_1 = 1$, it is clear that y_1 and y_2 are individually solutions of Eq. (11). Notice that y_1 satisfies the initial conditions $y_1(0) = 1$, $y_1'(0) = 0$ and y_2 satisfies the initial conditions $y_2(0) = 0$, $y_2'(0) = 1$. Thus $W(y_1, y_2)(0) = 1 \neq 0$, and consequently y_1 and y_2 are linearly independent. Hence the general solution of Airy's equation is

$$y = a_0 y_1(x) + a_1 y_2(x), \qquad -\infty < x < \infty.$$

Example 3. Find a solution of Airy's equation in powers of $x - 1$.

The point $x = 1$ is an ordinary point of Eq. (11), and thus we look for a solution of the form

$$y = \sum_{n=0}^{\infty} a_n (x - 1)^n.$$

Then

$$y' = \sum_{n=1}^{\infty} n a_n (x - 1)^{n-1} = \sum_{n=0}^{\infty} (n + 1) a_{n+1} (x - 1)^n,$$

and

$$y'' = \sum_{n=2}^{\infty} n(n-1)a_n(x-1)^{n-2} = \sum_{n=0}^{\infty} (n+2)(n+1)a_{n+2}(x-1)^n.$$

Substituting for y and y'' in Eq. (11) gives

$$\sum_{n=0}^{\infty} (n+2)(n+1)a_{n+2}(x-1)^n = x \sum_{n=0}^{\infty} a_n(x-1)^n. \tag{17}$$

Now in order to equate the coefficients of like powers of $(x-1)$ we must express x, the coefficient of y in Eq. (11), in powers of $x-1$; that is, we write $x = 1 + (x-1)$. Note that this is precisely the Taylor series for x about $x = 1$. Then Eq. (17) takes the form

$$\sum_{n=0}^{\infty} (n+2)(n+1)a_{n+2}(x-1)^n = [1 + (x-1)] \sum_{n=0}^{\infty} a_n(x-1)^n$$

$$= \sum_{n=0}^{\infty} a_n(x-1)^n + \sum_{n=0}^{\infty} a_n(x-1)^{n+1}.$$

Shifting the index of summation in the second series on the right gives

$$\sum_{n=0}^{\infty} (n+2)(n+1)a_{n+2}(x-1)^n = \sum_{n=0}^{\infty} a_n(x-1)^n + \sum_{n=1}^{\infty} a_{n-1}(x-1)^n.$$

Equating coefficients of like powers of $x-1$ we obtain

$$2a_2 = a_0,$$
$$(3 \cdot 2)a_3 = a_1 + a_0,$$
$$(4 \cdot 3)a_4 = a_2 + a_1,$$
$$(5 \cdot 4)a_5 = a_3 + a_2,$$

$$\cdot$$
$$\cdot$$
$$\cdot$$

The general recurrence relation is

$$(n+2)(n+1)a_{n+2} = a_n + a_{n-1} \qquad \text{for} \qquad n \geq 1. \tag{18}$$

Solving for the first few a_n in terms of a_0 and a_1 gives

$$a_2 = \frac{a_0}{2}, \qquad a_3 = \frac{a_1}{6} + \frac{a_0}{6},$$

$$a_4 = \frac{a_2}{12} + \frac{a_1}{12} = \frac{a_0}{24} + \frac{a_1}{12},$$

$$a_5 = \frac{a_3}{20} + \frac{a_2}{20} = \frac{a_0}{30} + \frac{a_1}{120},$$

$$\cdot$$
$$\cdot$$
$$\cdot$$

Hence

$$y = a_0 \left[1 + \frac{(x-1)^2}{2} + \frac{(x-1)^3}{6} + \frac{(x-1)^4}{24} + \frac{(x-1)^5}{30} + \cdots \right]$$
$$+ a_1 \left[(x-1) + \frac{(x-1)^3}{6} + \frac{(x-1)^4}{12} + \frac{(x-1)^5}{120} + \cdots \right]. \quad (19)$$

In general, when the recurrence relation has three terms, such as the one given in Eq. (18), the determination of the coefficients in the series solution will be fairly complicated. In this example the formula for a_n in terms of a_0 and a_1 is not readily apparent. However, even without knowing the formula for a_n, we shall see in Section 4.2.1 that it is possible to establish that the series in Eq. (19) converge for all x, and further define functions y_1 and y_2, which are linearly independent solutions of the Airy equation (11). Thus

$$y = a_0 y_1(x) + a_1 y_2(x)$$

is the general solution of Airy's equation for $-\infty < x < \infty$.

It is worth emphasizing, as we saw in Example 3, that if we look for a solution of Eq. (1) of the form $y = \sum_{n=0}^{\infty} a_n(x - x_0)^n$, then the coefficients $P(x)$, $Q(x)$, and $R(x)$ in Eq. (1) must also be expressed in powers of $(x - x_0)$. Alternatively we can make the change of variable $x - x_0 = t$, obtaining a new differential equation for y as a function of t and then look for solutions of this new equation of the form $\sum_{n=0}^{\infty} a_n t^n$. When we have finished the calculations we replace t by $x - x_0$ (see Problem 3).

Another point of interest is the following. The functions y_1 and y_2 defined by the series in Eq. (16) are linearly independent solutions of Eq. (11) for all x, and this is similarly true for the functions y_1 and y_2 defined by the series in Eq. (19). According to the general theory of second order linear equations each of the first two functions can be expressed as a linear combination of the latter two functions and vice versa—a result that is certainly not obvious from an examination of the series alone.

Example 4. The Hermite* equation is

$$y'' - 2xy' + \lambda y = 0, \qquad -\infty < x < \infty, \qquad (20)$$

where λ is a constant. This equation is important in many branches of mathematical physics; for example, in quantum mechanics the Hermite equation arises in the investigation of the Schrödinger (1887–1961) equation for a harmonic oscillator.

* Charles Hermite (1822–1901) was a distinguished French mathematician whose work in the field of algebra was particularly elegant.

To find a solution of Eq. (20) in the neighborhood of the ordinary point $x = 0$, we substitute the series (4), (5), and (13) for y, y', and y'' in Eq. (20), obtaining

$$\sum_{n=0}^{\infty} (n+2)(n+1)a_{n+2}x^n - \sum_{n=1}^{\infty} 2na_n x^n + \sum_{n=0}^{\infty} \lambda a_n x^n = 0$$

or

$$(2a_2 + \lambda a_0) + \sum_{n=1}^{\infty} [(n+2)(n+1)a_{n+2} - 2na_n + \lambda a_n]x^n = 0. \qquad (21)$$

Hence $a_2 = -\lambda a_0/2$, and the general recurrence relation is

$$a_{n+2} = \frac{2n - \lambda}{(n+2)(n+1)} a_n, \qquad n \geq 1. \qquad (22)$$

Note that the recurrence relation (22) actually includes the result $a_2 = -\lambda a_0/2$ corresponding to $n = 0$. It is clear from Eq. (22) that a_0 determines a_2, which in turn determines a_4, and so on. Similarly, the coefficients of the odd powers of x are determined in terms of a_1. The formal series solution for the Hermite equation is

$$y = a_0 \left[1 - \frac{\lambda}{2!} x^2 - \frac{(4-\lambda)\lambda}{4!} x^4 - \frac{(8-\lambda)(4-\lambda)\lambda}{6!} x^6 - \cdots \right]$$

$$+ a_1 \left[x + \frac{(2-\lambda)}{3!} x^3 + \frac{(6-\lambda)(2-\lambda)}{5!} x^5 + \frac{(10-\lambda)(6-\lambda)(2-\lambda)}{7!} x^7 + \cdots \right]$$

$$= a_0 y_1(x) + a_1 y_2(x). \qquad (23)$$

Again, the results of Section 4.2.1 will establish the convergence of these series for all x. If λ is a nonnegative even integer, one or the other of the above series terminates. In particular for $\lambda = 0, 2, 4, 6$, respectively, one solution of the Hermite equation is 1, x, $1 - 2x^2$, and $x - 2x^3/3$. The polynomial solution corresponding to $\lambda = 2n$, after being multiplied by a proper constant,* is known as the Hermite polynomial $H_n(x)$.

PROBLEMS

1. Proceeding formally, find two linearly independent power series solutions in powers of $x - x_0$ for each of the following differential equations. What is the recurrence relation?

    (a) $y'' - y = 0$, $x_0 = 0$

    (b) $y'' - xy' - y = 0$, $x_0 = 0$

    (c) $y'' - xy' - y = 0$, $x_0 = 1$

---

* This process is usually referred to as normalization. The constant is often chosen so that the solution has a specified value at a particular point, or so that a certain integral of the solution over a prescribed range has a definite value. For example, the Hermite polynomial of order n is uniquely specified by the statement that it is a polynomial solution of degree n of the Hermite equation with $\lambda = 2n$, and with the coefficient of x^n equal to 2^n.

(d) $y'' + k^2x^2y = 0$; $x_0 = 0$, k a constant

(e) $(1 - x)y'' + y = 0$, $x_0 = 0$

✗ (f) $(2 + x^2)y'' - xy' + 4y = 0$, $x_0 = 0$

2. Find the solution of Problem 1(b) that satisfies the initial conditions $y(0) = 0$, $y'(0) = 1$; the initial conditions $y(0) = 2$, $y'(0) = 1$.

3. By making the change of variable $x - 1 = t$, find two linearly independent series solutions of

$$y'' + (x - 1)^2 y' + (x^2 - 1)y = 0$$

in powers of $x - 1$. Show that you can obtain the same result by expanding $x^2 - 1$ in a Taylor series about $x = 1$, and then substituting

$$y = \sum_{n=0}^{\infty} a_n(x - 1)^n.$$

4. Show directly, using the ratio test, that the series solutions of Airy's equation about $x = 0$ converge for all x. See Eq. (16) of the text.

5. Consider the initial value problem $(y')^2 = 1 - y^2$, $y(0) = 0$ with the additional condition $y'(0) > 0$. This additional condition is necessary to determine a unique solution since only the squares of y and y' appear in the differential equation; hence if $\phi(x)$ is a solution of the differential equation, then so is $-\phi(x)$. Look for a series solution of the initial value problem of the form $y = a_0 + a_1x + a_2x^2 + \ldots$, and show that $a_0 = 0$, $a_1 = 1$, $a_2 = 0$, and $a_3 = -1/3!$. Note that the first two nonzero terms in the series solution are the first two terms in the Taylor series expansion of $\sin x$ about $x = 0$.

4.2.1   SERIES SOLUTIONS NEAR AN ORDINARY POINT, PART II

In the previous section we considered the problem of finding solutions of

$$P(x)y'' + Q(x)y' + R(x)y = 0, \tag{1}$$

where P, Q, and R are polynomials, in the neighborhood of an ordinary point x_0. Assuming that Eq. (1) does have a solution $y = \phi(x)$, and that ϕ has a Taylor series

$$y = \phi(x) = \sum_{n=0}^{\infty} a_n(x - x_0)^n \tag{2}$$

which converges for $|x - x_0| < \rho$, $\rho > 0$, we found that the a_n can be determined by directly substituting the series (2) for y in Eq. (1).

Let us now consider how we might justify the statement that if x_0 is an ordinary point of Eq. (1), then there exist solutions of the form (2). We will also consider the question of the radius of convergence of such a series. In doing this we will be led to a generalization of the definition of an ordinary point.

Suppose, then, that there is a solution of Eq. (1) of the form (2). It is clear on differentiating Eq. (2) m times and setting x equal to x_0 that

$$m!\, a_m = \phi^{(m)}(x_0).$$

Hence, in order to compute the a_n in the series (2), we must show that we can determine $\phi^{(n)}(x_0)$ for $n = 0, 1, 2, \ldots$ from the differential equation (1).

Suppose that $y = \phi(x)$ is a solution of Eq. (1) satisfying the initial conditions $y(x_0) = y_0$, $y'(x_0) = y_0'$. Then $a_0 = y_0$ and $a_1 = y_0'$. If we are solely interested in finding a solution of Eq. (1) without specifying any initial conditions, then a_0 and a_1 will be arbitrary. To determine $\phi^{(n)}(x_0)$ and the corresponding a_n for $n = 2, 3, \ldots$, we turn to Eq. (1). Since ϕ is a solution of Eq. (1) we have

$$P(x)\phi''(x) + Q(x)\phi'(x) + R(x)\phi(x) = 0.$$

For the interval about x_0 for which P is nonvanishing we can write this equation in the form

$$\phi''(x) = -p(x)\phi'(x) - q(x)\phi(x) \tag{3}$$

where $p(x) = Q(x)/P(x)$ and $q(x) = R(x)/P(x)$. Setting x equal to x_0 in Eq. (3) gives

$$\phi''(x_0) = -p(x_0)\phi'(x_0) - q(x_0)\phi(x_0).$$

Hence a_2 is given by

$$2! \, a_2 = \phi''(x_0) = -p(x_0)a_1 - q(x_0)a_0. \tag{4}$$

To determine a_3 we differentiate Eq. (3) and then set x equal to x_0, obtaining

$$3! \, a_3 = \phi'''(x_0) = -[p\phi'' + (p' + q)\phi' + q'\phi]_{x=x_0}$$
$$= -2! \, p(x_0)a_2 - [p'(x_0) + q(x_0)]a_1 - q'(x_0)a_0. \tag{5}$$

Substituting for a_2 from Eq. (4) gives a_3 in terms of a_1 and a_0. Since P, Q, and R are polynomials and $P(x_0) \neq 0$, all of the derivatives of p and q exist at x_0. Hence we can continue to differentiate Eq. (3) indefinitely, determining after each differentiation the successive coefficients $a_4, \ldots$, etc., by setting x equal to x_0.

Notice that the important property that we used in determining the a_n was that we could compute infinitely many derivatives of the functions p and q. It might seem reasonable to relax our assumption that the functions p and q are ratios of polynomials, and simply require that they be infinitely differentiable in the neighborhood of x_0. Unfortunately, this condition is too weak to insure that we can prove the convergence of the resulting series expansion for $y = \phi(x)$. However, there is a more general condition we can require of p and q. It allows us to carry out the above computations for the a_n and also to prove the convergence of the series solution. This condition is that the functions p and q be *analytic* at x_0; that is, that they have Taylor series expansions that converge in some interval about the point x_0:

$$p(x) = p_0 + p_1(x - x_0) + \cdots + p_n(x - x_0)^n + \cdots = \sum_{n=0}^{\infty} p_n(x - x_0)^n, \tag{6}$$

$$q(x) = q_0 + q_1(x - x_0) + \cdots + q_n(x - x_0)^n + \cdots = \sum_{n=0}^{\infty} q_n(x - x_0)^n. \tag{7}$$

As mentioned in Section 4.1 if each of the functions p and q is the quotient of polynomials, then they have series expansions* of the form (6) and (7). With this idea

---

* The careful reader has probably reached the conclusion (which is correct) that if p is of the form (6) then p has infinitely many derivatives at x_0, but that the converse does not follow. An example of a function with infinitely many derivatives at $x = 0$, but which does not have a Taylor series expansion, is given in Problem 8.

in mind we can generalize the definition of an ordinary point and a singular point of Eq. (1) as follows: If the functions $p = Q/P$ and $q = R/P$ are analytic at x_0, then the point x_0 is said to be an *ordinary point* of the differential equation (1); otherwise it is a *singular point*.

Now let us turn to the question of the interval of convergence of the series solution. One possibility (but not an appealing one) is actually to compute the series solution for each problem and then to apply one of the tests for convergence of an infinite series to determine the radius of convergence of the series solution. Fortunately, the question can be answered at once for a wide class of problems by the following theorem.

Theorem 4.1. *If x_0 is an ordinary point of the differential equation* (1)

$$P(x)y'' + Q(x)y' + R(x)y = 0;$$

that is, if $p = Q/P$ and $q = R/P$ are analytic at x_0, then the general solution of Eq. (1) *is*

$$y = \sum_{n=0}^{\infty} a_n(x - x_0)^n = a_0 y_1(x) + a_1 y_2(x), \tag{8}$$

where a_0 and a_1 are arbitrary, and y_1 and y_2 are linearly independent series solutions which are analytic at x_0. Further the radius of convergence for each of the series solutions y_1 and y_2 is at least as large as the minimum of the radii of convergence for the series for p and q. The coefficients in the series solutions are determined by substituting the series (2) for y in Eq. (1).

Notice from the form of the series solution that $y_1(x) = 1 + b_2(x - x_0)^2 + \dots$, and $y_2(x) = (x - x_0) + c_2(x - x_0)^2 + \dots$. Hence y_1 is a solution satisfying the initial conditions $y_1(x_0) = 1$, $y_1'(x_0) = 0$, and y_2 is a solution satisfying the initial conditions $y_2(x_0) = 0$, $y_2'(x_0) = 1$. Also note that while the calculation of the coefficients by successively differentiating the differential equation is excellent in theory, it is not in general a practical computational procedure. Rather, one should substitute the series (2) for y in the differential equation (1) and determine the coefficients so that the differential equation is satisfied as in the examples in the previous section.

While this theorem, which in a slightly more general form is due to Fuchs (1833–1902), is not too difficult to prove, it is beyond the scope and intent of this book to do so. What is important for our purposes is that there is a series solution of the form (2), and that the radius of convergence of the series solution cannot be less than the smaller of the radii of convergence of the series for p and q; hence we need only determine these.

This can be done in one of two ways. Again, one possibility is simply to compute the power series for p and q, and then to determine the radii of convergence by using one of the convergence tests for infinite series. However, there is an easier way when P, Q, and R are polynomials. It is shown in the theory of functions of a complex variable that the ratio of two polynomials, say Q/P, has a convergent power series expansion about a point $x = x_0$ if $P(x_0) \neq 0$. Further, assuming that any factors common to Q and P have been canceled, the radius of convergence of the power series for Q/P about the point x_0 is precisely the distance from x_0 to the nearest zero of P. In determining this distance we must remember that $P(x) = 0$ may have complex roots, and these must also be considered.

Example 1. What is the radius of convergence of the Taylor series for $(1 + x^2)^{-1}$ about $x = 0$?

For $x^2 < 1$ we know that a convergent power series expansion for $(1 + x^2)^{-1}$ is

$$\frac{1}{1 + x^2} = 1 - x^2 + x^4 - x^6 + \cdots + (-1)^n x^{2n} + \cdots.$$

Clearly this series diverges for $x^2 \geq 1$ since then the nth term does not approach zero as $n \to \infty$. Hence the radius of convergence is $\rho = 1$. (This can be verified by the ratio test.)

To determine the radius of convergence using the theory that was just discussed, we note that the polynomial $1 + x^2$ has zeros at $x = \pm i$; since the distance from 0 to i in the complex plane is 1, the radius of convergence of the power series about $x = 0$ is 1.

Example 2. What is the radius of convergence of the Taylor series for $(x^2 - 2x + 2)^{-1}$ about $x = 0$? About $x = 1$?

First notice that

$$x^2 - 2x + 2 = 0$$

has solutions $x = 1 \pm i$. The distance from $x = 0$ to either $x = 1 + i$ or $x = 1 - i$ in the complex plane is $\sqrt{2}$; hence the radius of convergence of the Taylor series expansion $\sum_{n=0}^{\infty} a_n x^n$ about $x = 0$ is $\sqrt{2}$.

The distance from $x = 1$ to either $x = 1 + i$ or $x = 1 - i$ is 1; hence the radius of convergence of the Taylor series expansion $\sum_{n=0}^{\infty} b_n (x - 1)^n$ about $x = 1$ is 1.

According to Theorem 4.1 the series solutions of the Airy equation in Examples 2 and 3, and of the Hermite equation in Example 4 of the previous section, converge for all values of x, $x - 1$, and x, respectively, since in each problem $P(x) = 1$ and hence never vanishes.

It should be emphasized that a series solution may converge for a wider range of x than indicated by Theorem 4.1, so the theorem actually gives only a lower bound on the radius of convergence of the series solution. This is illustrated by the Legendre polynomial solution of the Legendre equation given in the next example.

Example 3. Determine a lower bound for the radius of convergence of series solutions about $x = 0$ for the Legendre equation

$$(1 - x^2)y'' - 2xy' + \alpha(\alpha + 1)y = 0,$$

where α is a constant.

Note that $P(x) = 1 - x^2$, $Q(x) = -2x$, and $R(x) = \alpha(\alpha + 1)$ are polynomials, and that the zeros of P, $x = \pm 1$, are a distance 1 from $x = 0$. Hence a series solution of the form $\sum\limits_{n=0}^{\infty} a_n x^n$ will converge for $|x| < 1$ at least, and possibly for larger values of x. Indeed, it can be shown that if α is a positive integer, one of the series solutions terminates after a finite number of terms and hence converges not just for $|x| < 1$ but for all x. For example, if $\alpha = 1$ the polynomial solution is $y = x$. See Problems 10 through 16 at the end of this section for a more complete discussion of the Legendre equation.

Example 4. Determine a lower bound for the radius of convergence of series solutions of the differential equation

$$(1 + x^2)y'' + 2xy' + 4x^2 y = 0 \tag{9}$$

about the point $x = 0$; about the point $x = -\frac{1}{2}$.

Again, P, Q, and R are polynomials, and P has zeros at $x = \pm i$. The distance in the complex plane from 0 to $\pm i$ is 1, and from $-\frac{1}{2}$ to $\pm i$ is $\sqrt{1 + \frac{1}{4}} = \sqrt{5}/2$. Hence in the first case the series $\sum\limits_{n=0}^{\infty} a_n x^n$ converges at least for $|x| < 1$, and in the second case the series $\sum\limits_{n=0}^{\infty} b_n(x + \frac{1}{2})^n$ converges at least for $|x + \frac{1}{2}| < \sqrt{5}/2$.

An interesting observation that we can make about Eq. (9) follows from the existence and uniqueness Theorem 3.2 and Theorem 4.1 that was just stated. Suppose that initial conditions $y(0) = y_0$ and $y'(0) = y_0'$ are given. Since $1 + x^2 \neq 0$ for all x, we know from Theorem 3.2 that there exists a unique solution of the initial value problem on $-\infty < x < \infty$. On the other hand, Theorem 4.1 only guarantees a series solution of the form $\sum\limits_{n=0}^{\infty} a_n x^n$ ($a_0 = y_0$, $a_1 = y_0'$) for $-1 < x < 1$. The unique solution on the interval $-\infty < x < \infty$ may not have a power series about $x = 0$ that converges for all x.

Example 5. Can we determine a series solution about $x = 0$ for the differential equation

$$y'' + (\sin x)y' + (1 + x^2)y = 0,$$

and if so what is the radius of convergence?

For this differential equation, $p(x) = \sin x$ and $q(x) = 1 + x^2$. Recall from Section 4.1 that $\sin x$ has a Taylor series expansion about $x = 0$, which converges for all x. Further, q also has a Taylor series expansion about $x = 0$, namely

$q(x) = 1 + x^2$, which converges for all x. Thus there is a series solution of the form $y = \sum\limits_{n=0}^{\infty} a_n x^n$ with a_0 and a_1 arbitrary, and the series converges for all x.

PROBLEMS

1. Determine $\phi''(x_0)$, $\phi'''(x_0)$, and $\phi^{iv}(x_0)$ for the given point x_0 if $y = \phi(x)$ is a solution of the given initial value problem.

(a) $y'' + xy' + y = 0$; $y(0) = 1$, $y'(0) = 0$

(b) $y'' + (\sin x)y' + (\cos x)y = 0$; $y(0) = 0$, $y'(0) = 1$

(c) $x^2 y'' + (1 + x)y' + 3(\ln x)y = 0$; $y(1) = 2$, $y'(1) = 0$

(d) $y'' + x^2 y' + (\sin x)y = 0$; $y(0) = a_0$, $y'(0) = a_1$

In part (d) express $\phi''(0)$, $\phi'''(0)$, and $\phi^{iv}(0)$ in terms of a_0 and a_1.

2. Determine a lower bound for the radius of convergence of series solutions about each given point x_0 for each of the following differential equations.

(a) $y'' + 4y' + 6xy = 0$; $x_0 = 0$, $x_0 = 4$

(b) $(x^2 - 2x - 3)y'' + xy' + 4y = 0$; $x_0 = 4$, $x_0 = -4$, $x_0 = 0$

(c) $(1 + x^3)y'' + 4xy' + y = 0$; $x_0 = 0$, $x_0 = 2$

(d) $xy'' + y = 0$; $x_0 = 1$

3. Determine a lower bound for the radius of convergence of series solutions for each of the differential equations in Problem 1 of Section 4.2.

4. The Tchebycheff (1821–1894) differential equation is

$$(1 - x^2)y'' - xy' + \alpha^2 y = 0,$$

where α is a constant.

(a) Determine two linearly independent solutions in powers of x for $|x| < 1$.

(b) Show that if α is a nonnegative integer n, then there is a polynomial solution of degree n. These polynomials, when properly normalized, are called the Tchebycheff polynomials. They are very useful in problems requiring a polynomial approximation to a function defined on $-1 \leq x \leq 1$.

(c) Find a polynomial solution for each of the cases $\alpha = n = 0, 1, 2$, and 3.

5. Find the first three terms in each of two linearly independent power series solutions in powers of x of

$$y'' + (\sin x)y = 0.$$

Hint: Expand $\sin x$ in a Taylor series about $x = 0$, and retain a sufficient number of terms to compute the necessary coefficients in $y = \sum\limits_{n=0}^{\infty} a_n x^n$.

6. Find the first three terms in each of two linearly independent power series solutions in powers of x of

$$e^x y'' + xy = 0.$$

What is the radius of convergence of each series solution?

Hint: Expand e^x or xe^{-x} in a power series about $x = 0$.

7. Suppose that you are told that x and x^2 are solutions of a differential equation $P(x)y'' + Q(x)y' + R(x)y = 0$. What can you say about the point $x = 0$: Is it an ordinary point or a singular point?

Hint: Use Theorem 3.2, and note the value of x and x^2 at $x = 0$.

*8. To show that it is possible to have a function that is infinitely differentiable at a point but does not have a Taylor series expansion about the point, consider the function

$$f(x) = \begin{cases} e^{-1/x^2}, & x \neq 0, \\ 0, & x = 0. \end{cases}$$

Show, using the definition of a derivative as the limit of a difference quotient, that $f'(0) = f''(0) = 0$. It can be shown that f is infinitely differentiable at the origin, and that $f^{(n)}(0) = 0$ for all n. Hence the series

$$\sum_{n=0}^{\infty} \frac{f^{(n)}(0)}{n!} x^n$$

vanishes for all x, and converges to $f(x)$ only at $x = 0$.

9. The series methods discussed in this section are directly applicable to the first order linear differential equation $P(x)y' + Q(x)y = 0$ at a point x_0, if the function $p = Q/P$ has a Taylor series expansion about that point. Such a point is called an ordinary point, and further, the radius of convergence of the series $y = \sum_{n=0}^{\infty} a_n(x - x_0)^n$ is at least as large as the radius of convergence of the series for Q/P. Solve the following differential equations by series in powers of x and verify that a_0 is arbitrary in each case. Problems (e) and (f) involve nonhomogeneous differential equations to which the series methods can be easily extended.

(a) $y' - y = 0$
(b) $y' - xy = 0$
(c) $y' = e^{x^2}y$, three terms
(d) $(1 - x)y' = y$
*(e) $y' - y = x^2$
*(f) $y' + xy = 1 + x$

Where possible, compare the series solution with the solution obtained by using the methods of Chapter 2.

Legendre Equation.   Problems 10 through 16 deal with the Legendre equation

$$(1 - x^2)y'' - 2xy' + \alpha(\alpha + 1)y = 0.$$

As indicated in Example 3, the point $x = 0$ is an ordinary point of this equation, and the distance from the origin to the nearest zero of $P(x) = 1 - x^2$ is 1. Hence the radius of convergence of series solutions about $x = 0$ is at least 1. Also notice that it is only necessary to consider $\alpha > -1$ because if $\alpha \leq -1$, then the substitution $\alpha = -(1 + \gamma)$ where $\gamma \geq 0$ leads to the Legendre equation $(1 - x^2)y'' - 2xy' + \gamma(\gamma + 1)y = 0$.

10. Show that two linearly independent solutions of the Legendre equation for $|x| < 1$ are

$$y_1(x) = 1 + \sum_{m=1}^{\infty} (-1)^m$$

$$\times \frac{\alpha(\alpha - 2)(\alpha - 4) \cdots (\alpha - 2m + 2)(\alpha + 1)(\alpha + 3) \cdots (\alpha + 2m - 1)}{(2m)!} x^{2m},$$

$$y_2(x) = x + \sum_{m=1}^{\infty} (-1)^m$$

$$\times \frac{(\alpha - 1)(\alpha - 3) \cdots (\alpha - 2m + 1)(\alpha + 2)(\alpha + 4) \cdots (\alpha + 2m)}{(2m + 1)!} x^{2m+1}.$$

11. Show that if α is zero or a positive even integer $2n$ the series solution for y_1 reduces to a polynomial of degree $2n$ containing only even powers of x. Show that corresponding to $\alpha = 0, 2$, and 4 these polynomials are

$$1, 1 - 3x^2, 1 - 10x^2 + \tfrac{35}{3}x^4.$$

Show that if α is a positive odd integer $2n + 1$ then the series solution for y_2 reduces to a polynomial of degree $2n + 1$ containing only odd powers of x and that corresponding to $\alpha = 1, 3, 5$ these polynomials are

$$x, x - \tfrac{5}{3}x^3, x - \tfrac{14}{3}x^3 + \tfrac{21}{5}x^5.$$

12. The Legendre polynomial P_n is defined as the polynomial solution of the Legendre equation with $\alpha = n$ satisfying the condition $P_n(1) = 1$. Using the results of Problem 11, show that

$$P_0(x) = 1, \quad P_1(x) = x, \quad P_2(x) = \tfrac{1}{2}(3x^2 - 1), \quad P_3(x) = \tfrac{1}{2}(5x^3 - 3x).$$

It can be shown that the general formula is

$$P_n(x) = \frac{1}{2^n} \sum_{k=0}^{[n/2]} \frac{(-1)^k (2n - 2k)!}{k!\,(n - k)!\,(n - 2k)!} x^{n-2k},$$

where $[n/2]$ denotes the greatest integer less than or equal to $n/2$. By observing the form of $P_n(x)$ for n even and n odd show that $P_n(-1) = (-1)^n$.

13. The Legendre polynomials play an important role in mathematical physics. For example, in solving Laplace's equation (the potential equation) in spherical coordinates we encounter the equation

$$\frac{d^2F(\varphi)}{d\varphi^2} + \cot \varphi \, \frac{dF(\varphi)}{d\varphi} + n(n + 1)F(\varphi) = 0 \qquad 0 < \varphi < \pi,$$

where n is a positive integer. Show that the change of variable $x = \cos \varphi$ leads to the Legendre equation with $\alpha = n$ for $y = f(x) = F(\cos^{-1} x)$.

14. Show that for $n = 0, 1, 2, 3$ the corresponding Legendre polynomial is given by

$$P_n(x) = \frac{1}{2^n n!} \frac{d^n}{dx^n} (x^2 - 1)^n.$$

This formula, known as Rodrigues' (1794–1851) formula, is true for all positive integers n.

15. Show that the Legendre equation can also be written as

$$[(1 - x^2)y']' = -\alpha(\alpha + 1)y.$$

Then it follows that $[(1 - x^2)P_n'(x)]' = -n(n + 1)P_n(x)$ and $[(1 - x^2)P_m'(x)]' = -m(m + 1)P_m(x)$. Show, by multiplying the first equation by $P_m(x)$ and the second equation by $P_n(x)$, and then integrating by parts, that

$$\int_{-1}^{1} P_n(x)P_m(x)\,dx = 0 \qquad \text{if} \qquad n \neq m.$$

This property of the Legendre polynomials is known as the orthogonality property. If $m = n$, it can be shown that the value of the above integral is $2/(2n + 1)$.

16. Given a polynomial f of degree n, it is possible to express f as a linear combination of $P_0, P_1, P_2, \ldots, P_n$:

$$f(x) = \sum_{k=0}^{n} a_k P_k(x).$$

Using the result of Problem 15, show that

$$a_k = \frac{2k+1}{2} \int_{-1}^{1} f(x) P_k(x) \, dx.$$

4.3 REGULAR SINGULAR POINTS

In this section we will consider the equation

$$P(x)y'' + Q(x)y' + R(x)y = 0 \tag{1}$$

in the neighborhood of a singular point x_0. Recall that if the functions P, Q, and R are polynomials having no common factors, the singular points of Eq. (1) are the points for which $P(x) = 0$.

Example 1. Determine the singular points and ordinary points of the Bessel equation of order ν

$$x^2 y'' + xy' + (x^2 - \nu^2)y = 0. \tag{2}$$

The point $x = 0$ is clearly a singular point since $P(x) = x^2$ vanishes there. Furthermore, if we divide Eq. (2) by x^2, the coefficients of y' and y will be undefined at $x = 0$. All other points are ordinary points of Eq. (2).

Example 2. Determine the singular points and ordinary points of the Legendre equation

$$(1 - x^2)y'' - 2xy' + \alpha(\alpha + 1)y = 0, \tag{3}$$

where α is a constant.

The singular points are the zeros of $P(x) = 1 - x^2$, namely the points $x = \pm 1$. All other points are ordinary points.

The solution of Eq. (1) in the neighborhood of a singular point x_0 frequently becomes large in magnitude, experiences rapid changes in magnitude, or is peculiar in some manner. Thus, the behavior of a physical system governed by such an equation is frequently most interesting in the neighborhood of a singular point. Often geometrical singularities in a physical problem, such as corners or sharp edges, lead to singular points in the corresponding differential equation. Mathematically, the singular points of a differential equation, while usually few in number, determine the principal features of the solution to a much larger extent than one might at first suspect. Thus, while we might want to avoid the few points where a differential equation is singular, it is precisely at these points that it is necessary to study the solution most carefully.

Unfortunately, if x_0 is a singular point the methods of the previous section fail, and it is necessary to consider a more general type of series expansion than a Taylor series. This is illustrated by the differential equation in Example 3 below, for which one of the solutions does not have a Taylor series expansion about $x = 0$. Without any additional information about the behavior of Q/P and R/P in the neighborhood of the singular point, it is impossible to describe the behavior of the solutions of Eq. (1) near $x = x_0$. It may be that there are two linearly independent solutions of Eq. (1) that remain bounded as $x \to x_0$, or there may be only one with the other becoming unbounded as $x \to x_0$, or they may both become unbounded as $x \to x_0$. To illustrate this consider the following examples.

Example 3. The differential equation

$$x^2 y'' - 2y = 0 \tag{4}$$

has a singular point at $x = 0$. It can be easily verified by direct substitution that for $x > 0$ or $x < 0$, $y = x^2$ and $y = 1/x$ are linearly independent solutions of Eq. (4). Thus in any interval not containing the origin the general solution of Eq. (4) is $y = c_1 x^2 + c_2 x^{-1}$. The only solution of Eq. (4) that is bounded as $x \to 0$ is $y = c_1 x^2$. Indeed, this solution is analytic at the origin even though if Eq. (2) is put in the standard form, $y'' - (2/x^2)y = 0$, the function $q(x) = -2/x^2$ is not analytic at $x = 0$, and Theorem 4.1 is not applicable. On the other hand, notice that the solution $y = x^{-1}$ does not have a Taylor series expansion about the origin (is not analytic at $x = 0$); therefore, the method of Section 4.2 would fail in this case.

Example 4. The differential equation

$$x^2 y'' - 2xy' + 2y = 0 \tag{5}$$

also has a singular point at $x = 0$. It can be verified that $y_1(x) = x$ and $y_2(x) = x^2$ are linearly independent solutions of Eq. (5), and both are analytic at $x = 0$. Nevertheless, it is still not proper to pose an initial value problem with initial conditions at $x = 0$. It is impossible to prescribe arbitrary initial conditions at $x = 0$, since any linear combination of x and x^2 is zero at $x = 0$.

It is also possible to construct a differential equation with a singular point x_0, such that every solution of the differential equation becomes unbounded as $x \to x_0$. Even if Eq. (1) does not have a solution that remains bounded as $x \to x_0$, it is often important to determine how the solutions of Eq. (1) behave as $x \to x_0$; for example, does $y \to \infty$ in the same manner as $(x - x_0)^{-1}$ or $|x - x_0|^{-1/2}$, or in some other manner?

In order to develop a reasonably simple mathematical theory for solving Eq. (1) in the neighborhood of a singular point x_0, it is necessary to restrict ourselves to cases in which the singularities in the functions Q/P and R/P at $x = x_0$

are not too severe, that is, to so-called "weak singularities." A priori, it is not clear exactly what is an acceptable singularity. However, we will show later (see Section 4.5.1, Problems 5 and 6) that if P, Q, and R are polynomials, the conditions that distinguish "weak singularities" are

$$\lim_{x \to x_0} (x - x_0) \frac{Q(x)}{P(x)} \quad \text{is finite} \tag{6}$$

and

$$\lim_{x \to x_0} (x - x_0)^2 \frac{R(x)}{P(x)} \quad \text{is finite.} \tag{7}$$

This means that the singularity in Q/P can be no worse than $(x - x_0)^{-1}$ and the singularity in R/P can be no worse than $(x - x_0)^{-2}$. Such a point is called a *regular singular point* of Eq. (1). For more general functions than polynomials, x_0 is a *regular singular point* of Eq. (1) if it is a singular point, and if both*

$$(x - x_0) \frac{Q(x)}{P(x)} \quad \text{and} \quad (x - x_0)^2 \frac{R(x)}{P(x)} \tag{8}$$

have convergent Taylor series about x_0; that is, if the functions in Eq. (8) are analytic at $x = x_0$. Equations (6) and (7) imply that this will be the case when P, Q, and R are polynomials. Any singular point of Eq. (1) that is not a regular singular point is called an *irregular singular point* of Eq. (1).

In the following sections we will discuss how to solve Eq. (1) in the neighborhood of a regular singular point. A discussion of the solutions of differential equations in the neighborhood of irregular singular points is beyond the scope of an elementary text.

Example 5. In Example 2 we observed that the singular points of the Legendre equation

$$(1 - x^2)y'' - 2xy' + \alpha(\alpha + 1)y = 0$$

are $x = \pm 1$. Determine whether these singular points are regular or irregular singular points.

We consider the point $x = 1$ first and also observe that on dividing by $1 - x^2$ the coefficients of y' and y are $-2x/(1 - x^2)$ and $\alpha(\alpha + 1)/(1 - x^2)$, respectively. Thus we calculate

$$\lim_{x \to 1} (x - 1) \frac{-2x}{1 - x^2} = \lim_{x \to 1} \frac{(x - 1)(-2x)}{(1 - x)(1 + x)} = \lim_{x \to 1} \frac{2x}{1 + x} = 1$$

and

$$\lim_{x \to 1} (x - 1)^2 \frac{\alpha(\alpha + 1)}{1 - x^2} = \lim_{x \to 1} \frac{(x - 1)^2 \alpha(\alpha + 1)}{(1 - x)(1 + x)} = \lim_{x \to 1} \frac{(x - 1)(-\alpha)(\alpha + 1)}{1 + x} = 0.$$

---

* The functions given in Eq. (8) may not be defined at x_0, in which case their values at x_0 are to be taken equal to their limits as $x \to x_0$.

Since these limits are finite, the point $x = 1$ is a regular singular point. It can be shown in a similar manner that $x = -1$ is also a regular singular point.

Example 6. Determine the singular points of the differential equation

$$2(x - 2)^2 xy'' + 3xy' + (x - 2)y = 0,$$

and classify them as regular or irregular.

Dividing by $2(x - 2)^2 x$ we have

$$y'' + \frac{3}{2(x - 2)^2} y' + \frac{1}{2(x - 2)x} y = 0,$$

so $p(x) = Q(x)/P(x) = 3/2(x - 2)^2$ and $q(x) = R(x)/P(x) = 1/2x(x - 2)$. The singular points are $x = 0$ and $x = 2$. Consider $x = 0$. We have

$$\lim_{x \to 0} xp(x) = \lim_{x \to 0} x \frac{3}{2(x - 2)^2} = 0$$

$$\lim_{x \to 0} x^2 q(x) = \lim_{x \to 0} x^2 \frac{1}{2x(x - 2)} = 0.$$

Since these limits are finite, $x = 0$ is a regular singular point. For $x = 2$ we have

$$\lim_{x \to 2} (x - 2)p(x) = \lim_{x \to 2} (x - 2) \frac{3}{2(x - 2)^2}$$

does not exist; hence $x = 2$ is an irregular singular point.

PROBLEMS

In Problems 1 through 8 determine whether each of the points -1, 0, and 1 is an ordinary point, a regular singular point, or an irregular singular point for the given differential equation.

1. $xy'' + (1 - x)y' + xy = 0$
2. $x^2(1 - x^2)y'' + 2xy' + 4y = 0$
3. $2x^4(1 - x^2)y'' + 2xy' + 3x^2 y = 0$
4. $x^2(1 - x^2)y'' + \dfrac{2}{x} y' + 4y = 0$
5. $(1 - x^2)^2 y'' + x(1 - x)y' + (1 + x)y = 0$
6. $y'' + \left(\dfrac{x}{1 - x}\right)^2 y' + 3(1 + x)^2 y = 0$
7. $(x + 3)y'' - 2xy' + (1 - x^2)y = 0$
8. $x(1 - x^2)^3 y'' + (1 - x^2)^2 y' + 2(1 + x)y = 0$
9. Determine whether the singular point $x = 0$ of the Bessel equation

$$x^2 y'' + xy' + (x^2 - v^2)y = 0$$

is a regular or irregular singular point.

10. For the following differential equation classify the point $x = 0$ as an ordinary point, a regular singular point, or an irregular singular point:

$$xy'' + e^x y' + (3 \cos x)y = 0.$$

11. For the following differential equations classify the point $x = 0$ as an ordinary point, a regular singular point, or an irregular singular point:

(a) $x^2 y'' + 2(e^x - 1)y' + (e^{-x} \cos x)y = 0$

(b) $x^2 y'' - 3(\sin x)y' + (1 + x^2)y = 0$

Hint: Consider the power series for e^x and $\sin x$ in parts (a) and (b), respectively.

12. Assuming that the function f defined by $f(x) = x^{-1} \sin x$ for $x \neq 0$ and $f(0) = 1$ is analytic at $x = 0$, show that $x = 0$ is a regular singular point of the differential equation

$$(\sin x)y'' + xy' + 4y = 0,$$

and an irregular singular point of the differential equation

$$(x \sin x)y'' + 3y' + xy = 0.$$

*13. The definitions of an ordinary point and a regular singular point given in the preceding sections apply only if the point x_0 is finite. In more advanced work in differential equations it is often necessary to discuss the point at infinity. This is done by making the change of variable $\xi = 1/x$ and studying the resulting equation at $\xi = 0$. Show that for the differential equation $P(x)y'' + Q(x)y' + R(x)y = 0$ the point at infinity is an ordinary point if

$$\frac{1}{P(1/\xi)}\left[\frac{2P(1/\xi)}{\xi} - \frac{Q(1/\xi)}{\xi^2}\right] \quad \text{and} \quad \frac{R(1/\xi)}{\xi^4 P(1/\xi)}$$

have Taylor series expansions about $\xi = 0$. Show also that the point at infinity is a regular singular point if one or the other of the above functions does not have a Taylor series expansion, but both

$$\frac{\xi}{P(1/\xi)}\left[\frac{2P(1/\xi)}{\xi} - \frac{Q(1/\xi)}{\xi^2}\right] \quad \text{and} \quad \frac{R(1/\xi)}{\xi^2 P(1/\xi)}$$

do have such expansions.

*14. For the following differential equations determine whether the point at infinity is an ordinary point, a regular singular point, or an irregular singular point.

(a) The Legendre equation

$$(1 - x^2)y'' - 2xy' + \alpha(\alpha + 1)y = 0$$

(b) The Bessel equation

$$x^2 y'' + xy' + (x^2 - \nu^2)y = 0$$

(c) The Hermite equation

$$y'' - 2xy' + \lambda y = 0$$

(d) The Airy equation

$$y'' - xy = 0$$

4.4 EULER EQUATIONS

One of the simplest examples of a differential equation having a regular singular point is the Euler equation or equidimensional equation

$$L[y] = x^2 \frac{d^2y}{dx^2} + \alpha x \frac{dy}{dx} + \beta y = 0, \tag{1}$$

where α and β are constants. Unless otherwise stated we shall consider α and β to be real. It is easy to see that $x = 0$ is a regular singular point of Eq. (1). Because the solution of the Euler equation is typical of the solution of all differential equations with a regular singular point, it is worthwhile considering this equation in detail before discussing the more general problem.

In any interval not including the origin, Eq. (1) has a general solution of the form $y = c_1 y_1(x) + c_2 y_2(x)$ where y_1 and y_2 are linearly independent. For convenience we will first consider the interval $x > 0$, extending our results later to the interval $x < 0$. First, we note that $(x^r)' = rx^{r-1}$ and $(x^r)'' = r(r-1)x^{r-2}$. Hence, if we assume that Eq. (1) has a solution of the form*

$$y = x^r, \tag{2}$$

we obtain

$$L[x^r] = x^2(x^r)'' + \alpha x(x^r)' + \beta x^r$$
$$= x^r F(r), \tag{3}$$

where

$$F(r) = r(r-1) + \alpha r + \beta. \tag{4}$$

If r is a root of the quadratic equation $F(r) = 0$, then $L[x^r]$ is zero, and $y = x^r$ is a solution of Eq. (1). The roots of $F(r) = 0$ are

$$r_1 = \frac{-(\alpha - 1) + \sqrt{(\alpha - 1)^2 - 4\beta}}{2},$$

$$r_2 = \frac{-(\alpha - 1) - \sqrt{(\alpha - 1)^2 - 4\beta}}{2}, \tag{5}$$

and $F(r) = (r - r_1)(r - r_2)$. Just as in the case of the second order linear differential equation with constant coefficients (see Section 3.5), we must examine separately the cases $(\alpha - 1)^2 - 4\beta$ positive, zero, and negative. Indeed, the entire theory presented in this section is similar to the theory for second order linear equations with constant coefficients, e^{rx} being replaced by x^r (see Problems 4 and 5). Also, we note that in solving a specific Euler equation it is much easier to substitute $y = x^r$, obtain the quadratic equation (4), and calculate the roots r_1 and r_2, rather than to memorize the general formula (5).

---

* This assumption is also suggested by the fact that the first order Euler equation $xy' + ky = 0$ has the solution $y = x^{-k}$.

Case 1. $(\alpha - 1)^2 - 4\beta > 0$. In this case Eqs. (5) give two real unequal roots r_1 and r_2. Since $W(x^{r_1}, x^{r_2})$ is nonvanishing for $r_1 \neq r_2$ and $x > 0$, it follows that the general solution of Eq. (1) is

$$y = c_1 x^{r_1} + c_2 x^{r_2}, \qquad x > 0. \tag{6}$$

Note that if r_1 is not a rational number, then x^{r_1} is defined by $x^{r_1} = e^{r_1 \ln x}$.

Example 1. Solve

$$2x^2 y'' + 3xy' - y = 0, \qquad x > 0. \tag{7}$$

Substituting $y = x^r$ gives

$$x^r(2r^2 + r - 1) = 0,$$

$$x^r(2r - 1)(r + 1) = 0.$$

Hence

$$r_1 = \tfrac{1}{2}, \qquad r_2 = -1,$$

and

$$y = c_1 x^{1/2} + c_2 x^{-1}, \qquad x > 0. \tag{8}$$

Case 2. $(\alpha - 1)^2 - 4\beta = 0$. In this case, according to Eqs. (5), $r_1 = r_2 = -(\alpha - 1)/2$, and we have only one solution, $y_1(x) = x^{r_1}$, of the differential equation. A second solution can be obtained by the method of reduction of order, but for the purpose of our future discussion we will consider an alternative method. Since $r_1 = r_2$, it follows that $F(r) = (r - r_1)^2$. Thus in this case not only does $F(r_1) = 0$, but also $F'(r_1) = 0$. This suggests differentiating Eq. (3) with respect to r and then setting r equal to r_1. Differentiating Eq. (3) with respect to r gives

$$\frac{\partial}{\partial r} L[x^r] = \frac{\partial}{\partial r} [x^r F(r)].$$

Substituting for $F(r)$, interchanging differentiation with respect to x and r, and noting that $\partial(x^r)/\partial r = x^r \ln x$, we obtain

$$L[x^r \ln x] = (r - r_1)^2 x^r \ln x + 2(r - r_1)x^r. \tag{9}$$

The right-hand side of Eq. (9) vanishes for $r = r_1$; consequently,

$$y_2(x) = x^{r_1} \ln x, \qquad x > 0, \tag{10}$$

is a second solution of Eq. (1). Since x^{r_1} and $x^{r_1} \ln x$ are linearly independent for $x > 0$, the general solution of Eq. (1) is

$$y = (c_1 + c_2 \ln x)x^{r_1}, \qquad x > 0. \tag{11}$$

Example 2. Solve

$$x^2 y'' + 5xy' + 4y = 0, \qquad x > 0. \tag{12}$$

Substituting $y = x^r$ gives

$$x^r(r^2 + 4r + 4) = 0.$$

Hence $r_1 = r_2 = -2$, and

$$y = x^{-2}(c_1 + c_2 \ln x), \qquad x > 0. \tag{13}$$

Case 3. $(\alpha - 1)^2 - 4\beta < 0$. In this case the roots r_1 and r_2 are complex conjugates, say $r_1 = \lambda + i\mu$ and $r_2 = \lambda - i\mu$. We must now explain what is meant by x^r when r is complex. Remembering that when $x > 0$ and r is real we can write

$$x^r = e^{r \ln x}, \tag{14}$$

we can use this equation to *define* x^r when r is complex. Then

$$x^{(\lambda + i\mu)} = e^{(\lambda + i\mu) \ln x} = e^{\lambda \ln x} e^{i\mu \ln x}$$

$$= x^\lambda e^{i\mu \ln x}$$

$$= x^\lambda [\cos(\mu \ln x) + i \sin(\mu \ln x)], \qquad x > 0. \tag{15}$$

With this definition of x^r for complex values of r, it can be verified that the usual laws of algebra and the differential calculus hold, and hence x^{r_1} and x^{r_2} are indeed solutions of Eq. (1). The general solution of Eq. (1) is

$$y = c_1 x^{\lambda + i\mu} + c_2 x^{\lambda - i\mu}. \tag{16}$$

The disadvantage of this expression is that the functions $x^{\lambda + i\mu}$ and $x^{\lambda - i\mu}$ are complexed-valued. Recall that we had a similar situation for the second order differential equation with constant coefficients when the roots of the characteristic equation were complex. In the same way as we did then we observe that the real and imaginary parts of $x^{\lambda + i\mu}$,

$$x^\lambda \cos(\mu \ln x) \qquad \text{and} \qquad x^\lambda \sin(\mu \ln x), \tag{17}$$

are linearly independent solutions of Eq. (1). Hence for the case of complex roots, the general solution of Eq. (1) is

$$y = c_1 x^\lambda \cos(\mu \ln x) + c_2 x^\lambda \sin(\mu \ln x), \qquad x > 0. \tag{18}$$

Example 3. Solve

$$x^2 y'' + xy' + y = 0. \tag{19}$$

Substituting $y = x^r$ gives

$$x^r[r(r - 1) + r + 1] = 0.$$

Hence $r = \pm i$, and the general solution is

$$y = c_1 \cos(\ln x) + c_2 \sin(\ln x), \qquad x > 0. \tag{20}$$

Now let us consider the interval $x < 0$. The difficulty here is in explaining what is meant by x^r when x is negative, and r is not an integer; similarly $\ln x$ is not defined for $x < 0$. The solutions of the Euler equation that we have given for $x > 0$ can be shown to be valid for $x < 0$ but they are in general complex-valued. Thus in Example 1 the solution $x^{1/2}$ is imaginary for $x < 0$.

It is always possible to obtain real-valued solutions of the Euler equation (1) in the interval $x < 0$ by making the following change of variable. Let $x = -\xi$ where $\xi > 0$, and let $y = u(\xi)$. Further, we have

$$\frac{dy}{dx} = \frac{du}{d\xi}\frac{d\xi}{dx} = -\frac{du}{d\xi}, \qquad \frac{d^2y}{dx^2} = \frac{d}{d\xi}\left[-\frac{du}{d\xi}\right]\frac{d\xi}{dx} = \frac{d^2u}{d\xi^2}. \tag{21}$$

Thus Eq. (1) takes the form

$$\xi^2\frac{d^2u}{d\xi^2} + \alpha\xi\frac{du}{d\xi} + \beta u = 0, \qquad \xi > 0. \tag{22}$$

But this is exactly the problem that we have just solved; from Eqs. (6), (11), and (18) we have

$$u(\xi) = \begin{cases} c_1\xi^{r_1} + c_2\xi^{r_2} \\ (c_1 + c_2\ln\xi)\xi^{r_1} \\ c_1\xi^\lambda\cos(\mu\ln\xi) + c_2\xi^\lambda\sin(\mu\ln\xi) \end{cases} \tag{23}$$

depending on whether $(\alpha - 1)^2 - 4\beta$ is positive, zero, or negative. To obtain u in terms of x we replace ξ by $-x$ in Eqs. (23).

We can combine our results for $x > 0$ and $x < 0$ as follows. We have $|x| = x$ for $x > 0$ and $|x| = -x$ for $x < 0$. Thus we need only replace x by $|x|$ in Eqs. (6), (11), and (18) to obtain real-valued solutions valid in any interval not containing the origin (also see Problems 6 and 7). These results are summarized in the following theorem.

Theorem 4.2 *To solve the Euler equation* (1)

$$x^2y'' + \alpha xy' + \beta y = 0$$

in any interval not containing the origin, substitute $y = x^r$, *and compute the roots* r_1 *and* r_2 *of the equation*

$$F(r) = r^2 + (\alpha - 1)r + \beta = 0.$$

If the roots are real and unequal,

$$y = c_1|x|^{r_1} + c_2|x|^{r_2}. \tag{24}$$

If the roots are real and equal,

$$y = (c_1 + c_2\ln|x|)|x|^{r_1}. \tag{25}$$

If the roots are complex,

$$y = |x|^\lambda[c_1\cos(\mu\ln|x|) + c_2\sin(\mu\ln|x|)], \tag{26}$$

where $r_1, r_2 = \lambda \pm i\mu$.

For an Euler equation of the form

$$(x - x_0)^2 y'' + \alpha(x - x_0)y' + \beta y = 0, \tag{27}$$

the situation is exactly the same. One looks for solutions of the form $y = (x - x_0)^r$. The general solution is given by either Eq. (24), (25), or (26) with x replaced by $(x - x_0)$. Alternatively, we can reduce Eq. (27) to the form of Eq. (1) by making the change of independent variable $t = x - x_0$.

The situation for a general second order differential equation with a regular singular point is similar to that for an Euler equation. We will consider that problem in the next section.

PROBLEMS

1. Determine the general solution of each of the following second order differential equations that is valid in any interval not including the singular point.

(a) $x^2 y'' + 4xy' + 2y = 0$ (b) $(x + 1)^2 y'' + 3(x + 1)y' + \frac{3}{4}y = 0$

(c) $x^2 y'' - 3xy' + 4y = 0$ (d) $x^2 y'' + 3xy' + 5y = 0$

(e) $x^2 y'' - xy' + y = 0$ (f) $(x - 1)^2 y'' + 8(x - 1)y' + 12y = 0$

(g) $x^2 y'' + 6xy' - y = 0$ (h) $2x^2 y'' - 4xy' + 6y = 0$

(i) $x^2 y'' - 5xy' + 9y = 0$ (j) $(x - 2)^2 y'' + 5(x - 2)y' + 8y = 0$

(k) $x^2 y'' + 2xy' + 4y = 0$ (l) $x^2 y'' - 4xy' + 4y = 0$

2. Using the method of reduction of order, show that if r_1 is a repeated root of $r(r - 1) + \alpha r + \beta = 0$, then x^{r_1} and $x^{r_1} \ln x$ are solutions of

$$x^2 y'' + \alpha xy' + \beta y = 0, \qquad x > 0.$$

3. Find all values of α for which all solutions of $x^2 y'' + \alpha xy' + \frac{5}{2}y = 0$ approach zero as $x \to 0$.

4. The second order Euler equation $x^2 y'' + \alpha xy' + \beta y = 0$ can be reduced to a second order linear equation with constant coefficients by an appropriate change of the independent variable. Remembering that x^r plays the same role for the Euler equation as $e^{rx} = (e^x)^r$ plays for the equation with constant coefficients, it would appear reasonable to make the change of variable $x = e^z$, or $z = \ln x$. Consider only the interval $x > 0$.

(a) Show that

$$\frac{dy}{dx} = \frac{1}{x}\frac{dy}{dz} \qquad \text{and} \qquad \frac{d^2y}{dx^2} = \frac{1}{x^2}\frac{d^2y}{dz^2} - \frac{1}{x^2}\frac{dy}{dz}$$

(b) Show that the Euler equation becomes

$$\frac{d^2y}{dz^2} + (\alpha - 1)\frac{dy}{dz} + \beta y = 0.$$

Letting r_1 and r_2 denote the roots of $r^2 + (\alpha - 1)r + \beta = 0$, show that
    (c) if r_1 and r_2 are real and unequal, then

$$y = c_1 e^{r_1 z} + c_2 e^{r_2 z} = c_1 x^{r_1} + c_2 x^{r_2},$$

(d) if r_1 and r_2 are equal, then

$$y = (c_1 + c_2 z)e^{r_1 z} = (c_1 + c_2 \ln x)x^{r_1},$$

(e) if r_1 and r_2 are complex conjugates, $r_1 = \lambda + i\mu$, then

$$y = e^{\lambda z}(c_1 \cos \mu z + c_2 \sin \mu z) = x^\lambda[c_1 \cos (\mu \ln x) + c_2 \sin (\mu \ln x)].$$

5. Using the method of Problem 4, solve the following differential equations for $x > 0$.

(a) $x^2 y'' - 2y = 0$

(b) $x^2 y'' - 3xy' + 4y = \ln x$

(c) $x^2 y'' + 7xy' + 5y = x$

(d) $x^2 y'' - 2xy' + 2y = 3x^2 + 2 \ln x$

(e) $x^2 y'' + xy' + 4y = \sin (\ln x)$

(f) $3x^2 y'' + 12xy' + 9y = 0$

6. Show that if $L[y] = x^2 y'' + \alpha xy' + \beta y$, then

$$L[(-x)^r] = (-x)^r F(r)$$

for $x < 0$, where $F(r) = r(r - 1) + \alpha r + \beta$. Hence conclude that if $r_1 \neq r_2$ are roots of $F(r) = 0$, then linearly independent solutions of $L[y] = 0$ for $x < 0$ are $(-x)^{r_1}$ and $(-x)^{r_2}$.

7. Suppose that x^{r_1} and x^{r_2} are solutions of an Euler equation, where $r_1 \neq r_2$, and r_1 is an integer. According to Eq. (24) the general solution in any interval not containing the origin is $y = c_1 |x|^{r_1} + c_2 |x|^{r_2}$. Show that the general solution can also be written as $y = k_1 x^{r_1} + k_2 |x|^{r_2}$.

Hint: Show by a proper choice of constants that the expressions are identical for $x > 0$; and by a different choice of constants that they are identical for $x < 0$.

8. If the constants α and β in the Euler equation $x^2 y'' + \alpha xy' + \beta y = 0$ are complex numbers, it is still possible to obtain solutions of the form x^r. However, in general, the solutions are no longer real-valued. Determine the general solution of each of the following equations for the interval $x > 0$.

(a) $x^2 y'' + 2ixy' - iy = 0$

(b) $x^2 y'' + (1 - i)xy' + 2y = 0$

(c) $x^2 y'' + xy' - 2iy = 0$

Hint: See Problems 13, 14, and 15 of Section 3.5.1.

4.5   SERIES SOLUTIONS NEAR A REGULAR SINGULAR POINT, PART I

We will now consider the question of solving the general second order linear equation

$$P(x)y'' + Q(x)y' + R(x)y = 0 \tag{1}$$

in the neighborhood of a regular singular point $x = x_0$. For convenience we will assume that $x_0 = 0$. If $x_0 \neq 0$ the equation can be transformed into one for which the regular singular point is at the origin by letting $x - x_0$ equal z.

The fact that $x = 0$ is a regular singular point of Eq. (1) means that $xQ(x)/P(x) = xp(x)$ and $x^2 R(x)/P(x) = x^2 q(x)$ have finite limits as $x \to 0$, and are analytic at $x = 0$. Thus they have power series expansions of the form

$$xp(x) = \sum_{n=0}^{\infty} p_n x^n, \qquad x^2 q(x) = \sum_{n=0}^{\infty} q_n x^n, \tag{2}$$

which are convergent for some interval $|x| < \rho$, $\rho > 0$, about the origin. It is convenient for theoretical purposes to divide Eq. (1) by $P(x)$ and then to multiply by x^2, obtaining

$$x^2 y'' + x[xp(x)]y' + [x^2 q(x)]y = 0, \tag{3a}$$

or

$$x^2 y'' + x(p_0 + p_1 x + \cdots + p_n x^n + \cdots)y'$$
$$+ (q_0 + q_1 x + \cdots + q_n x^n + \cdots)y = 0. \tag{3b}$$

If all of the p_n and q_n are zero except

$$p_0 = \lim_{x \to 0} \frac{xQ(x)}{P(x)}, \qquad q_0 = \lim_{x \to 0} \frac{x^2 R(x)}{P(x)},$$

then Eq. (3) reduces to the Euler equation

$$x^2 y'' + p_0 xy' + q_0 y = 0, \tag{4}$$

which was discussed in the previous section. In general, of course, some of the p_n and q_n, $n \geq 1$, are not zero. However, the essential character of solutions of Eq. (3) is identical to that of solutions of the Euler equation. The presence of the terms $p_1 x + \cdots + p_n x^n + \cdots$ and $q_1 x + \cdots + q_n x^n + \cdots$ merely complicates the calculations.

We shall primarily restrict our discussion to the interval $x > 0$. The interval $x < 0$ can be treated, just as for the Euler equation, by making the change of variable $x = -\xi$ and then solving the resulting equation for $\xi > 0$.

Since the coefficients in Eq. (3) are "Euler coefficients" times power series, it is natural to seek solutions of the form of "Euler solutions" times power series:

$$y = x^r(a_0 + a_1 x + \cdots + a_n x^n + \cdots) = x^r \sum_{n=0}^{\infty} a_n x^n = \sum_{n=0}^{\infty} a_n x^{r+n}. \tag{5}$$

As part of our problem we have to determine:

1. The values of r for which Eq. (1) has a solution of the form (5).

2. The recurrence relation for the a_n.

3. The radius of convergence of the series $\sum_{n=0}^{\infty} a_n x^n$.

We shall find that Eq. (1) always has at least one solution of the form (5) with a_0 arbitrary, and possibly a second solution corresponding to a second value of r. If there is not a second solution of the form (5), the second solution will involve a logarithmic term just as did the second solution of the Euler equation when the roots were equal. In theory, of course, once we obtain the first solution, which is of the form (5), a second solution can be obtained by the method of reduction of order. Unfortunately, this is not always a convenient method for obtaining a second solution; we will give an alternate method in Sections 4.6 and 4.7.

The general theory is due to the German mathematician Frobenius (1849–1917) and is fairly complicated. Rather than trying to present this theory we shall simply assume in this and the next three sections that there does exist a solution of the stated form. In particular we assume that any power series in an expression for a solution has a nonzero radius of convergence, and shall concentrate on showing how to determine the coefficients in such a series.

Let us first illustrate the method of Frobenius for a specific differential equation with two solutions of the form (5). Consider the equation

$$2x^2y'' - xy' + (1 + x)y = 0. \tag{6}$$

By comparison with Eq. (3), it is clear that $x = 0$ is a regular singular point of Eq. (6). Also by comparison with Eqs. (3a) and (3b) we see that $xp(x) = -\frac{1}{2}$ and $x^2q(x) = (1 + x)/2$. Thus $p_0 = -\frac{1}{2}$, $q_0 = \frac{1}{2}$, $q_1 = \frac{1}{2}$, and all other p's and q's are zero. The corresponding Euler equation is

$$2x^2y'' - xy' + y = 0. \tag{7}$$

We will formally try to find a solution of Eq. (6) of the form (5) for $x > 0$. Then y' and y'' are given by

$$y' = \sum_{n=0}^{\infty} a_n(r + n)x^{r+n-1}, \tag{8}$$

and

$$y'' = \sum_{n=0}^{\infty} a_n(r + n)(r + n - 1)x^{r+n-2}. \tag{9}$$

Hence

$$2x^2y'' - xy' + (1 + x)y = \sum_{n=0}^{\infty} 2a_n(r + n)(r + n - 1)x^{r+n}$$

$$- \sum_{n=0}^{\infty} a_n(r + n)x^{r+n} + \sum_{n=0}^{\infty} a_n x^{r+n} + \sum_{n=0}^{\infty} a_n x^{r+n+1}.$$

Noting that the last sum can be written in the form $\sum_{n=1}^{\infty} a_{n-1}x^{r+n}$ and combining terms gives

$$2x^2y'' - xy' + (1 + x)y = a_0[2r(r - 1) - r + 1]x^r$$

$$+ \sum_{n=1}^{\infty} \{[2(r + n)(r + n - 1) - (r + n) + 1]a_n + a_{n-1}\}x^{r+n} = 0. \tag{10}$$

If Eq. (10) is to be identically satisfied, the coefficient of each power of x in Eq. (10) must be zero. Corresponding to x^r we obtain, since $a_0 \neq 0$,*

$$2r(r - 1) - r + 1 = 2r^2 - 3r + 1 = 0. \tag{11}$$

---

* Nothing is gained by taking $a_0 = 0$, as this would simply mean that the series would start with the term $a_1 x^{r+1}$ or, if $a_1 = 0$, the term $a_2 x^{r+2}$, and so on. Suppose the first nonzero term is $a_m x^{r+m}$; this is equivalent to using the series (5) with r replaced by $r + m$ and a_0 replaced by a_m. Indeed r is chosen to be the lowest power of x appearing in the series and we take $a_0 \neq 0$ as its coefficient.

From the coefficient of x^{r+n} we have

$$[2(r + n)(r + n - 1) - (r + n) + 1]a_n + a_{n-1} = 0,$$

or

$$a_n = -\frac{1}{2(r + n)^2 - 3(r + n) + 1} a_{n-1}, \qquad n \geq 1. \tag{12}$$

Equation (11) is called the *indicial equation* for Eq. (6). Note that it is exactly the equation we would obtain for the Euler equation (7) associated with Eq. (6). Its roots determine the two possible values of r for which there may be solutions of Eq. (6) of the form (5). Factoring Eq. (11) gives $(2r - 1)(r - 1) = 0$; hence the roots of the indicial equation are

$$r_1 = 1, \qquad r_2 = \tfrac{1}{2}. \tag{13}$$

In the following, whenever the roots of the indicial equation are real and unequal, the larger one will be denoted by r_1. The roots of the indicial equation are often referred to as the *exponents of the singularity* at the regular singular point.

Equation (12) gives a recurrence relation for the coefficients a_n. Corresponding to the factored form of the indicial equation we can write the recurrence relation as

$$a_n = \frac{-1}{[2(r + n) - 1][(r + n) - 1]} a_{n-1}, \qquad n \geq 1. \tag{14}$$

For each root r_1 and r_2 of the indicial equation we use the recurrence relation to determine the corresponding a_n. For $r = r_1 = 1$ we have

$$a_n = \frac{-1}{(2n + 1)n} a_{n-1}, \qquad n \geq 1.$$

Thus

$$a_n = \frac{-1}{(2n + 1)n} \frac{-1}{(2n - 1)(n - 1)} a_{n-2}$$

$$\cdot$$
$$\cdot$$
$$\cdot$$

$$= \frac{(-1)^n}{(2n + 1)n(2n - 1)(n - 1) \cdots (5 \cdot 2)(3 \cdot 1)} a_0.$$

Upon grouping the factors $n, n - 1, \ldots, 1$, we have

$$a_n = \frac{(-1)^n}{[(2n + 1)(2n - 1) \cdots 5 \cdot 3]n!} a_0, \qquad n \geq 1. \tag{15}$$

Hence one solution of Eq. (6), omitting the constant multiplier a_0, is

$$y_1(x) = x\left[1 + \sum_{n=1}^{\infty} \frac{(-1)^n x^n}{[(2n + 1)(2n - 1) \cdots 5 \cdot 3]n!}\right], \qquad x > 0. \tag{16}$$

To determine the radius of convergence of the series in Eq. (16) we use the ratio test:

$$\lim_{n \to \infty} \left| \frac{a_{n+1}x^{n+1}}{a_n x^n} \right| = \lim_{n \to \infty} \frac{|x|}{(2n + 3)(n + 1)} = 0$$

for all x. Thus the series converges for all x.

Corresponding to the second root $r = r_2 = \frac{1}{2}$, we proceed similarly. From Eq. (14) we have

$$a_n = \frac{-1}{2n(n - \frac{1}{2})} a_{n-1} = \frac{-1}{n(2n - 1)} a_{n-1}$$

$$= \frac{-1}{n(2n - 1)} \frac{-1}{(n - 1)(2n - 3)} a_{n-2}$$

$$\vdots$$

$$= \frac{(-1)^n}{n(2n - 1)(n - 1)(2n - 3) \cdots (2 \cdot 3)(1 \cdot 1)} a_0$$

$$= \frac{(-1)^n}{[(2n - 1)(2n - 3) \cdots 3 \cdot 1]n!} a_0, \qquad n \geq 1. \tag{17}$$

Again omitting the constant multiplier a_0, we obtain a second solution

$$y_2(x) = x^{1/2}\left[1 + \sum_{n=1}^{\infty} \frac{(-1)^n x^n}{[(2n - 1)(2n - 3) \cdots 3 \cdot 1]n!} \right], \qquad x > 0. \tag{18}$$

As before, we can show that the series in Eq. (18) converges for all x. Since the leading terms in the series solutions y_1 and y_2 are x and $x^{1/2}$, respectively, it is clear that the solutions are linearly independent. Hence the general solution of Eq. (6) is

$$y = c_1 y_1(x) + c_2 y_2(x), \qquad x > 0.$$

The preceding example is illuminating in several respects. Let us reconsider our calculations from a more general viewpoint. Let

$$F(r) = 2r(r - 1) - r + 1$$
$$= (2r - 1)(r - 1); \tag{19}$$

then the indicial equation is $F(r) = 0$, and the recurrence relation (12) becomes

$$F(r + n)a_n + a_{n-1} = 0, \qquad n \geq 1 \tag{20}$$

Equation (20) is typical of the general case except that the term a_{n-1} is, in general, replaced by a linear combination of $a_0, a_1, a_2, \ldots, a_{n-1}$. However, we can use Eq. (20) as a model for the discussion that follows with the understanding that the discussion is not limited to the specific example just completed.

There are two possible sources of trouble. First, if the two roots r_1 and r_2 of the indicial equation are equal, then we will obtain only one series solution of the form (5). Second, there is the possibility that for a given root r_1 or r_2 we may not be able to solve for one of the a_n from Eq. (20) because $F(r + n)$ may be zero. When can this happen? The only zeros of $F(r)$ are r_1 and r_2, and if $r_1 > r_2$, then clearly $r_1 + n > r_1 > r_2$ for all n and $F(r_1 + n) \neq 0$ for all n. Thus there will never be any difficulty in computing the series solution corresponding to the larger of the roots of the indicial equation. On the other hand consider $F(r_2 + n)$, and suppose that $r_1 - r_2 = N$, a positive integer. Then for $n = N$ we have

$$F(r_2 + N)a_N + a_{N-1} = 0;$$

but $F(r_2 + N) = F(r_1) = 0$ and we cannot solve for a_N unless also $a_{N-1} = 0$, and hence cannot obtain a second series solution. In more general problems, as we shall see in the next section, the recurrence relation will be more complicated, and even though $F(r_2 + N)$ may vanish, it may be possible to determine a second solution of the form (5) corresponding to $r = r_2$. In any event the cases $r_1 = r_2$ and $r_1 - r_2 = N$, a positive integer, require special attention.

Finally, we note that if the roots of the indicial equation are complex, then this type of difficulty cannot occur; on the other hand, the solutions corresponding to r_1 and r_2 will be complex-valued functions of x. However, just as for the Euler equation, it is possible to obtain real-valued solutions by taking the real and imaginary parts of the complex-valued solutions.

Before turning to a general discussion and the special cases $r_1 = r_2$ and $r_1 - r_2 = N$, a positive integer, a practical point should be mentioned. If P, Q, and R are polynomials, it is much better to work directly with Eq. (1) than with Eq. (3). This avoids the necessity of expressing $xQ(x)/P(x)$ and $x^2R(x)/P(x)$ as power series. For example, it is more convenient to consider the equation

$$x(1 + x)y'' + 2y' + xy = 0$$

than to write it in the form

$$x^2y'' + x\,\frac{2}{1 + x}\,y' + \frac{x^2}{1 + x}\,y = 0,$$

and then to expand $2/(1 + x)$ and $x^2/(1 + x)$, which gives

$$x^2y'' + x[2(1 - x + x^2 - \cdots)]y' + (x^2 - x^3 + x^4 - \cdots)y = 0.$$

PROBLEMS

1. Show that each of the following differential equations has a regular singular point at $x = 0$. Determine the indicial equation, the recurrence relation, and the roots of the indicial equation. Find the series solution ($x > 0$) corresponding to the larger root. If

the roots are unequal and do not differ by an integer, find the series solution corresponding to the smaller root.

(a) $2xy'' + y' + xy = 0$
(b) $x^2y'' + xy' + (x^2 - \frac{1}{9})y = 0$
(c) $xy'' + y = 0$
(d) $xy'' + y' - y = 0$
(e) $3x^2y'' + 2xy' + x^2y = 0$
(f) $x^2y'' + xy' + (x - 2)y = 0$

2. The Legendre equation of order α is

$$(1 - x^2)y'' - 2xy' + \alpha(\alpha + 1)y = 0.$$

The solution of this equation near the ordinary point $x = 0$ was discussed in Problems 10 and 11 of Section 4.2.1. In Example 5 of Section 4.3 it was shown that $x = \pm 1$ are regular singular points. Determine the indicial equation and its roots for the point $x = 1$. Find a series solution in powers of $x - 1$ for $x - 1 > 0$.
Hint: Write $1 + x = 2 + (x - 1)$, and $x = 1 + (x - 1)$. Alternatively, make the change of variable $x - 1 = z$ and determine a series solution in powers of z.

3. The Laguerre (1834–1886) differential equation is

$$xy'' + (1 - x)y' + \lambda y = 0.$$

Show that $x = 0$ is a regular singular point. Determine the indicial equation, its roots, the recurrence relation, and one solution ($x > 0$). Show that if $\lambda = m$, a positive integer, this solution reduces to a polynomial. When properly normalized this polynomial is known as the Laguerre polynomial, $L_m(x)$.

4. The Bessel equation of order zero is

$$x^2y'' + xy' + x^2y = 0.$$

Show that $x = 0$ is a regular singular point; that the roots of the indicial equation are $r_1 = r_2 = 0$; and that one solution for $x > 0$ is

$$J_0(x) = 1 + \sum_{n=1}^{\infty} \frac{(-1)^n x^{2n}}{2^{2n}(n!)^2}.$$

Notice that the series converges for all x, not just $x > 0$. In particular, $J_0(x)$ is bounded as $x \to 0$. The function J_0 is known as the Bessel function of the first kind of order zero.

5. Referring to Problem 4 show, using the method of reduction of order, that the second solution of the Bessel equation of order zero contains a logarithmic term.
Hint: If $y_2(x) = J_0(x)v(x)$, then

$$y_2(x) = J_0(x) \int^x \frac{ds}{s[J_0(s)]^2}.$$

Find the first term in the series expansion for $1/s[J_0(s)]^2$.

6. The Bessel equation of order one is

$$x^2y'' + xy' + (x^2 - 1)y = 0.$$

(a) Show that $x = 0$ is a regular singular point; that the roots of the indicial equation are $r_1 = 1$ and $r_2 = -1$; and that one solution for $x > 0$ is

$$J_1(x) = \frac{x}{2} \sum_{n=0}^{\infty} \frac{(-1)^n x^{2n}}{(n + 1)!\, n!\, 2^{2n}}.$$

Notice as in Problem 4 that the series actually converges for all x and that $J_1(x)$ is bounded as $x \to 0$. The function J_1 is known as the Bessel function of the first kind of order one.

(b) Show that it is impossible to determine a second solution of the form

$$x^{-1} \sum_{n=0}^{\infty} b_n x^n, \qquad x > 0.$$

4.5.1 SERIES SOLUTIONS NEAR A REGULAR SINGULAR POINT, PART II

Now let us consider the general problem of determining a solution of the equation

$$x^2 y'' + x[xp(x)]y' + [x^2 q(x)]y = 0 \tag{1}$$

where

$$xp(x) = \sum_{n=0}^{\infty} p_n x^n \quad \text{and} \quad x^2 q(x) = \sum_{n=0}^{\infty} q_n x^n,$$

and both series have nonzero radii of convergence [see Eq. (3) of Section 4.5]. The point $x = 0$ is a regular singular point. For convenience, we first suppose that $x > 0$, and look for a solution of the form

$$y = \sum_{n=0}^{\infty} a_n x^{r+n}. \tag{2}$$

Substituting in Eq. (1) gives

$$a_0 r(r-1)x^r + a_1(r+1)rx^{r+1} + \cdots + a_n(r+n)(r+n-1)x^{r+n} + \cdots$$
$$+ (p_0 + p_1 x + \cdots + p_n x^n + \cdots)$$
$$\times [a_0 r x^r + a_1(r+1)x^{r+1} + \cdots + a_n(r+n)x^{r+n} + \cdots]$$
$$+ (q_0 + q_1 x + \cdots + q_n x^n + \cdots)$$
$$\times (a_0 x^r + a_1 x^{r+1} + \cdots + a_n x^{r+n} + \cdots) = 0. \tag{3}$$

Multiplying the infinite series together and then collecting terms, we obtain

$$a_0 F(r)x^r + [a_1 F(r+1) + a_0(p_1 r + q_1)]x^{r+1}$$
$$+ \{a_2 F(r+2) + a_0(p_2 r + q_2) + a_1[p_1(r+1) + q_1]\}x^{r+2}$$
$$+ \cdots + \{a_n F(r+n) + a_0(p_n r + q_n) + a_1[p_{n-1}(r+1) + q_{n-1}]$$
$$+ \cdots + a_{n-1}[p_1(r+n-1) + q_1]\}x^{r+n} \cdots = 0,$$

or in a more compact form

$$a_0 F(r)x^r + \sum_{n=1}^{\infty} \left\{ F(r+n)a_n + \sum_{k=0}^{n-1} a_k[(r+k)p_{n-k} + q_{n-k}] \right\}x^{r+n} = 0, \tag{4}$$

where

$$F(r) = r(r-1) + p_0 r + q_0. \tag{5}$$

If Eq. (4) is to be satisfied identically, the coefficient of each power of x must be zero.

Since $a_0 \neq 0$, the term involving x^r yields the indicial equation $F(r) = 0$. We emphasize again that the indicial equation $F(r) = 0$ is exactly the equation we would obtain in looking for solutions x^r of the Euler equation $x^2y'' + xp_0y' + q_0y = 0$ corresponding to Eq. (1). Let us denote the roots of the indicial equation by r_1 and r_2 with $r_1 \geq r_2$ if the roots are real. If the roots are complex the ordering is immaterial. Only for these values of r can we expect to find solutions of Eq. (1) of the form (2).

Setting the coefficient of x^{r+n} in Eq. (4) equal to zero gives the recurrence relation

$$F(r + n)a_n + \sum_{k=0}^{n-1} a_k[(r + k)p_{n-k} + q_{n-k}] = 0, \qquad n \geq 1. \tag{6}$$

Equation (6) shows that, in general, a_n depends on the value of r and all of the preceding coefficients $a_0, a_1, \ldots, a_{n-1}$. It also shows that we can successively compute $a_1, a_2, \ldots, a_n, \ldots$, in terms of the coefficients in the series for $xp(x)$ and $x^2q(x)$ provided that $F(r + 1), F(r + 2), \ldots, F(r + n), \ldots$ do not vanish. The only values of r for which $F(r)$ vanishes are $r = r_1$ and $r = r_2$; since $r_1 + n$ is not equal to r_1 or r_2 for $n \geq 1$, it follows that $F(r_1 + n) \neq 0$ for $n \geq 1$. Hence we can always determine one solution

$$y_1(x) = x^{r_1}\left[1 + \sum_{n=1}^{\infty} a_n(r_1)x^n\right]. \tag{7}$$

Here we have introduced the notation $a_n(r_1)$ to indicate that a_n has been determined from Eq. (6) with $r = r_1$. Also for definiteness we have taken a_0 to be one.

If r_2 is not equal to r_1, and $r_1 - r_2$ is not a positive integer, then $r_2 + n$ is not equal to r_1 for any value of $n \geq 1$; hence $F(r_2 + n) \neq 0$, and we can obtain a second solution

$$y_2(x) = x^{r_2}\left[1 + \sum_{n=1}^{\infty} a_n(r_2)x^n\right]. \tag{8}$$

Before continuing, we will make several remarks about the series solutions (7) and (8). Within their radii of convergence, the power series $\sum_{n=0}^{\infty} a_n(r_1)x^n$ and $\sum_{n=0}^{\infty} a_n(r_2)x^n$ define functions that are analytic at $x = 0$. Thus the singular behavior, if there is any, of the solutions y_1 and y_2 is determined by the terms x^{r_1} and x^{r_2} which multiply these two analytic functions, respectively. Next, to obtain real valued solutions for $x < 0$, we can make the substitution $x = -\xi$ with $\xi > 0$. As we might expect from our discussion of the Euler equation it turns out that we need only replace x^{r_1} in Eq. (7) and x^{r_2} in Eq. (8) by $|x|^{r_1}$ and $|x|^{r_2}$, respectively. Finally, we note that if r_1 and r_2 are complex numbers, then they are necessarily complex conjugates and $r_2 \neq r_1 + N$. Thus we can compute two series solutions of the form (2); however, they are complex-valued functions of x. Real-valued solutions can be obtained by taking the real and imaginary parts of the complex-valued solutions.

If $r_2 = r_1$, we obtain only one solution of the form (2). The second solution contains a logarithmic term. A procedure for determining a second solution is discussed in Section 4.6 and an example is given in Section 4.7.

If $r_1 - r_2 = N$, where N is a positive integer, then $F(r_2 + N) = F(r_1) = 0$, and we cannot determine $a_N(r_2)$ from Eq. (6) unless it happens that the sum in Eq. (6) also vanishes for $n = N$. If this is so, it can be shown that a_N is arbitrary and we can obtain a second series solution; if not, there is only one series solution of the form (2). If the latter is the case, then the second solution contains a logarithmic term. The procedure is discussed briefly in Section 4.6, and examples of both cases are given in Section 4.7.

The next question to consider is the convergence of the infinite series in the formal solution. As we might expect, this is related to the radii of convergence of the power series for $xp(x)$ and $x^2q(x)$. We summarize the results of our discussion and the additional conclusions about the convergence of the series solution in the following theorem.

Theorem 4.3. *Consider the differential equation* (1)

$$x^2 y'' + x[xp(x)]y' + [x^2 q(x)]y = 0,$$

where $x = 0$ is a regular singular point. Then $xp(x)$ and $x^2 q(x)$ are analytic at $x = 0$ with convergent power series expansions

$$xp(x) = \sum_{n=0}^{\infty} p_n x^n, \qquad x^2 q(x) = \sum_{n=0}^{\infty} q_n x^n$$

for $|x| < \rho$, where $\rho > 0$ is the minimum of the radii of convergence of the power series for $xp(x)$ and $x^2q(x)$. Let r_1 and r_2 be the roots of the indicial equation

$$F(r) = r(r - 1) + p_0 r + q_0 = 0,$$

with $r_1 \geq r_2$ if r_1 and r_2 are real. Then in either of the intervals $-\rho < x < 0$ or $0 < x < \rho$, there exists a solution of the form

$$y_1(x) = |x|^{r_1} \left[1 + \sum_{n=1}^{\infty} a_n(r_1) x^n \right] \qquad (9)$$

where the $a_n(r_1)$ are given by the recurrence relation (6) with $a_0 = 1$ and $r = r_1$. If $r_1 - r_2$ is not zero or a positive integer, then in either of the intervals $-\rho < x < 0$ or $0 < x < \rho$, there exists a second linearly independent solution of the form

$$y_2(x) = |x|^{r_2} \left[1 + \sum_{n=1}^{\infty} a_n(r_2) x^n \right]. \qquad (10)$$

The $a_n(r_2)$ are also determined by the recurrence relation (6) with $a_0 = 1$ and $r = r_2$. The power series in Eqs. (9) and (10) converge at least for $|x| < \rho$.

It is possible that the series in Eqs. (9) and (10) may have radii of convergence greater than ρ; if so, the solutions (9) and (10) will be valid in correspondingly larger intervals. While it is beyond the scope of this text to prove this theorem, we can make several observations. To determine the roots r_1 and r_2 of the indicial equation, which play such an important role in the present theory, we need only determine p_0 and q_0 and then solve the equation $r(r - 1) + p_0 r + q_0 = 0$. These coefficients are given by

$$p_0 = \lim_{x \to 0} x p(x), \qquad q_0 = \lim_{x \to 0} x^2 q(x). \tag{11}$$

In practice, rather than remember the recurrence formula (6) for the a_n we usually substitute the series (2) for y in Eq. (1), determine the values of r, and then explicitly calculate the a_n.

Finally we note that if $x = 0$ is a regular singular point of the equation

$$P(x)y'' + Q(x)y' + R(x)y = 0, \tag{12}$$

where the functions P, Q, and R are polynomials, then $xp(x) = xQ(x)/P(x)$ and $x^2 q(x) = x^2 R(x)/P(x)$. Thus

$$p_0 = \lim_{x \to 0} x \frac{Q(x)}{P(x)}, \qquad q_0 = \lim_{x \to 0} x^2 \frac{R(x)}{P(x)}. \tag{13}$$

Further, the radius of convergence for the series in Eqs. (9) and (10) is at least equal to the distance from the origin to the nearest zero of P.

Example.   Discuss the nature of the solutions of the equation

$$2x(1 + x)y'' + (3 + x)y' - xy = 0$$

in the neighborhood of the singular points.

This equation is of the form (12) with $P(x) = 2x(1 + x)$, $Q(x) = 3 + x$, and $R(x) = -x$. Clearly the points $x = 0$ and $x = -1$ are singular points. The point $x = 0$ is a regular singular point, since

$$\lim_{x \to 0} x \frac{Q(x)}{P(x)} = \lim_{x \to 0} x \frac{3 + x}{2x(1 + x)} = \frac{3}{2},$$

$$\lim_{x \to 0} x^2 \frac{R(x)}{P(x)} = \lim_{x \to 0} x^2 \frac{-x}{2x(1 + x)} = 0.$$

Further from Eq. (13), $p_0 = \frac{3}{2}$ and $q_0 = 0$. Thus the indicial equation is $r(r - 1) + \frac{3}{2}r = 0$, and the roots are $r_1 = 0$, $r_2 = -\frac{1}{2}$. Since these roots are not equal and do not differ by an integer there are two linearly independent solutions of the form

$$y_1(x) = 1 + \sum_{n=1}^{\infty} a_n(0)x^n \quad \text{and} \quad y_2(x) = |x|^{-\frac{1}{2}}\left[1 + \sum_{n=1}^{\infty} a_n(-\tfrac{1}{2})x^n\right]$$

for $0 < |x| < \rho$. A lower bound for the radius of convergence of the series is 1, the distance from $x = 0$ to $x = -1$, the other zero of $P(x)$. Notice that the solution y_1 is bounded as $x \to 0$, indeed is analytic there. On the other hand, the second solution y_2 is unbounded as $x \to 0$.

The point $x = -1$ is also a regular singular point, since

$$\lim_{x \to -1} (x + 1) \frac{Q(x)}{P(x)} = \lim_{x \to -1} \frac{(x + 1)(3 + x)}{2x(1 + x)} = -1,$$

$$\lim_{x \to -1} (x + 1)^2 \frac{R(x)}{P(x)} = \lim_{x \to -1} \frac{(x + 1)^2(-x)}{2x(1 + x)} = 0.$$

In this case $p_0 = -1$, $q_0 = 0$ so the indicial equation is $r(r - 1) - r = 0$. The roots of the indicial equation are $r_1 = 2$ and $r_2 = 0$. Corresponding to the larger root there is a solution of the form

$$y_1(x) = (x + 1)^2 \left[1 + \sum_{n=1}^{\infty} a_n(2)(x + 1)^n \right].$$

The series converges at least for $|x + 1| < 1$. Since the two roots differ by a positive integer there may or may not be a second solution of the form

$$y_2(x) = \left[1 + \sum_{n=1}^{\infty} a_n(0)(x + 1)^n \right].$$

We cannot say more without further analysis.

PROBLEMS

1. Find all the regular singular points of each of the following differential equations. Determine the indicial equation and the exponents of the singularity at each regular singular point.

(a) $xy'' + 2xy' + 6e^x y = 0$

(b) $x^2 y'' - x(2 + x)y' + (2 + x^2)y = 0$

(c) $x(x - 1)y'' + 6x^2 y' + 3y = 0$

(d) $y'' + 4xy' + 6y = 0$

(e) $x^2 y'' + 3(\sin x)y' - 2y = 0$

(f) $2x(x + 2)y'' + y' - xy = 0$

(g) $x^2 y'' + \frac{1}{2}(x + \sin x)y' + y = 0$

(h) $(x + 1)^2 y'' + 3(x^2 - 1)y' + 3y = 0$

2. Show that

$$x^2 y'' + (\sin x)y' - (\cos x)y = 0$$

has a regular singular point at $x = 0$, and that the roots of the indicial equation are ± 1. Determine the first three nonzero terms in the series corresponding to the larger root.

3. Show that

$$(\ln x)y'' + \frac{1}{2}y' + y = 0$$

has a regular singular point at $x = 1$. Determine the roots of the indicial equation at $x = 1$. Determine the first three nonzero terms in the series $\sum_{n=0}^{\infty} a_n(x - 1)^{r+n}$ corresponding to the larger root. Take $x - 1 > 0$. What would you expect the radius of convergence of the series to be?

4. In several problems in mathematical physics (for example, the Schrödinger equation for a hydrogen atom) it is necessary to study the differential equation

$$x(1 - x)y'' + [\gamma - (1 + \alpha + \beta)x]y' - \alpha\beta y = 0, \qquad (i)$$

where α, β, and γ are constants. This equation is known as the *hypergeometric* equation.

(a) Show that $x = 0$ is a regular singular point, and that the roots of the indicial equation are 0 and $1 - \gamma$.

(b) Show that $x = 1$ is a regular singular point, and that the roots of the indicial equation are 0 and $\gamma - \alpha - \beta$.

(c) Assuming that $1 - \gamma$ is not a positive integer, show that in the neighborhood of $x = 0$ one solution of (i) is

$$y_1(x) = 1 + \frac{\alpha\beta}{\gamma \cdot 1!} x + \frac{\alpha(\alpha + 1)\beta(\beta + 1)}{\gamma(\gamma + 1)2!} x^2 + \cdots.$$

What would you expect the radius of convergence of this series to be?

(d) Assuming that $1 - \gamma$ is not an integer or zero, show that a second solution for $0 < x < 1$ is

$$y_2(x) = x^{(1-\gamma)}\left[1 + \frac{(\alpha - \gamma + 1)(\beta - \gamma + 1)}{(2 - \gamma)1!} x \right.$$
$$\left. + \frac{(\alpha - \gamma + 1)(\alpha - \gamma + 2)(\beta - \gamma + 1)(\beta - \gamma + 2)}{(2 - \gamma)(3 - \gamma)2!} x^2 + \cdots \right].$$

*(e) Show that the point at infinity is a regular singular point, and that the roots of the indicial equation are α and β. See Problem 13 of Section 4.3.

5. Consider the differential equation

$$x^3y'' + \alpha xy' + \beta y = 0$$

where $\alpha \neq 0$ and β are real constants.

(a) Show that $x = 0$ is an irregular singular point.

(b) Show that if we attempt to determine a solution of the form $x^r \sum\limits_{n=0}^{\infty} a_n x^n$, the indicial equation for r will be linear, and as a consequence there will be only one formal solution of this form.

6. Consider the differential equation

$$y'' + \frac{\alpha}{x^s} y' + \frac{\beta}{x^t} y = 0, \qquad (i)$$

where $\alpha \neq 0$ and $\beta \neq 0$ are real numbers, and s and t are positive integers that for the moment are arbitrary.

(a) Show that if $s > 1$ or $t > 2$, then the point $x = 0$ is an irregular singular point.

(b) Suppose we try to find a solution of Eq. (i) of the form

$$y = \sum_{n=0}^{\infty} a_n x^{r+n}, \qquad x > 0. \qquad (ii)$$

Show that if $s = 2$ and $t = 2$, then there is only one possible value of r for which there is a formal solution of Eq. (i) of the form (ii).

(c) Show that if $s = 1$ and $t = 3$ there are no solutions of Eq. (i) of the form (ii).

(d) Show that the maximum values of s and t for which the indicial equation is quadratic in r [and hence we can hope to find two solutions of the form (ii)] are $s = 1$ and $t = 2$. These are precisely the conditions that distinguish a "weak singularity," or a regular singular point from an irregular singular point, as we defined them in Section 4.3.

As a note of caution we should point out that while it is sometimes possible to obtain a formal series solution of the form (ii) at an irregular singular point, the series may not converge.

*4.6   SERIES SOLUTIONS NEAR A REGULAR SINGULAR POINT; $r_1 = r_2$ AND $r_1 - r_2 = N$

We now consider the cases in which the roots r_1 and r_2 of the indicial equation are equal or differ by a positive integer, $r_1 - r_2 = N$. If $r_1 = r_2$, we would expect by analogy with the Euler equation that the second solution will contain a logarithmic term. This may also be the case when the roots differ by an integer. The general situation, including the case in which r_1 and r_2 do not differ by an integer, is summarized in the following theorem.

Theorem 4.4.   *Consider the differential equation*

$$L[y] = x^2 y'' + x[xp(x)]y' + [x^2 q(x)]y = 0, \tag{1}$$

where $x = 0$ is a regular singular point. Then the functions $xp(x)$ and $x^2 q(x)$ are analytic at $x = 0$ with power series representations

$$xp(x) = \sum_{n=0}^{\infty} p_n x^n, \qquad x^2 q(x) = \sum_{n=0}^{\infty} q_n x^n, \tag{2}$$

which converge for $|x| < \rho$, where $\rho > 0$ is the minimum of the radii of convergence of the power series for $xp(x)$ and $x^2 q(x)$. Let r_1 and r_2, where $r_1 \geq r_2$ if they are real, be the roots of the indicial equation

$$F(r) = r(r - 1) + p_0 r + q_0 = 0. \tag{3}$$

Then in either of the intervals $-\rho < x < 0$ or $0 < x < \rho$, Eq. (1) has two linearly independent solutions y_1 and y_2 of the following form.

1. *If $r_1 - r_2$ is not an integer or zero, then*

$$y_1(x) = |x|^{r_1} \left[1 + \sum_{n=1}^{\infty} a_n(r_1) x^n \right], \tag{4a}$$

$$y_2(x) = |x|^{r_2} \left[1 + \sum_{n=1}^{\infty} a_n(r_2) x^n \right]. \tag{4b}$$

2. *If $r_1 = r_2$, then*

$$y_1(x) = |x|^{r_1} \left[1 + \sum_{n=1}^{\infty} a_n(r_1) x^n \right], \tag{5a}$$

$$y_2(x) = y_1(x) \ln |x| + |x|^{r_2} \sum_{n=1}^{\infty} b_n(r_1) x^n. \tag{5b}$$

3. If $r_1 - r_2 = N$, a positive integer, then

$$y_1(x) = |x|^{r_1}\left[1 + \sum_{n=1}^{\infty} a_n(r_1)x^n\right], \tag{6a}$$

$$y_2(x) = ay_1(x)\ln|x| + |x|^{r_2}\left[1 + \sum_{n=1}^{\infty} c_n(r_2)x^n\right]. \tag{6b}$$

The coefficients $a_n(r_1)$, $a_n(r_2)$, $b_n(r_1)$, $c_n(r_2)$, and the constant a can be determined by substituting the form of the series solutions for y in Eq. (1). The constant a may turn out to be zero. Each of the series in Eq. (4), (5), and (6) converges at least for $|x| < \rho$ and defines a function that is analytic at $x = 0$.

The case in which $r_1 - r_2$ is not an integer was discussed in Section 4.5.1. We turn now to the cases $r_1 = r_2$ and $r_1 - r_2 = N$, and assume that $x > 0$. Real-valued solutions for $x < 0$ can be obtained, as before, by letting $x = -\xi$.

$r_1 = r_2$. The method of finding the second solution is essentially the same as that which we used in finding the second solution of the Euler equation (see Section 4.4) when the roots of the indicial equation were equal. We seek a solution of Eq. (1) of the form

$$y = \phi(r, x) = x^r \sum_{n=0}^{\infty} a_n x^n = x^r \left(a_0 + \sum_{n=1}^{\infty} a_n x^n\right), \tag{7}$$

where we have written $y = \phi(r, x)$ to emphasize the fact that ϕ depends on the choice of r. If we substitute the series (2) for $xp(x)$ and $x^2q(x)$, compute and substitute for y' and y'' from Eq. (7), and finally collect terms, we obtain (see Eqs. (3) and (4) of Section 4.5.1)

$$L[\phi](r, x) = x^r a_0 F(r) + \sum_{n=1}^{\infty}\left\{a_n F(r + n) + \sum_{k=0}^{n-1} a_k[(r + k)p_{n-k} + q_{n-k}]\right\}x^{r+n}, \tag{8}$$

where

$$F(r) = r(r - 1) + p_0 r + q_0. \tag{9}$$

If the roots of the indicial equation are both equal to r_1, we can obtain one solution by setting r equal to r_1 and choosing the a_n so that each term of the series in Eq. (8) is zero.

To determine a second solution we will think of r as a continuous variable, and determine a_n as a function of r by requiring that the coefficient of x^{r+n}, $n \geq 1$, in Eq. (8) be zero. Then assuming $F(r + n) \neq 0$ we have

$$a_n(r) = -\frac{\sum_{k=0}^{n-1} a_k[(r + k)p_{n-k} + q_{n-k}]}{F(r + n)}, \quad n \geq 1. \tag{10}$$

With this choice for $a_n(r)$ for $n \geq 1$, Eq. (8) reduces to

$$L[\phi](r, x) = a_0 x^r F(r). \tag{11}$$

Since r_1 is a repeated root of $F(r)$, it follows from Eq. (9) that $F(r) = (r - r_1)^2$. Setting $r = r_1$ in Eq. (11) we find that $L[\phi](r_1, x) = 0$; hence, as we already know,

$$y_1(x) = \phi(r_1, x) = x^{r_1}\left[a_0 + \sum_{n=1}^{\infty} a_n(r_1)x^n\right], \qquad x > 0, \qquad (12)$$

is one solution of Eq. (1). But more important, it also follows from Eq. (11), just as for the Euler equation, that

$$L\left[\frac{\partial\phi}{\partial r}\right](r_1, x) = a_0 \frac{\partial}{\partial r}\left[x^r(r - r_1)^2\right]\bigg|_{r=r_1}$$

$$= a_0[(r - r_1)^2 x^r \ln x + 2(r - r_1)x^r]|_{r=r_1}$$

$$= 0. \qquad (13)$$

Hence, a second solution of Eq. (1) is

$$y_2(x) = \frac{\partial\phi(r, x)}{\partial r}\bigg|_{r=r_1} = \frac{\partial}{\partial r}\left\{x^r\left[a_0 + \sum_{n=1}^{\infty} a_n(r)x^n\right]\right\}\bigg|_{r=r_1}$$

$$= (x^{r_1} \ln x)\left[a_0 + \sum_{n=1}^{\infty} a_n(r_1)x^n\right] + x^{r_1}\sum_{n=1}^{\infty} a'_n(r_1)x^n$$

$$= y_1(x) \ln x + x^{r_1}\sum_{n=1}^{\infty} a'_n(r_1)x^n, \qquad x > 0, \qquad (14)$$

where $a'_n(r_1)$ denotes da_n/dr evaluated at $r = r_1$ and $a_n(r)$ is given by Eq. (10).

Note that in addition to the assumptions we have already made, we have also assumed that it is permissible to differentiate the series $\sum_{n=1}^{\infty} a_n(r)x^n$ term by term with respect to r. Further, it is certainly not a trivial matter to determine $a_n(r_1)$ and $a'_n(r_1)$; however, the determination of two solutions of a second order differential equation with a regular singular point is not a trivial problem.

In summary, there are three ways that we can proceed to find a second solution when $r_1 = r_2$. First, we can compute $b_n(r_1)$ directly by substituting the expression (5b) for y in Eq. (1). Second, we can compute $b_n(r_1) = a'_n(r_1)$ by first determining $a_n(r)$ and then calculating $a'_n(r_1)$. Notice that in calculating the first solution it is necessary to know $a_n(r_1)$ so it may be just as easy to calculate the general expression for $a_n(r)$ from which we can calculate both $a_n(r_1)$ and $a'_n(r_1)$. The first procedure may be more convenient if only a few terms in the series for $y_2(x)$ are needed or if the formula for $a_n(r)$ is very complicated or difficult to obtain. The third alternative is to use the method of reduction of order, but this will not avoid the use of series.

$r_1 - r_2 = N$. For this case the detailed arguments to derive the form (6b) of the second solution are considerably more complicated and will not be given.

However, we note that if we assume $y_2(x) = \sum\limits_{n=0}^{\infty} a_n x^{r_2+n}$, then there may be difficulty in computing $a_N(r_2)$ from Eq. (10) since $F(r + N) = (r + N - r_1)(r + N - r_2) = (r - r_2)(r + N - r_2)$ vanishes for $r = r_2$. This difficulty is circumvented by choosing a_0, which can be chosen arbitrarily, to be $r - r_2$. Then each of the a's will be proportional to $r - r_2$, so there will be a factor of $r - r_2$ in the numerator of Eq. (10) to cancel the corresponding factor in the denominator that occurs when $n = N$. Following an analysis similar to that for the case $r_1 = r_2$, it turns out that the solution is of the form (6b) with

$$c_n(r_2) = \frac{d}{dr} [(r - r_2)a_n(r)]\bigg|_{r=r_2}, \qquad n = 1, 2, \ldots \tag{15}$$

where $a_n(r)$ is given by Eq. (10) with $a_0 = r - r_2$, and with

$$a = \lim_{r \to r_2} (r - r_2)a_N(r). \tag{16}$$

If $a_N(r_2)$ is finite, then $a = 0$ and there is no logarithmic term in y_2. The procedure is discussed in Coddington (Chapter 4).

In practice the best way to determine whether a is zero in the second solution is simply to try to compute the a_n corresponding to the root r_2 and to see whether it is possible to determine $a_N(r_2)$. If so, there is no further problem. If not, we must use the form (6b) with $a \neq 0$.

Again, in summary, there are three ways to find a second solution when $r_1 - r_2 = N$. First, we can calculate a and $c_n(r_2)$ directly by substituting the expression (6b) for y in Eq. (1). Second, we can calculate $c_n(r_2)$ and a of Eq. (6b) using the formulas (15) and (16). If this is the planned procedure, then in calculating the solution corresponding to $r = r_1$ be sure to obtain the general formula for $a_n(r)$ rather than just $a_n(r_1)$. The third alternative is to use the method of reduction of order.

In the next section we will give three examples that illustrate the cases $r_1 = r_2$, $r_1 - r_2 = N$ with a zero, and $r_1 - r_2 = N$ with a nonzero.

*4.7 BESSEL'S EQUATION

In this section we consider three special cases of Bessel's equation,

$$x^2 y'' + xy' + (x^2 - \nu^2)y = 0, \tag{1}$$

where ν is a constant, which illustrate the theory discussed in Section 4.6. Clearly $x = 0$ is a regular singular point. For simplicity we consider only the case $x > 0$.

Bessel Equation of Order Zero.   This example illustrates the situation in which the roots of the indicial equation are equal. Setting $\nu = 0$ in Eq. (1) gives

$$L[y] = x^2 y'' + xy' + x^2 y = 0. \tag{2}$$

Substituting

$$y = \phi(r, x) = a_0 x^r + \sum_{n=1}^{\infty} a_n x^{r+n}, \tag{3}$$

we obtain

$$L[\phi](r, x) = \sum_{n=0}^{\infty} a_n[(n + r)(n + r - 1) + (n + r)]x^{r+n} + \sum_{n=0}^{\infty} a_n x^{r+n+2}$$

$$= a_0[r(r - 1) + r]x^r + a_1[(r + 1)r + (r + 1)]x^{r+1}$$

$$+ \sum_{n=2}^{\infty} \{a_n[(n + r)(n + r - 1) + (n + r)] + a_{n-2}\}x^{r+n} = 0. \tag{4}$$

The roots of the indicial equation $F(r) = r(r - 1) + r = 0$ are $r_1 = 0$ and $r_2 = 0$; hence we have the case of equal roots. The recurrence relation is

$$a_n(r) = \frac{-a_{n-2}(r)}{(n + r)(n + r - 1) + (n + r)} = -\frac{a_{n-2}(r)}{(n + r)^2}, \qquad n \geq 2. \tag{5}$$

To determine $y_1(x)$ we set r equal to 0. Then from Eq. (4) it follows that for the coefficient of x^{r+1} to be zero we must choose $a_1 = 0$. Hence from Eq. (5), $a_3 = a_5 = a_7 = \cdots = a_{2n+1} = \cdots = 0$. Further

$$a_n(0) = -a_{n-2}(0)/n^2, \qquad n = 2, 4, 6, 8, \ldots,$$

or letting $n = 2m$,

$$a_{2m}(0) = -a_{2m-2}(0)/(2m)^2, \qquad m = 1, 2, 3, \ldots.$$

Thus

$$a_2(0) = -\frac{a_0}{2^2},$$

$$a_4(0) = -\frac{a_2(0)}{(2 \cdot 2)^2} = +\frac{a_0}{(2 \cdot 2)^2 2^2} = \frac{a_0}{2^4 2^2}$$

$$a_6(0) = -\frac{a_4(0)}{(2 \cdot 3)^2} = -\frac{a_0}{(2 \cdot 3)^2 2^4 2^2} = -\frac{a_0}{2^6(3 \cdot 2)^2}$$

$$\vdots$$

$$a_{2m}(0) = \frac{(-1)^m a_0}{2^{2m}(m!)^2}, \qquad m = 1, 2, 3, \ldots. \tag{6}$$

Hence

$$y_1(x) = a_0\left[1 + \sum_{m=1}^{\infty} \frac{(-1)^m x^{2m}}{2^{2m}(m!)^2}\right], \qquad x > 0. \tag{7}$$

The function in brackets is known as the *Bessel function of the first kind of order zero*, and is denoted by $J_0(x)$. It follows from Theorem 4.4 that the series converges for all x, and that J_0 is analytic at $x = 0$. Some of the important properties of J_0 are discussed in the problems.

In this example we will determine $y_2(x)$ by computing $a'_n(0)$. The alternate procedure in which we simply substitute the form (5b) of Section 4.6 in Eq. (2), and then determine the b_n is discussed in Problem 7. First we note from the coefficient of x^{r+1} in Eq. (4) that $(r + 1)^2 a_1(r) = 0$. It follows that not only does $a_1(0) = 0$, but also $a'_1(0) = 0$. It is easy to deduce from the recurrence relation (5) that $a'_3(0) = a'_5(0) = \ldots = a'_{2n+1}(0) = \ldots = 0$; hence we need only compute $a'_{2m}(0)$, $m = 1, 2, 3, \ldots$. From Eq. (5)

$$a_{2m}(r) = -a_{2m-2}(r)/(2m + r)^2, \qquad m = 1, 2, 3, \ldots.$$

Hence

$$a_2(r) = -\frac{a_0}{(2 + r)^2}$$

$$a_4(r) = -\frac{a_2(r)}{(4 + r)^2} = +\frac{a_0}{(4 + r)^2(2 + r)^2}$$

$$a_6(r) = -\frac{a_4}{(6 + r)^2} = -\frac{a_0}{(6 + r)^2(4 + r)^2(2 + r)^2}$$

$$\cdot$$
$$\cdot$$
$$\cdot$$

$$a_{2m}(r) = \frac{(-1)^m a_0}{(2m + r)^2(2m - 2 + r)^2 \cdots (4 + r)^2(2 + r)^2}, \qquad m = 1, 2, 3, \ldots. \quad (8)$$

The computation of $a'_{2m}(r)$ can be carried out most conveniently by noting that if

$$f(x) = (x - \alpha_1)^{\beta_1}(x - \alpha_2)^{\beta_2}(x - \alpha_3)^{\beta_3} \cdots (x - \alpha_n)^{\beta_n},$$

then

$$f'(x) = \beta_1(x - \alpha_1)^{\beta_1 - 1}[(x - \alpha_2)^{\beta_2} \cdots (x - \alpha_n)^{\beta_n}]$$
$$\qquad + \beta_2(x - \alpha_2)^{\beta_2 - 1}[(x - \alpha_1)^{\beta_1}(x - \alpha_3)^{\beta_3} \cdots (x - \alpha_n)^{\beta_n}] + \cdots.$$

Hence for x not equal to $\alpha_1, \alpha_2, \ldots, \alpha_n$

$$\frac{f'(x)}{f(x)} = \frac{\beta_1}{x - \alpha_1} + \frac{\beta_2}{x - \alpha_2} + \cdots + \frac{\beta_n}{x - \alpha_n}.$$

Thus, from Eq. (8)

$$\frac{a'_{2m}(r)}{a_{2m}(r)} = -2\left(\frac{1}{2m + r} + \frac{1}{2m - 2 + r} + \cdots + \frac{1}{2 + r}\right),$$

and setting r equal to 0 we obtain

$$a'_{2m}(0) = -2\left[\frac{1}{2m} + \frac{1}{2(m - 1)} + \frac{1}{2(m - 2)} + \cdots + \frac{1}{2}\right]a_{2m}(0).$$

Substituting for $a_{2m}(0)$ from Eq. (6), and letting

$$H_m = \frac{1}{m} + \frac{1}{m-1} + \cdots + \frac{1}{2} + 1, \tag{9}$$

we obtain finally

$$a'_{2m}(0) = -H_m \frac{(-1)^m a_0}{2^{2m}(m!)^2}, \qquad m = 1, 2, 3, \ldots . \tag{10}$$

The second solution of the Bessel equation of order zero is obtained by setting $a_0 = 1$, and substituting for $y_1(x)$ and $a'_{2m}(0) = b_{2m}(0)$ in Eq. (5b) of Section 4.6. We obtain

$$y_2(x) = J_0(x) \ln x + \sum_{m=1}^{\infty} \frac{(-1)^{m+1} H_m}{2^{2m}(m!)^2} x^{2m}, \qquad x > 0. \tag{11}$$

In place of y_2, the second solution is usually taken to be a certain linear combination of J_0 and y_2. It is known as the Bessel function of the second kind of order zero, and is denoted by Y_0. Following Copson (Chapter 12), we define*

$$Y_0(x) = \frac{2}{\pi} [y_2(x) + (\gamma - \ln 2)J_0(x)]. \tag{12}$$

Here γ is a constant, known as the Euler-Máscheroni (1750–1800) constant; it is defined by the equation

$$\gamma = \lim_{n \to \infty} (H_n - \ln n) \cong 0.5772. \tag{13}$$

Substituting for $y_2(x)$ in Eq. (12), we obtain

$$Y_0(x) = \frac{2}{\pi} \left[\left(\gamma + \ln \frac{x}{2} \right) J_0(x) + \sum_{m=1}^{\infty} \frac{(-1)^{m+1} H_m}{2^{2m}(m!)^2} x^{2m} \right], \qquad x > 0. \tag{14}$$

The general solution of the Bessel equation of order zero for $x > 0$ is

$$y = c_1 J_0(x) + c_2 Y_0(x).$$

Notice that since $J_0(x) \to 1$ as $x \to 0$, $Y_0(x)$ has a logarithmic singularity at $x = 0$; that is, $Y_0(x)$ behaves as $(2/\pi) \ln x$ when $x \to 0$ through positive values. Thus if we are interested in solutions of Bessel's equation of order zero that are finite at the origin, which is often the case, we must discard Y_0. The graphs of the functions J_0 and Y_0 are shown in Figure 4.2.

It is interesting to note from Figure 4.2 that for x large both $J_0(x)$ and $Y_0(x)$ are oscillatory. Such a behavior might be anticipated from the original equation; indeed it is true for the solutions of the Bessel equation of order ν. If we divide Eq. (1) by x^2, we obtain

$$y'' + \frac{1}{x} y' + \left(1 - \frac{\nu^2}{x^2} \right) y = 0.$$

---

* Other authors use other definitions for Y_0. The present choice for Y_0 is also known as the Weber (1842–1913) function.

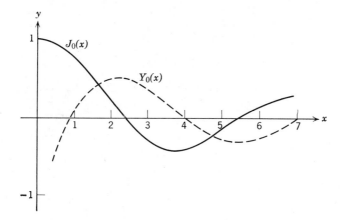

FIGURE 4.2   The Bessel functions of order zero.

For x very large it is reasonable to suspect that the terms $(1/x)y'$ and $(v^2/x^2)y$ are small and hence can be neglected. If this is true, then the Bessel equation of order v can be approximated by

$$y'' + y = 0.$$

The solutions of this equation are $\sin x$ and $\cos x$; thus we might anticipate that the solutions of Bessel's equation for x large are similar to linear combinations of $\sin x$ and $\cos x$. This is correct insofar as the Bessel functions are oscillatory; however, it is only partly correct. For x large the functions J_0 and Y_0 also decay as x increases; in particular, it can be shown that $J_0(x)$ behaves as $(2/\pi x)^{1/2} \cos (x - \pi/4)$ as $x \to \infty$. This cannot be deduced from the equation $y'' + y = 0$. Thus the equation $y'' + y = 0$ does not provide an adequate approximation to the Bessel equation for x large. The solutions of the two equations have some features in common but they also have significant differences. Also notice the solutions of Bessel's equation of order one-half that are discussed next.

Bessel Equation of Order One-Half.   This example illustrates the situation in which the roots of the indicial equation differ by a positive integer, but there is no logarithmic term in the second solution. Setting $v = \frac{1}{2}$ in Eq. (1) gives

$$L[y] = x^2 y'' + xy' + (x^2 - \tfrac{1}{4})y = 0. \tag{15}$$

If we substitute the series (3) for $y = \phi(r, x)$, we obtain

$$L[\phi](r, x) = \sum_{n=0}^{\infty} [(r + n)(r + n - 1) + (r + n) - \tfrac{1}{4}]a_n x^{r+n} \pm \sum_{n=0}^{\infty} a_n x^{r+n+2}$$

$$= (r^2 - \tfrac{1}{4})a_0 x^r + [(r + 1)^2 - \tfrac{1}{4}]a_1 x^{r+1}$$

$$+ \sum_{n=2}^{\infty} \{[(r + n)^2 - \tfrac{1}{4}]a_n + a_{n-2}\}x^{r+n} = 0. \tag{16}$$

The roots of the indicial equation are $r_1 = \frac{1}{2}$, $r_2 = -\frac{1}{2}$; hence the roots differ by an integer. The recurrence relation is

$$[(r + n)^2 - \tfrac{1}{4}]a_n = -a_{n-2}, \qquad n \geq 2. \tag{17}$$

Corresponding to the larger root $r_1 = \frac{1}{2}$ we find from the coefficient of x^{r+1} in Eq. (16) that $a_1 = 0$. Hence, from Eq. (17), $a_3 = a_5 = \cdots = a_{2n+1} = \cdots = 0$. Further, for $r = \frac{1}{2}$

$$a_n = -\frac{a_{n-2}}{n(n + 1)}, \qquad n = 2, 4, 6, \ldots,$$

or letting $n = 2m$,

$$a_{2m} = -\frac{a_{2m-2}}{(2m + 1)2m}, \qquad m = 1, 2, 3, \ldots.$$

Thus

$$a_2 = -\frac{a_0}{3 \cdot 2} = -\frac{a_0}{3!}$$

$$a_4 = -\frac{a_2}{5 \cdot 4} = +\frac{a_0}{5 \cdot 4 \cdot 3!} = \frac{a_0}{5!}$$

$$\cdot$$
$$\cdot$$
$$\cdot$$

$$= \frac{(-1)^m a_0}{(2m + 1)!}, \qquad m = 1, 2, 3, \ldots. \tag{18}$$

Hence, taking $a_0 = 1$, we obtain

$$y_1(x) = x^{1/2}\left[1 + \sum_{m=1}^{\infty} \frac{(-1)^m x^{2m}}{(2m + 1)!}\right]$$

$$= x^{-1/2} \sum_{m=0}^{\infty} \frac{(-1)^m x^{2m+1}}{(2m + 1)!}, \qquad x > 0. \tag{19}$$

The power series in Eq. (19) is precisely the Taylor series for $\sin x$; hence one solution of the Bessel equation of order one-half is $x^{-1/2} \sin x$. The Bessel function of the first kind of order one-half, $J_{1/2}$, is defined as $(2/\pi)^{1/2} y_1$. Thus

$$J_{1/2}(x) = \left(\frac{2}{\pi x}\right)^{1/2} \sin x, \qquad x > 0. \tag{20}$$

Corresponding to the root $r_2 = -\frac{1}{2}$ it is possible that we may have difficulty in computing a_1 since $N = r_1 - r_2 = 1$. However, it is clear from Eq. (16) that for $r = -\frac{1}{2}$ the coefficients of a_0 and a_1 are zero, and hence a_0 and a_1 can be chosen arbitrarily. Then corresponding to a_0 we obtain $a_2, a_4, \ldots$ from the recurrence relation (17); and corresponding to a_1 we obtain $a_3, a_5, a_7, \ldots$. Hence the second solution will not involve a logarithmic term. It is left as an exercise for the student to show from Eq. (17) that for $r = -\frac{1}{2}$,

$$a_{2n} = \frac{(-1)^n a_0}{(2n)!}, \qquad n = 1, 2, \ldots,$$

and

$$a_{2n+1} = \frac{(-1)^n a_1}{(2n+1)!}, \qquad n = 1, 2, \ldots.$$

Hence

$$y_2(x) = x^{-\frac{1}{2}}\left[a_0 \sum_{n=0}^{\infty} \frac{(-1)^n x^{2n}}{(2n)!} + a_1 \sum_{n=0}^{\infty} \frac{(-1)^n x^{2n+1}}{(2n+1)!} \right]$$

$$= a_0 \frac{\cos x}{x^{\frac{1}{2}}} + a_1 \frac{\sin x}{x^{\frac{1}{2}}}, \qquad x > 0. \tag{21}$$

The constant a_1 simply introduces a multiple of $y_1(x)$. The second linearly independent solution of the Bessel equation of order one-half is usually taken to be the solution generated by a_0 with $a_0 = (2/\pi)^{\frac{1}{2}}$. It is denoted by $J_{-\frac{1}{2}}$. Then

$$J_{-\frac{1}{2}}(x) = \left(\frac{2}{\pi x} \right)^{\frac{1}{2}} \cos x, \qquad x > 0. \tag{22}$$

The general solution of Eq. (15) is $y = c_1 J_{\frac{1}{2}}(x) + c_2 J_{-\frac{1}{2}}(x)$.

Bessel Equation of Order One. This example illustrates the situation in which the roots of the indicial equation differ by a positive integer and the second solution involves a logarithmic term. Setting $\nu = 1$ in Eq. (1) gives

$$L[y] = x^2 y'' + xy' + (x^2 - 1)y = 0. \tag{23}$$

If we substitute for $y = \phi(r, x)$ the series (3), and collect terms as in the previous examples, we obtain

$$L[\phi](r, x) = a_0(r^2 - 1)x^r + a_1[(r+1)^2 - 1]x^{r+1}$$

$$+ \sum_{n=2}^{\infty} \{[(r+n)^2 - 1]a_n + a_{n-2}\}x^{r+n} = 0. \tag{24}$$

The roots of the indicial equation are $r_1 = 1$ and $r_2 = -1$. The recurrence relation is

$$[(r+n)^2 - 1]a_n(r) = -a_{n-2}(r), \qquad n \geq 2. \tag{25}$$

Corresponding to the larger root $r = 1$ the recurrence relation is

$$a_n = -\frac{a_{n-2}}{(n+2)n}, \qquad n = 2, 3, 4, \ldots.$$

We also find from the coefficient of x^{r+1} in Eq. (24) that $a_1 = 0$, and hence from the recurrence relation $a_3 = a_5 = \cdots = 0$. For even values of n, let $n = 2m$; then

$$a_{2m} = -\frac{a_{2m-2}}{(2m+2)(2m)} = -\frac{a_{2m-2}}{2^2(m+1)m}, \qquad m = 1, 2, 3, \ldots.$$

Hence

$$a_2 = -\frac{a_0}{2^2(2)(1)}$$

$$a_4 = -\frac{a_2}{2^2(3)(2)} = +\frac{a_0}{2^4(3 \cdot 2)(2 \cdot 1)} = \frac{a_0}{2^4 3! 2!}$$

$$a_6 = -\frac{a_4}{2^2(4)(3)} = -\frac{a_0}{2^6 4! 3!}$$

$$\vdots$$

$$a_{2m} = \frac{(-1)^m a_0}{2^{2m}(m+1)! m!}, \qquad m = 1, 2, 3, \ldots. \tag{26}$$

With $a_0 = 1$ we have

$$y_1(x) = x \sum_{m=0}^{\infty} \frac{(-1)^m x^{2m}}{2^{2m}(m+1)! \, m!}. \tag{27}$$

The Bessel function of the first kind of order one is usually taken to be $\frac{1}{2} y_1$ and is denoted by J_1:

$$J_1(x) = \tfrac{1}{2} y_1(x) = \frac{x}{2} \sum_{m=0}^{\infty} \frac{(-1)^m x^{2m}}{2^{2m}(m+1)! \, m!}. \tag{28}$$

The series converges absolutely for all x; hence the function J_1 is defined for all x.

In determining a second solution of Bessel's equation of order one, we will illustrate the method of direct substitution. The calculation is difficult, and inexperienced students are not expected to perform similar calculations in their entirety. However, it is feasible for a student to obtain a and the first few c_n of Eq. (29) given below. According to Theorem 4.4 we assume that

$$y_2(x) = a J_1(x) \ln x + x^{-1} \left[1 + \sum_{n=1}^{\infty} c_n x^n \right], \qquad x > 0. \tag{29}$$

Computing $y_2'(x)$, $y_2''(x)$, substituting in Eq. (23), and making use of the fact that J_1 is a solution of Eq. (23) gives

$$2a x J_1'(x) + \sum_{n=0}^{\infty} [(n-1)(n-2)c_n + (n-1)c_n - c_n] x^{n-1} + \sum_{n=0}^{\infty} c_n x^{n+1} = 0 \tag{30}$$

where $c_0 = 1$. Substituting for $J_1(x)$ from Eq. (28), shifting the indices of summation in the two series, and carrying out several steps of algebra gives

$$-c_1 + [0 \cdot c_2 + c_0]x + \sum_{n=2}^{\infty} [(n^2 - 1)c_{n+1} + c_{n-1}]x^n$$

$$= -a\left[x + \sum_{m=1}^{\infty} \frac{(-1)^m(2m + 1)x^{2m+1}}{2^{2m}(m + 1)!\, m!}\right]. \quad (31)$$

From Eq. (31) we observe first that $c_1 = 0$, and $a = -c_0 = -1$. Next since there are only odd powers of x on the right, the coefficient of each even power of x on the left must be zero. Thus, since $c_1 = 0$, we have $c_3 = c_5 = \cdots = 0$. Corresponding to the odd powers of x we obtain the recurrence relation (let $n = 2m + 1$ in the series on the left-hand side of Eq. (31))

$$[(2m + 1)^2 - 1]c_{2m+2} + c_{2m} = \frac{(-1)(-1)^m(2m + 1)}{2^{2m}(m + 1)!\, m!}, \quad m = 1, 2, 3, \ldots . \quad (32)$$

When we set $m = 1$ in Eq. (32) we obtain

$$(3^2 - 1)c_4 + c_2 = (-1)3/(2^2 \cdot 2!).$$

Notice that c_2 can be selected *arbitrarily*, and then this equation determines c_4. Also notice that in the equation for the coefficient of x, c_2 appeared multiplied by 0, and that equation was used to determine a. That c_2 is arbitrary is not surprising, since c_2 is the coefficient of x in the expression $x^{-1}\left[1 + \sum_{n=1}^{\infty} c_n x^n\right]$. Consequently, c_2 simply generates a multiple of J_1, and y_2 is only determined up to an additive multiple of J_1. In accord with the usual practice we choose $c_2 = 1/2^2$. Then we obtain

$$c_4 = \frac{-1}{2^4 \cdot 2}\,[\tfrac{3}{2} + 1] = \frac{-1}{2^4 2!}\,[(1 + \tfrac{1}{2}) + 1]$$

$$= \frac{(-1)}{2^4 \cdot 2!}\,(H_2 + H_1).$$

Although it is not an easy task to show (and we do not expect the student to perform this calculation), the solution of the recurrence relation (32) is

$$c_{2m} = \frac{(-1)(-1)^m(H_m + H_{m-1})}{2^{2m}m!\,(m - 1)!}, \quad m = 1, 2, \ldots$$

with the understanding that $H_0 = 0$. Thus

$$y_2(x) = -J_1(x)\ln x + \frac{1}{x}\left[1 - \sum_{m=1}^{\infty} \frac{(-1)^m(H_m + H_{m-1})}{2^{2m}m!\,(m - 1)!}\,x^{2m}\right], \quad x > 0. \quad (33)$$

The calculation of $y_2(x)$ using the alternate procedure [see Eqs. (15) and (16) of Section 4.6] in which we determine the $c_n(r_2)$ is slightly easier. In particular the latter procedure yields the general formula for c_{2m} without the necessity of solving a recurrence relation of the form (32) (see Problem 8). In this regard the reader may also wish to compare the calculations of the second solution of Bessel's equation of order zero in the text and in Problem 7.

The second solution of Eq. (23), the Bessel function of the second kind of order one, Y_1, is usually taken to be a certain linear combination of J_1 and y_2. Following Copson (Chapter 12), Y_1 is defined as

$$Y_1(x) = \frac{2}{\pi} [-y_2(x) + (\gamma - \ln 2)J_1(x)], \tag{34}$$

where γ is defined in Eq. (13). The general solution of Eq. (23) for $x > 0$ is

$$y = c_1 J_1(x) + c_2 Y_1(x).$$

Notice that while J_1 is analytic at $x = 0$, the second solution Y_1 becomes unbounded in the same manner as $1/x$ as $x \to 0$.

PROBLEMS

1. Show that each of the following differential equations has a regular singular point at $x = 0$, and determine two linearly independent solutions for $x > 0$.

(a) $x^2 y'' + 2xy' + xy = 0$ (b) $x^2 y'' + 3xy' + (1 + x)y = 0$

(c) $x^2 y'' + xy' + 2xy = 0$ (d) $x^2 y'' + 4xy' + (2 + x)y = 0$

2. Find two linearly independent solutions of the Bessel equation of order $\frac{3}{2}$,

$$x^2 y'' + xy' + (x^2 - \tfrac{9}{4})y = 0, \qquad x > 0.$$

3. Show that the Bessel equation of order one-half,

$$x^2 y'' + xy' + (x^2 - \tfrac{1}{4})y = 0, \qquad x > 0,$$

can be reduced to the equation

$$v'' + v = 0$$

by the change of dependent variable $y = x^{-1/2}v(x)$. From this conclude that $y_1(x) = x^{-1/2} \cos x$ and $y_2(x) = x^{-1/2} \sin x$ are solutions of the Bessel equation of order one-half.

4. Show directly that the series for $J_0(x)$, Eq. (7), converges absolutely for all x.

5. Show directly that the series for $J_1(x)$, Eq. (28), converges absolutely for all x and that $J_0'(x) = -J_1(x)$.

6. Consider the Bessel equation of order ν

$$x^2 y'' + xy' + (x^2 - \nu^2)y = 0, \qquad x > 0.$$

Take ν real and greater than zero.

(a) Show that $x = 0$ is a regular singular point, and that the roots of the indicial equation are ν and $-\nu$.

(b) Corresponding to the larger root ν, show that one solution is

$$y_1(x) = x^\nu \left[1 + \sum_{m=1}^{\infty} \frac{(-1)^m}{m! \, (m + \nu)(m + \nu - 1) \cdots (2 + \nu)(1 + \nu)} \left(\frac{x}{2}\right)^{2m} \right].$$

(c) If 2ν is not an integer show that a second solution is

$$y_2(x) = x^{-\nu}\left[1 + \sum_{m=1}^{\infty} \frac{(-1)^m}{m!\,(m-\nu)(m-\nu-1)\cdots(2-\nu)(1-\nu)}\left(\frac{x}{2}\right)^{2m}\right].$$

Note that $y_1(x)$ is analytic at $x = 0$, and that $y_2(x)$ is unbounded as $x \to 0$.

(d) Verify by direct methods that the power series in the expressions for $y_1(x)$ and $y_2(x)$ converge absolutely for all x. Also verify that y_2 is a solution provided only that ν is not an integer.

7. In this section we showed that one solution of Bessel's equation of order zero,

$$L[y] = x^2y'' + xy' + x^2y = 0$$

is J_0, where $J_0(x)$ is given by Eq. (7) with $a_0 = 1$. According to Theorem 4.4 a second solution has the form $(x > 0)$

$$y_2(x) = J_0(x) \ln x + \sum_{n=1}^{\infty} b_n x^n.$$

(a) Show that

$$L[y_2](x) = \sum_{n=2}^{\infty} n(n-1)b_n x^n + \sum_{n=1}^{\infty} nb_n x^n + \sum_{n=1}^{\infty} b_n x^{n+2} + 2xJ_0'(x). \qquad \text{(i)}$$

(b) Substituting the series representation for $J_0(x)$ in Eq. (i), show that

$$b_1 x + 2^2 b_2 x^2 + \sum_{n=3}^{\infty} (n^2 b_n + b_{n-2})x^n = -2\sum_{n=1}^{\infty} \frac{(-1)^n 2nx^{2n}}{2^{2n}(n!)^2}. \qquad \text{(ii)}$$

(c) Note that only even powers of x appear on the right-hand side of Eq. (ii). Show that $b_1 = b_3 = b_5 = \cdots = 0$, $b_2 = 1/2^2(1!)^2$, and that

$$(2n)^2 b_{2n} + b_{2n-2} = -2(-1)^n(2n)/2^{2n}(n!)^2, \qquad n = 2, 3, 4, \ldots.$$

Deduce that

$$b_4 = \frac{-1}{2^2 4^2}(1 + \tfrac{1}{2}) \qquad \text{and} \qquad b_6 = \frac{1}{2^2 4^2 6^2}(1 + \tfrac{1}{2} + \tfrac{1}{3}).$$

The general solution of the recurrence relation is $b_{2n} = (-1)^{n+1}H_n/2^{2n}(n!)^2$. Substituting for b_n in the expression for $y_2(x)$ we obtain the solution given in Eq. (11).

8. Find a second solution of Bessel's equation of order one by computing the $c_n(r_2)$ and a of Eq. (6b) of Section 4.6 according to the formulas (15) and (16) of that section. Some guidelines along the way of this calculation are the following. First, use Eq. (24) to show that $a_1(-1)$ and $a_1'(-1) = 0$. Then show that $c_1(-1) = 0$ and, from the recurrence relation, that $c_n(-1) = 0$ for $n = 3, 5, \ldots$. Finally, use Eq. (25) to show that

$$a_{2m}(r) = \frac{(-1)^m a_0}{(2m+r+1)(2m+r-1)^2(2m+r-3)^2 \cdots (r+3)^2(r+1)}$$

$m = 1, 2, 3, \ldots,$ and calculate $c_{2m}(-1) = (-1)^{m+1}(H_m + H_{m-1})/2^{2m}m!(m-1)!$.

9. By a suitable change of variables it is often possible to transform a differential

equation with variable coefficients into a Bessel equation of a certain order. For example, show that a solution of

$$x^2 y'' + (\alpha^2 \beta^2 x^{2\beta} + \tfrac{1}{4} - \nu^2 \beta^2) y = 0, \qquad x > 0,$$

is given by $y = x^{\frac{1}{2}} f(\alpha x^\beta)$ where $f(\xi)$ is a solution of the Bessel equation of order ν.

10. Using the result of Problem 9 show that the general solution of the Airy equation

$$y'' - xy = 0, \qquad x > 0,$$

is $y = x^{\frac{1}{2}} [c_1 f_1(\tfrac{2}{3} i x^{\frac{3}{2}}) + c_2 f_2(\tfrac{2}{3} i x^{\frac{3}{2}})]$ where $f_1(\xi)$ and $f_2(\xi)$ are linearly independent solutions of the Bessel equation of order one-third.

11. It can be shown that J_0 has infinitely many zeros for $x > 0$. In particular the first three zeros are approximately 2.405, 5.520, and 8.653 (see Figure 4.2). Letting $\lambda_j, j = 1, 2, \ldots,$ denote the zeros of J_0, it follows that

$$J_0(\lambda_j x) = \begin{cases} 1, & x = 0, \\ 0, & x = 1. \end{cases}$$

Verify that $y = J_0(\lambda_j x)$ satisfies the differential equation

$$y'' + \frac{1}{x} y' + \lambda_j^2 y = 0, \qquad x > 0.$$

Hence show that

$$\int_0^1 x J_0(\lambda_i x) J_0(\lambda_j x) \, dx = 0 \qquad \text{if} \qquad \lambda_i \neq \lambda_j.$$

This important property of $J_0(\lambda_i x)$ is known as the orthogonality property. *Hint:* Write the differential equation for $J_0(\lambda_i x)$. Multiply it by $x J_0(\lambda_j x)$ and subtract it from $x J_0(\lambda_i x)$ times the differential equation for $J_0(\lambda_j x)$. Then integrate from 0 to 1.

REFERENCES

Coddington, E. A., *An Introduction to Ordinary Differential Equations*, Prentice-Hall, Englewood Cliffs, N.J., 1961.

Copson, E. T., *An Introduction to the Theory of Functions of a Complex Variable*, Oxford University, Oxford, 1935.

Proofs of Theorems 4.1, 4.3, and 4.4 can be found in intermediate or advanced books; for example, see Chapters 3 and 4 of Coddington, or Chapters 3 and 4 of:

Rainville, E. D., *Intermediate Differential Equations*, Second ed., Macmillan, New York, 1964.

Also see these texts for a discussion of the point at infinity, which was mentioned in Problem 13 of Section 4.3. The behavior of solutions near an irregular singular point is an even more advanced topic; a brief discussion can be found in Chapter 5 of:

Coddington, E. A. and Levinson, N., *Theory of Ordinary Differential Equations*, McGraw-Hill, New York, 1955.

More complete discussions of the Bessel equation, the Legendre equation, and many of the other named equations can be found in advanced books on differential equations, methods of applied mathematics, and special functions. A text dealing with special functions such as the Legendre polynomials, the Bessel functions, etc., is:

Hochstadt, H., *Special Functions of Mathematical Physics*, Holt, New York, 1961.

CHAPTER 5

Higher Order Linear Equations

5.1 INTRODUCTION

In this chapter we will extend the theory of second order linear equations developed in Chapter 3 to higher order linear equations. An nth order linear differential equation is an equation of the form

$$P_0(x)\frac{d^n y}{dx^n} + P_1(x)\frac{d^{n-1}y}{dx^{n-1}} + \cdots + P_{n-1}(x)\frac{dy}{dx} + P_n(x)y = G(x). \qquad (1)$$

We will assume, unless otherwise stated, that the functions $P_0, \ldots, P_n$, and G are continuous real-valued functions on some interval $\alpha < x < \beta$, and that P_0 is nowhere zero in the interval. Then dividing Eq. (1) by $P_0(x)$ we obtain

$$L[y] = \frac{d^n y}{dx^n} + p_1(x)\frac{d^{n-1}y}{dx^{n-1}} + \cdots + p_{n-1}(x)\frac{dy}{dx} + p_n(x)y = g(x), \qquad (2)$$

where, using the same notation as in Chapter 3, we have introduced the linear differential operator L. The mathematical theory associated with Eq. (2) is completely analogous to that for the second order linear equation; for this reason we shall, for the most part, simply state the results for the nth order linear equation. The proofs of most of the results are also similar to those for the second order linear equation and are usually left as exercises.

First we note that since Eq. (2) involves the nth derivative of y with respect to x, it will, so to speak, require n integrations to solve Eq. (2). Each of these integrations introduces an arbitrary constant. Hence we can expect that to obtain a unique solution it is necessary to specify n initial conditions, say

$$y(x_0) = y_0, y'(x_0) = y_0', \ldots, y^{(n-1)}(x_0) = y_0^{(n-1)}, \qquad (3)$$

where x_0 may be any point in the interval $\alpha < x < \beta$ and $y_0, y_0', \ldots, y_0^{(n-1)}$ is any set of prescribed real constants. That there does exist such a solution and that it is unique is assured by the following existence and uniqueness theorem.

Theorem 5.1. *If the functions $p_1, p_2, \ldots, p_n$, and g are continuous on the open interval $\alpha < x < \beta$, then there exists one and only one function $y = \phi(x)$ satisfying Eq. (2) on the interval $\alpha < x < \beta$ and the prescribed initial conditions (3).*

The proof of this theorem will not be given here; however, we note that if the functions $p_1, p_2, \ldots, p_n$ are constants we will actually construct the solution of Eq. (2) satisfying the initial conditions (3) (see Sections 5.3–5.5). Even though we know a solution in this case, we do not know it is unique without the use of Theorem 5.1. A proof of the complete theorem can be found in Ince (Section 3.32) or Coddington (Chapter 6).

As in the corresponding second order problem, we will first discuss the problem of solving the homogeneous or complementary equation

$$y^{(n)} + p_1(x)y^{(n-1)} + \cdots + p_{n-1}(x)y' + p_n(x)y = 0. \tag{4}$$

The general theory of the homogeneous equation (4) is discussed in Section 5.2, and a method of solving Eq. (4) when the p_j are constants is given in Section 5.3. Next we will consider the problem of finding a particular solution y_p of the non-homogeneous equation (2). The method of undetermined coefficients and the method of variation of parameters for determining a particular solution of Eq. (2) are discussed in Sections 5.4 and 5.5, respectively.

PROBLEMS

1. Assuming that the functions $p_1, \ldots, p_n$ are continuous on an interval including the origin, and that $y = \phi(x)$ is a solution of the initial value problem

$$y^{(n)} + p_1(x)y^{(n-1)} + \cdots + p_n(x)y = 0, \qquad y(0) = y_0, \ldots, y^{(n-1)}(0) = y_0^{(n-1)},$$

determine $\phi^{(n)}(0)$. Show that if $p_1, \ldots, p_n$ are differentiable at $x = 0$, then $\phi^{(n+1)}(0)$ can be determined in terms of the initial data.

2. We can expect by analogy with the first order and second order differential equations that under suitable conditions on f the nth order differential equation $y^{(n)} = f(x, y, y', y'', \ldots, y^{(n-1)})$ has a solution involving n arbitrary constants. Conversely, a family of functions involving n arbitrary constants can be shown to be the solution of an nth order differential equation. By eliminating the constants $c_1, \ldots, c_n$, determine the differential equation satisfied by each of the following functions.

(a) $y = c_1 + c_2 x + c_3 x^2 + \sin x$ (b) $y = c_1 + c_2 \cos x + c_3 \sin x$

(c) $y = c_1 e^x + c_2 e^{-x} + c_3 e^{2x}$ (d) $y = c_1 x + c_2 x^2 + c_3 x^3$

(e) $y = x + c_1 + c_2 \cos x + c_3 \sin x$ (f) $y = c_1 + c_2 x + c_3 \sinh x + c_4 \cosh x$

3. For each of the following linear equations determine intervals in which a solution is sure to exist.

(a) $y^{\text{iv}} + 4y''' + 3y = x$ (b) $xy''' + (\sin x)y'' + 3y = \cos x$

(c) $x(x - 1)y^{\text{iv}} + e^x y'' + 4x^2 y = 0$ (d) $y''' + xy'' + x^2 y' + x^3 y = \ln x$

5.2   GENERAL THEORY OF nth ORDER LINEAR EQUATIONS

If the functions $y_1, y_2, \ldots, y_n$ are solutions of the nth order linear homogeneous differential equation

$$L[y] = y^{(n)} + p_1(x)y^{(n-1)} + \cdots + p_n(x)y = 0, \tag{1}$$

it follows by direct computation that the linear combination

$$y = c_1 y_1(x) + c_2 y_2(x) + \cdots + c_n y_n(x), \tag{2}$$

where $c_1, \ldots, c_n$ are arbitrary constants, is also a solution of Eq. (1). It is natural to ask whether every solution of Eq. (1) can be expressed as a linear combination of $y_1, y_2, \ldots, y_n$. This will be true if, regardless of the initial conditions

$$y(x_0) = y_0, \, y'(x_0) = y_0', \, \ldots, \, y^{(n-1)}(x_0) = y_0^{(n-1)} \tag{3}$$

that are specified, it is possible to choose the constants $c_1, \ldots, c_n$ so that the linear combination (2) satisfies the initial conditions. Specifically, for any choice of the point x_0 in $\alpha < x < \beta$, and for any choice of $y_0, y_0', y_0'', \ldots, y_0^{(n-1)}$, we must be able to determine $c_1, \ldots, c_n$ so that the equations

$$
\begin{aligned}
c_1 y_1(x_0) \quad &+ \cdots + c_n y_n(x_0) \quad = y_0 \\
c_1 y_1'(x_0) \quad &+ \cdots + c_n y_n'(x_0) \quad = y_0'
\end{aligned}
$$

$$\vdots \tag{4}$$

$$c_1 y_1^{(n-1)}(x_0) + \cdots + c_n y_n^{(n-1)}(x_0) = y_0^{(n-1)}$$

are satisfied. Equations (4) can always be solved for the constants $c_1, \ldots, c_n$, provided that the determinant of the coefficients does not vanish. On the other hand, if the determinant of the coefficients does vanish, it is always possible to choose values of $y_0, y_0', \ldots, y_0^{(n-1)}$ such that Eqs. (4) do not have a solution. Hence a necessary and sufficient condition for the existence of a solution of Eqs. (4) for arbitrary values of $y_0, y_0', \ldots, y_0^{(n-1)}$ is that the Wronskian

$$
W(y_1, y_2, \ldots, y_n) =
\begin{vmatrix}
y_1 & y_2 & \cdots & y_n \\
y_1' & y_2' & \cdots & y_n' \\
\vdots & \vdots & & \vdots \\
y_1^{(n-1)} & y_2^{(n-1)} & \cdots & y_n^{(n-1)}
\end{vmatrix}
\tag{5}
$$

does not vanish at $x = x_0$. Since x_0 can be any point in the interval $\alpha < x < \beta$ it is necessary and sufficient that $W(y_1, y_2, \ldots, y_n)$ be nonzero at every point in the interval. Just as for the second order linear equation, it can be shown that if $y_1, y_2, \ldots, y_n$ are solutions of Eq. (1), then $W(y_1, y_2, \ldots, y_n)$ is either identically zero on the interval $\alpha < x < \beta$ or else never vanishes (see Problem 2). Hence we have the following theorem.

Theorem 5.2.   *If the functions $p_1, p_2, \ldots, p_n$ are continuous on the open interval $\alpha < x < \beta$, if the functions $y_1, y_2, \ldots, y_n$ are solutions of Eq. (1), and if $W(y_1, \ldots, y_n)(x) \neq 0$ at least at one point in $\alpha < x < \beta$, then any solution of Eq. (1) can be expressed as a linear combination of the solutions $y_1, y_2, \ldots, y_n$.*

Such a set of solutions $y_1, y_2, \ldots, y_n$ of Eq. (1) is referred to as a *fundamental set of solutions* of Eq. (1). That a fundamental set of solutions exists can be shown in precisely the same way as for the second order linear equation (see Theorem 3.7). Since all solutions of Eq. (1) are of the form (2), it is customary to use the term *general solution* to refer to an arbitrary linear combination of any fundamental set of solutions of Eq. (1).

The discussion of linear independence given in Section 3.3 can also be generalized. The functions $y_1, \ldots, y_n$ are said to be *linearly independent* on $\alpha < x < \beta$ if there is no set of constants $c_1, c_2, \ldots, c_n$ (except $c_1 = c_2 = \cdots = c_n = 0$) such that

$$c_1 y_1(x) + c_2 y_2(x) + \cdots + c_n y_n(x) = 0 \tag{6}$$

for all x in $\alpha < x < \beta$. If $y_1, \ldots, y_n$ are solutions of Eq. (1), it can be shown that a necessary and sufficient condition for them to be linearly independent is that $W(y_1, \ldots, y_n)$ be nonzero on $\alpha < x < \beta$ (see Problem 3). Hence the functions forming a fundamental set of solutions of Eq. (1) are linearly independent, and a linearly independent set of n solutions of Eq. (1) forms a fundamental set of solutions of Eq. (1).

Nonhomogeneous Problem.   Now consider the nonhomogeneous equation

$$L[y] = y^{(n)} + p_1(x)y^{(n-1)} + \cdots + p_n(x)y = g(x). \tag{7}$$

If y_{p_1} and y_{p_2} are any two particular solutions of Eq. (7), it follows immediately from the linearity of the operator L that $L[y_{p_1} - y_{p_2}] = g - g = 0$. Hence the difference of any two solutions of the nonhomogeneous equation (7) is a solution of the homogeneous equation (1). Since any solution of the homogeneous equation can be expressed as a linear combination of a fundamental set of solutions $y_1, y_2, \ldots y_n$, it follows that any solution of Eq. (7) can be written as

$$\begin{aligned}
y &= y_c(x) + y_p(x) \\
&= c_1 y_1(x) + c_2 y_2(x) + \cdots + c_n y_n(x) + y_p(x),
\end{aligned} \tag{8}$$

where y_p is any solution of the nonhomogeneous equation (7). The linear combination (8) is usually referred to as the *general solution* of the nonhomogeneous equation (7).

The primary problem is to determine a fundamental set of solutions $y_1, y_2, \ldots, y_n$. If the coefficients in the differential equation are constants this is a fairly simple problem; it is discussed in the next section. If the coefficients are not constants, it is usually necessary to use numerical methods (Chapter 8) or series methods similar to those used for the second order linear differential equation with variable coefficients.*

In conclusion, it can be shown that the method of reduction of order also applies to nth order linear differential equations. Thus if y_1 is one solution of Eq. (1),

---

* A discussion of series methods for higher order equations can be found in Ince (Chapter 16) or Coddington and Levinson (Chapter 4).

the substitution $y = y_1(x)v(x)$ leads to a linear differential equation of order $n - 1$ for v' (see Problems 7 and 8). Corresponding to y_1 and the $n - 1$ linearly independent solutions $v_1, \ldots, v_{n-1}$ of the reduced equation we obtain the fundamental set of solutions $y_1, y_1v_1, \ldots, y_1v_{n-1}$ of Eq. (1). If one solution of the equation for v' is known, the method of reduction of order can again be used to obtain a linear differential equation of order $n - 2$ and so on until a first order equation is obtained. However, in practice the method of reduction of order is seldom useful for equations of higher than second order. If $n \geq 3$ the reduced equation is itself at least of second order, and only rarely will it be simpler to solve than the original equation. On the other hand, as we have seen in Section 3.4, if $n = 2$ the reduced equation is of first order, and hence considerable simplification may be achieved.

PROBLEMS

1. Verify that the differential operator L defined by $L[y] = y^{(n)} + p_1(x)y^{(n-1)} + \cdots + p_n(x)y$ is a linear differential operator. That is, show that

$$L[c_1y_1 + c_2y_2] = c_1L[y_1] + c_2L[y_2],$$

where y_1 and y_2 are n times differentiable functions and c_1 and c_2 are arbitrary constants. Hence, show that if $y_1, y_2, \ldots, y_n$ are solutions of $L[y] = 0$, then the linear combination $c_1y_1 + \cdots + c_ny_n$ is also a solution of $L[y] = 0$.

2. In Section 3.2 it was shown that the Wronskian $W(y_1, y_2)$ of two solutions of $y'' + p_1(x)y' + p_2(x)y = 0$ can be written as $W(y_1, y_2)(x) = c \exp\left[-\int^x p_1(t)\, dt\right]$, where c is a constant. More generally, it can be shown that if $y_1, y_2, \ldots, y_n$ are solutions of $y^{(n)} + p_1(x)y^{(n-1)} + \cdots + p_n(x)y = 0$ for $\alpha < x < \beta$, then

$$W(y_1, y_2, \ldots, y_n)(x) = c \exp\left[-\int^x p_1(t)\, dt\right].$$

This is known as Abel's identity. To show this result for $n = 3$ we can proceed as follows.

(a) Show that

$$W' = \begin{vmatrix} y_1 & y_2 & y_3 \\ y_1' & y_2' & y_3' \\ y_1''' & y_2''' & y_3''' \end{vmatrix},$$

where $W(y_1, y_2, y_3) = W$.

Hint: The derivative of a 3 by 3 determinant is the sum of three 3 by 3 determinants with the first, second, and third rows differentiated, respectively.

(b) Substitute for y_1''', y_2''', and y_3''' from the differential equation; multiply the first row by p_3, the second by p_2, and add these to the last row to obtain

$$W' = -p_1W.$$

The desired result follows from this equation. The proof for the general case is similar. Since the exponential function never vanishes, this result shows that $W(y_1, y_2, \ldots, y_n)$ either is identically zero or else is nowhere zero on $\alpha < x < \beta$.

3. The purpose of this problem is to show that if $W(y_1, y_2, \ldots, y_n)$ is not identically zero on $\alpha < x < \beta$, then $y_1, y_2, \ldots, y_n$ are linearly independent; and if they are linearly independent and solutions of

$$L[y] = y^{(n)} + p_1(x)y^{(n-1)} + \cdots + p_n(x)y = 0, \qquad \alpha < x < \beta, \qquad \text{(i)}$$

then $W(y_1, y_2, \ldots, y_n)$ is nowhere zero on $\alpha < x < \beta$.

(a) Suppose $W(y_1, y_2, \ldots, y_n)$ is not identically zero on $\alpha < x < \beta$. To show that $y_1, y_2, \ldots, y_n$ are linearly independent we must show that it is impossible to find constants $c_1, c_2, \ldots, c_n$ (not all zero) such that

$$c_1 y_1(x) + c_2 y_2(x) + \cdots + c_n y_n(x) = 0 \qquad \text{(ii)}$$

for all x in $\alpha < x < \beta$. By writing the equations for the first, second, $\ldots$, and $(n-1)$st derivatives of Eq. (ii) at a point at which $W(y_1, y_2, \ldots, y_n) \neq 0$, show that the c's must all be zero if Eq. (ii) is true. Hence the $y_1, y_2, \ldots, y_n$ are linearly independent.

(b) Suppose that $y_1, y_2, \ldots, y_n$ are linearly independent solutions of Eq. (i). To show that $W(y_1, y_2, \ldots, y_n)$ is nowhere zero on $\alpha < x < \beta$, assume that $W(y_1, y_2, \ldots, y_n)(x_0) = 0$ and show that this leads to a contradiction.
Hint: If $W(y_1, y_2, \ldots, y_n)(x_0) = 0$, there exists a nonzero solution of Eq. (i) satisfying the initial conditions $y = y' = y'' = \cdots = y^{(n-1)} = 0$ at x_0; then use the existence and uniqueness theorem.

4. Verify that the given functions are solutions of the differential equation, and compute the Wronskian of the solutions.

(a) $y''' + y' = 0$; 1, $\cos x$, $\sin x$

(b) $y^{iv} + y'' = 0$; 1, x, $\cos x$, $\sin x$

(c) $y''' + 2y'' - y' - 2y = 0$; e^x, e^{-x}, e^{-2x}

(d) $y^{iv} + 2y''' + y'' = 0$; 1, x, e^{-x}, xe^{-x}

(e) $xy''' - y'' = 0$; 1, x, x^3

5. Show that $W(5, \sin^2 x, \cos 2x) = 0$ for all x. Can you establish this result directly without evaluating the Wronskian?

6. Let the linear differential operator L be defined by

$$L[y] = a_0 y^{(n)} + a_1 y^{(n-1)} + a_2 y^{(n-2)} + \cdots + a_n y,$$

where $a_0, a_1, \ldots, a_n$ are real constants. Compute (a) $L[x^n]$; (b) $L[e^{rx}]$.

(c) Determine four solutions of the equation $y^{iv} - 5y'' + 4y = 0$. Do you think the four solutions form a fundamental set of solutions? Why?

7. Show that if y_1 is a solution of

$$y''' + p_1(x)y'' + p_2(x)y' + p_3(x)y = 0,$$

then the substitution $y = y_1(x)v(x)$ leads to the following second order linear equation for v':

$$y_1 v''' + (3y_1' + p_1 y_1)v'' + (3y_1'' + 2p_1 y_1' + p_2 y_1)v' = 0.$$

8. Using the method of reduction of order, solve the following differential equations.

(a) $x^3 y''' - 3x^2 y'' + 6xy' - 6y = 0$, $x > 0$; $y_1(x) = x$

(b) $x^2(x + 3)y''' - 3x(x + 2)y'' + 6(1 + x)y' - 6y = 0$, $x > 0$;
$y_1(x) = x^2$, $y_2(x) = x^3$

5.3   HOMOGENEOUS EQUATIONS WITH CONSTANT COEFFICIENTS

Consider the nth order linear homogeneous differential equation

$$L[y] = a_0 y^{(n)} + a_1 y^{(n-1)} + \cdots + a_{n-1} y' + a_n y = 0, \tag{1}$$

where $a_0, a_1, \ldots, a_n$ are real constants. It is natural to anticipate from our knowledge of second order linear equations with constant coefficients that for suitable values of r, $y = e^{rx}$ is a solution of Eq. (1). Indeed,

$$L[e^{rx}] = e^{rx}(a_0 r^n + a_1 r^{n-1} + \cdots + a_{n-1} r + a_n)$$
$$= e^{rx} Z(r) \tag{2}$$

for all r. For those values of r for which $Z(r) = 0$, it follows that $L[e^{rx}] = 0$ and $y = e^{rx}$ is a solution of Eq. (1). The polynomial $Z(r)$ is referred to as the *characteristic polynomial* or *auxiliary polynomial* and the equation $Z(r) = 0$ as the *characteristic equation* or *auxiliary equation* of the differential equation (1). A polynomial of degree n has n zeros, say $r_1, r_2, \ldots, r_n$; hence we can write the characteristic polynomial in the form

$$Z(r) = a_0(r - r_1)(r - r_2) \cdots (r - r_n). \tag{3}$$

Real and Unequal Roots.   If the roots of the characteristic equation are real and no two are equal, then we have n distinct solutions $e^{r_1 x}, e^{r_2 x}, \ldots, e^{r_n x}$ of Eq. (1). To show in this case that the general solution of Eq. (1) is of the form

$$y = c_1 e^{r_1 x} + c_2 e^{r_2 x} + \cdots + c_n e^{r_n x}, \tag{4}$$

we must show that the functions $e^{r_1 x}, \ldots, e^{r_n x}$ are linearly independent on the interval $-\infty < x < \infty$. Let us assume that they are linearly dependent and show that this leads to a contradiction. Then there exist constants $c_1, c_2, \ldots, c_n$, not all zero, such that $c_1 e^{r_1 x} + c_2 e^{r_2 x} + \cdots + c_n e^{r_n x} = 0$ for all x in $-\infty < x < \infty$. Multiplying by $e^{-r_1 x}$ gives $c_1 + c_2 e^{(r_2 - r_1)x} + \cdots + c_n e^{(r_n - r_1)x} = 0$ and, differentiating, we obtain

$$(r_2 - r_1)c_2 e^{(r_2 - r_1)x} + (r_3 - r_1)c_3 e^{(r_3 - r_1)x} + \cdots + (r_n - r_1)c_n e^{(r_n - r_1)x} = 0$$

for $-\infty < x < \infty$. Multiplying this last result by $e^{-(r_2 - r_1)x}$ and then differentiating gives

$$(r_3 - r_2)(r_3 - r_1)c_3 e^{(r_3 - r_2)x} + \cdots + (r_n - r_2)(r_n - r_1)c_n e^{(r_n - r_2)x} = 0$$

for $-\infty < x < \infty$. Continuing in this manner, we finally obtain

$$(r_n - r_{n-1})(r_n - r_{n-2}) \cdots (r_n - r_2)(r_n - r_1)c_n e^{(r_n - r_{n-1})x} = 0 \tag{5}$$

for $-\infty < x < \infty$. Since the exponential function does not vanish and the r_i are unequal, we have $c_n = 0$. Thus $c_1 e^{r_1 x} + c_2 e^{r_2 x} + \cdots + c_{n-1} e^{r_{n-1} x} = 0$ and, proceeding as above, we obtain $c_{n-1} = 0$. Similarly $c_{n-2}, \ldots, c_1 = 0$, and this is a contradiction of the assumption that $e^{r_1 x}, \ldots, e^{r_n x}$ are linearly dependent.

Complex Roots.   If the characteristic equation has complex roots, they must occur in conjugate pairs, $\lambda \pm i\mu$, since the coefficients $a_0, \ldots, a_n$ are real numbers. Provided that none of the roots are repeated, the general solution of Eq. (1) is still of the form* (4). However, following the same procedure as for the second order equation (Section 3.5.1) in place of the complex-valued solutions $e^{(\lambda+i\mu)x}$ and $e^{(\lambda-i\mu)x}$ we will normally use the real-valued solutions

$$e^{\lambda x} \cos \mu x, \qquad e^{\lambda x} \sin \mu x \tag{6}$$

obtained as the real and imaginary parts of $e^{(\lambda+i\mu)x}$. Thus, even though some of the roots of the characteristic equation are complex, it is still possible to express the general solution of Eq. (1) as a linear combination of real-valued solutions.

Example 1.   Find the general solution of

$$y^{\text{iv}} - y = 0. \tag{7}$$

Substituting e^{rx} for y, we find that the characteristic equation is

$$r^4 - 1 = (r^2 - 1)(r^2 + 1) = 0.$$

The roots are $r = 1, -1, i, -i$; hence the general solution of Eq. (7) is

$$y = c_1 e^x + c_2 e^{-x} + c_3 \cos x + c_4 \sin x.$$

Repeated Roots.   If the roots of the characteristic equation are not distinct, that is, if some of the roots are repeated, then the solution (4) is clearly not the general solution of Eq. (1). Recalling that if r_1 was a repeated root for the second order linear equation $a_0 y'' + a_1 y' + a_2 y = 0$, then the two linearly independent solutions were $e^{r_1 x}$ and $xe^{r_1 x}$, it seems reasonable to expect that if a root of $Z(r) = 0$, say $r = r_1$, is repeated s times ($s \le n$) then

$$e^{r_1 x}, \ xe^{r_1 x}, \ x^2 e^{r_1 x}, \ldots, \ x^{s-1} e^{r_1 x} \tag{8}$$

are solutions of Eq. (1). To prove this, we observe that if r_1 is an s-fold zero of $Z(r)$, then Eq. (2) can be written as

$$L[e^{rx}] = e^{rx} a_0 (r - r_1)^s (r - r_{s+1}) \ldots (r - r_n)$$

$$= e^{rx}(r - r_1)^s H(r) \tag{9}$$

for all values of r, where $H(r_1) \ne 0$. Next, we use the facts that $\partial e^{rx}/\partial r = xe^{rx}$ and that differentiation of e^{rx} with respect to x and r can be interchanged. Then on differentiating Eq. (9) with respect to r we obtain

$$L[xe^{rx}] = e^{rx}[x(r - r_1)^s H(r) + s(r - r_1)^{s-1} H(r) + (r - r_1)^s H'(r)]. \tag{10}$$

Since $s \ge 2$ the right-hand side of Eq. (10) vanishes for $r = r_1$ and hence $xe^{r_1 x}$ is also a solution of Eq. (1). If $s \ge 3$, differentiating Eq. (10) again with respect to r

---

* The linear independence of the solutions $e^{r_1 x}, \ldots, e^{r_n x}$, in the case that some of the r's are complex numbers, follows by a generalization of the argument just given for the case of the r's real.

and setting r equal to r_1 shows that $x^2 e^{r_1 x}$ is also a solution of Eq. (1). This process can be continued through $s - 1$ differentiations, which gives the desired result. Notice that the sth derivative of the right-hand side of Eq. (9) will not vanish for $r = r_1$ since the sth derivative of $(r - r_1)^s$ is a constant and $H(r_1) \neq 0$. It is reasonable to expect that $e^{r_1 x}, x e^{r_1 x}, \ldots, x^{s-1} e^{r_1 x}$ are linearly independent, and we will accept this fact without proof.

Finally, if a complex root $\lambda + i\mu$ is repeated s times the complex conjugate $\lambda - i\mu$ is also repeated s times. Corresponding to these $2s$ complex-valued solutions we can find $2s$ real-valued solutions by noting that the real and imaginary parts of $e^{(\lambda+i\mu)x}, x e^{(\lambda+i\mu)x}, \ldots, x^{s-1} e^{(\lambda+i\mu)x}$ are also linearly independent solutions:

$$e^{\lambda x} \cos \mu x, \quad e^{\lambda x} \sin \mu x, \quad x e^{\lambda x} \cos \mu x, \quad x e^{\lambda x} \sin \mu x,$$
$$\ldots, x^{s-1} e^{\lambda x} \cos \mu x, \quad x^{s-1} e^{\lambda x} \sin \mu x.$$

Hence the general solution of Eq. (1) can always be expressed as a linear combination of n real-valued solutions. Consider the following example.

Example 2. Find the general solution of

$$y^{\text{iv}} + 2y'' + y = 0. \tag{11}$$

The characteristic equation is

$$r^4 + 2r^2 + 1 = (r^2 + 1)(r^2 + 1) = 0.$$

The roots are $r = i, i, -i, -i$, and the general solution of Eq. (11) is

$$y = c_1 \cos x + c_2 \sin x + c_3 x \cos x + c_4 x \sin x.$$

In determining the roots of the characteristic equation it is often necessary to compute the cube roots, or fourth roots, or even higher roots of a (possibly complex) number. This can usually be done most conveniently by using Euler's formula $e^{ix} = \cos x + i \sin x$, and the algebraic laws given in Section 3.5.1. This is illustrated in the following example, and is also discussed in Problems 1 and 2.

Example 3. Find the general solution of

$$y^{\text{iv}} + y = 0. \tag{12}$$

The characteristic equation is

$$r^4 + 1 = 0.$$

In this case the polynomial is not readily factored. We must compute the fourth roots of -1. Now -1, thought of as a complex number, is $-1 + 0i$. It has magnitude 1 and polar angle π. Thus

$$-1 = \cos \pi + i \sin \pi = e^{i\pi}.$$

Moreover, the angle is determined only up to a multiple of 2π. Thus

$$-1 = \cos(\pi + 2m\pi) + i \sin(\pi + 2m\pi) = e^{i(\pi + 2m\pi)},$$

where m is zero or any positive or negative integer. Thus

$$(-1)^{1/4} = e^{i(\pi/4 + m\pi/2)} = \cos\left(\frac{\pi}{4} + \frac{m\pi}{2}\right) + i \sin\left(\frac{\pi}{4} + \frac{m\pi}{2}\right).$$

The four fourth roots of -1 are obtained by setting $m = 0, 1, 2,$ and 3; they are

$$\frac{1+i}{\sqrt{2}}, \qquad \frac{-1+i}{\sqrt{2}}, \qquad \frac{-1-i}{\sqrt{2}}, \qquad \frac{-i}{\sqrt{2}}.$$

It is easy to verify that for any other value of m we obtain one of these four roots. For example, corresponding to $m = 4$ we obtain $(1 + i)/\sqrt{2}$. The general solution of Eq. (12) is

$$y = e^{x/\sqrt{2}}\left(c_1 \cos\frac{x}{\sqrt{2}} + c_2 \sin\frac{x}{\sqrt{2}}\right) + e^{-x/\sqrt{2}}\left(c_3 \cos\frac{x}{\sqrt{2}} + c_4 \sin\frac{x}{\sqrt{2}}\right).$$

In conclusion we mention that the problem of determining the roots of $Z(r) = 0$ for $n > 2$ may require the use of numerical methods if they cannot be found by inspection or by a simple process of trial and error. Although there are formulas similar to the quadratic formula for the roots of cubic and quartic polynomial equations, there is no formula* for $n > 4$; even for third and fourth degree polynomial equations it is usually more efficient to use numerical methods, such as Newton's method, rather than the exact formula for determining the roots.

If the constants $a_0, a_1, \ldots, a_n$ in Eq. (1) are complex numbers, the solution of Eq. (1) is still of the form (4). In this case, however, the roots of the characteristic equation are, in general, complex numbers; and it is no longer true that the complex conjugate of a root is also a root. The corresponding solutions are complex-valued.

PROBLEMS

1. Express each of the following complex numbers in the form

$$R(\cos\theta + i \sin\theta) = Re^{i\theta}.$$

---

* The formula for solving the cubic equation is usually attributed to Cardano (1501–1576), and that for the quartic equation to his pupil Ferrari (1522–1565). That it is impossible to express the roots of a general algebraic equation of degree higher than four by a formula involving only rational operations (addition, multiplication, etc.) and root extractions was established by Abel and Galois (1811–1832). A discussion of methods of solving algebraic equations can be found in Uspensky.

Note that $e^{i(\theta+2m\pi)} = e^{i\theta}$, $m = 0, \pm1, \pm2, \ldots$.

(a) $1 + i$ (b) $-1 + i\sqrt{3}$ (c) -1

(d) $-i$ (e) $\sqrt{3} - i$ (f) $-1 - i$

2. Observing that $e^{i(\theta+2m\pi)} = e^{i\theta}$ for m an integer, and that (one form of DeMoivre's formula, Problem 12, Section 3.5.1)

$$[e^{i(\theta+2m\pi)}]^{1/n} = e^{i[(\theta+2m\pi)/n]} = \cos\left(\frac{\theta}{n} + \frac{2m\pi}{n}\right) + i\sin\left(\frac{\theta}{n} + \frac{2m\pi}{n}\right),$$

determine the indicated roots of each of the following complex numbers.

(a) $1^{1/3}$ (b) $(1 - i)^{1/2}$

(c) $1^{1/4}$ (d) $\left[2\left(\cos\dfrac{\pi}{3} + i\sin\dfrac{\pi}{3}\right)\right]^{1/2}$

In each of Problems 3 through 13 determine the general solution of the given differential equation.

3. $y''' - y'' - y' + y = 0$ 4. $y''' - 3y'' + 3y' - y = 0$

5. $2y''' - 4y'' - 2y' + 4y = 0$ 6. $y^{iv} - 4y''' + 4y'' = 0$

7. $y^{vi} + y = 0$ 8. $y^{iv} - 5y'' + 4y = 0$

9. $y^{vi} - 3y^{iv} + 3y'' - y = 0$ 10. $y^{vi} - y'' = 0$

11. $y^{v} - 3y^{iv} + 3y''' - 3y'' + 2y' = 0$ 12. $y^{iv} - 8y' = 0$

13. $y^{viii} + 8y^{iv} + 16y = 0$

14. Solve the initial value problem

$$y''' + y' = 0; \quad y(0) = 0, \quad y'(0) = 1, \quad y''(0) = 2.$$

15. Show that the general solution of the differential equation

$$y^{iv} - y = 0$$

can be written as

$$y = c_1 \cos x + c_2 \sin x + c_3 \cosh x + c_4 \sinh x.$$

Determine the solution satisfying the initial conditions $y(0) = 0$, $y'(0) = 0$, $y''(0) = 1$, $y'''(0) = 1$. Why is it convenient to use $\cosh x$ and $\sinh x$ rather than e^x and e^{-x}?

Problems 16 through 19 deal with the nth order Euler equation.

*16. The nth order Euler or equidimensional equation is

$$L[y] = x^n y^{(n)} + a_1 x^{n-1} y^{(n-1)} + \cdots + a_{n-1} x y' + a_n y = 0 \tag{i}$$

where $a_1, a_2, \ldots, a_n$ are real constants. Consider only the interval $x > 0$.

(a) Show that $L[x^r] = x^r F(r)$ where

$$F(r) = r(r - 1) \cdots (r - n + 1) + a_1[r(r - 1) \cdots (r - n + 2)] + \cdots + a_{n-1}r + a_n$$

is a polynomial of degree n in r. The functions $x^{r_1}, x^{r_2}, \ldots, x^{r_n}$ corresponding to the roots $r_1, r_2, \ldots, r_n$ of $F(r) = 0$ are solutions of Eq. (i).

(b) Show that if r_1 is an s-fold root of $F(r) = 0$ then $x^{r_1}, x^{r_1} \ln x, x^{r_1}(\ln x)^2, \ldots, x^{r_1}(\ln x)^{s-1}$ are solutions of Eq. (i).

When the roots of $F(r) = 0$ are complex, real-valued solutions can be obtained in the manner described for the second order Euler equation. See Section 4.4.

*17. Using the results of Problem 16 determine the general solution of each of the following differential equations. Consider only the interval $x > 0$.

(a) $x^3y''' + x^2y'' - 2xy' + 2y = 0$ (b) $x^3y''' + xy' - y = 0$

(c) $x^3y''' + 2x^2y'' + xy' - y = 0$ (d) $x^4y'''' + 5x^3y''' + 7x^2y'' + 8xy' = 0$

(e) $x^3y''' + x^2y'' = 0$ (f) $x^3y''' + 4x^2y'' - 2xy' - 4y = 0$

*18. In determining the drag on a very small sphere of radius a placed in a uniform viscous flow it is necessary to solve the differential equation

$$\rho^3 f''''(\rho) + 8\rho^2 f'''(\rho) + 8\rho f''(\rho) - 8f'(\rho) = 0, \qquad \rho > a,$$

where ρ is the distance from the center of the sphere. Note that this is an Euler equation for f' and show, using the results of Problem 16, that the general solution is

$$f(\rho) = \frac{A}{\rho^3} + \frac{B}{\rho} + C + D\rho^2$$

where A, B, C, and D are constants. Show that the solution satisfying the boundary conditions $f(a) = 0$, $f'(a) = 0$, and $f(\rho) \to U$ (the velocity at ∞) as $\rho \to \infty$ is $f(\rho) = Ua^3/2\rho^3 - 3Ua/2\rho + U$. The formula for the drag (Stokes' formula) on the sphere turns out to be $6\pi\mu aU$ where μ is the viscosity of the fluid. This result was used by R. A. Millikan (1868–1953) in his famous experiment to measure the charge on an electron. Also see Problem 9 of Section 2.10.

*19. Show, for $x > 0$, that the change of variable $x = e^z$ reduces the third order Euler equation $x^3y''' + a_1x^2y'' + a_2xy' + a_3y = 0$ to a third order linear equation with constant coefficients. This transformation also reduces the nth order Euler equation to an nth order linear equation with constant coefficients. Solve Problem 17b by this method.

*20. It is often of importance in engineering applications to know whether all solutions of a linear homogeneous equation approach zero as x approaches infinity. If so, the equation is said to be *asymptotically stable*. If x represents time and y the response of a physical system, the condition that the differential equation is asymptotically stable means that regardless of the initial conditions the response of the system to the initial conditions eventually decays to zero as x becomes large. For the second order equation $ay'' + by' + cy = 0$, it was shown in Section 3.7.1 (also see Section 3.5.1, Problem 10) that a sufficient condition for asymptotic stability is that a, b, and c be positive. More generally the equation

$$a_0y^{(n)} + a_1y^{(n-1)} + \cdots + a_ny = 0, \tag{i}$$

where $a_0, a_1, \ldots, a_n$ are real, is asymptotically stable if all the roots of the corresponding characteristic equation have negative real parts. A necessary and sufficient condition for this to be true has been given by Hurwitz (1859–1919). For $n = 4$ the Hurwitz stability criterion can be stated as follows: Eq. (i) is asymptotically stable if and only if for $a_0 > 0$,

$$a_1, \quad \begin{vmatrix} a_1 & a_0 \\ a_3 & a_2 \end{vmatrix}, \quad \begin{vmatrix} a_1 & a_0 & 0 \\ a_3 & a_2 & a_1 \\ 0 & a_4 & a_3 \end{vmatrix}, \quad \begin{vmatrix} a_1 & a_0 & 0 & 0 \\ a_3 & a_2 & a_1 & a_0 \\ 0 & a_4 & a_3 & a_2 \\ 0 & 0 & 0 & a_4 \end{vmatrix}$$

are all positive. For $n = 3$ the condition just applies to the first three expressions with $a_4 = 0$, and for $n = 2$ to the first two expressions with $a_3 = 0$. A complete discussion of the Hurwitz stability criterion can be found in Guillemin (Chapter 6, Article 26).

Determine whether each of the following equations is asymptotically stable, and verify your result if possible by actually computing the general solution.

(a) $y''' + 3y'' + 3y' + y = 0$ (b) $y''' - y = 0$

(c) $y''' + y'' - y' + y = 0$ (d) $y^{iv} + 2y'' + y = 0$

(e) $y''' + 0.1y'' + 1.2y' - 0.4y = 0$ (f) $y''' + 3.2y'' + 2.41y' + 0.21y = 0$

5.4 THE METHOD OF UNDETERMINED COEFFICIENTS

A particular solution of the nonhomogeneous nth order linear equation with constant coefficients

$$L[y] = a_0 y^{(n)} + a_1 y^{(n-1)} + \cdots + a_{n-1} y' + a_n y = g(x) \tag{1}$$

can be obtained by the method of undetermined coefficients provided that $g(x)$ is of an appropriate form. While the method of undetermined coefficients is not as general as the method of variation of parameters described in the next section, it is usually much easier to use when applicable.

Just as for the second order linear equation, it is clear that when the constant coefficient linear differential operator L is applied to a polynomial $A_0 x^m + A_1 x^{m-1} + \cdots + A_m$, an exponential function $e^{\alpha x}$, or a sine function $\sin \beta x$, or a cosine function $\cos \beta x$, the result is a polynomial, an exponential function, or a linear combination of sine and cosine functions, respectively. Hence if $g(x)$ is a sum of polynomials, exponentials, sines and cosines, or even products of such functions, we can expect that it is possible to find $y_p(x)$ by choosing a suitable combination of polynomials, exponentials etc., with a number of undetermined constants. The constants are then determined so that Eq. (1) is satisfied.

First consider the case that $g(x)$ is a polynomial of degree m

$$g(x) = b_0 x^m + b_1 x^{m-1} + \cdots + b_m, \tag{2}$$

where $b_0, b_1, \ldots, b_m$ are given constants. It is natural to look for a particular solution of the form

$$y_p(x) = A_0 x^m + A_1 x^{m-1} + \cdots + A_m. \tag{3}$$

Substituting for y in Eq. (1), and equating the coefficients of like powers of x, we find from the terms in x^m that $a_n A_0 = b_0$. Provided that $a_n \neq 0$ we have $A_0 = b_0/a_n$. The constants $A_1, \ldots, A_m$ are determined from the coefficients of the terms x^{m-1}, $x^{m-2}, \ldots, x^0$.

If $a_n = 0$, that is, if a constant is a solution of the homogeneous equation, we cannot solve for A_0; in this case it is necessary to assume for $y_p(x)$ a polynomial of degree $m + 1$ in order to obtain a term in $L[y_p](x)$ to balance against $b_0 x^m$. However, it is not necessary to carry the constant in the assumed form for $y_p(x)$. More generally it is easy to verify that if zero is an s-fold root of the characteristic polynomial, in which case $1, x, x^2, \ldots, x^{s-1}$ are solutions of the homogeneous equation, then a suitable form for $y_p(x)$ is

$$y_p(x) = x^s(A_0x^m + A_1x^{m-1} + \cdots + A_m). \tag{4}$$

As a second problem suppose that $g(x)$ is of the form

$$g(x) = e^{\alpha x}(b_0x^m + b_1x^{m-1} + \cdots + b_m). \tag{5}$$

Then we would expect $y_p(x)$ to be of the form

$$y_p(x) = e^{\alpha x}(A_0x^m + A_1x^{m-1} + \cdots + A_m), \tag{6}$$

provided that $e^{\alpha x}$ is not a solution of the homogeneous equation. If α is an s-fold root of the auxiliary equation, a suitable form for $y_p(x)$ is

$$y_p(x) = x^s e^{\alpha x}(A_0x^m + A_1x^{m-1} + \cdots + A_m). \tag{7}$$

These results can be proved, as for the second order linear nonhomogeneous equation, by reducing this problem to the previous one by the substitution $y = e^{\alpha x}u(x)$. The function u will satisfy an nth order linear nonhomogeneous equation with constant coefficients—the nonhomogeneous term being precisely the polynomial (2) (see Problem 18).

Similarly, if $g(x)$ is of the form

$$g(x) = e^{\alpha x}(b_0x^m + b_1x^{m-1} + \cdots + b_m)\begin{cases}\sin \beta x, \\ \cos \beta x,\end{cases} \tag{8}$$

then a suitable form for $y_p(x)$, provided that $\alpha + i\beta$ is not a root of the characteristic equation, is

$$y_p(x) = e^{\alpha x}(A_0x^m + A_1x^{m-1} + \cdots + A_m) \cos \beta x$$
$$+ e^{\alpha x}(B_0x^m + B_1x^{m-1} + \cdots + B_m) \sin \beta x. \tag{9}$$

If $\alpha + i\beta$ is an s-fold root of the characteristic equation, then it is necessary to multiply the right-hand side of Eq. (9) by x^s.

These results are summarized in Table 5.1.

TABLE 5.1

$g(x)$	$y_p(x)$
$P_m(x) = b_0x^m + b_1x^{m-1} + \cdots + b_m$	$x^s(A_0x^m + \cdots + A_m)$
$P_m(x)e^{\alpha x}$	$x^s(A_0x^m + \cdots + A_m)e^{\alpha x}$
$P_m(x)e^{\alpha x}\begin{cases}\sin \beta x \\ \cos \beta x\end{cases}$	$x^s[(A_0x^m + \cdots + A_m)e^{\alpha x} \cos \beta x$ $+ (B_0x^m + \cdots + B_m)e^{\alpha x} \sin \beta x]$

Here s is the smallest nonnegative integer for which every term in $y_p(x)$ differs from every term in the complementary function $y_c(x)$.

If $g(x)$ is a sum of terms of the form (2), (5), and (8), it is usually easier in practice to compute separately the particular solution corresponding to each term in $g(x)$. From the principle of superposition (since the differential equation is linear), the particular solution of the complete problem is the sum of the particular solutions of the individual problems. This is illustrated in the following example.

Example.   Find a particular solution of

$$y''' - 4y' = x + 3 \cos x + e^{-2x}. \tag{10}$$

First we solve the homogeneous equation. The characteristic equation is $r^3 - 4r = 0$, and the roots are $0, \pm 2$; hence

$$y_c(x) = c_1 + c_2 e^{2x} + c_3 e^{-2x}.$$

Using the superposition principle, we can write a particular solution of Eq. (10) as the sum of particular solutions corresponding to the differential equations

$$y''' - 4y' = x, \qquad y''' - 4y' = 3 \cos x, \qquad y''' - 4y' = e^{-2x}.$$

Our initial guess for a particular solution, y_{p_1}, of the first equation is $A_0 x + A_1$; but since a constant is a solution of the homogeneous equation we multiply by x. Thus

$$y_{p_1}(x) = x(A_0 x + A_1).$$

For the second equation we guess

$$y_{p_2}(x) = B \cos x + C \sin x,$$

and there is no need to modify this initial guess since $\cos x$ and $\sin x$ are not solutions of the homogeneous equation. Finally, for the third equation, since e^{-2x} is a solution of the homogeneous equation, we assume that

$$y_{p_3}(x) = Exe^{-2x}.$$

The constants are determined by substituting into the individual differential equations; they are $A_0 = -\frac{1}{8}$, $A_1 = 0$, $B = 0$, $C = -\frac{3}{5}$, and $E = \frac{1}{8}$. Hence a particular solution of Eq. (10) is

$$y_p(x) = -\tfrac{1}{8}x^2 - \tfrac{3}{5} \sin x + \tfrac{1}{8}xe^{-2x}.$$

The method of undetermined coefficients can be used whenever it is possible to guess the correct form for $y_p(x)$. However, this is usually impossible for other than constant coefficient differential equations and for other than nonhomogeneous terms of the type described above. For more complicated problems we can use the method of variation of parameters, which is discussed in the next section.

PROBLEMS

In each of Problems 1 through 11 determine the general solution of the given differential equation. Where specified find the solution satisfying the given initial conditions.

1. $y''' + 4y' = x, \quad y(0) = y'(0) = 0, \quad y''(0) = 1$

2. $y''' - y'' - y' + y = 2e^{-x} + 3$

3. $y''' + y'' + y' + y = e^{-x} + 4x \qquad$ 4. $y''' - y' = 2 \sin x$

5. $y^{iv} + 2y'' + y = 3x + 4, \quad y(0) = y'(0) = 0, \quad y''(0) = y'''(0) = 1$

6. $y^{iv} - 4y'' = x^2 + e^x \qquad\qquad$ 7. $y^{iv} + 2y'' + y = 3 + \cos 2x$

8. $y''' - 3y'' + 2y' = x + e^x, \quad y(0) = 1, \quad y'(0) = -\frac{1}{4}, \quad y''(0) = -\frac{3}{2}$

9. $y^{vi} + y''' = x \qquad\qquad$ 10. $y^{iv} + y''' = \sin 2x$

11. $y^{iv} - y = 3x + \cos x$

In each of Problems 12 through 17 determine a suitable form for $y_p(x)$ if the method of undetermined coefficients is to be used. Do not evaluate the constants.

12. $y''' - 2y'' + y' = x^3 + 2e^x$

13. $y''' - y' = xe^{-x} + 2 \cos x$

14. $y^{iv} - 2y'' + y = e^x + \sin x$

15. $y^{iv} - y''' - y'' + y' = x^2 + 4 + x \sin x$

16. $y^{iv} + 4y'' = \sin 2x + xe^x + 4$

17. $y^{iv} + 2y''' + 2y'' = 3e^x + 2xe^{-x} + e^{-x} \sin x$

18. Consider the nonhomogeneous nth order linear differential equation

$$a_0 y^{(n)} + a_1 y^{(n-1)} + \cdots + a_n y = g(x)$$

where $a_0, \ldots, a_n$ are constants. Verify that if $g(x)$ is of the form

$$e^{\alpha x}(b_0 x^m + \cdots + b_m),$$

then the substitution $y = e^{\alpha x} u(x)$ reduces the above equation to the form

$$t_0 u^{(n)} + t_1 u^{(n-1)} + \cdots + t_n u = b_0 x^m + \cdots + b_m,$$

where $t_0, \ldots, t_n$ are constants. Determine t_0 and t_n in terms of the a's and α. Thus the problem of determining a particular solution of the original equation is reduced to the simpler problem of determining a particular solution of an equation with constant coefficients and a polynomial for the nonhomogeneous term.

5.5 THE METHOD OF VARIATION OF PARAMETERS

The method of variation of parameters for determining a particular solution of the nonhomogeneous nth order linear differential equation

$$L[y] = y^{(n)} + p_1(x)y^{(n-1)} + \cdots + p_{n-1}(x)y' + p_n(x)y = g(x) \qquad (1)$$

is a direct extension of the theory for the second order differential equation (see Section 3.6.2). As before, in order to use the method of variation of parameters, it is first necessary to solve the corresponding homogeneous differential equation.

In general, this may be difficult unless the coefficients are constants. However, the method of variation of parameters is still more general than the method of undetermined coefficients in the following sense. The method of undetermined coefficients is usually applicable only for constant coefficient equations and a limited class of functions g; for constant coefficient equations the homogeneous equation can be solved, and hence a particular solution for *any* continuous function g can be determined by the method of variation of parameters. Suppose then that we know a fundamental set of solutions $y_1, y_2, \ldots, y_n$ of the homogeneous equation. Then

$$y_c(x) = c_1 y_1(x) + c_2 y_2(x) + \cdots + c_n y_n(x). \tag{2}$$

The method of variation of parameters for determining a particular solution of Eq. (1) rests on the possibility of determining n functions $u_1, u_2, \ldots, u_n$ such that $y_p(x)$ is of the form

$$y_p(x) = u_1(x)y_1(x) + u_2(x)y_2(x) + \cdots + u_n(x)y_n(x). \tag{3}$$

Since we have n functions to determine, we will have to specify n conditions. One of these is clearly that y_p satisfy Eq. (1). The other $n - 1$ conditions are chosen so as to facilitate the calculations. Since we can hardly expect a simplification in determining y_p if we must solve high order differential equations for the u_i, $i = 1, 2, \ldots, n$, it is natural to impose conditions to suppress the terms that lead to higher derivatives of the u_i. From Eq. (3) we obtain

$$y_p' = (u_1 y_1' + u_2 y_2' + \cdots + u_n y_n') + (u_1' y_1 + u_2' y_2 + \cdots + u_n' y_n). \tag{4}$$

Thus the first condition that we impose on the u_i is that

$$u_1' y_1 + u_2' y_2 + \cdots + u_n' y_n = 0. \tag{5}$$

Continuing this process in a similar manner through $n - 1$ derivatives of y_p gives

$$y_p^{(m)} = u_1 y_1^{(m)} + u_2 y_2^{(m)} + \cdots + u_n y_n^{(m)}, \qquad m = 0, 1, 2, \ldots, n - 1, \tag{6}$$

and the following $n - 1$ conditions on the functions $u_1, \ldots, u_n$:

$$u_1' y_1^{(m-1)} + u_2' y_2^{(m-1)} + \cdots + u_n' y_n^{(m-1)} = 0, \qquad m = 1, 2, \ldots, n - 1. \tag{7}$$

The nth derivative of y_p is

$$y_p^{(n)} = (u_1 y_1^{(n)} + \cdots + u_n y_n^{(n)}) + (u_1' y_1^{(n-1)} + \cdots + u_n' y_n^{(n-1)}). \tag{8}$$

Finally, we impose the condition that y_p be a solution of Eq. (1). On substituting for the derivatives of y_p from Eqs. (6) and (8), collecting terms, and making use of the fact that $L[y_i] = 0$, $i = 1, 2, \ldots, n$, we obtain

$$u_1' y_1^{(n-1)} + u_2' y_2^{(n-1)} + \cdots + u_n' y_n^{(n-1)} = g. \tag{9}$$

Equation (9), coupled with the $n - 1$ equations (7), give n simultaneous linear nonhomogeneous equations for $u_1', u_2', \ldots, u_n'$:

$$y_1 u_1' + y_2 u_2' + \cdots + y_n u_n' = 0,$$
$$y_1' u_1' + y_2' u_2' + \cdots + y_n' u_n' = 0,$$
$$y_1'' u_1' + y_2'' u_2' + \cdots + y_n'' u_n' = 0,$$

$$\cdot$$
$$\cdot \qquad\qquad\qquad\qquad (10)$$
$$\cdot$$

$$y_1^{(n-1)} u_1' + \cdots + y_n^{(n-1)} u_n' = g.$$

A sufficient condition for the existence of a solution of the system of equations (10) is that the determinant of the coefficients is nonzero for each value of x. However, the determinant of the coefficients is precisely $W(y_1, y_2, \ldots, y_n)$, and it is nowhere zero since $y_1, \ldots, y_n$ are linearly independent solutions of the homogeneous equation. Hence it is possible to determine $u_1', \ldots, u_n'$. Using Cramer's rule, we find that the solution of the system of equations (10) is

$$u_m'(x) = \frac{g(x) W_m(x)}{W(x)}, \qquad m = 1, 2, \ldots, n. \qquad (11)$$

Here $W(x) = W(y_1, y_2, \ldots, y_n)(x)$ and W_m is the determinant obtained from $W(y_1, y_2, \ldots, y_n)$ by replacing the mth column by the column $(0, 0, \ldots, 0, 1)$. With this notation a particular solution of Eq. (1) is given by

$$y_p(x) = \sum_{m=1}^{n} y_m(x) \int^x \frac{g(t) W_m(t)}{W(t)} dt. \qquad (12)$$

While the procedure is straightforward, the computational problem of evaluating $y_p(x)$ from Eq. (12) for $n > 2$ is not a trivial one. The calculation may be simplified to some extent by using Abel's indentity:

$$W(x) = W(y_1, y_2, \ldots, y_n)(x) = c \exp\left[-\int^x p_1(t) \, dt \right].$$

(See Problem 2 of Section 5.2.) The constant c can be determined by evaluating $W(y_1, y_2, \ldots, y_n)$ at a conveniently chosen point.

PROBLEMS

In each of Problems 1 through 3 use the method of variation of parameters to determine a particular solution of the given differential equation.

1. $y''' + y' = \tan x, \qquad 0 < x < \pi/2$

2. $y''' - y' = x$

3. $y''' - 2y'' - y' + 2y = e^{4x}$

4. Given that x, x^2, and $1/x$ are solutions of the homogeneous equation corresponding to

$$x^3 y''' + x^2 y'' - 2xy' + 2y = 2x^4, \qquad x > 0,$$

determine a particular solution.

5. Find a formula involving integrals for a particular solution of the differential equation

$$y''' - y'' + y' - y = g(x).$$

6. Find a formula involving integrals for a particular solution of the differential equation

$$y^{iv} - y = g(x).$$

Hint: The functions $\sin x$, $\cos x$, $\sinh x$, $\cosh x$ form a fundamental set of solutions of the homogeneous equation.

7. Find a formula involving integrals for a particular solution of the differential equation

$$x^3 y''' - 3x^2 y'' + 6xy' - 6y = g(x), \qquad x > 0.$$

Hint: Verify that x, x^2, and x^3 are solutions of the homogeneous equation.

REFERENCES

Coddington, E. A., *An Introduction to Ordinary Differential Equations*, Prentice-Hall, Englewood Cliffs, N.J., 1961.

Coddington, E. A. and Levinson, N., *Theory of Ordinary Differential Equations*, McGraw-Hill New York, 1955.

Guillemin, E. A., *The Mathematics of Circuit Analysis*, Wiley, New York, 1949; MIT Press, Cambridge, 1969.

Ince, E. L., *Ordinary Differential Equations*, Longmans, Green, London, 1927; Dover, New York, 1956.

Uspensky, J. V., *Theory of Equations*, McGraw-Hill, New York, 1948.

CHAPTER 6

The Laplace Transform

6.1 INTRODUCTION. DEFINITION OF THE LAPLACE TRANSFORM

Among the tools that are very useful in solving linear differential equations are *integral transforms.* An integral transform is a relation of the form

$$F(s) = \int_\alpha^\beta K(s, t)f(t)\, dt, \tag{1}$$

wherein a given function f is transformed into another function F by means of an integral. The function F is said to be the transform of f, and the function K is called the *kernel* of the transformation. The general idea is to use the relation (1) to transform a problem for f into a simpler problem for F, to solve this simpler problem, and then to recover the desired function f from its transform F. By making a suitable choice of the kernel K and the integration limits α and β, it is often possible to simplify substantially a problem involving a linear differential equation. Several integral transformations are widely used, each being appropriate for certain types of problems.

In this chapter we will discuss the properties and some of the applications of the Laplace* transform. This transform is defined in the following way. Let $f(t)$ be given for $t \geq 0$, and suppose that f satisfies certain conditions to be stated a little later. Then the Laplace transform of f, which we will denote by $\mathcal{L}\{f(t)\}$ or by $F(s)$, is defined by the equation

$$\mathcal{L}\{f(t)\} = F(s) = \int_0^\infty e^{-st}f(t)\, dt. \tag{2}$$

This transform makes use of the kernel $K(s, t) = e^{-st}$ and is associated particularly with linear differential equations with constant coefficients. It is especially useful in solving problems with nonhomogeneous terms of a discontinuous or impulsive

* P. S. Laplace (1749–1827), one of the greatest of all French mathematicians, is especially famous for his work in astronomy, probability, and mechanics. He studied the relation (2) in 1782, but the techniques described in this chapter were not developed until much later, mainly by Oliver Heaviside (1850–1925).

nature. Such problems are relatively awkward to handle by the methods previously discussed, which involve piecing together solutions valid in different intervals. The Laplace transform is particularly valuable in circuit analysis, where discontinuous or impulsive forcing terms are common, but it is also important in other applications.

Since the Laplace transform is defined by an integral over the range from zero to infinity, it is useful to mention first some basic facts about such integrals. In the first place, an integral over an unbounded interval is called an *improper integral*, and is defined as a limit of integrals over finite intervals; thus

$$\int_a^\infty f(t)\,dt = \lim_{A\to\infty} \int_a^A f(t)\,dt, \tag{3}$$

where A is a positive real number. If the integral from a to A exists for each $A > a$, and if the limit as $A \to \infty$ exists, then the improper integral is said to *converge* to that limiting value. Otherwise the integral is said to *diverge*, or to fail to exist. The following examples illustrate both possibilities.

Example 1.   Let $f(t) = e^{ct}$, $t \geq 0$, where c is a real nonzero constant. Then

$$\int_0^\infty e^{ct}\,dt = \lim_{A\to\infty} \int_0^A e^{ct}\,dt = \lim_{A\to\infty} \frac{e^{ct}}{c}\bigg|_0^A$$

$$= \lim_{A\to\infty} \frac{1}{c}(e^{cA} - 1).$$

It follows that the improper integral converges if $c < 0$, and diverges if $c > 0$. If $c = 0$ the integrand is unity, and the integral again diverges.

Example 2.   Let $f(t) = 1/t$, $t \geq 1$. Then

$$\int_1^\infty \frac{dt}{t} = \lim_{A\to\infty} \int_1^A \frac{dt}{t} = \lim_{A\to\infty} \ln A.$$

Since $\lim_{A\to\infty} \ln A = \infty$, the improper integral diverges.

Example 3.   Let $f(t) = t^{-p}$, $t \geq 1$, where p is a real constant and $p \neq 1$; the case $p = 1$ was considered in Example 2. Then

$$\int_1^\infty t^{-p}\,dt = \lim_{A\to\infty} \int_1^A t^{-p}\,dt = \lim_{A\to\infty} \frac{1}{1-p}(A^{1-p} - 1).$$

As $A \to \infty$, $A^{1-p} \to 0$ if $p > 1$, but $A^{1-p} \to \infty$ if $p < 1$. Hence $\int_1^\infty t^{-p}\,dt$ converges for $p > 1$, but (incorporating the result of Example 2) diverges for $p \leq 1$. These results are analogous to those for the infinite series $\sum_{n=1}^\infty n^{-p}$.

Before discussing the possible existence of $\int_a^\infty f(t)\,dt$, it is helpful to define certain terms. A function f is said to be *piecewise continuous* on an interval $\alpha \le t \le \beta$ if the interval can be partitioned by a finite number of points $\alpha = t_0 < t_1 < \cdots < t_n = \beta$ so that

1. f is continuous on each open subinterval $t_{i-1} < t < t_i$;

2. f approaches a finite limit as the end points of each subinterval are approached from within the subinterval.

In other words, f is piecewise continuous on $\alpha \le t \le \beta$ if it is continuous there except for a finite number of jump discontinuities. If f is piecewise continuous on $\alpha \le t \le \beta$ for every $\beta > \alpha$, then f is said to be piecewise continuous on $t \ge \alpha$. An example of a piecewise continuous function is shown in Figure 6.1.

If f is piecewise continuous on the interval $a \le t \le A$, then it can be shown that $\int_a^A f(t)\,dt$ exists. Hence, if f is piecewise continuous for $t \ge a$, then $\int_a^A f(t)\,dt$ exists for each $A > a$. However, piecewise continuity is not enough to insure convergence of the improper integral $\int_a^\infty f(t)\,dt$, as the above examples show.

If f cannot be easily integrated in terms of elementary functions, the definition of convergence of $\int_a^\infty f(t)\,dt$ may be difficult to apply. Frequently, the most convenient way to test the convergence or divergence of an improper integral is by the following comparison theorem, which is analogous to a similar theorem for infinite series.

Theorem 6.1. *If f is piecewise continuous for $t \ge a$, if $|f(t)| \le g(t)$ when $t \ge M$ for some positive constant M, and if $\int_M^\infty g(t)\,dt$ converges, then $\int_a^\infty f(t)\,dt$ also converges. On the other hand, if $f(t) \ge g(t) \ge 0$ for $t \ge M$, and if $\int_M^\infty g(t)\,dt$ diverges, then $\int_a^\infty f(t)\,dt$ also diverges.*

FIGURE 6.1   A piecewise continuous function.

The proof of this result from the calculus will not be given here. It is made plausible, however, by comparing the areas represented by $\int_M^\infty g(t)\, dt$ and $\int_M^\infty |f(t)|\, dt$. The functions most useful for comparison purposes are e^{ct} and t^{-p}, which were considered in Examples 1, 2, and 3.

We now return to a consideration of the Laplace transform $\mathscr{L}\{f(t)\}$ or $F(s)$, which is defined by Eq. (2) whenever this improper integral converges. In general, the parameter s may be complex, but for our discussion we need consider only real values of s. According to the foregoing discussion of integrals the function f must satisfy certain conditions in order for its Laplace transform F to exist. The simplest and most useful of these conditions are stated in the following theorem.

Theorem 6.2.   *Suppose that*
  (i) *f is piecewise continuous on the interval $0 \le t \le A$ for any positive A;*
  (ii) *$|f(t)| \le Ke^{at}$ when $t \ge M$. In this inequality K, a, and M are real constants, K and M necessarily positive.*
  Then the Laplace transform $\mathscr{L}\{f(t)\} = F(s)$, defined by Eq. (2), exists for $s > a$.

To establish this theorem it is necessary to show only that the integral in Eq. (2) converges for $s > a$. Splitting the improper integral into two parts, we have

$$\int_0^\infty e^{-st}f(t)\, dt = \int_0^M e^{-st}f(t)\, dt + \int_M^\infty e^{-st}f(t)\, dt. \tag{4}$$

The first integral on the right side of Eq. (4) exists by hypothesis (i) of the theorem; hence the existence of $F(s)$ depends on the convergence of the second integral. By hypothesis (ii) we have, for $t \ge M$,

$$|e^{-st}f(t)| \le Ke^{-st}e^{at} = Ke^{(a-s)t},$$

and thus, by Theorem 6.1, $F(s)$ exists provided $\int_M^\infty e^{(a-s)t}\, dt$ converges. Referring to Example 1 with c replaced by $a - s$, we see that this latter integral converges when $a - s < 0$, which establishes Theorem 6.2.

Unless the contrary is specifically stated, we will, in this chapter, deal only with functions satisfying the conditions of Theorem 6.2. Such functions are described as piecewise continuous, and of *exponential order* as $t \to \infty$. The Laplace transforms of some important elementary functions are given in the examples below.

Example 4.   Let $f(t) = 1$, $t \ge 0$. Then

$$\mathscr{L}\{1\} = \int_0^\infty e^{-st}\, dt = \frac{1}{s}, \qquad s > 0.$$

Example 5. Let $f(t) = e^{at}$, $t \geq 0$. Then

$$\mathscr{L}\{e^{at}\} = \int_0^\infty e^{-st} e^{at} \, dt = \int_0^\infty e^{-(s-a)t} \, dt$$

$$= \frac{1}{s - a}, \qquad s > a.$$

Example 6. Let $f(t) = \sin at$, $t \geq 0$. Then

$$\mathscr{L}\{\sin at\} = F(s) = \int_0^\infty e^{-st} \sin at \, dt, \qquad s > 0.$$

Since

$$F(s) = \lim_{A \to \infty} \int_0^A e^{-st} \sin at \, dt,$$

upon integrating by parts we obtain

$$F(s) = \lim_{A \to \infty} \left[-\frac{e^{-st} \cos at}{a} \bigg|_0^A - \frac{s}{a} \int_0^A e^{-st} \cos at \, dt \right]$$

$$= \frac{1}{a} - \frac{s}{a} \int_0^\infty e^{-st} \cos at \, dt.$$

A second integration by parts then yields

$$F(s) = \frac{1}{a} - \frac{s^2}{a^2} \int_0^\infty e^{-st} \sin at \, dt$$

$$= \frac{1}{a} - \frac{s^2}{a^2} F(s).$$

Hence, solving for $F(s)$, we have

$$F(s) = \frac{a}{s^2 + a^2}, \qquad s > 0.$$

Now let us suppose that f_1 and f_2 are two functions whose Laplace transforms exist for $s > a_1$ and $s > a_2$, respectively. Then, for s greater than the maximum of a_1 and a_2,

$$\mathscr{L}\{c_1 f_1(t) + c_2 f_2(t)\} = \int_0^\infty e^{-st} [c_1 f_1(t) + c_2 f_2(t)] \, dt$$

$$= c_1 \int_0^\infty e^{-st} f_1(t) \, dt + c_2 \int_0^\infty e^{-st} f_2(t) \, dt,$$

and hence

$$\mathscr{L}\{c_1 f_1(t) + c_2 f_2(t)\} = c_1 \mathscr{L}\{f_1(t)\} + c_2 \mathscr{L}\{f_2(t)\}. \tag{5}$$

Equation (5) is a statement of the fact that the Laplace transform is a *linear operator*. This property is of paramount importance, and we will make frequent use of it later.

PROBLEMS

1. Sketch the graph of each of the following functions. In each case determine whether f is continuous, piecewise continuous, or neither on the interval $0 \leq t \leq 3$.

(a) $f(t) = \begin{cases} t^2, & 0 \leq t \leq 1 \\ 2 + t, & 1 < t \leq 2 \\ 6 - t, & 2 < t \leq 3 \end{cases}$

(b) $f(t) = \begin{cases} t^2, & 0 \leq t \leq 1 \\ (t - 1)^{-1}, & 1 < t \leq 2 \\ 1, & 2 < t \leq 3 \end{cases}$

(c) $f(t) = \begin{cases} t^2, & 0 \leq t \leq 1 \\ 1, & 1 < t \leq 2 \\ 3 - t, & 2 < t \leq 3 \end{cases}$

(d) $f(t) = \begin{cases} t, & 0 \leq t \leq 1 \\ 3 - t, & 1 < t \leq 2 \\ 1, & 2 < t \leq 3 \end{cases}$

2. Find the Laplace transform of each of the following functions.

(a) t (b) t^2 (c) t^n, where n is a positive integer

3. Find the Laplace transform of $f(t) = \cos at$, where a is a real constant.

4. Recalling that $\cosh bt = (e^{bt} + e^{-bt})/2$ and $\sinh bt = (e^{bt} - e^{-bt})/2$, find the Laplace transform of each of the following functions; a and b are real constants.

(a) $\cosh bt$ (b) $\sinh bt$

(c) $e^{at} \cosh bt$ (d) $e^{at} \sinh bt$

5. Recall that $\cos bt = (e^{ibt} + e^{-ibt})/2$ and $\sin bt = (e^{ibt} - e^{-ibt})/2i$. Assuming that the necessary elementary integration formulas extend to this case, find the Laplace transform of each of the following functions; a and b are real constants.

(a) $\sin bt$ (b) $\cos bt$

(c) $e^{at} \sin bt$ (d) $e^{at} \cos bt$

6. Using integration by parts, find the Laplace transform of each of the following functions; n is a positive integer and a is a real constant.

(a) te^{at} (b) $t \sin at$ (c) $t \cosh at$

(d) $t^n e^{at}$ (e) $t^n \sin at$ (f) $t^n \sinh at$

7. Determine whether each of the following integrals converges or diverges.

(a) $\displaystyle\int_0^\infty (t^2 + 1)^{-1} \, dt$ (b) $\displaystyle\int_0^\infty te^{-t} \, dt$ (c) $\displaystyle\int_1^\infty t^{-2}e^t \, dt$

8. Suppose that f and f' are continuous for $t \geq 0$, and of exponential order as $t \to \infty$. Show by integration by parts that if $F(s) = \mathcal{L}\{f(t)\}$, then $\lim_{s \to \infty} F(s) = 0$. The result is actually true under less restrictive conditions, such as those of Theorem 6.2.

9. **The Gamma Function.**   The Gamma function is denoted by $\Gamma(p)$ and is defined by the integral

$$\Gamma(p + 1) = \int_0^\infty e^{-x}x^p \, dx. \tag{i}$$

This integral converges at infinity for all p. For $p < 0$ it is also improper because the integrand becomes unbounded as $x \to 0$. However, the integral can be shown to converge at $x = 0$ for $p > -1$.

(a) Show that for $p > 0$

$$\Gamma(p + 1) = p\Gamma(p).$$

(b) If p is a positive integer n, show that

$$\Gamma(n + 1) = n!.$$

Since $\Gamma(p)$ is also defined when p is not an integer, this function provides an extension of the factorial function to nonintegral values of the independent variable.

(c) Show that for $p > 0$

$$p(p + 1)(p + 2) \cdots (p + n - 1) = \frac{\Gamma(p + n)}{\Gamma(p)}.$$

Thus $\Gamma(p)$ can be determined for all positive values of p if $\Gamma(p)$ is known in a single interval of unit length, say $0 < p \leq 1$. It is possible to show that $\Gamma(\frac{1}{2}) = \sqrt{\pi}$. Find $\Gamma(\frac{3}{2})$ and $\Gamma(\frac{11}{2})$.

(d) Show that $\Gamma(1) = 1$. Thus it is consistent to define $0! = 1$.

10. Consider the Laplace transform of t^p, where $p > -1$.

(a) Referring to Problem 9, show that

$$\mathscr{L}\{t^p\} = \int_0^\infty e^{-st} t^p \, dt = \frac{1}{s^{p+1}} \int_0^\infty e^{-x} x^p \, dx$$

$$= \frac{\Gamma(p + 1)}{s^{p+1}}, \qquad s > 0.$$

(b) If p is a positive integer n, show that

$$\mathscr{L}\{t^n\} = \frac{n!}{s^{n+1}}, \qquad s > 0.$$

(c) Show that

$$\mathscr{L}\{t^{-\frac{1}{2}}\} = \frac{2}{\sqrt{s}} \int_0^\infty e^{-x^2} \, dx, \qquad s > 0.$$

It is possible to show that

$$\int_0^\infty e^{-x^2} \, dx = \frac{\sqrt{\pi}}{2};$$

hence

$$\mathscr{L}\{t^{-\frac{1}{2}}\} = \sqrt{\pi/s}, \qquad s > 0.$$

(d) Show that

$$\mathscr{L}\{t^{\frac{1}{2}}\} = \sqrt{\pi}/2s^{\frac{3}{2}}, \qquad s > 0.$$

6.2   SOLUTION OF INITIAL VALUE PROBLEMS

In this section we will show how the Laplace transform can be used to solve initial value problems for differential equations with constant coefficients. The usefulness of the Laplace transform in this connection rests primarily on the fact that the transform of f' is related in a simple way to the transform of f. The relationship is made clear in the following theorem.

Theorem 6.3.   *Suppose that f is continuous and that f' is piecewise continuous on any interval $0 \leq t \leq A$. Suppose further that there exist constants K, a, and M such that $|f(t)| \leq Ke^{at}$ for $t \geq M$. Then $\mathcal{L}\{f'(t)\}$ exists for $s > a$, and moreover*

$$\mathcal{L}\{f'(t)\} = s\mathcal{L}\{f(t)\} - f(0). \tag{1}$$

In order to prove this theorem we consider the integral

$$\int_0^A e^{-st} f'(t)\, dt.$$

Letting $t_1, t_2, \ldots, t_n$ be the points in the interval $0 \leq t \leq A$ where f' is discontinuous, we obtain

$$\int_0^A e^{-st}f'(t)\,dt = \int_0^{t_1} e^{-st}f'(t)\,dt + \int_{t_1}^{t_2} e^{-st}f'(t)\,dt + \cdots + \int_{t_n}^A e^{-st}f'(t)\,dt.$$

Integrating each term on the right by parts yields

$$\int_0^A e^{-st}f'(t)\,dt = e^{-st}f(t)\Big|_0^{t_1} + e^{-st}f(t)\Big|_{t_1}^{t_2} + \cdots + e^{-st}f(t)\Big|_{t_n}^A$$
$$+ s\left[\int_0^{t_1} e^{-st}f(t)\,dt + \int_{t_1}^{t_2} e^{-st}f(t)\,dt + \cdots + \int_{t_n}^A e^{-st}f(t)\,dt\right].$$

Since f is continuous, the contributions of the integrated terms at $t_1, t_2, \ldots, t_n$ cancel. Combining the integrals gives

$$\int_0^A e^{-st}f'(t)\,dt = e^{-sA}f(A) - f(0) + s\int_0^A e^{-st}f(t)\,dt.$$

As $A \to \infty$, $e^{-sA}f(A) \to 0$ whenever $s > a$. Hence, for $s > a$,

$$\mathcal{L}\{f'(t)\} = s\mathcal{L}\{f(t)\} - f(0),$$

which establishes the theorem.

If f' and f'' satisfy the same conditions that are imposed on f and f' respectively in Theorem 6.3, then it follows that the Laplace transform of f'' also exists for $s > a$ and is given by

$$\mathcal{L}\{f''(t)\} = s^2\mathcal{L}\{f(t)\} - sf(0) - f'(0). \tag{2}$$

Indeed, provided the function f and its derivatives satisfy suitable conditions, an expression for the transform of the nth derivative $f^{(n)}$ can be derived by successive applications of this theorem. The result is given in the following corollary.

Corollary. Suppose that the functions $f, f', \ldots, f^{(n-1)}$ are continuous, and that $f^{(n)}$ is piecewise continuous on any interval $0 \leq t \leq A$. Suppose further that there exist constants K, a, and M such that $|f(t)| \leq Ke^{at}$, $|f'(t)| \leq Ke^{at}, \ldots$, $|f^{(n-1)}(t)| \leq Ke^{at}$ for $t \geq M$. Then $\mathcal{L}\{f^{(n)}(t)\}$ exists for $s > a$ and is given by

$$\mathcal{L}\{f^{(n)}(t)\} = s^n \mathcal{L}\{f(t)\} - s^{n-1}f(0) - \cdots - sf^{(n-2)}(0) - f^{(n-1)}(0). \qquad (3)$$

We will now show how the Laplace transform can be used to solve initial value problems. First we consider the differential equation

$$y'' - y' - 2y = 0 \qquad (4)$$

and the initial conditions

$$y(0) = 1, \qquad y'(0) = 0. \qquad (5)$$

This simple problem is easily solved by the methods of Section 3.5. The characteristic equation is

$$r^2 - r - 2 = (r - 2)(r + 1) = 0, \qquad (6)$$

and consequently the general solution of Eq. (4) is

$$y = c_1 e^{-t} + c_2 e^{2t}. \qquad (7)$$

The initial conditions (5) require that $c_1 + c_2 = 1$, $-c_1 + 2c_2 = 0$; hence $c_1 = \frac{2}{3}$ and $c_2 = \frac{1}{3}$, so that the solution of the initial value problem (4) and (5) is

$$y = \phi(t) = \tfrac{2}{3}e^{-t} + \tfrac{1}{3}e^{2t}. \qquad (8)$$

Now let us solve the given problem by means of the Laplace transform. To do this we must assume that the problem has a solution $y = \phi(t)$, which with its first two derivatives satisfies the conditions of the Corollary. Then, taking the Laplace transform of the differential equation (4), we obtain

$$\mathcal{L}\{y''\} - \mathcal{L}\{y'\} - 2\mathcal{L}\{y\} = 0, \qquad (9)$$

where we have used the linearity of the transform to write the transform of a sum as the sum of the separate transforms. Upon using the corollary to express $\mathcal{L}\{y''\}$ and $\mathcal{L}\{y'\}$ in terms of $\mathcal{L}\{y\}$, we find that Eq. (9) then takes the form

$$s^2\mathcal{L}\{y\} - sy(0) - y'(0) - [s\mathcal{L}\{y\} - y(0)] - 2\mathcal{L}\{y\} = 0,$$

or

$$(s^2 - s - 2)Y(s) + (1 - s)y(0) - y'(0) = 0, \qquad (10)$$

where $Y(s) = \mathcal{L}\{y\}$. Substituting for $y(0)$ and $y'(0)$ in Eq. (10) from the initial conditions (5), and then solving for $Y(s)$, we obtain

$$Y(s) = \frac{s - 1}{s^2 - s - 2} = \frac{s - 1}{(s - 2)(s + 1)}. \qquad (11)$$

We have thus obtained an expression for the Laplace transform $Y(s)$ of the solution $y = \phi(t)$ of the given initial value problem. To determine the function ϕ we must find the function whose Laplace transform is $Y(s)$, as given by Eq. (11).

In the present case this can be done most easily by expanding the right side of Eq. (11) in partial fractions. Thus we find that

$$Y(s) = \frac{\frac{1}{3}}{s - 2} + \frac{\frac{2}{3}}{s + 1}. \tag{12}$$

Finally, using the result of Example 5 of Section 6.1, it follows that $\frac{1}{3}e^{2t}$ has the transform $\frac{1}{3}(s - 2)^{-1}$; similarly $\frac{2}{3}e^{-t}$ has the transform $\frac{2}{3}(s + 1)^{-1}$. Hence, by the linearity of the Laplace transform,

$$y = \phi(t) = \tfrac{1}{3}e^{2t} + \tfrac{2}{3}e^{-t}$$

has the transform (12). This is the same solution, of course, as that obtained earlier in a different way.

While in this simple problem the use of the Laplace transform offers no particular advantages, we can nevertheless point out some of the essential features of the transform method. In the first place, the transform $Y(s)$ of the unknown function $y = \phi(t)$ is found by solving an *algebraic equation*, Eq. (10), rather than a *differential equation*. Next, the solution satisfying the initial conditions (5) as well as the differential equation (4) is automatically found. Thus the task of determining appropriate values for the arbitrary constants appearing in the general solution does not arise. The reader should also note that the coefficient of $Y(s)$ in Eq. (10) is precisely the polynomial that also appears in the characteristic equation (6); in fact, this is always the case. Since the use of the partial fraction expansion of $Y(s)$ in order to determine $\phi(t)$ requires that this polynomial be factored, the use of Laplace transforms does not avoid the necessity of finding the roots of the characteristic equation. For equations of higher than second order this may be a difficult algebraic problem, particularly if the roots are irrational or complex.

The main difficulty that occurs in solving initial value problems by the transform technique lies in the problem of determining the function $y = \phi(t)$ corresponding to the transform $Y(s)$. This problem is known as the inversion problem for the Laplace transform; $\phi(t)$ is called the inverse transform corresponding to $Y(s)$, and the process of finding $\phi(t)$ from $Y(s)$ is known as "inverting the transform." We will also use the notation $\mathcal{L}^{-1}\{Y(s)\}$ to denote the inverse transform of $Y(s)$. There is a general formula for the inverse Laplace transform, but its use requires a knowledge of the theory of functions of a complex variable, and we will not consider it in this book. However, it is still possible to develop many important properties of the Laplace transform, and to solve many interesting problems, without the use of complex variables.

In solving the initial value problem (4), (5) we did not consider the question of whether there may be functions other than that given by Eq. (8) that also have the transform (12). In fact, it can be shown that if f is a continuous function with the Laplace transform F, then there is no other continuous function having the same transform. In other words, there is essentially a one-to-one correspondence between functions and their Laplace transforms. This fact suggests the compilation of a table (somewhat analogous to a table of integrals) giving the transforms of

TABLE 6.1 Elementary Laplace Transforms

$f(t) = \mathcal{L}^{-1}\{F(s)\}$	$F(s) = \mathcal{L}\{f(t)\}$	Notes
1	$\dfrac{1}{s}, \quad s > 0$	Sec. 6.1; Ex. 4
e^{at}	$\dfrac{1}{s-a}, \quad s > a$	Sec. 6.1; Ex. 5
$\sin at$	$\dfrac{a}{s^2+a^2}, \quad s > 0$	Sec. 6.1; Ex. 6
$t^n, \ n = $ positive integer	$\dfrac{n!}{s^{n+1}}, \quad s > 0$	Sec. 6.1; Prob. 10
$t^p, \ p > -1$	$\dfrac{\Gamma(p+1)}{s^{p+1}}, \quad s > 0$	
$\cos at$	$\dfrac{s}{s^2+a^2}, \quad s > 0$	Sec. 6.1; Prob. 3
$\sinh at$	$\dfrac{a}{s^2-a^2}, \quad s > |a|$	Sec. 6.1; Prob. 4
$\cosh at$	$\dfrac{s}{s^2-a^2}, \quad s > |a|$	
$e^{at} \sin bt$	$\dfrac{b}{(s-a)^2+b^2}, \quad s > a$	Sec. 6.1; Prob. 5
$e^{at} \cos bt$	$\dfrac{s-a}{(s-a)^2+b^2}, \quad s > a$	
$t^n e^{at}, \ n = $ positive integer	$\dfrac{n!}{(s-a)^{n+1}}, \quad s > a$	Sec. 6.1; Prob. 6
$u_c(t)$	$\dfrac{e^{-cs}}{s}, \quad s > 0$	Sec. 6.3
$u_c(t)f(t-c)$	$e^{-cs}F(s)$	Sec. 6.3
$e^{ct}f(t)$	$F(s-c)$	Sec. 6.3
$f(ct)$	$\dfrac{1}{c} F\left(\dfrac{s}{c}\right), \quad c > 0$	Sec. 6.3; Prob. 4
$\displaystyle\int_0^t f(t-\tau)g(\tau)\,d\tau$	$F(s)G(s)$	Sec. 6.5
$\delta(t-c)$	e^{-cs}	Sec. 6.4
$f^{(n)}(t)$	$s^n F(s) - s^{n-1}f(0) - \cdots - f^{(n-1)}(0)$	Sec. 6.2
$(-t)^n f(t)$	$F^{(n)}(s)$	Sec. 6.2; Prob. 14

functions frequently encountered, and vice versa (see Table 6.1).* The entries in the second column are the transforms of those in the first column. Perhaps more important, the functions in the first column are the inverse transforms of those in the second column. Thus, for example, if the transform of the solution of a differential equation is known, the solution itself can often be found merely by looking it up in the table. Some of the entries in Table 6.1 have been used as examples, or appear as problems in Section 6.1, while others will be developed later in the chapter. The third column of the table indicates where the derivation of some of the given transforms may be found.

Frequently, a Laplace transform $F(s)$ is expressible as a sum of several terms,

$$F(s) = F_1(s) + F_2(s) + \cdots + F_n(s). \tag{13}$$

Suppose that $f_1(t) = \mathscr{L}^{-1}\{F_1(s)\}, \ldots, f_n(t) = \mathscr{L}^{-1}\{F_n(s)\}$. Then the function

$$f(t) = f_1(t) + \cdots + f_n(t)$$

has the Laplace transform $F(s)$. By the uniqueness property stated above there is no other continuous function f having the same transform. Thus

$$\mathscr{L}^{-1}\{F(s)\} = \mathscr{L}^{-1}\{F_1(s)\} + \cdots + \mathscr{L}^{-1}\{F_n(s)\}; \tag{14}$$

that is, the inverse Laplace transform is also a linear operator.

In many problems it is convenient to make use of this property by decomposing a given transform into a sum of functions whose inverse transforms are already known or can be found in the table. Partial fraction expansions are particularly useful in this connection, and a general result covering many cases is given in Problem 19. Other useful properties of Laplace transforms are derived later in this chapter.

As a further illustration of the technique of solving initial value problems by means of the Laplace transform, consider the following example.

Example.   Find the solution of the differential equation

$$y'' + y = \sin 2t, \tag{15}$$

satisfying the initial conditions

$$y(0) = 0, \qquad y'(0) = 1. \tag{16}$$

We will assume that this initial value problem has a solution $y = \phi(t)$, which with its first two derivatives satisfies the conditions of the Corollary to Theorem 6.3. Then, taking the Laplace transform of the differential equation, we have

$$s^2 Y(s) - s y(0) - y'(0) + Y(s) = \frac{2}{s^2 + 4},$$

---

* Larger tables are also available. See the references at the end of the chapter.

where the transform of sin $2t$ has been obtained from Table 6.1. Substituting for $y(0)$ and $y'(0)$ from the initial conditions and solving for $Y(s)$, we obtain

$$Y(s) = \frac{s^2 + 6}{(s^2 + 1)(s^2 + 4)}.$$

Using partial fractions we can write $Y(s)$ in the form

$$Y(s) = \frac{as + b}{s^2 + 1} + \frac{cs + d}{s^2 + 4}.$$

Upon determining the constants a, b, c, and d, we find that

$$Y(s) = \frac{\frac{5}{3}}{s^2 + 1} - \frac{\frac{2}{3}}{s^2 + 4}.$$

Referring to Example 6 of Section 6.1, or to Table 6.1, it follows that

$$y = \phi(t) = \tfrac{5}{3} \sin t - \tfrac{1}{3} \sin 2t. \tag{17}$$

The reader should note that in solving a differential equation, such as Eq. (15), by means of the Laplace transform, we *assume* at the beginning that the solution $y = \phi(t)$ *has* a transform $Y(s)$. Once the solution has been found, this assumption should be justified. For example, the expression on the right side of Eq. (17) clearly satisfies the conditions of the Corollary to Theorem 6.3 for $n = 2$.

This example also illustrates that nonhomogeneous problems are solved in exactly the same manner as homogeneous ones; it is not necessary to solve the homogeneous and nonhomogeneous equations separately, as in Chapter 3. In this respect the Laplace transform has a definite advantage. Again, the solution satisfying the initial conditions is found directly, without first finding the general solution of the differential equation.

The most important elementary applications of the Laplace transform are in the study of mechanical vibrations and in the analysis of electric circuits; the governing equations were derived in Sections 3.7 and 3.8, respectively. A vibrating spring-mass system has the equation of motion

$$m \frac{d^2u}{dt^2} + c \frac{du}{dt} + ku = F(t), \tag{18}$$

where m is the mass, c the damping coefficient, k the spring constant, and $F(t)$ the applied external force. The equation describing an electric circuit containing a resistance R, an inductance L, and a capacitance C (an *RLC* circuit) is

$$L \frac{d^2Q}{dt^2} + R \frac{dQ}{dt} + \frac{1}{C} Q = E(t), \tag{19}$$

where $Q(t)$ is the charge on the condenser and $E(t)$ is the applied voltage. In terms of the current $I(t) = dQ(t)/dt$ we can write Eq. (19) in the form

$$L\frac{d^2I}{dt^2} + R\frac{dI}{dt} + \frac{1}{C}I = \frac{dE}{dt}(t). \tag{20}$$

Suitable initial conditions on u, Q, or I must also be prescribed.

We have noted previously in Section 3.8 that Eq. (18) for the spring-mass system and Eqs. (19) or (20) for the electric circuit are identical mathematically, differing only in the interpretation of the constants and variables appearing in them. Other physical problems also lead to the same differential equation. Thus once the mathematical problem is solved, its solution can be interpreted in terms of whichever corresponding physical problem is of immediate interest.

In the problem lists following this and other sections in this chapter are numerous initial value problems for second order linear differential equations with constant coefficients. Many can be interpreted as models of particular physical systems, but usually we do not point this out explicitly.

PROBLEMS

In each of Problems 1 through 10 use the Laplace transform to solve the given initial value problem.

1. $y'' - y' - 6y = 0$; $y(0) = 1$, $y'(0) = -1$
2. $y'' + 3y' + 2y = 0$; $y(0) = 1$, $y'(0) = 0$
3. $y'' - 2y' + 2y = 0$; $y(0) = 0$, $y'(0) = 1$
4. $y'' - 4y' + 4y = 0$; $y(0) = 1$, $y'(0) = 1$
5. $y^{iv} - 4y''' + 6y'' - 4y' + y = 0$;

$\quad\quad\quad\quad y(0) = 0$, $y'(0) = 1$, $y''(0) = 0$, $y'''(0) = 1$
6. $y^{iv} - y = 0$; $y(0) = 1$, $y'(0) = 0$, $y''(0) = 1$, $y'''(0) = 0$
7. $y'' - 2y' - 2y = 0$; $y(0) = 2$, $y'(0) = 0$
8. $y'' + \omega^2 y = \cos 2t$, $\omega^2 \neq 4$; $y(0) = 1$, $y'(0) = 0$
9. $y'' - 2y' + 2y = \cos t$; $y(0) = 1$, $y'(0) = 0$
10. $y'' - 2y' + 2y = e^{-t}$; $y(0) = 0$, $y'(0) = 1$

In each of Problems 11 and 12 find the Laplace transform $Y(s) = \mathscr{L}\{y\}$ of the solution of the given initial value problem. A method for determining the inverse transforms will be developed in Section 6.3.

11. $y'' + 4y = \begin{cases} 1, & 0 \leq t < \pi, \\ 0, & \pi \leq t < \infty; \end{cases}$ $y(0) = 1$, $y'(0) = 0$

12. $y'' + y = \begin{cases} t, & 0 \leq t < 1, \\ 0, & 1 \leq t < \infty; \end{cases}$ $y(0) = 0$, $y'(0) = 0$

*13. The Laplace transform of certain functions can be conveniently found from their Taylor series expansions.

(a) Using the Taylor series for $\sin t$,

$$\sin t = \sum_{n=0}^{\infty} \frac{(-1)^n t^{2n+1}}{(2n + 1)!},$$

and assuming that the Laplace transform of this series can be computed term by term, verify that

$$\mathscr{L}\{\sin t\} = \frac{1}{s^2 + 1}, \qquad s > 1.$$

(b) Let

$$f(t) = \begin{cases} (\sin t)/t, & t \neq 0, \\ 1, & t = 0. \end{cases}$$

Find the Taylor series for f about $t = 0$. Assuming that the Laplace transform of this function can be computed term by term, verify that

$$\mathscr{L}\{f(t)\} = \arctan 1/s, \qquad s > 1.$$

(c) The Bessel function of the first kind of order zero J_0 has the Taylor series (see Section 4.7)

$$J_0(t) = \sum_{n=0}^{\infty} \frac{(-1)^n t^{2n}}{2^{2n}(n!)^2}.$$

Assuming that the following Laplace transforms can be computed term by term, verify that

$$\mathscr{L}\{J_0(t)\} = (s^2 + 1)^{-\frac{1}{2}}, \qquad s > 1,$$

and that

$$\mathscr{L}\{J_0(\sqrt{t})\} = \frac{1}{s} e^{-1/4s}, \qquad s > 0.$$

Problems 14 through 17 are concerned with differentiation of the Laplace transform.

14. Let

$$F(s) = \int_0^{\infty} e^{-st} f(t) \, dt.$$

It is possible to show that as long as f satisfies the conditions of Theorem 6.2, it is legitimate to differentiate under the integral sign with respect to the parameter s when $s > a$.

(a) Show that

$$F'(s) = \mathscr{L}\{-tf(t)\}.$$

(b) Show that

$$F^{(n)}(s) = \mathscr{L}\{(-t)^n f(t)\};$$

hence differentiating the Laplace transform corresponds to multiplying the original function by $-t$.

15. Use the result of Problem 14 to find the Laplace transform of each of the following functions.

(a) $f(t) = t^n e^{at}$, where a is a real constant and n is a positive integer

(b) $f(t) = t^2 \sin t$ (c) $f(t) = t^n$, where n is a positive integer

*16. Consider Bessel's equation of order zero

$$ty'' + y' + ty = 0.$$

Recall from Section 4.3 that $t = 0$ is a regular singular point for this equation, and therefore solutions may become unbounded as $t \to 0$. However, let us try to determine whether there are any solutions that remain finite at $t = 0$ and have finite derivatives there. Assuming that there is such a solution $y = \phi(t)$, let $Y(s) = \mathcal{L}\{\phi(t)\}$.

(a) Show that $Y(s)$ satisfies

$$(1 + s^2)Y'(s) + sY(s) = 0.$$

(b) Show that $Y(s) = c(1 + s^2)^{-\frac{1}{2}}$, where c is an arbitrary constant.

(c) Expanding $(1 + s^2)^{-\frac{1}{2}}$ in a binomial series valid for $s > 1$, and assuming that it is permissible to take the inverse transform term by term, show that

$$y = c \sum_{n=0}^{\infty} \frac{(-1)^n t^{2n}}{2^{2n}(n!)^2} = cJ_0(t),$$

where J_0 is the Bessel function of the first kind of order zero. Note that $J_0(0) = 1$, and that J_0 has finite derivatives of all orders at $t = 0$. It was shown in Section 4.7 that the second solution of this equation becomes unbounded as $t \to 0$.

17. For each of the following initial value problems use the results of Problem 14 to find the differential equation satisfied by $Y(s) = \mathcal{L}\{\phi(t)\}$, where $y = \phi(t)$ is the solution of the given initial value problem.

(a) $y'' + ty = 0;$ $y(0) = 1,$ $y'(0) = 0$ (Airy's equation)

(b) $(1 - t^2)y'' - 2ty' + \alpha(\alpha + 1)y = 0;$ $y(0) = 0,$ $y'(0) = 1$

(Legendre's equation)

Note that the differential equation for $Y(s)$ is of first order in part (a), but of second order in part (b). This is due to the fact that t appears at most to the first power in the equation of part (a), whereas it appears to the second power in that of part (b). This illustrates that the Laplace transform is not often useful in solving differential equations with variable coefficients, unless all of the coefficients are at most linear functions of the independent variable.

18. Suppose that

$$g(t) = \int_0^t f(\tau)\, d\tau.$$

If $G(s)$ and $F(s)$ are the Laplace transforms of $g(t)$ and $f(t)$, respectively, show that

$$G(s) = \frac{F(s)}{s}.$$

19. In this problem we show how a general partial fraction expansion can be used to calculate many inverse Laplace transforms. Suppose that

$$F(s) = \frac{P(s)}{Q(s)};$$

where $Q(s)$ is a polynomial of degree n with distinct zeros $r_1, \ldots, r_n$ and $P(s)$ is a polynomial of degree less than n. In this case it is possible to show that $P(s)/Q(s)$ has a partial fraction expansion of the form

$$\frac{P(s)}{Q(s)} = \frac{A_1}{s - r_1} + \cdots + \frac{A_n}{s - r_n}, \tag{i}$$

where the coefficients $A_1, \ldots, A_n$ must be determined.

(a) Show that

$$A_k = \frac{P(r_k)}{Q'(r_k)}, \qquad k = 1, \ldots, n. \tag{ii}$$

Hint: One way to do this is to multiply Eq. (i) by $s - r_k$ and then to take the limit as $s \to r_k$.

(b) Show that

$$\mathcal{L}^{-1}\{F(s)\} = \sum_{k=1}^{n} \frac{P(r_k)}{Q'(r_k)} e^{r_k t}. \tag{iii}$$

6.3 STEP FUNCTIONS

In Section 6.2 the general procedure involved in solving initial value problems by means of the Laplace transform was outlined. Some of the most interesting elementary applications of the transform method occur in the solution of linear differential equations with discontinuous or impulsive forcing functions. Equations of this type frequently arise in the analysis of the flow of current in electric circuits or the vibrations of mechanical systems. In this section and the following ones we will develop some additional properties of the Laplace transform that are useful in the solution of such problems. Unless a specific statement is made to the contrary, all functions appearing below will be assumed to be piecewise continuous and of exponential order, so that their Laplace transforms exist, at least for s sufficiently large.

In order to deal effectively with functions having jump discontinuities, it is very helpful to introduce a function known as the *unit step function*. This function will be denoted by u_c, and is defined by

$$u_c(t) = \begin{cases} 0, & t < c, \\ 1, & t \geq c, \end{cases} \qquad c \geq 0. \tag{1}$$

The graph of $y = u_c(t)$ is shown in Figure 6.2.

FIGURE 6.2 $y = u_c(t)$.

FIGURE 6.3 $y = u_\pi(t) - u_{2\pi}(t)$.

Example 1.   Sketch the graph of $y = h(t)$, where

$$h(t) = u_\pi(t) - u_{2\pi}(t), \qquad t \geq 0.$$

From the definition of $u_c(t)$ in Eq. (1) we have

$$h(t) = \begin{cases} 0 - 0 = 0, & 0 \leq t < \pi, \\ 1 - 0 = 1, & \pi \leq t < 2\pi, \\ 1 - 1 = 0, & 2\pi \leq t < \infty. \end{cases}$$

Thus the equation $y = h(t)$ has the graph shown in Figure 6.3.

The unit step function u_c can also be used to translate a given function f, with domain $t \geq 0$, a distance c to the right. For example, the function defined by

$$y = u_c(t)f(t - c) = \begin{cases} 0, & 0 \leq t < c, \\ f(t - c), & t \geq c, \end{cases}$$

represents a translation of the function f a distance c in the positive t direction (see Figure 6.4).

(a) $y = f(t)$.

(b) $y = u_c(t)f(t - c)$.

FIGURE 6.4

The Laplace transform of u_c is easily determined:

$$\mathcal{L}\{u_c(t)\} = \int_0^\infty e^{-st} u_c(t)\, dt$$

$$= \int_c^\infty e^{-st}\, dt$$

$$= \frac{e^{-cs}}{s}, \qquad s > 0. \tag{2}$$

The unit step function is particularly important in transform theory because of the following relation between the transform of $f(t)$ and that of its translation $u_c(t)f(t - c)$.

Theorem 6.4.   *If $F(s) = \mathcal{L}\{f(t)\}$ exists for $s > a \geq 0$, and if c is a positive constant, then*

$$\mathcal{L}\{u_c(t)f(t - c)\} = e^{-cs}\mathcal{L}\{f(t)\} = e^{-cs}F(s), \qquad s > a. \tag{3}$$

Conversely, if $f(t) = \mathcal{L}^{-1}\{F(s)\}$, then

$$u_c(t)f(t - c) = \mathcal{L}^{-1}\{e^{-cs}F(s)\}. \tag{4}$$

To prove Theorem 6.4 it is sufficient to compute the transform of $u_c(t)f(t - c)$:

$$\mathcal{L}\{u_c(t)f(t - c)\} = \int_0^\infty e^{-st} u_c(t) f(t - c)\, dt$$

$$= \int_c^\infty e^{-st} f(t - c)\, dt.$$

Introducing a new integration variable $\xi = t - c$, we have

$$\mathcal{L}\{u_c(t)f(t - c)\} = \int_0^\infty e^{-(\xi + c)s} f(\xi)\, d\xi = e^{-cs} \int_0^\infty e^{-s\xi} f(\xi)\, d\xi$$

$$= e^{-cs}F(s).$$

Thus Eq. (3) is established; Eq. (4) follows by taking the inverse transform of both sides of Eq. (3).

A simple example of this theorem occurs if we take $f(t) = 1$. Recalling that $\mathcal{L}\{1\} = 1/s$, we immediately have from Eq. (3) that $\mathcal{L}\{u_c(t)\} = e^{-cs}/s$. This result agrees with that of Eq. (2). Examples 2 and 3 illustrate further how Theorem 6.4 can be used in the calculation of transforms and inverse transforms.

───────────────

Example 2.   If the function f is defined by

$$f(t) = \begin{cases} \sin t, & 0 \leq t < 2\pi, \\ \sin t + \cos t, & t \geq 2\pi, \end{cases}$$

find $\mathcal{L}\{f(t)\}$.

Note that $f(t) = \sin t + g(t)$, where

$$g(t) = \begin{cases} 0, & t < 2\pi, \\ \cos t, & t \ge 2\pi. \end{cases}$$

Thus

$$g(t) = u_{2\pi}(t) \cos (t - 2\pi),$$

and

$$\mathscr{L}\{f(t)\} = \mathscr{L}\{\sin t\} + \mathscr{L}\{u_{2\pi}(t) \cos (t - 2\pi)\}$$
$$= \mathscr{L}\{\sin t\} + e^{-2\pi s}\mathscr{L}\{\cos t\}.$$

Introducing the transforms of $\sin t$ and $\cos t$, we obtain

$$\mathscr{L}\{f(t)\} = \frac{1}{s^2 + 1} + e^{-2\pi s} \frac{s}{s^2 + 1}$$
$$= \frac{1 + se^{-2\pi s}}{s^2 + 1},$$

The readers should compare this method with the calculation of $\mathscr{L}\{f(t)\}$ directly from the definition.

Example 3.   Find the inverse transform of

$$F(s) = \frac{1 - e^{-2s}}{s^2}.$$

From the linearity of the inverse transform we have

$$f(t) = \mathscr{L}^{-1}\{F(s)\} = \mathscr{L}^{-1}\left\{\frac{1}{s^2}\right\} - \mathscr{L}^{-1}\left\{\frac{e^{-2s}}{s^2}\right\}$$
$$= t - u_2(t)(t - 2).$$

The function f may also be written as

$$f(t) = \begin{cases} t, & 0 \le t < 2, \\ 2, & t \ge 2. \end{cases}$$

The following theorem contains another very useful property of Laplace transforms that is somewhat analogous to that given in Theorem 6.4.

Theorem 6.5.   *If $F(s) = \mathscr{L}\{f(t)\}$ exists for $s > a \ge 0$, and if c is a constant,* then

$$\mathscr{L}\{e^{ct}f(t)\} = F(s - c), \qquad s > a + c. \tag{5}$$

Conversely, if $f(t) = \mathscr{L}^{-1}\{F(s)\}$, then

$$e^{ct}f(t) = \mathscr{L}^{-1}\{F(s - c)\}. \tag{6}$$

According to Theorem 6.5, multiplication of $f(t)$ by e^{ct} results in a translation of the transform $F(s)$ a distance c in the positive s-direction, and conversely. The proof of this theorem requires merely the evaluation of $\mathcal{L}\{e^{ct}f(t)\}$. Thus

$$\mathcal{L}\{e^{ct}f(t)\} = \int_0^\infty e^{-st}e^{ct}f(t)\ dt = \int_0^\infty e^{-(s-c)t}f(t)\ dt$$

$$= F(s - c),$$

which is Eq. (5). The restriction $s > a + c$ follows from the observation that, according to hypothesis (ii) of Theorem 6.2, $|f(t)| \leq Ke^{at}$; hence $|e^{ct}f(t)| \leq Ke^{(a+c)t}$. Equation (6) follows by taking the inverse transform of Eq. (5), and the proof is complete.

The principal application of Theorem 6.5 is in the evaluation of certain inverse transforms, as illustrated by Example 4.

Example 4.   Find the inverse transform of

$$G(s) = \frac{1}{s^2 - 4s + 5}.$$

By completing the square in the denominator we can write

$$G(s) = \frac{1}{(s - 2)^2 + 1} = F(s - 2),$$

where $F(s) = (s^2 + 1)^{-1}$. Since $\mathcal{L}^{-1}\{F(s)\} = \sin t$, it follows from Theorem 6.5 that

$$g(t) = \mathcal{L}^{-1}\{G(s)\} = e^{2t}\sin t.$$

The results of this section are often useful in solving differential equations, particularly those having discontinuous forcing functions. The next section is devoted to an example illustrating this fact.

PROBLEMS

1. Sketch the graph of each of the following functions on the interval $t \geq 0$.

(a) $u_1(t) + 2u_3(t) - 6u_4(t)$

(b) $(t - 3)u_2(t) - (t - 2)u_3(t)$

(c) $f(t - \pi)u_\pi(t)$,   where $f(t) = t^2$

(d) $f(t - 3)u_3(t)$,   where $f(t) = \sin t$

(e) $f(t - 1)u_2(t)$,   where $f(t) = 2t$

(f) $f(t) = (t - 1)u_1(t) - 2(t - 2)u_2(t) + (t - 3)u_3(t)$

2. Find the Laplace transform of each of the following functions.

(a) $f(t) = \begin{cases} 0, & t < 2 \\ (t-2)^2, & t \geq 2 \end{cases}$ (b) $f(t) = \begin{cases} 0, & t < 1 \\ t^2 - 2t + 2, & t \geq 1 \end{cases}$

(c) $f(t) = \begin{cases} 0, & t < \pi \\ t - \pi, & \pi \leq t < 2\pi \\ 0, & t \geq 2\pi \end{cases}$ (d) $f(t) = u_1(t) + 2u_3(t) - 6u_4(t)$

(e) $f(t) = (t - 3)u_2(t) - (t - 2)u_3(t)$ (f) $f(t) = t - u_1(t)(t - 1)$, $t \geq 0$

3. Find the inverse Laplace transform of each of the following functions.

(a) $F(s) = \dfrac{3!}{(s - 2)^4}$ (b) $F(s) = \dfrac{e^{-2s}}{s^2 + s - 2}$ $(s+2)(s-1)$

(c) $F(s) = \dfrac{2(s - 1)e^{-2s}}{s^2 - 2s + 2}$ (d) $F(s) = \dfrac{2e^{-2s}}{s^2 - 4}$

(e) $F(s) = \dfrac{(s - 2)e^{-s}}{s^2 - 4s + 3}$ (f) $F(s) = \dfrac{e^{-s} + e^{-2s} - e^{-3s} - e^{-4s}}{s}$

4. Suppose that $F(s) = \mathcal{L}\{f(t)\}$ exists for $s > a \geq 0$.

(a) Show that if c is a positive constant, then

$$\mathcal{L}\{f(ct)\} = \frac{1}{c}F\left(\frac{s}{c}\right), \qquad s > ca.$$

$(s-2)^2 - 1$

(b) Show that if k is a positive constant, then

$$\mathcal{L}^{-1}\{F(ks)\} = \frac{1}{k}f\left(\frac{t}{k}\right).$$

(c) Show that if a and b are constants with $a > 0$, then

$$\mathcal{L}^{-1}\{F(as + b)\} = \frac{1}{a}e^{-bt/a}f\left(\frac{t}{a}\right).$$

5. Using the results of Problem 4, find the inverse Laplace transform of each of the following functions.

(a) $F(s) = \dfrac{2^{n+1}n!}{s^{n+1}}$ (b) $F(s) = \dfrac{2s + 1}{4s^2 + 4s + 5}$

(c) $F(s) = \dfrac{1}{9s^2 - 12s + 3}$ (d) $F(s) = \dfrac{e^2 e^{-4s}}{2s - 1}$

6. Find the Laplace transform of each of the following functions. In part (d) assume that term by term integration of the infinite series is permissible.

(a) $f(t) = \begin{cases} 1, & 0 \leq t < 1 \\ 0, & t \geq 1 \end{cases}$ (b) $f(t) = \begin{cases} 1, & 0 \leq t < 1 \\ 0, & 1 \leq t < 2 \\ 1, & 2 \leq t < 3 \\ 0, & t \geq 3 \end{cases}$

FIGURE 6.5 Square wave.

FIGURE 6.6 Square wave.

FIGURE 6.7 Sawtooth wave.

FIGURE 6.8 Rectified sine wave.

(c) $f(t) = 1 - u_1(t) + \cdots + u_{2n}(t) - u_{2n+1}(t) = 1 + \sum_{k=1}^{2n+1} (-1)^k u_k(t)$

(d) $f(t) = 1 + \sum_{k=1}^{\infty} (-1)^k u_k(t)$.   See Figure 6.5.

*7. Let f satisfy $f(t + T) = f(t)$ for all $t \geq 0$ and for some fixed positive number T; f is said to be periodic with period T on $0 \leq t < \infty$. Show that

$$\mathcal{L}\{f(t)\} = \frac{\int_0^T e^{-st} f(t)\, dt}{1 - e^{-sT}}.$$

*8. Use the result of Problem 7 to find the Laplace transform of each of the following functions.

(a) $f(t) = \begin{cases} 1, & 0 \leq t < 1, \\ 0, & 1 \leq t < 2; \end{cases}$

$f(t + 2) = f(t)$.

Compare with Problem 6(d).

(b) $f(t) = \begin{cases} 1, & 0 \leq t < 1, \\ -1, & 1 \leq t < 2; \end{cases}$

$f(t + 2) = f(t)$.

See Figure 6.6.

(c) $f(t) = t,$ $0 \leq t < 1$;

$f(t + 1) = f(t)$.

See Figure 6.7.

(d) $f(t) = \sin t,\ 0 \leq t < \pi$;

$f(t + \pi) = f(t)$.

See Figure 6.8.

*9. (a) If $f(t) = 1 - u_1(t)$, find $\mathcal{L}\{f(t)\}$; compare with Problem 6(a). Sketch the graph of $y = f(t)$.

(b) Let $g(t) = \int_0^t f(\xi)\, d\xi$, where the function f is defined in part (a). Sketch the graph of $y = g(t)$, and find $\mathcal{L}\{g(t)\}$.

(c) Let $h(t) = g(t) - u_1(t)g(t - 1)$, where g is defined in part (b). Sketch the graph of $y = h(t)$, and find $\mathcal{L}\{h(t)\}$.

*10. Consider the function p defined by

$$p(t) = \begin{cases} t, & 0 \le t < 1 \\ 2 - t, & 1 \le t < 2; \end{cases} \qquad p(t + 2) = p(t).$$

(a) Sketch the graph of $y = p(t)$.

(b) Find $\mathscr{L}\{p(t)\}$ by noting that p is the periodic extension of the function h in Problem 9(c) and then using the result of Problem 7.

(c) Find $\mathscr{L}\{p(t)\}$ by noting that

$$p(t) = \int_0^t f(t) \, dt,$$

where f is the function in Problem 8(b), and then using Theorem 6.3.

6.3.1   A DIFFERENTIAL EQUATION WITH A DISCONTINUOUS FORCING FUNCTION

Let us now turn our attention to a typical problem involving a discontinuous forcing function (nonhomogeneous term). Consider the differential equation

$$y'' + 4y = h(t), \tag{1}$$

where (see Figure 6.3 on page 243)

$$h(t) = \begin{cases} 1, & \pi \le t < 2\pi, \\ 0, & 0 \le t < \pi \text{ and } t \ge 2\pi. \end{cases} \tag{2}$$

Assume that the initial conditions are

$$y(0) = 1, \qquad y'(0) = 0. \tag{3}$$

This problem governs, for example, the charge on the condenser in a simple electric circuit containing no resistor and with an impressed voltage of the form (2). Alternatively, y may represent the response of an undamped harmonic oscillator subject to the applied force $h(t)$.

Before taking the transform of Eq. (1) it is convenient to write $h(t)$ in the form (see Example 1 of Section 6.3)

$$h(t) = u_\pi(t) - u_{2\pi}(t). \tag{4}$$

Then the Laplace transform of Eq. (1) is

$$(s^2 + 4)Y(s) - sy(0) - y'(0) = \mathscr{L}\{u_\pi(t)\} - \mathscr{L}\{u_{2\pi}(t)\}$$

$$= \frac{e^{-\pi s}}{s} - \frac{e^{-2\pi s}}{s}.$$

Introducing the initial values (3) and solving for $Y(s)$, we obtain

$$Y(s) = \frac{s}{s^2 + 4} + \frac{e^{-\pi s}}{s(s^2 + 4)} - \frac{e^{-2\pi s}}{s(s^2 + 4)}. \tag{5}$$

To find $y = \phi(t)$ we must obtain the inverse transform of each term on the right side of Eq. (5). From Table 6.1 the inverse transform of the first term is given by

$$\mathscr{L}^{-1}\left\{\frac{s}{s^2 + 4}\right\} = \cos 2t. \tag{6}$$

To compute the inverse transform of the second term we note that, according to Theorem 6.4,

$$\mathscr{L}^{-1}\left\{\frac{e^{-\pi s}}{s(s^2 + 4)}\right\} = u_\pi(t)g(t - \pi),$$

where

$$g(t) = \mathscr{L}^{-1}\left\{\frac{1}{s(s^2 + 4)}\right\}.$$

Using partial fractions, we obtain

$$\frac{1}{s(s^2 + 4)} = \frac{1}{4}\left(\frac{1}{s} - \frac{s}{s^2 + 4}\right),$$

and it follows from Table 6.1 that

$$g(t) = \tfrac{1}{4}(1 - \cos 2t).$$

Hence

$$\mathscr{L}^{-1}\left\{\frac{e^{-\pi s}}{s(s^2 + 4)}\right\} = \tfrac{1}{4}u_\pi(t)[1 - \cos 2(t - \pi)]. \tag{7}$$

Similarly

$$\mathscr{L}^{-1}\left\{\frac{e^{-2\pi s}}{s(s^2 + 4)}\right\} = \tfrac{1}{4}u_{2\pi}(t)[1 - \cos 2(t - 2\pi)]. \tag{8}$$

Combining Eqs. (6), (7), and (8) we obtain finally

$$\phi(t) = \mathscr{L}^{-1}\{Y(s)\}$$

$$= \cos 2t + \tfrac{1}{4}u_\pi(t)[1 - \cos 2(t - \pi)] - \tfrac{1}{4}u_{2\pi}(t)[1 - \cos 2(t - 2\pi)]. \tag{9}$$

The graph of $y = \phi(t)$ is sketched in Figure 6.9.

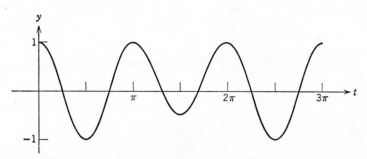

FIGURE 6.9   Solution of initial value problem (1), (2), (3).

Note that the second and third terms on the right side of Eq. (9) are zero for $t < \pi$ and $t < 2\pi$, respectively. If $\phi(t)$ is written without use of the unit step functions u_π and $u_{2\pi}$, then

$$\phi(t) = \begin{cases} \cos 2t, & 0 \le t < \pi, \\ \tfrac{3}{4} \cos 2t + \tfrac{1}{4}, & \pi \le t < 2\pi, \\ \cos 2t, & 2\pi \le t < \infty. \end{cases} \qquad (10)$$

The initial value problem we have just solved involves a differential equation with a forcing function having jump discontinuities. It therefore does not quite fall within the scope of the fundamental existence and uniqueness Theorem 3.2. Thus it is not surprising that while the solution $y = \phi(t)$ given by Eq. (10) is continuous and has a continuous first derivative for all $t > 0$, it does not have a second derivative at the points $t = \pi$ and $t = 2\pi$ where the forcing function h has jump discontinuities. From Eq. (10) we have

$$\phi''(t) = \begin{cases} -4 \cos 2t, & 0 < t < \pi, \\ -3 \cos 2t, & \pi < t < 2\pi, \\ -4 \cos 2t, & 2\pi < t < \infty. \end{cases}$$

In the neighborhood of $t = \pi$, for example, $\phi''(t) \to -4$ as $t \to \pi$ from the left, but $\phi''(t) \to -3$ as $t \to \pi$ from the right. Thus ϕ'' has a jump discontinuity at $t = \pi$, and the situation is the same at $t = 2\pi$. These discontinuities reflect the discontinuities in the forcing function at the same points.

Consider now the general second order linear equation,

$$y'' + p(t)y' + q(t)y = g(t), \qquad (11)$$

where p and q are continuous on some interval $\alpha < t < \beta$, but g is only piecewise continuous there. If $y = \psi(t)$ is a solution of Eq. (11) then ψ and ψ' will be continuous on $\alpha < t < \beta$, but ψ'' will have jump discontinuities at the same points as g. Similar remarks apply to higher order equations; the highest derivative of the solution appearing in the differential equation will have jump discontinuities at the same points as the forcing function, but the solution itself and its lower derivatives will be continuous even at those points.

An alternate way to solve the initial value problem (1), (2), and (3) is to solve the differential equation in the three separate intervals $0 < t < \pi$, $\pi < t < 2\pi$, and $t > 2\pi$. In the first of these intervals the arbitrary constants in the general solution are chosen so as to satisfy the given initial conditions. In the second interval the constants are chosen so as to make the solution ϕ and its derivative ϕ' continuous at $t = \pi$. Note that there are not enough constants available to require that ϕ'' also be continuous there. Similarly, in the third interval the constants are chosen so that ϕ and ϕ' will be continuous at $t = 2\pi$. The same general procedure can be used in other problems having piecewise continuous forcing functions. However, once the necessary theory is developed, it is usually easier to use the Laplace transform to solve such problems all at once.

PROBLEMS

In each of Problems 1 through 7 find the solution of the given initial value problem.

1. $y'' + y = f(t);$ $\qquad f(t) = \begin{cases} 1, & 0 \leq t < \pi/2 \\ 0, & \pi/2 \leq t < \infty \end{cases}$
   $y(0) = 0, \qquad y'(0) = 1$

2. $y'' + 2y' + 2y = h(t);$ $\qquad h(t) = \begin{cases} 1, & \pi \leq t < 2\pi \\ 0, & 0 \leq t < \pi \text{ and } t \geq 2\pi \end{cases}$
   $y(0) = 0, \qquad y'(0) = 1$

3. $y'' + 4y = \sin t - u_{2\pi}(t) \sin (t - 2\pi);$ $\qquad y(0) = 0, \qquad y'(0) = 0$

4. $y'' + 4y = \sin t + u_\pi(t) \sin (t - \pi);$ $\qquad y(0) = 0, \qquad y'(0) = 0$

5. $y'' + 2y' + y = f(t);$ $\qquad f(t) = \begin{cases} 1, & 0 \leq t < 1 \\ 0, & t \geq 1 \end{cases}$
   $y(0) = 1, \qquad y'(0) = 0$

6. $y'' + 3y' + 2y = u_2(t);$ $\qquad y(0) = 0, \qquad y'(0) = 1$

7. $y'' + y = u_\pi(t);$ $\qquad y(0) = 1, \qquad y'(0) = 0$

*8. Let the function f be defined by

$$f(t) = \begin{cases} 1, & 0 \leq t < \pi, \\ 0, & \pi \leq t < 2\pi, \end{cases}$$

and for all other positive values of t so that $f(t + 2\pi) = f(t)$. That is, f is periodic with period 2π (see Problem 7 in Section 6.3). Find the solution of the initial value problem

$$y'' + y = f(t); \qquad y(0) = 1, \qquad y'(0) = 0.$$

*9. Suppose that the function f is given by

$$f(t) = \begin{cases} t, & 0 \leq t < 1, \\ 1, & t \geq 1. \end{cases}$$

(a) Find $\mathscr{L}\{f(t)\}$.
(b) Solve the initial value problem

$$y'' + y = f(t); \qquad y(0) = 0, \qquad y'(0) = 1.$$

6.4 IMPULSE FUNCTIONS

In some applications it is necessary to deal with phenomena of an impulsive nature; for example, voltages or forces of large magnitude that act over very short time intervals. Such problems often lead to differential equations of the form

$$ay'' + by' + cy = g(t), \tag{1}$$

where $g(t)$ is large during a short interval $t_0 - \tau < t < t_0 + \tau$ and is otherwise zero.

FIGURE 6.10 $y = d_r(t)$.

The integral $I(\tau)$, defined by

$$I(\tau) = \int_{t_0-\tau}^{t_0+\tau} g(t)\, dt, \tag{2}$$

or, since $g(t) = 0$ outside of the interval $(t_0 - \tau, t_0 + \tau)$,

$$I(\tau) = \int_{-\infty}^{\infty} g(t)\, dt, \tag{3}$$

is a measure of the strength of the forcing function. In a mechanical system, where $g(t)$ is a force, then $I(\tau)$ is the total *impulse* of the force $g(t)$ over the time interval $(t_0 - \tau, t_0 + \tau)$. Similarly, if y is the current in an electric circuit and $g(t)$ is the time derivative of the voltage, then $I(\tau)$ represents the total voltage impressed on the circuit during the interval $(t_0 - \tau, t_0 + \tau)$.

In particular, let us suppose that t_0 is zero, and that $g(t)$ is given by

$$g(t) = d_r(t) = \begin{cases} 1/2\tau, & -\tau < t < \tau, \\ 0, & t \le -\tau \text{ or } t \ge \tau, \end{cases} \tag{4}$$

where τ is a small positive constant (see Figure 6.10). According to Eq. (2) or Eq. (3) it follows immediately that in this case $I(\tau) = 1$ independent of the value of τ, as long as $\tau \ne 0$. Now let us idealize the forcing function d_r by prescribing it to act over shorter and shorter time intervals; that is, we will require that $\tau \to 0$, as indicated in Figure 6.11. As a result of this limiting operation we obtain

$$\lim_{\tau \to 0} d_r(t) = 0, \qquad t \ne 0. \tag{5}$$

Further, since $I(\tau) = 1$ for each $\tau \ne 0$, it follows that

$$\lim_{\tau \to 0} I(\tau) = 1. \tag{6}$$

Equations (5) and (6) can be used to define an idealized "unit impulse function" δ, which imparts an impulse of magnitude one at $t = 0$, but is zero for all values of t other than zero. That is, the "function" δ is defined to have the properties

$$\delta(t) = 0, \qquad t \ne 0; \tag{7}$$

$$\int_{-\infty}^{\infty} \delta(t)\, dt = 1. \tag{8}$$

FIGURE 6.11

Clearly, there is no ordinary function of the kind studied in elementary calculus that satisfies both of Eqs. (7) and (8). The "function" δ, defined by those equations, is an example of what are known as *generalized functions* and is usually called the Dirac* "delta function". It is frequently convenient to represent impulsive forcing functions by means of the delta function. Since $\delta(t)$ corresponds to a unit impulse at $t = 0$, a unit impulse at an arbitrary point $t = t_0$ is given by $\delta(t - t_0)$. From Eqs. (7) and (8) it follows that

$$\delta(t - t_0) = 0, \qquad t \neq t_0; \tag{9}$$

$$\int_{-\infty}^{\infty} \delta(t - t_0)\, dt = 1. \tag{10}$$

The delta function does not satisfy the conditions of Theorem 6.2, but its Laplace transform can nevertheless be formally defined. Since $\delta(t)$ is defined as the limit of $d_\tau(t)$ as $\tau \to 0$, it is natural to define the Laplace transform of δ as a similar limit of the transform of d_τ. In particular, we will assume that $t_0 > 0$, and define $\mathscr{L}\{\delta(t - t_0)\}$ by the equation

$$\mathscr{L}\{\delta(t - t_0)\} = \lim_{\tau \to 0} \mathscr{L}\{d_\tau(t - t_0)\}. \tag{11}$$

To evaluate the limit in Eq. (11) we first observe that if $\tau < t_0$, which must eventually be the case as $\tau \to 0$, then $t_0 - \tau > 0$. Since $d_\tau(t - t_0)$ is nonzero only in the interval from $t_0 - \tau$ to $t_0 + \tau$, we have

$$\mathscr{L}\{d_\tau(t - t_0)\} = \int_0^\infty e^{-st}\, d_\tau(t - t_0)\, dt$$

$$= \int_{t_0 - \tau}^{t_0 + \tau} e^{-st}\, d_\tau(t - t_0)\, dt.$$

* Paul A. M. Dirac (1902–   ) was awarded the Nobel Prize (with Erwin Schrödinger) in 1933 for his work in quantum mechanics.

Substituting for $d_\tau(t - t_0)$ from Eq. (4), we obtain

$$\mathscr{L}\{d_\tau(t - t_0)\} = \frac{1}{2\tau} \int_{t_0-\tau}^{t_0+\tau} e^{-st}\, dt = - \frac{1}{2\tau s}\, e^{-st}\Big|_{t=t_0-\tau}^{t=t_0+\tau}$$

$$= \frac{1}{2\tau s}\, e^{-st_0}(e^{s\tau} - e^{-s\tau})$$

or

$$\mathscr{L}\{d_\tau(t - t_0)\} = \frac{\sinh s\tau}{s\tau}\, e^{-st_0}. \tag{12}$$

The quotient $(\sinh s\tau)/s\tau$ is indeterminate as $\tau \to 0$, but its limit can be evaluated by L'Hospital's rule. We obtain

$$\lim_{\tau \to 0} \frac{\sinh s\tau}{s\tau} = \lim_{\tau \to 0} \frac{s \cosh s\tau}{s} = 1.$$

Then from Eq. (11) it follows that

$$\mathscr{L}\{\delta(t - t_0)\} = e^{-st_0}. \tag{13}$$

Since Eq. (13) holds for all $t_0 > 0$, we can define $\mathscr{L}\{\delta(t)\}$ by letting $t_0 \to 0$ in Eq. (13); thus

$$\mathscr{L}\{\delta(t)\} = \lim_{t_0 \to 0} e^{-st_0} = 1. \tag{14}$$

   In a similar way it is possible to define the integral of the product of the delta function and any continuous function f. We have

$$\int_{-\infty}^{\infty} \delta(t - t_0)f(t)\, dt = \lim_{\tau \to 0} \int_{-\infty}^{\infty} d_\tau(t - t_0)f(t)\, dt. \tag{15}$$

Using the definition (4) of $d_\tau(t)$ and the mean value theorem for integrals, we find that

$$\int_{-\infty}^{\infty} d_\tau(t - t_0)f(t)\, dt = \frac{1}{2\tau} \int_{t_0-\tau}^{t_0+\tau} f(t)\, dt$$

$$= \frac{1}{2\tau} \cdot 2\tau \cdot f(t^*) = f(t^*),$$

where $t_0 - \tau < t^* < t_0 + \tau$. Hence, $t^* \to t_0$ as $\tau \to 0$, and it follows from Eq. (15) that

$$\int_{-\infty}^{\infty} \delta(t - t_0)f(t)\, dt = f(t_0). \tag{16}$$

   It is often convenient to introduce the delta function when working with impulse problems, and to operate formally on it as though it were a function of the ordinary kind. This is illustrated in the example below. It is important to realize, however, that the ultimate justification of such procedures must rest on a careful analysis of the limiting operations involved. Such a rigorous mathematical theory has been developed, but it is too difficult to discuss here.

Example. Find the solution of the initial value problem

$$y'' + 2y' + 2y = \delta(t - \pi) \tag{17}$$

$$y(0) = 0, \qquad y'(0) = 0. \tag{18}$$

This initial value problem might arise from the study of a certain electric circuit to which a unit voltage impulse is applied at time $t = \pi$.

To solve the given problem we take the Laplace transform of the differential equation, obtaining

$$(s^2 + 2s + 2)Y(s) = e^{-\pi s},$$

where the initial conditions (18) have been used. Thus

$$Y(s) = \frac{e^{-\pi s}}{s^2 + 2s + 2} = e^{-\pi s} \frac{1}{(s + 1)^2 + 1}. \tag{19}$$

By Theorem 6.5 or from Table 6.1

$$\mathcal{L}^{-1}\left\{\frac{1}{(s + 1)^2 + 1}\right\} = e^{-t} \sin t. \tag{20}$$

Hence, by Theorem 6.4 we have

$$y = \mathcal{L}^{-1}\{Y(s)\} = u_\pi(t)e^{-(t-\pi)} \sin (t - \pi), \tag{21}$$

which is the formal solution of the given problem. It is also possible to write y in the form

$$y = \begin{cases} 0, & t < \pi, \\ e^{-(t-\pi)} \sin (t - \pi), & t \geq \pi. \end{cases} \tag{22}$$

The graph of Eq. (22) is shown in Figure 6.12.

FIGURE 6.12   Solution of initial value problem (17), (18).

As the initial conditions at $t = 0$ are homogeneous, and there is no external excitation until $t = \pi$, there is no response in the interval $0 < t < \pi$. The impulse at $t = \pi$ produces a response that persists indefinitely, although it decays exponentially in the absence of any further external excitation. It is also interesting to note that the response is continuous at $t = \pi$ in spite of the singular nature of the forcing function at that point; however, the slope of the solution has a jump discontinuity at this point, and its second derivative is not defined there. This is required by the differential equation (17) since a singularity on one side of the equation must be balanced by one on the other.

PROBLEMS

In each of Problems 1 through 7 find the solution of the given initial value problem by means of the Laplace transform.

1. $y'' + 2y' + 2y = \delta(t - \pi);$ $y(0) = 1,$ $y'(0) = 0$
2. $y'' + 4y = \delta(t - \pi) - \delta(t - 2\pi);$ $y(0) = 0,$ $y'(0) = 0$
3. $y'' + 2y' + y = \delta(t) + u_{2\pi}(t);$ $y(0) = 0,$ $y'(0) = 1$
4. $y'' - y = 2\delta(t - 1);$ $y(0) = 1,$ $y'(0) = 0$
5. $y'' + 2y' + 3y = \sin t + \delta(t - \pi);$ $y(0) = 0,$ $y'(0) = 1$
6. $y'' + \omega^2 y = \delta(t - \pi/\omega);$ $y(0) = 1,$ $y'(0) = 0$
7. $y'' + y = \delta(t - \pi) \cos t;$ $y(0) = 0;$ $y'(0) = 1$

*8. (a) Show by the method of variation of parameters that the solution of the initial value problem
$$y'' + 2y' + 2y = f(t);   y(0) = 0,   y'(0) = 0$$
is
$$y = \int_0^t e^{-(t-\tau)} f(\tau) \sin(t - \tau)\, d\tau.$$

(b) Show that if $f(t) = \delta(t - \pi)$, then the solution of part (a) reduces to
$$y = u_\pi(t) e^{-(t-\pi)} \sin(t - \pi),$$

which agrees with Eq. (21) of the text.

6.5 THE CONVOLUTION INTEGRAL

Sometimes it is possible to identify a Laplace transform $H(s)$ as the product of two other transforms $F(s)$ and $G(s)$, the latter transforms corresponding to known functions f and g, respectively. In this event, we might anticipate that $H(s)$ would be the transform of the product of f and g. However, this is not the case; in other words, the Laplace transform cannot be commuted with ordinary multiplication. On the other hand, if an appropriately defined "generalized product" is introduced, then the situation changes, as stated in the following theorem.

Theorem 6.6. *If* $F(s) = \mathcal{L}\{f(t)\}$ *and* $G(s) = \mathcal{L}\{g(t)\}$ *both exist for* $s > a \geq 0$, *then*

$$H(s) = F(s)G(s) = \mathcal{L}\{h(t)\}, \qquad s > a, \tag{1}$$

where

$$h(t) = \int_0^t f(t - \tau)g(\tau)\, d\tau = \int_0^t f(\tau)g(t - \tau)\, d\tau. \tag{2}$$

The function h is known as the convolution of f and g; the integrals in Eqs. (2) are known as convolution integrals.

The equality of the two integrals in Eq. (2) follows by making the change of variable $t - \tau = \xi$ in the first integral. Before giving the proof of this theorem let us make some observations about the convolution integral. According to this theorem, the transform of the convolution of two functions, rather than the transform of their ordinary product, is given by the product of the separate transforms. It is conventional to emphasize that the convolution integral can be thought of as a "generalized product" by writing

$$h(t) = (f * g)(t). \tag{3}$$

In particular, the notation $(f * g)(t)$ serves to indicate the first integral appearing in Eq. (2).

The convolution $f * g$ has many of the properties of ordinary multiplication. For example, it is relatively simple to show that

$$f * g = g * f \qquad \text{(commutative law)} \tag{4}$$
$$f * (g_1 + g_2) = f * g_1 + f * g_2 \qquad \text{(distributive law)} \tag{5}$$
$$(f * g) * h = f * (g * h) \qquad \text{(associative law)} \tag{6}$$
$$f * 0 = 0 * f = 0. \tag{7}$$

The proofs of these properties are left to the reader. However, there are other properties of ordinary multiplication which the convolution integral does not have. For example, it is not true in general that $f * 1$ is equal to f. To see this, note that

$$(f * 1)(t) = \int_0^t f(t - \tau) \cdot 1 \, d\tau = \int_0^t f(t - \tau)\, d\tau.$$

If, for example, $f(t) = \cos t$, then

$$(f * 1)(t) = \int_0^t \cos (t - \tau)\, d\tau = -\sin (t - \tau) \Big|_{\tau=0}^{\tau=t}$$
$$= -\sin 0 + \sin t$$
$$= \sin t.$$

Clearly, $(f * 1)(t) \neq f(t)$. Similarly, it may not be true that $f * f$ is nonnegative. See Problem 3 for an example.

Convolution integrals arise in various applications in which the behavior of the system at time t depends not only on its state at time t, but on its past history as well. Systems of this kind are sometimes called hereditary systems, and occur in such diverse fields as neutron transport, viscoelasticity, and population dynamics.

Turning now to the proof of Theorem 6.6, we note first that if

$$F(s) = \int_0^\infty e^{-s\xi} f(\xi)\, d\xi$$

and

$$G(s) = \int_0^\infty e^{-s\eta} g(\eta)\, d\eta,$$

then

$$F(s)G(s) = \int_0^\infty e^{-s\xi} f(\xi)\, d\xi \int_0^\infty e^{-s\eta} g(\eta)\, d\eta. \tag{8}$$

Since the integrand of the first integral does not depend on the integration variable of the second, we can write $F(s)G(s)$ as an iterated integral,

$$F(s)G(s) = \int_0^\infty g(\eta)\, d\eta \int_0^\infty e^{-s(\xi+\eta)} f(\xi)\, d\xi. \tag{9}$$

This expression can be put into a more convenient form by introducing new variables of integration. First let $\xi = t - \eta$, for fixed η. Then the integral with respect to ξ in Eq. (9) is transformed into one with respect to t; hence

$$F(s)G(s) = \int_0^\infty g(\eta)\, d\eta \int_\eta^\infty e^{-st} f(t - \eta)\, dt. \tag{10}$$

Next let $\eta = \tau$; then Eq. (10) becomes

$$F(s)G(s) = \int_0^\infty g(\tau)\, d\tau \int_\tau^\infty e^{-st} f(t - \tau)\, dt. \tag{11}$$

The integral on the right side of Eq. (11) is carried out over the shaded wedge-shaped region extending to infinity in the $t\tau$ plane shown in Figure 6.13. Assuming that the order of integration can be reversed, we finally obtain

$$F(s)G(s) = \int_0^\infty e^{-st}\, dt \int_0^t f(t - \tau) g(\tau)\, d\tau, \tag{12}$$

or

$$F(s)G(s) = \int_0^\infty e^{-st} h(t)\, dt$$

$$= \mathcal{L}\{h(t)\}, \tag{13}$$

where $h(t)$ is defined by Eq. (2). This completes the proof of Theorem 6.6.

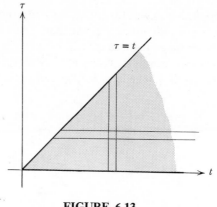

FIGURE 6.13

Example.   Find the inverse transform of

$$H(s) = \frac{a}{s^2(s^2 + a^2)}.$$

It is convenient to think of $H(s)$ as the product of s^{-2} and $a/(s^2 + a^2)$, which, according to Table 6.1, are the transforms of t and $\sin at$, respectively. Hence, by Theorem 6.6, the inverse transform of $H(s)$ is

$$h(t) = \int_0^t (t - \tau) \sin a\tau \, d\tau = \frac{at - \sin at}{a^2}.$$

The reader may show that the same result is obtained if $h(t)$ is written in the alternate form

$$h(t) = \int_0^t \tau \sin a(t - \tau) \, d\tau,$$

which verifies Eq. (2) in this case.

PROBLEMS

1. Establish the commutative, distributive, and associative properties of the convolution integral.

(a) $f * g = g * f$ (b) $f * (g_1 + g_2) = f * g_1 + f * g_2$

(c) $f * (g * h) = (f * g) * h$

2. Find an example different from the one in the text showing that $(f * 1)(t)$ need not be equal to $f(t)$.

3. Show, by means of the example $f(t) = \sin t$, that $f * f$ is not necessarily non-negative.

4. Find the Laplace transform of each of the following functions.

(a) $f(t) = \displaystyle\int_0^t (t - \tau)^2 \cos 2\tau \, d\tau$

(b) $f(t) = \displaystyle\int_0^t e^{-(t-\tau)} \sin \tau \, d\tau$

(c) $f(t) = \displaystyle\int_0^t (t - \tau)e^\tau \, d\tau$

(d) $f(t) = \displaystyle\int_0^t \sin (t - \tau) \cos \tau \, d\tau$

5. Find the inverse Laplace transform of each of the following functions by using the convolution theorem.

(a) $F(s) = \dfrac{1}{s^4(s^2 + 1)}$

(b) $F(s) = \dfrac{s}{(s + 1)(s^2 + 4)}$

(c) $F(s) = \dfrac{1}{(s + 1)^2(s^2 + 4)}$

(d) $F(s) = \dfrac{G(s)}{s^2 + 1}$

6. Find the solution of the initial value problem

$$y'' + 2y' + 2y = f(t); \qquad y(0) = 0, \qquad y'(0) = 0,$$

where $f(t) = 1 - u_{2\pi}(t)$.

7. Find the solution of the initial value problem

$$y'' + 2y' + 2y = \sin \alpha t; \qquad y(0) = 0, \qquad y'(0) = 0,$$

where α is a constant.

8. Consider the equation

$$\phi(t) + \int_0^t k(t - \xi)\phi(\xi) \, d\xi = f(t),$$

in which f and k are known functions, and ϕ is to be determined. Since the unknown function ϕ appears under an integral sign, the given equation is called an *integral equation*; in particular, it belongs to a class of integral equations known as Volterra (1860–1940) integral equations. Take the Laplace transform of the given integral equation, and obtain an expression for $\mathcal{L}\{\phi(t)\}$ in terms of the transforms $\mathcal{L}\{f(t)\}$ and $\mathcal{L}\{k(t)\}$ of the given functions f and k. The inverse transform of $\mathcal{L}\{\phi(t)\}$ is the solution of the original integral equation.

9. Consider the Volterra integral equation (see Problem 8)

$$\phi(t) + \int_0^t (t - \xi)\phi(\xi) \, d\xi = \sin 2t.$$

(a) Show that if u is a function such that $u''(t) = \phi(t)$, then

$$u''(t) + u(t) - tu'(0) - u(0) = \sin 2t.$$

(b) Show that the given integral equation is equivalent to the initial value problem

$$u''(t) + u(t) = \sin 2t; \qquad u(0) = 0, \qquad u'(0) = 0.$$

(c) Solve the given integral equation by using the Laplace transform.

(d) Solve the initial value problem of part (b) and verify that the solution is the same as that obtained in (c).

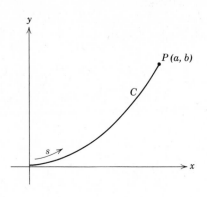

FIGURE 6.14

*10. **The Tautochrone.** A problem of interest in the history of mathematics is that of finding the *tautochrone*—the curve down which a particle will slide freely under gravity alone, reaching the bottom in the same time regardless of its starting point on the curve. This problem arose in the construction of a clock pendulum whose period is independent of the amplitude of its motion. The tautochrone was found by Christian Huygens (1629–1695) in 1673 by geometrical methods, and later by Leibniz and Jakob Bernoulli using analytical arguments. Bernoulli's solution (in 1690) was one of the first occasions on which a differential equation was explicitly solved.

The geometrical configuration is shown in Figure 6.14. The starting point $P(a, b)$ is joined to the terminal point $(0, 0)$ by the arc C. Arc length s is measured from the origin, and $f(y)$ denotes the rate of change of s with respect to y:

$$f(y) = \frac{ds}{dy} = \left[1 + \left(\frac{dx}{dy} \right)^2 \right]^{1/2}. \tag{i}$$

Then it follows from the principle of conservation of energy that the time $T(b)$ required for a particle to slide from P to the origin is

$$T(b) = \frac{1}{\sqrt{2g}} \int_0^b \frac{f(y)}{\sqrt{b - y}} \, dy. \tag{ii}$$

(a) Assume that $T(b) = T_0$, a constant, for each b. By taking the Laplace transform of Eq. (ii) in this case, and using the convolution theorem, show that

$$F(s) = \sqrt{\frac{2g}{\pi}} \frac{T_0}{\sqrt{s}} ; \tag{iii}$$

then show that

$$f(y) = \frac{\sqrt{2g} \, T_0}{\pi \sqrt{y}}. \tag{iv}$$

Hint: See Problem 10 of Section 6.1.

(b) Combining Eqs. (i) and (iv), show that

$$\frac{dx}{dy} = \sqrt{\frac{2\alpha - y}{y}}, \tag{v}$$

where $\alpha = gT_0^2/\pi^2$.

(c) Use the substitution $y = 2\alpha \sin^2 (\theta/2)$ to solve Eq. (v), and show that

$$x = \alpha(\theta + \sin \theta), \qquad y = \alpha(1 - \cos \theta). \tag{vi}$$

Equations (vi) can be identified as parametric equations of a cycloid. Thus the tautochrone is an arc of a cycloid.

6.6 GENERAL DISCUSSION AND SUMMARY

In concluding this chapter, we will briefly discuss some more general aspects of the transform method. Consider the initial value problem consisting of the differential equation

$$ay'' + by' + cy = f(t), \tag{1}$$

where a, b, and c are real constants and f is a given function, together with the initial conditions

$$y(0) = y_0, \qquad y'(0) = y_0'. \tag{2}$$

The Laplace transform of Eq. (1) is

$$(as^2 + bs + c)Y(s) - (as + b)y(0) - ay'(0) = F(s), \tag{3}$$

where $Y(s)$ and $F(s)$ are the transforms of y and $f(t)$, respectively. Using the initial conditions (2) and solving for $Y(s)$, we obtain

$$Y(s) = \frac{F(s) + (as + b)y_0 + ay_0'}{as^2 + bs + c}. \tag{4}$$

Thus $Y(s)$ can be thought of as the product of the functions

$$H(s) = (as^2 + bs + c)^{-1} \tag{5}$$

and

$$G(s) = F(s) + (as + b)y_0 + ay_0'. \tag{6}$$

The first function H is characterized entirely by the properties of the system under consideration, and may be called the *system function*. For an electric circuit it is common to call the function $Z(s) = 1/sH(s)$ the *impedance* of the circuit. The second function G is dependent on the forcing function and initial conditions, and may be called the *excitation function*. Further, the excitation function also consists of two parts: the function F, which is the transform of the forcing function f, and the function $(as + b)y_0 + ay_0'$, which depends on the initial conditions.

At this point it is tempting to write

$$Y(s) = H(s)G(s) \tag{7}$$

and, by invoking the convolution theorem, to obtain

$$y = \int_0^t h(t - \tau)g(\tau) \, d\tau \tag{8}$$

as the solution of Eqs. (1) and (2). However, the function $(as + b)y_0 + ay_0'$ cannot be a Laplace transform, at least not of an ordinary function.* Consequently, it is necessary to write $Y(s)$ as

$$Y(s) = H(s)F(s) + \frac{(as + b)y_0 + ay_0'}{as^2 + bs + c}. \tag{9}$$

---

* If it were it would approach zero as $s \to \infty$ (see Problem 8 of Section 6.1). It is possible, however, to interpret this function as the transform of a certain combination of the delta function and its "derivative."

It then follows that

$$y = \int_0^t h(t - \tau)f(\tau)\,d\tau + \mathscr{L}^{-1}\left\{\frac{(as + b)y_0 + ay_0'}{as^2 + bs + c}\right\}. \tag{10}$$

Once specific values of a, b, and c are given, the inverse transform indicated in Eq. (10) can be found from Table 6.1 or by the method of partial fractions. Using the linearity of the inverse transform we can write Eq. (10) as

$$y = \int_0^t h(t - \tau)f(\tau)\,d\tau + y_0\mathscr{L}^{-1}\left\{\frac{as + b}{as^2 + bs + c}\right\} + y_0'\mathscr{L}^{-1}\left\{\frac{a}{as^2 + bs + c}\right\}. \tag{11}$$

Then, if we let

$$y_1(t) = \mathscr{L}^{-1}\left\{\frac{as + b}{as^2 + bs + c}\right\}, \qquad y_2(t) = \mathscr{L}^{-1}\left\{\frac{a}{as^2 + bs + c}\right\}, \tag{12}$$

we have

$$y = \int_0^t h(t - \tau)f(\tau)\,d\tau + y_0 y_1(t) + y_0' y_2(t). \tag{13}$$

The first term on the right side of Eq. (13) is the response of the system to the forcing function f. In the terminology of Chapter 3 it is a particular solution of the nonhomogeneous equation (1). The remaining terms on the right side of Eq. (13) represent the response of the system to the initial conditions. If we think of y_0 and y_0' as arbitrary constants, these terms comprise the complementary function, or the general solution of the homogeneous equation corresponding to Eq. (1).

REFERENCES

The books listed below contain additional information on the Laplace transform and its applications:

Churchill, R. V., *Operational Mathematics*, Second Edition, McGraw-Hill, New York, 1958.

Doetsch, Gustave, *Guide to the Applications of Laplace Transforms*, Van Nostrand, London, 1967.

Kaplan, W., *Operational Methods for Linear Systems*, Addison-Wesley, Reading, Mass., 1962.

McLachlan, N. W., *Modern Operational Calculus*, Dover, New York, 1962.

Miles, J. W., *Integral Transforms in Applied Mathematics*, Cambridge University Press, 1971.

Each of the books mentioned above contains a table of transforms. Very extensive tables are also available; see, for example:

Erdelyi, A. (editor), *Tables of Integral Transforms*, Vol. 1, McGraw-Hill, New York, 1954.

Roberts, G. E. and Kaufman, H., *Table of Laplace Transforms*, Saunders, Philadelphia, 1966.

A further discussion of generalized functions can be found in:

Lighthill, M. J., *Fourier Analysis and Generalized Functions*, Cambridge University Press, 1958.

CHAPTER 7

Systems of First Order Equations

7.1 INTRODUCTION

This chapter is devoted to a discussion of systems of simultaneous ordinary differential equations. Such systems arise naturally in problems involving several dependent variables, each of which is a function of a single independent variable. We will denote the independent variable by t, and let x_1, x_2, x_3, ... represent dependent variables which are functions of t. Differentiation with respect to t will be denoted by a prime.

For example, the motion of a particle in space is governed by Newton's law in three dimensions,

$$m \frac{d^2 x_1}{dt^2} = F_1\left(t, x_1, x_2, x_3, \frac{dx_1}{dt}, \frac{dx_2}{dt}, \frac{dx_3}{dt}\right),$$

$$m \frac{d^2 x_2}{dt^2} = F_2\left(t, x_1, x_2, x_3, \frac{dx_1}{dt}, \frac{dx_2}{dt}, \frac{dx_3}{dt}\right), \tag{1}$$

$$m \frac{d^2 x_3}{dt^2} = F_3\left(t, x_1, x_2, x_3, \frac{dx_1}{dt}, \frac{dx_2}{dt}, \frac{dx_3}{dt}\right),$$

where m is the mass of the particle; x_1, x_2, and x_3 are its spatial coordinates; and F_1, F_2, and F_3 are the forces acting on the particle in the x_1, x_2, and x_3 directions, respectively. If the particle is an idealization of a space vehicle, for instance, then F_1, F_2, and F_3 include the gravitational forces exerted on the vehicle by neighboring celestial bodies, any forces produced by the vehicle's own propulsive system, and also drag forces if the vehicle is within the earth's atmosphere.

As another example, consider the spring-mass system shown in Figure 7.1. The two masses move on a frictionless surface under the influence of external forces $F_1(t)$ and $F_2(t)$, and they are also constrained by the three springs whose constants are k_1, k_2, and k_3, respectively. Using arguments similar to those in Section 3.7,

FIGURE 7.1 A two-degree-of-freedom spring-mass system.

we find the following equations for the coordinates x_1 and x_2 of the two masses:

$$m_1 \frac{d^2 x_1}{dt^2} = k_2(x_2 - x_1) - k_1 x_1 + F_1(t)$$

$$= -(k_1 + k_2)x_1 + k_2 x_2 + F_1(t),$$

$$m_2 \frac{d^2 x_2}{dt^2} = - k_3 x_2 - k_2(x_2 - x_1) + F_2(t)$$

$$= k_2 x_1 - (k_2 + k_3)x_2 + F_2(t). \tag{2}$$

A derivation of Eqs. (2) is outlined in Problem 6.

Next, consider the parallel RLC-circuit shown in Figure 7.2. Let V be the voltage drop across the capacitor and I the current through the inductance. Then, referring to Section 3.8 and to Problem 7 of this section, we can show that the voltage and current in this circuit are governed by the system of equations

$$\frac{dI}{dt} = \frac{V}{L},$$

$$\frac{dV}{dt} = -\frac{I}{C} - \frac{V}{RC}, \tag{3}$$

FIGURE 7.2 A parallel resistance-capacitance-inductance circuit.

where L is the inductance, C the capacitance, and R the resistance.

As a final example we mention the predator-prey problem, one of the fundamental problems of mathematical ecology. Let $H(t)$ and $P(t)$ denote the populations at time t of two species, one of which (P) preys upon the other (H). For example, $P(t)$ and $H(t)$ may be the number of foxes and rabbits, respectively, in a wood, or the number of bass and redear (which are eaten by bass) in a pond. Without the prey the predators will decrease, and without the predator the prey will increase. A mathematical model showing how an ecological balance can be maintained when both are present was proposed about 1925 by Lotka and Volterra.* This model consists of the system of differential equations

$$\frac{dH}{dt} = a_1 H - b_1 HP,$$

$$\frac{dP}{dt} = -a_2 P + b_2 HP. \tag{4}$$

In Eqs. (4) the coefficient a_1 is the birthrate of the population H; similarly, a_2 is the death rate of the population P. The coefficients b_1 and b_2 are the coefficients of interaction between predator and prey. Equations of the form (4) are sometimes called predator-prey equations. They are discussed in some detail in Section 9.5.

Some examples of systems of differential equations for electric circuits are found in the problems at the end of this section.

There is an important connection between systems of equations and single equations of arbitrary order. In fact, an nth order equation

$$y^{(n)} = F(t, y, y', \ldots, y^{(n-1)}) \tag{5}$$

can always be reduced to a system of n first order equations having a rather special form. To show this we introduce the variables $x_1, x_2, \ldots, x_n$ defined by

$$x_1 = y, x_2 = y', x_3 = y'', \ldots, x_n = y^{(n-1)}. \tag{6}$$

It then follows immediately that

$$x_1' = x_2,$$
$$x_2' = x_3,$$
$$\vdots \tag{7}$$
$$x_{n-1}' = x_n,$$

and from Eq. (5)

$$x_n' = F(t, x_1, x_2, \ldots, x_n). \tag{8}$$

Equations (7) and (8) are a special case of the more general system

$$x_1' = F_1(t, x_1, \ldots, x_n),$$
$$x_2' = F_2(t, x_1, \ldots, x_n),$$
$$\vdots \tag{9}$$
$$x_n' = F_n(t, x_1, \ldots, x_n).$$

* Alfred J. Lotka (1880–1949) was an American biophysicist. Vito Volterra (1860–1940), an Italian mathematician, is best known for his work on integral equations (see Problem 8 of Section 6.5). He was also one of the pioneers of the theory of modern integration and functional analysis.

In a similar way the systems (1) and (2) can also be reduced to systems of first order equations of the form (9) containing, respectively, six and four equations. The systems (3) and (4) are already in the form (9). In fact, systems of the form (9) include almost all cases of interest, and as a consequence much of the more advanced theory of differential equations is devoted to such systems.

The system of equations (9) is said to have a *solution* on the interval $\alpha < t < \beta$ if there exists a set of n functions $x_1 = \phi_1(t), \ldots, x_n = \phi_n(t)$, which are differentiable at all points in $\alpha < t < \beta$ and which satisfy the system (9) at all points in this interval. In addition to the given system of differential equations, there may also be given initial conditions of the form

$$x_1(t_0) = x_1^0, x_2(t_0) = x_2^0, \ldots, x_n(t_0) = x_n^0, \tag{10}$$

where t_0 is a specified value of t in $\alpha < t < \beta$, and $x_1^0, \ldots, x_n^0$ are prescribed numbers. The differential equations (9) and initial conditions (10) together form an initial value problem.

In order to guarantee that the initial value problem (9) and (10) has a solution and that the solution is unique, it is necessary to impose certain conditions on the functions $F_1, F_2, \ldots, F_n$. The following theorem is analogous to the existence and uniqueness theorems for single equations of first and second order, Theorems 2.2 and 3.1, respectively.

Theorem 7.1.   *Let each of the functions $F_1, \ldots, F_n$ and the partial derivatives $\partial F_1/\partial x_1, \ldots, \partial F_1/\partial x_n, \ldots, \partial F_n/\partial x_1, \ldots, \partial F_n/\partial x_n$ be continuous in a region R of $t x_1 x_2 \cdots x_n$ space containing the point $(t_0, x_1^0, x_2^0, \ldots, x_n^0)$. Then there is an interval $|t - t_0| < h$ in which there exists a unique solution $x_1 = \phi_1(t), \ldots, x_n = \phi_n(t)$ of the system of differential equations (9), which also satisfies the initial conditions (10).*

The proof of this theorem can be constructed by generalizing the argument in Section 2.11, but we will not give it here. Note, however, that in the hypotheses of the theorem nothing is said about the partial derivatives of the functions $F_1, F_2, \ldots, F_n$ with respect to t. Also, in the conclusion, the length $2h$ of the interval in which the solution exists is not specified exactly, and in some cases may be very short. Finally, the theorem is not the most general one known, and the given conditions are sufficient, but not necessary, to ensure the conclusion.

If each of the functions $F_1, \ldots, F_n$ in Eqs. (9) is a linear function of the dependent variables $x_1, \ldots, x_n$, then the system of equations is said to be *linear;* otherwise it is *nonlinear.* Thus the most general system of n linear first order equations has the form

$$x_1' = p_{11}(t)x_1 + \cdots + p_{1n}(t)x_n + g_1(t),$$

$$\vdots \tag{11}$$

$$x_n' = p_{n1}(t)x_1 + \cdots + p_{nn}(t)x_n + g_n(t).$$

If each of the functions $g_1, \ldots, g_n$ is identically zero, then the system (11) is said to be *homogeneous;* otherwise it is *nonhomogeneous.* Observe that the systems (2) and (3) are both linear; the system (2) is nonhomogeneous unless $F_1(t) = F_2(t) = 0$, while the system (3) is homogeneous. The system (4) is nonlinear. The system (1) may be either linear or nonlinear, depending on the nature of the functions F_1, F_2, and F_3.

For the linear system (11) the existence and uniqueness theorem is simpler and also has a stronger conclusion. It is analogous to Theorems 2.1 and 3.2

Theorem 7.2. *If the functions $p_{11}, \ldots, p_{nn}, g_1, \ldots, g_n$ are continuous on an open interval $\alpha < t < \beta$, containing the point $t = t_0$, then there exists a unique solution $x_1 = \phi_1(t), \ldots, x_n = \phi_n(t)$ of the system of differential equations (11), which also satisfies the initial conditions (10). This solution is valid throughout the interval $\alpha < t < \beta$.*

Note that in contrast to the situation for a nonlinear system the existence and uniqueness of the solution of a linear system is guaranteed throughout the interval in which the hypotheses are satisfied. Furthermore, for a linear system the initial values $x_1{}^0, \ldots, x_n{}^0$ at $t = t_0$ are completely arbitrary, whereas in the nonlinear case the initial point must lie in the region R defined in Theorem 7.1.

The rest of this chapter is devoted to systems of linear first order equations (nonlinear systems are included in the discussion in Chapters 8 and 9). The exposition is most effectively phrased in terms of certain results from matrix theory and, beginning in Section 7.3, the rest of the chapter is based on the use of matrices. The necessary facts about matrices are summarized in Sections 7.3 and 7.4. For readers completely unfamiliar with matrix theory, Section 7.2 contains a discussion of elementary elimination techniques by which systems of equations can be replaced by a single equation of higher order. This procedure does not provide an adequate foundation for a general discussion of linear systems, but is sometimes quite efficient as a means of solving systems of two or three equations.

PROBLEMS

1. Reduce each of the systems (1) and (2) to a system of first order equations of the form (9).

2. Consider the initial value problem $u'' + p(t)u' + q(t)u = g(t)$, $u(0) = u_0$, $u'(0) = u_0'$. Transform this problem into an initial value problem for two first order linear equations.

3. Show that if a_{11}, a_{12}, a_{21}, and a_{22} are constants with a_{12} and a_{21} not both zero, and if the functions g_1 and g_2 are differentiable, then the initial value problem

$$x_1' = a_{11}x_1 + a_{12}x_2 + g_1(t), \qquad x_1(0) = x_1{}^0$$
$$x_2' = a_{21}x_1 + a_{22}x_2 + g_2(t), \qquad x_2(0) = x_2{}^0$$

can be transformed into an initial value problem for a single second order equation. Can the same procedure be carried out if $a_{11}, \ldots, a_{22}$ are functions of t?

4. Consider the linear homogeneous system

$$x' = p_{11}(t)x + p_{12}(t)y,$$
$$y' = p_{21}(t)x + p_{22}(t)y.$$

Show that if $x = x_1(t)$, $y = y_1(t)$ and $x = x_2(t)$, $y = y_2(t)$ are two solutions of the given system, then $x = c_1x_1(t) + c_2x_2(t)$, $y = c_1y_1(t) + c_2y_2(t)$ is also a solution for any constants c_1 and c_2. This is the principle of superposition.

5. Let $x = x_1(t)$, $y = y_1(t)$ and $x = x_2(t)$, $y = y_2(t)$ be any two solutions of the linear nonhomogeneous system

$$x' = p_{11}(t)x + p_{12}(t)y + g_1(t),$$
$$y' = p_{21}(t)x + p_{22}(t)y + g_2(t).$$

Show that $x = x_1(t) - x_2(t)$, $y = y_1(t) - y_2(t)$ is a solution of the corresponding homogeneous system.

6. Equations (2) can be derived by drawing a free-body diagram showing the forces acting on each mass. Figure 7.3a shows the situation when the displacements x_1 and x_2 of the two masses are both positive (to the right) and $x_2 > x_1$. Then springs 1 and 2 are elongated and spring 3 is compressed, giving rise to forces as shown in Figure 7.3b. Use Newton's law ($F = ma$) to derive Eqs. (2).

(a)

(b)

FIGURE 7.3

Electric Circuits. The theory of electric circuits, such as that shown in Figure 7.2, consisting of inductances, resistors, and capacitors, is based on Kirchhoff's laws: (1) The net flow of current through each node (or junction) is zero, (2) the net voltage drop around each closed loop is zero. In addition to Kirchhoff's laws we also have the relation between the current I in amperes through each circuit element and the voltage drop V in volts across the element; namely,

$$V = RI; \quad R = \text{resistance in ohms}$$

$$C\frac{dV}{dt} = I; \quad C = \text{capacitance in farads}$$

$$L\frac{dI}{dt} = V; \quad L = \text{inductance in henrys.}$$

Kirchhoff's laws and the current-voltage relation for each circuit element provide a system of algebraic and differential equations from which the voltage and current throughout the circuit can be determined. Problems 7 through 9 illustrate the procedure described above.

7. Consider the circuit shown in Figure 7.2. Let I_1, I_2, and I_3 be the current through the capacitor, resistor, and inductance, respectively. Likewise, let V_1, V_2, and V_3 be the corresponding voltage drops. The arrows denote the arbitrarily chosen directions in which currents and voltage drops will be taken to be positive.

(a) Applying Kirchhoff's second law to the upper loop in the circuit, show that

$$V_1 - V_2 = 0. \tag{i}$$

In a similar way, show that

$$V_2 - V_3 = 0. \tag{ii}$$

(b) Applying Kirchhoff's first law to either node in the circuit, show that

$$I_1 + I_2 + I_3 = 0. \tag{iii}$$

(c) Use the current-voltage relation through each element in the circuit to obtain the equations

$$CV_1' = I_1, \quad V_2 = RI_2, \quad LI_3' = V_3. \tag{iv}$$

(d) Eliminate V_2, V_3, I_1, and I_2 among Eqs. (i) through (iv) to obtain

$$CV_1' = -I_3 - \frac{V_1}{R}, \quad LI_3' = V_1. \tag{v}$$

Observe that if we omit the subscripts in Eqs. (v) then we have the system (3) of the text.

8. Consider the circuit shown in Figure 7.4. Use the method outlined in Problem 7 to show that the current I through the inductance and the voltage V across the capacitor satisfy the system of differential equations

$$\frac{dI}{dt} = -I - V,$$

$$\frac{dV}{dt} = 2I - V.$$

FIGURE 7.4

9. Consider the circuit shown in Figure 7.5. Use the method outlined in Problem 7 to show that the current I through the inductance and the voltage V across the capacitor satisfy the system of differential equations

$$L\frac{dI}{dt} = -R_1 I - V,$$

$$C\frac{dV}{dt} = I - \frac{V}{R_2}.$$

FIGURE 7.5

7.2 SOLUTION OF LINEAR SYSTEMS BY ELIMINATION

The most elementary way of solving a system of linear differential equations with constant coefficients is by the direct process of elimination. That is, dependent variables are successively eliminated with the object of eventually obtaining a single higher order equation containing only one dependent variable. After solving this equation, the other dependent variables can be found in turn.

For systems of only two or three first order equations such methods are quite efficient. They can be applied to nonhomogeneous systems as well as to homogeneous ones, and can also be used to advantage in some problems involving systems of higher order differential equations. Elimination methods do not, however, lend themselves readily to theoretical discussions; nor are they a convenient way to solve large systems of equations.

To illustrate the method of elimination let us consider the system of differential equations

$$
\begin{aligned}
x_1' &= x_1 + x_2, \\
x_2' &= 4x_1 + x_2.
\end{aligned}
\tag{1}
$$

Solving the first of Eqs. (1) for x_2 gives

$$
x_2 = x_1' - x_1.
\tag{2}
$$

Hence $x_2' = x_1'' - x_1'$, and from the second of Eqs. (1) we obtain, after some minor algebraic simplification,

$$
x_1'' - 2x_1' - 3x_1 = 0.
\tag{3}
$$

The general solution of Eq. (3) is

$$
x_1 = \phi_1(t) = c_1 e^{3t} + c_2 e^{-t},
\tag{4}
$$

where c_1 and c_2 are arbitrary constants. From Eq. (2) it follows at once that

$$
x_2 = \phi_2(t) = 2c_1 e^{3t} - 2c_2 e^{-t}.
\tag{5}
$$

The solution of the system (1) given by Eqs. (4) and (5) contains two arbitrary constants. It is easy to show that these constants can be chosen so as to satisfy any prescribed initial conditions at any given point. Therefore any solution of the system (1) can be written in the form (4), (5) for suitable choices of c_1 and c_2. Equations (4) and (5) are thus said to form the general solution of the system (1). The reader may verify that the same solution is obtained if x_1 is eliminated from Eqs. (1) instead of x_2.

The same procedure can be applied to the nonhomogeneous system

$$
\begin{aligned}
x_1' &= x_1 + x_2 + g_1(t), \\
x_2' &= 4x_1 + x_2 + g_2(t).
\end{aligned}
\tag{6}
$$

The only differences are that now $x_2 = x_1' - x_1 - g_1(t)$, and that the second order equation (3) for x_1 is replaced by a nonhomogeneous equation.

As mentioned earlier, the method of elimination can also be used to solve systems of higher order equations. Such systems may occur naturally in various applications; they also arise when the method of elimination just described is applied to a system of more than two first order equations.

Thus let us consider the system

$$L_1[x_1] + L_2[x_2] = g_1(t),$$
$$L_3[x_1] + L_4[x_2] = g_2(t),$$

(7)

where L_1, L_2, L_3, and L_4 are linear differential operators, not necessarily of first order, with constant coefficients, and g_1 and g_2 are given functions. For example, suppose that L_1 is the second order differential operator

$$L_1[x] = ax'' + bx' + cx.$$

(8)

It is convenient to let $D = d/dt$, and to write Eq. (8) in the form

$$L_1[x] = (aD^2 + bD + c)x.$$

(9)

Before proceeding further it is necessary to discuss the meaning of an expression such as $L_2L_1[x]$, which indicates the successive applications of the operators L_1 and L_2 to the function x. Suppose for the moment that

$$L_1[x] = (aD + b)x,$$

(10)

and

$$L_2[x] = (cD + d)x,$$

(11)

where a, b, c, and d are constants. Then $L_2L_1[x]$ is defined by

$$L_2L_1[x] = cD[(aD + b)x] + d[(aD + b)x]$$
$$= caD^2x + cbDx + daDx + dbx$$
$$= [caD^2 + (cb + da)D + db]x.$$

(12)

Note that Eq. (12) can be written as

$$L_2L_1[x] = [(cD + d)(aD + b)]x,$$

(13)

where D is treated as an algebraic quantity in expanding the right-hand side. Similarly

$$L_1L_2[x] = aD[(cD + d)x] + b[(cD + d)x]$$
$$= [acD^2 + (ad + bc)D + bd]x.$$

(14)

It is clear from Eqs. (12) and (14) that

$$L_2L_1[x] = L_1L_2[x];$$

(15)

that is, L_1 and L_2 are commutative. It is possible to show that the property of commutativity is possessed not only by the operators L_1 and L_2 defined by Eqs. (10) and (11), but by any two linear differential operators L_1 and L_2 having constant coefficients, regardless of their order. On the other hand, operators with variable coefficients do not in general possess the property of commutativity (see Problem 22).

Returning to the system (7), it is now a simple matter to eliminate either of the dependent variables x_1 or x_2. To eliminate x_1 we need only apply L_3 to the first equation, L_1 to the second equation, and then subtract the first from the second. Thus

$$L_3 L_1[x_1] + L_3 L_2[x_2] = L_3[g_1], \tag{16}$$

$$L_1 L_3[x_1] + L_1 L_4[x_2] = L_1[g_2], \tag{17}$$

and subtraction of Eq. (16) from Eq. (17) gives

$$L_1 L_4[x_2] - L_3 L_2[x_2] = L_1[g_2] - L_3[g_1]. \tag{18}$$

Equation (18) does not involve x_1 and can be solved by standard methods. Once x_2 is determined from Eq. (18), x_1 can be found by solving either of Eqs. (7). Alternatively, x_2 can be eliminated from the system in a similar way. Note that in using this method the functions g_1 and g_2 must possess sufficiently many derivatives so that it is possible to compute the quantities $L_1[g_2]$ and $L_3[g_1]$ in Eq. (18).

In general, a number of extraneous constants will be introduced during this procedure. In solving Eq. (18) for x_2 the number of arbitrary constants will be equal to the order* of the operator $L_1 L_4 - L_3 L_2$. Additional constants are introduced when x_1 is determined by solving either of Eqs. (7). The relations among the constants can be found, and the extraneous ones eliminated, by substituting for x_1 and x_2 in whichever of Eqs. (7) was not used in solving for x_1. This is illustrated in the example below.

The question of the number of arbitrary constants that should appear in the general solution of the system (7) can be answered in the following way. If the system (7) were reduced to a system of first order equations, then the number of independent constants in the general solution would be equal to the number of equations in the system. This will be established in Section 7.5. By eliminating x_1 we obtain a single equation whose order is that of the operator $L_1 L_4 - L_3 L_2$. This latter equation could in turn be transformed into a system of first order equations equal in number to the order of $L_1 L_4 - L_3 L_2$. Thus it is reasonable to expect, and it can be proved to be true, that the number of independent arbitrary constants in the general solution of the system (7) is equal to the order of $L_1 L_4 - L_3 L_2$, provided that $L_1 L_4 - L_3 L_2$ is not zero. This fact can be used to check that the correct number of arbitrary constants appears in the final solution. The result can also be extended to larger systems of equations (see Problem 23).

---

* By the order of $L_1 L_4 - L_3 L_2$ we mean the order of the highest derivative appearing in it. If $L_1 L_4 - L_3 L_2$ is zero the system (7) is said to be *degenerate*. In this case there may be either infinitely many solutions or else none, depending on whether the right side of Eq. (18) does, or does not, vanish (see Problems 18 through 21). The situation here is similar to that for a system of linear algebraic equations whose determinant of coefficients vanishes.

Example.   Find the general solution of

$$(x_1'' + x_1' - x_1) + (x_2'' - 3x_2' + 2x_2) = 0, \tag{19}$$

$$(x_1' + 2x_1) + \qquad (2x_2' - 4x_2) = 0. \tag{20}$$

In this case $L_1[x_1] = (D^2 + D - 1)x_1$, $L_2[x_2] = (D^2 - 3D + 2)x_2$, $L_3[x_1] = (D + 2)x_1$, and $L_4[x_2] = (2D - 4)x_2$. Applying L_3 and L_1 to Eqs. (19) and (20) respectively, and subtracting the results, yields

$$(L_1 L_4 - L_3 L_2)[x_2]$$
$$= [(D^2 + D - 1)(2D - 4) - (D + 2)(D^2 - 3D + 2)]x_2 = 0$$

or

$$(D^3 - D^2 - 2D)x_2 = 0. \tag{21}$$

Thus the general solution of the system (19), (20) should contain three arbitrary constants. The general solution of Eq. (21) is

$$x_2 = \phi_2(t) = c_1 + c_2 e^{2t} + c_3 e^{-t}. \tag{22}$$

Substituting this expression for x_2 in Eq. (20) gives

$$x_1' + 2x_1 = 4c_1 + 6c_3 e^{-t} \tag{23}$$

whose general solution is

$$x_1 = \phi_1(t) \quad 2c_1 + 6c_3 e^{-t} + c_4 e^{-2t}. \tag{24}$$

To eliminate the extraneous constant we substitute for x_1 and x_2 in Eq. (19), and find that $c_4 = 0$. Note that if we had substituted for x_2 in Eq. (19) and then solved this second order equation for x_1, we would have introduced two extraneous constants instead of one. These would have been eliminated by substituting for x_1 and x_2 in Eq. (20).

PROBLEMS

In each of Problems 1 through 11 solve the given system of equations.

1. $x_1' = x_1 + x_2$
   $x_2' = 4x_1 - 2x_2$

2. $x_1' = x_1 + x_2 + 2e^t$
   $x_2' = 4x_1 + x_2 - e^t$

3. $x_1' = 2x_1 - 5x_2 - \sin 2t,$ $x_1(0) = 0$
   $x_2' = x_1 - 2x_2 + t,$ $x_2(0) = 1$

4. $x_1' = x_1 - x_2 - t^2$
   $x_2' = x_1 + 3x_2 + 2t$

5. $x_1' = 3x_1 - 4x_2 + e^t,$ $x_1(0) = 1$
   $x_2' = x_1 - x_2 - e^t,$ $x_2(0) = -1$

6. $x_1' = 4x_1 - 2x_2 + 2t$
   $x_2' = 8x_1 - 4x_2 + 1$

7. $x_1' = 3x_1 - 2x_2 - e^{-t} \sin t$
   $x_2' = 4x_1 - x_2 + 2e^{-t} \cos t$

8. $x_1' = x_1 - 5x_2,$ $x_1(0) = 1$
   $x_2' = 2x_1 - 5x_2,$ $x_2(0) = 0$

9. $x_1' = x_1$
   $x_2' = -x_2 + \sqrt{2}x_3$
   $x_3' = \sqrt{2}x_2$

10. $x_1' = x_1 - x_2 + 4x_3$
    $x_2' = 3x_1 + 2x_2 - x_3$
    $x_3' = 2x_1 + x_2 - x_3$

11. $x_1' = x_1 + x_2 + x_3$
    $x_2' = 2x_1 + x_2 - x_3$
    $x_3' = -8x_1 - 5x_2 - 3x_3$

12. Solve the example problem, Eqs. (19) and (20), by substituting for x_2 from Eq. (22) in Eq. (19).

In each of Problems 13 through 17 solve the given system of equations. Make sure that the proper number of arbitrary constants appears in the general solution.

13. $(D^2 - 3D + 2)x_1 + (D - 1)x_2 = 0$
    $(D - 2)x_1 + (D + 1)x_2 = 0$

14. $(2D + 1)x_1 + Dx_2 = t$
    $(D - 1)x_1 + Dx_2 = 2$

15. $(D^2 - 4D + 4)x_1 + 3Dx_2 = 1$
    $(D - 2)x_1 + (D + 2)x_2 = 0$

16. $D^2x_1 + (D + 1)x_2 = 0$
    $(D - 1)x_1 + x_2 = \sin t$

17. $(D^2 - 4D + 4)x_1 + (D^2 + 2D)x_2 = 0$
    $(D^2 - 2D)x_1 + (D^2 + 4D + 4)x_2 = 0$

In each of Problems 18 through 21 show that the given system of equations is degenerate. Determine by attempting to solve the system whether there are no solutions or infinitely many.

18. $Dx_1 + (D + 1)x_2 = t$
    $Dx_1 + (D + 1)x_2 = t + 2$

19. $(D + 2)x_1 + (D + 2)x_2 = e^{2t}$
    $(D - 2)x_1 + (D - 2)x_2 = e^{-2t}$ Hint: Let $u = x_1 + x_2$

20. $(D^2 - 1)x_1 + (D - 1)x_2 = \sin t$
    $(D + 1)x_1 + x_2 = 2e^t$

21. $(D^2 + 3D + 2)x_1 + (D^2 + 2D)x_2 = 0$
    $(D + 1)x_1 + Dx_2 = 0$

22. Linear differential operators with variable coefficients may not be commutative. Consider the operators

$$L_1[x] = x' + tx, \qquad L_2[x] = x' + t^2x.$$

Show that

$$L_1L_2[x] = L_1\{L_2[x]\} = x'' + (t^2 + t)x' + (2t + t^3)x,$$

and that

$$L_2L_1[x] = L_2\{L_1[x]\} = x'' + (t + t^2)x' + (1 + t^3)x.$$

Hence

$$L_1L_2[x] \neq L_2L_1[x].$$

23. Consider the system of equations

$$L_{11}[x_1] + L_{12}[x_2] + L_{13}[x_3] = g_1(t),$$
$$L_{21}[x_1] + L_{22}[x_2] + L_{23}[x_3] = g_2(t),$$
$$L_{31}[x_1] + L_{32}[x_2] + L_{33}[x_3] = g_3(t),$$

where $L_{11}, \ldots, L_{33}$ are linear differential operators with constant coefficients, but are not necessarily of first order. It can be shown that the order of this system, that is, the number of arbitrary constants appearing in its general solution, is equal to the order of the operator L defined by

$$L = \begin{vmatrix} L_{11} & L_{12} & L_{13} \\ L_{21} & L_{22} & L_{23} \\ L_{31} & L_{32} & L_{33} \end{vmatrix}.$$

This determinant is evaluated by treating $L_{11}, \ldots, L_{33}$ as polynomials in the operator D and formally using the customary methods of expanding determinants. Determine the order of each of the following systems.

(a) $(D^2 + 1)x_1 + (D^2 + 2)x_2 \qquad\qquad = 0$
$\qquad\qquad (D - 1)x_2 + (D^2 - 2)x_3 = 0$
$\quad D^2 x_1 \qquad\qquad - (D^3 + 1)x_3 = 0$

(b) $\qquad (D + 1)x_1 + \qquad Dx_2 + \quad (D^2 + 1)x_3 = 0$
$(D^2 + 2D + 2)x_1 + (D - 1)x_2 + \quad (D + 2)x_3 = 2te^t$
$\quad (D - 2)x_1 + (D + 3)x_2 + (D^2 + 3D)x_3 = te^{2t}$

7.3   REVIEW OF MATRICES

Whereas the process of elimination discussed in the last section is often a satisfactory way of solving systems of linear differential equations when the number of equations is small, it rapidly becomes unwieldy when larger systems are to be considered. More important, it does not provide an adequate basis for a systematic development of the theory associated with such equations. For both of these reasons it is advisable to bring to bear some of the results of matrix theory on the initial value problem for a system of linear differential equations. For reference purposes this section and the next are devoted to a brief summary* of some of the facts that will be needed later. We will assume, however, that the reader is familiar with determinants and how to evaluate them.

We will designate matrices by bold-faced capitals $\mathbf{A}, \mathbf{B}, \mathbf{C}, \ldots$, occasionally using bold-faced Greek capitals $\mathbf{\Phi}, \mathbf{\Psi}, \ldots$. A matrix $\mathbf{A}$ consists of a rectangular array of numbers, or elements, arranged in m rows and n columns, that is,

$$\mathbf{A} = \begin{pmatrix} a_{11} & a_{12} & \cdots & a_{1n} \\ a_{21} & a_{22} & \cdots & a_{2n} \\ \cdot & \cdot & & \cdot \\ \cdot & \cdot & & \cdot \\ \cdot & \cdot & & \cdot \\ a_{m1} & a_{m2} & \cdots & a_{mn} \end{pmatrix}. \tag{1}$$

* A representative list of books on matrix algebra is given at the end of the chapter.

We speak of $\mathbf{A}$ as an $m \times n$ matrix. Although later in the chapter we will often assume that the elements of certain matrices are real numbers, in this section we will assume that the elements of matrices may be complex numbers. The element lying in the ith row and jth column is designated by a_{ij}, the first subscript identifying its row and the second its column. Sometimes the notation (a_{ij}) is used to denote the matrix whose generic element is a_{ij}.

Associated with each matrix $\mathbf{A}$ is the matrix $\mathbf{A}^T$, known as the *transpose* of $\mathbf{A}$, and obtained from $\mathbf{A}$ by interchanging the rows and columns of $\mathbf{A}$. Thus, if $\mathbf{A} = (a_{ij})$, then $\mathbf{A}^T = (a_{ji})$. Also, we will denote by $\bar{a}_{ij}$ the complex conjugate of a_{ij}, and by $\bar{\mathbf{A}}$ the matrix obtained from $\mathbf{A}$ by replacing each element a_{ij} by its conjugate $\bar{a}_{ij}$. The matrix $\bar{\mathbf{A}}$ is called the *conjugate* of $\mathbf{A}$. It will also be necessary to consider the transpose of the conjugate matrix $\bar{\mathbf{A}}^T$. This matrix is called the *adjoint* of $\mathbf{A}$ and will be denoted by $\mathbf{A}^*$.

For example, let

$$\mathbf{A} = \begin{pmatrix} 3 & 2 - i \\ 4 + 3i & -5 + 2i \end{pmatrix}.$$

Then

$$\mathbf{A}^T = \begin{pmatrix} 3 & 4 + 3i \\ 2 - i & -5 + 2i \end{pmatrix}, \quad \bar{\mathbf{A}} = \begin{pmatrix} 3 & 2 + i \\ 4 - 3i & -5 - 2i \end{pmatrix},$$

$$\mathbf{A}^* = \begin{pmatrix} 3 & 4 - 3i \\ 2 + i & -5 - 2i \end{pmatrix}.$$

We will be particularly interested in two somewhat special kinds of matrices: square matrices, which have the same number of rows and columns, that is, $m = n$; and vectors (or column vectors), which can be thought of as $n \times 1$ matrices, or matrices having only one column. Square matrices having n rows and n columns are said to be of order n. We will denote (column) vectors by bold-faced lowercase letters $\mathbf{x}, \mathbf{y}, \boldsymbol{\xi}, \boldsymbol{\eta}, \ldots$. The transpose $\mathbf{x}^T$ of an $n \times 1$ column vector is a $1 \times n$ row vector; that is, the matrix consisting of one row whose elements are the same as the elements in the corresponding position of $\mathbf{x}$.

Algebraic Properties

1. *Equality.* Two $m \times n$ matrices $\mathbf{A}$ and $\mathbf{B}$ are said to be equal if corresponding elements are equal; that is, if $a_{ij} = b_{ij}$ for each i and j.

2. *Zero.* The symbol $\mathbf{0}$ will be used to denote the matrix (or vector) each of whose elements is zero.

3. *Addition.* The sum of two $m \times n$ matrices is defined as the matrix obtained by adding corresponding elements:

$$\mathbf{A} + \mathbf{B} = (a_{ij}) + (b_{ij}) = (a_{ij} + b_{ij}). \tag{2}$$

With this definition it follows that matrix addition is commutative and associative, so that

$$\mathbf{A} + \mathbf{B} = \mathbf{B} + \mathbf{A}, \quad \mathbf{A} + (\mathbf{B} + \mathbf{C}) = (\mathbf{A} + \mathbf{B}) + \mathbf{C}. \tag{3}$$

4. *Multiplication by a Number.* The product of a matrix $\mathbf{A}$ by a complex number α is defined as follows:

$$\alpha\mathbf{A} = \alpha(a_{ij}) = (\alpha a_{ij}). \tag{4}$$

The distributive laws

$$\alpha(\mathbf{A} + \mathbf{B}) = \alpha\mathbf{A} + \alpha\mathbf{B}, \quad (\alpha + \beta)\mathbf{A} = \alpha\mathbf{A} + \beta\mathbf{A} \tag{5}$$

are satisfied for this type of multiplication. In particular, the negative of $\mathbf{A}$, denoted by $-\mathbf{A}$, is defined by

$$-\mathbf{A} = (-1)\mathbf{A}. \tag{6}$$

5. *Subtraction.* The difference $\mathbf{A} - \mathbf{B}$ of two $m \times n$ matrices is defined by

$$\mathbf{A} - \mathbf{B} = \mathbf{A} + (-\mathbf{B}). \tag{7}$$

Thus

$$(a_{ij}) - (b_{ij}) = (a_{ij} - b_{ij}), \tag{8}$$

which is similar to Eq. (2).

6. *Multiplication.* The product $\mathbf{AB}$ of two matrices is defined whenever the number of columns of the first factor is the same as the number of rows in the second. If $\mathbf{A}$ and $\mathbf{B}$ are $m \times n$ and $n \times r$ matrices, respectively, then the product $\mathbf{C} = \mathbf{AB}$ is an $m \times r$ matrix. The element in the ith row and jth column of $\mathbf{C}$ is found by multiplying each element of the ith row of $\mathbf{A}$ by the corresponding element of the jth column of $\mathbf{B}$ and then adding the resulting products. Symbolically,

$$c_{ij} = \sum_{k=1}^{n} a_{ik}b_{kj}. \tag{9}$$

By direct calculation it can be shown that matrix multiplication satisfies the associative law

$$(\mathbf{AB})\mathbf{C} = \mathbf{A}(\mathbf{BC}) \tag{10}$$

and the distributive law

$$\mathbf{A}(\mathbf{B} + \mathbf{C}) = \mathbf{AB} + \mathbf{AC}. \tag{11}$$

However, in general, matrix multiplication is not commutative. In order for both products $\mathbf{AB}$ and $\mathbf{BA}$ to exist and to be of the same size, it is necessary that $\mathbf{A}$ and $\mathbf{B}$ be square matrices of the same order. Even in that case the two products need not be equal, so that, in general

$$\mathbf{AB} \neq \mathbf{BA}. \tag{12}$$

Example 1.   To illustrate the multiplication of matrices, and also the fact that matrix multiplication is not necessarily commutative, consider the matrices

$$\mathbf{A} = \begin{pmatrix} 1 & -2 & 1 \\ 0 & 2 & -1 \\ 2 & 1 & 1 \end{pmatrix}, \qquad \mathbf{B} = \begin{pmatrix} 2 & 1 & -1 \\ 1 & -1 & 0 \\ 2 & -1 & 1 \end{pmatrix}.$$

From the definition of multiplication given in Eq. (9) we have

$$\mathbf{AB} = \begin{pmatrix} 2-2+2 & 1+2-1 & -1+0+1 \\ 0+2-2 & 0-2+1 & 0+0-1 \\ 4+1+2 & 2-1-1 & -2+0+1 \end{pmatrix}$$

$$= \begin{pmatrix} 2 & 2 & 0 \\ 0 & -1 & -1 \\ 7 & 0 & -1 \end{pmatrix}.$$

Similarly, we find that

$$\mathbf{BA} = \begin{pmatrix} 0 & -3 & 0 \\ 1 & -4 & 2 \\ 4 & -5 & 4 \end{pmatrix}.$$

Clearly, $\mathbf{AB} \neq \mathbf{BA}$.

---

7. *Multiplication of Vectors.*   Matrix multiplication as described above also applies as a special case if the matrices $\mathbf{A}$ and $\mathbf{B}$ are $1 \times n$ and $n \times 1$ row and column vectors, respectively.   Denoting these vectors by $\mathbf{x}^T$ and $\mathbf{y}$ we have

$$\mathbf{x}^T \mathbf{y} = \sum_{i=1}^{n} x_i y_i. \tag{13}$$

The result of such an operation is a (complex) number, and it follows directly from Eq. (13) that

$$\mathbf{x}^T \mathbf{y} = \mathbf{y}^T \mathbf{x}, \qquad \mathbf{x}^T(\mathbf{y} + \mathbf{z}) = \mathbf{x}^T \mathbf{y} + \mathbf{x}^T \mathbf{z}, \qquad (\alpha \mathbf{x})^T \mathbf{y} = \alpha(\mathbf{x}^T \mathbf{y}) = \mathbf{x}^T(\alpha \mathbf{y}). \tag{14}$$

There is another very useful type of vector multiplication, which is also defined for any two vectors having the same number of components.   This product, denoted by $(\mathbf{x}, \mathbf{y})$, is called the *scalar* or *inner product*, and is defined by

$$(\mathbf{x}, \mathbf{y}) = \sum_{i=1}^{n} x_i \bar{y}_i. \tag{15}$$

The scalar product is also a (complex) number, and by comparing Eqs. (13) and (15) we see that

$$(\mathbf{x}, \mathbf{y}) = \mathbf{x}^T \bar{\mathbf{y}}. \tag{16}$$

From Eq. (15) it follows that

$$(\mathbf{x}, \mathbf{y}) = \overline{(\mathbf{y}, \mathbf{x})}, \qquad (\mathbf{x}, \mathbf{y} + \mathbf{z}) = (\mathbf{x}, \mathbf{y}) + (\mathbf{x}, \mathbf{z})$$
$$(\alpha\mathbf{x}, \mathbf{y}) = \alpha(\mathbf{x}, \mathbf{y}), \qquad (\mathbf{x}, \alpha\mathbf{y}) = \bar{\alpha}(\mathbf{x}, \mathbf{y}). \tag{17}$$

Note that even if the vector $\mathbf{x}$ has elements with nonzero imaginary parts, the scalar product of $\mathbf{x}$ with itself yields a nonnegative real number,

$$(\mathbf{x}, \mathbf{x}) = \sum_{i=1}^{n} x_i \bar{x}_i = \sum_{i=1}^{n} |x_i|^2. \tag{18}$$

If $(\mathbf{x}, \mathbf{y}) = 0$, then the two vectors $\mathbf{x}$ and $\mathbf{y}$ are said to be *orthogonal*. For example, the unit vectors $\mathbf{i}, \mathbf{j}, \mathbf{k}$ of three dimensional vector geometry form an orthogonal set.

On the other hand, the matrix product

$$\mathbf{x}^T \mathbf{x} = \sum_{i=1}^{n} x_i^{\,2} \tag{19}$$

may not be a real number. If all of the components of the second factor of Eqs. (13) and (15) are real, then the two products are identical, and both reduce to the dot product usually encountered in physical and geometrical contexts with $n = 3$.

For example, let

$$\mathbf{x} = \begin{pmatrix} i \\ -2 \\ 1+i \end{pmatrix}, \qquad \mathbf{y} = \begin{pmatrix} 2-i \\ i \\ 3 \end{pmatrix}.$$

Then

$$\mathbf{x}^T\mathbf{y} = (i)(2-i) + (-2)(i) + (1+i)(3) = 4 + 3i,$$
$$(\mathbf{x}, \mathbf{y}) = (i)(2+i) + (-2)(-i) + (1+i)(3) = 2 + 7i,$$
$$\mathbf{x}^T\mathbf{x} = (i)^2 + (-2)^2 + (1+i)^2 = 3 + 2i,$$
$$(\mathbf{x}, \mathbf{x}) = (i)(-i) + (-2)(-2) + (1+i)(1-i) = 7.$$

8. *Identity.* The multiplicative identity, or simply the identity matrix $\mathbf{I}$, is given by

$$\mathbf{I} = \begin{pmatrix} 1 & 0 \cdots 0 \\ 0 & 1 \cdots 0 \\ \cdot & \cdot \quad \cdot \\ \cdot & \cdot \quad \cdot \\ \cdot & \cdot \quad \cdot \\ 0 & 0 \cdots 1 \end{pmatrix}. \tag{20}$$

From the definition of matrix multiplication we have

$$\mathbf{AI} = \mathbf{IA} = \mathbf{A} \tag{21}$$

for any (square) matrix $\mathbf{A}$. Hence the commutative law does hold for square matrices if one of the matrices is the identity.

9. *Inverse.* In order to define an operation for square matrices analogous to division for numbers, we need to determine, for a given square matrix $\mathbf{A}$, another matrix $\mathbf{B}$ such that $\mathbf{AB} = \mathbf{I}$, where $\mathbf{I}$ is the identity. If $\mathbf{B}$ exists, it is called the multiplicative inverse, or simply the inverse, of $\mathbf{A}$, and we write $\mathbf{B} = \mathbf{A}^{-1}$. It is possible to show that if $\mathbf{A}^{-1}$ exists, then

$$\mathbf{AA}^{-1} = \mathbf{A}^{-1}\mathbf{A} = \mathbf{I}. \tag{22}$$

In other words, multiplication is commutative between any matrix and its inverse. If $\mathbf{A}$ has a multiplicative inverse $\mathbf{A}^{-1}$, then $\mathbf{A}$ is said to be *nonsingular*; otherwise $\mathbf{A}$ is called *singular*.

There are various ways to compute $\mathbf{A}^{-1}$ from $\mathbf{A}$, assuming that it exists. One way involves the use of determinants. Associated with each element a_{ij} of a given matrix is the minor M_{ij}, which is the determinant of the matrix obtained by deleting the ith row and jth column of the original matrix, that is, the row and column containing a_{ij}. Also associated with each element a_{ij} is the cofactor C_{ij} defined by the equation

$$C_{ij} = (-1)^{i+j}M_{ij}. \tag{23}$$

If $\mathbf{B} = \mathbf{A}^{-1}$, then it can be shown that the general element b_{ij} is given by

$$b_{ij} = \frac{C_{ji}}{\det \mathbf{A}}. \tag{24}$$

While Eq. (24) is not an efficient way* to calculate $\mathbf{A}^{-1}$, it does suggest a condition that $\mathbf{A}$ must satisfy in order for it to have an inverse. In fact, the condition is both necessary and sufficient: $\mathbf{A}$ is nonsingular if and only if $\det \mathbf{A} \neq 0$. If $\det \mathbf{A} = 0$, then $\mathbf{A}$ is singular.

Another and usually better way to compute $\mathbf{A}^{-1}$ is by means of elementary row operations. There are three such operations:

1. Interchange of two rows.
2. Multiplication of a row by a nonzero scalar.
3. Addition of any multiple of one row to another row.

Any nonsingular matrix $\mathbf{A}$ can be transformed into the identity $\mathbf{I}$ by a systematic sequence of these operations. It is possible to show that if the same sequence of operations is then performed upon $\mathbf{I}$, it is transformed into $\mathbf{A}^{-1}$. The transformation of a matrix by a sequence of elementary row operations is usually referred to as *row reduction*. The following example illustrates the process.

---

* For large n the number of multiplications required to evaluate $\mathbf{A}^{-1}$ by Eq. (24) is proportional to $n!$. If one uses more efficient methods, such as the row reduction procedure described below, the number of multiplications is proportional only to n^3. Even for small values of n (such as $n = 4$), determinants are not an economical tool in calculating inverses, and row reduction methods are preferred.

Example 2.   Find the inverse of $A = \begin{pmatrix} 1 & -1 & -1 \\ 3 & -1 & 2 \\ 2 & 2 & 3 \end{pmatrix}$.

The matrix **A** can be transformed into **I** by the following sequence of operations. The result of each step appears in the right hand column.

(a) Obtain zeros in the off-diagonal positions in the first column by adding (-3) times the first row to the second row and adding (-2) times the first row to the third row.

$\begin{pmatrix} 1 & -1 & -1 \\ 0 & 2 & 5 \\ 0 & 4 & 5 \end{pmatrix}$

(b) Obtain a one in the diagonal position in the second column by multiplying the second row by $\frac{1}{2}$.

$\begin{pmatrix} 1 & -1 & -1 \\ 0 & 1 & \frac{5}{2} \\ 0 & 4 & 5 \end{pmatrix}$

(c) Obtain zeros in the off-diagonal positions in the second column by adding the second row to the first row and adding (-4) times the second row to the third row.

$\begin{pmatrix} 1 & 0 & \frac{3}{2} \\ 0 & 1 & \frac{5}{2} \\ 0 & 0 & -5 \end{pmatrix}$

(d) Obtain a one in the diagonal position in the third column by multiplying the third row by $(-\frac{1}{5})$.

$\begin{pmatrix} 1 & 0 & \frac{3}{2} \\ 0 & 1 & \frac{5}{2} \\ 0 & 0 & 1 \end{pmatrix}$

(e) Obtain zeros in the off-diagonal positions in the third column by adding $(-\frac{3}{2})$ times the third row to the first row and adding $(-\frac{5}{2})$ times the third row to the first row.

$\begin{pmatrix} 1 & 0 & 0 \\ 0 & 1 & 0 \\ 0 & 0 & 1 \end{pmatrix}$

If we perform the same sequence of operations in the same order on **I**, we obtain the following sequence of matrices:

$\begin{pmatrix} 1 & 0 & 0 \\ 0 & 1 & 0 \\ 0 & 0 & 1 \end{pmatrix}$, $\begin{pmatrix} 1 & 0 & 0 \\ -3 & 1 & 0 \\ -2 & 0 & 1 \end{pmatrix}$, $\begin{pmatrix} 1 & 0 & 0 \\ -\frac{3}{2} & \frac{1}{2} & 0 \\ -2 & 0 & 1 \end{pmatrix}$, $\begin{pmatrix} -\frac{1}{2} & \frac{1}{2} & 0 \\ -\frac{3}{2} & \frac{1}{2} & 0 \\ 4 & -2 & 1 \end{pmatrix}$,

$\begin{pmatrix} -\frac{1}{2} & \frac{1}{2} & 0 \\ -\frac{3}{2} & \frac{1}{2} & 0 \\ -\frac{4}{5} & \frac{2}{5} & -\frac{1}{5} \end{pmatrix}$, $\begin{pmatrix} \frac{7}{10} & -\frac{1}{10} & \frac{3}{10} \\ \frac{1}{2} & -\frac{1}{2} & \frac{1}{2} \\ -\frac{4}{5} & \frac{2}{5} & -\frac{1}{5} \end{pmatrix}$.

The last of these matrices is A^{-1}, a result that can be verified by direct multiplication.

This example is made slightly simpler by the fact that the original matrix $\mathbf{A}$ had a one in the upper left corner ($a_{11} = 1$). If this is not the case, then the first step is to produce a one there by multiplying the first row by $1/a_{11}$, so long as $a_{11} \neq 0$. If $a_{11} = 0$, then the first row must be interchanged with some other row in order to bring a nonzero element into the upper left position before proceeding as above.

Matrix Functions.   We will sometimes need to consider vectors or matrices whose elements are functions of a real variable t. We write

$$
\mathbf{x}(t) = \begin{pmatrix} x_1(t) \\ \cdot \\ \cdot \\ \cdot \\ x_n(t) \end{pmatrix}, \qquad
\mathbf{A}(t) = \begin{pmatrix} a_{11}(t) \cdots a_{1n}(t) \\ \cdot \qquad \cdot \\ \cdot \qquad \cdot \\ \cdot \qquad \cdot \\ a_{n1}(t) \cdots a_{nn}(t) \end{pmatrix}, \tag{25}
$$

respectively.

The matrix $\mathbf{A}(t)$ is said to be continuous at $t = t_0$ or on an interval $\alpha < t < \beta$ if each element of $\mathbf{A}$ is a continuous function at the given point or on the given interval. Similarly, $\mathbf{A}(t)$ is said to be differentiable if each of its elements is differentiable, and its derivative $d\mathbf{A}/dt$ is defined by

$$
\frac{d\mathbf{A}}{dt} = \left(\frac{da_{ij}}{dt} \right); \tag{26}
$$

that is, each element of $d\mathbf{A}/dt$ is the derivative of the corresponding element of $\mathbf{A}$. In the same way the integral of a matrix function is defined as

$$
\int_a^b \mathbf{A}(t)\, dt = \left(\int_a^b a_{ij}(t)\, dt \right). \tag{27}
$$

For example, if

$$
\mathbf{A}(t) = \begin{pmatrix} \sin t & t \\ 1 & \cos t \end{pmatrix}
$$

then

$$
\mathbf{A}'(t) = \begin{pmatrix} \cos t & 1 \\ 0 & -\sin t \end{pmatrix}, \qquad
\int_0^\pi \mathbf{A}(t)\, dt = \begin{pmatrix} 2 & \pi^2/2 \\ \pi & 0 \end{pmatrix}.
$$

Many of the rules of elementary calculus can be easily extended to matrix functions; in particular,

$$
\frac{d}{dt}\,(\mathbf{CA}) = \mathbf{C}\,\frac{d\mathbf{A}}{dt}, \text{ where } \mathbf{C} \text{ is a constant matrix;} \tag{28}
$$

$$
\frac{d}{dt}\,(\mathbf{A} + \mathbf{B}) = \frac{d\mathbf{A}}{dt} + \frac{d\mathbf{B}}{dt}; \tag{29}
$$

$$
\frac{d}{dt}\,(\mathbf{AB}) = \mathbf{A}\,\frac{d\mathbf{B}}{dt} + \frac{d\mathbf{A}}{dt}\,\mathbf{B}. \tag{30}
$$

In Eqs. (28) and (30) care must be taken in each term to avoid carelessly interchanging the order of multiplication. The definitions expressed by Eqs. (26) and (27) also apply as special cases to vectors.

PROBLEMS

1. If $\mathbf{A} = \begin{pmatrix} 1 & -2 & 0 \\ 3 & 2 & -1 \\ -2 & 1 & 3 \end{pmatrix}$ and $\mathbf{B} = \begin{pmatrix} 4 & -2 & 3 \\ -1 & 5 & 0 \\ 6 & 1 & 2 \end{pmatrix}$,

find

(a) $2\mathbf{A} + \mathbf{B}$, (b) $\mathbf{A} - 4\mathbf{B}$, (c) $\mathbf{AB}$, (d) $\mathbf{BA}$.

2. If $\mathbf{A} = \begin{pmatrix} 1 + i & -1 + 2i \\ 3 + 2i & 2 - i \end{pmatrix}$ and $\mathbf{B} = \begin{pmatrix} i & 3 \\ 2 & -2i \end{pmatrix}$,

find

(a) $\mathbf{A} - 2\mathbf{B}$, (b) $3\mathbf{A} + \mathbf{B}$, (c) $\mathbf{AB}$, (d) $\mathbf{BA}$.

3. If $\mathbf{A} = \begin{pmatrix} -2 & 1 & 2 \\ 1 & 0 & -3 \\ 2 & -1 & 1 \end{pmatrix}$ and $\mathbf{B} = \begin{pmatrix} 1 & 2 & 3 \\ 3 & -1 & -1 \\ -2 & 1 & 0 \end{pmatrix}$,

find

(a) $\mathbf{A}^T$, (b) $\mathbf{B}^T$, (c) $\mathbf{A}^T + \mathbf{B}^T$, (d) $(\mathbf{A} + \mathbf{B})^T$.

4. If $\mathbf{A} = \begin{pmatrix} 3 - 2i & 1 + i \\ 2 - i & -2 + 3i \end{pmatrix}$,

find

(a) $\mathbf{A}^T$, (b) $\bar{\mathbf{A}}$, (c) $\mathbf{A}^*$.

5. If $\mathbf{A} = \begin{pmatrix} 3 & 2 & -1 \\ 2 & -1 & 2 \\ 1 & 2 & 1 \end{pmatrix}$ and $\mathbf{B} = \begin{pmatrix} 2 & 1 & -1 \\ -2 & 3 & 3 \\ 1 & 0 & 2 \end{pmatrix}$,

verify that $2(\mathbf{A} + \mathbf{B}) = 2\mathbf{A} + 2\mathbf{B}$.

6. If $\mathbf{A} = \begin{pmatrix} 1 & -2 & 0 \\ 3 & 2 & -1 \\ -2 & 0 & 3 \end{pmatrix}$, $\mathbf{B} = \begin{pmatrix} 2 & 1 & -1 \\ -2 & 3 & 3 \\ 1 & 0 & 2 \end{pmatrix}$, $\mathbf{C} = \begin{pmatrix} 2 & 1 & 0 \\ 1 & 2 & 2 \\ 0 & 1 & -1 \end{pmatrix}$,

verify that

(a) $(AB)C = A(BC)$,

(b) $(A + B) + C = A + (B + C)$,

(c) $A(B + C) = AB + AC$.

7. Prove each of the following laws of matrix algebra:

(a) $A + B = B + A$

(b) $A + (B + C) = (A + B) + C$

(c) $\alpha(A + B) = \alpha A + \alpha B$

(d) $(\alpha + \beta)A = \alpha A + \beta A$

(e) $A(BC) = (AB)C$

(f) $A(B + C) = AB + AC$

8. If $\mathbf{x} = \begin{pmatrix} 2 \\ 3i \\ 1 - i \end{pmatrix}$ and $\mathbf{y} = \begin{pmatrix} -1 + i \\ 2 \\ 3 - i \end{pmatrix}$, find

(a) $\mathbf{x}^T\mathbf{y}$, (b) $\mathbf{y}^T\mathbf{y}$, (c) $(\mathbf{x}, \mathbf{y})$, (d) $(\mathbf{y}, \mathbf{y})$.

9. If $\mathbf{x} = \begin{pmatrix} 1 - 2i \\ i \\ 2 \end{pmatrix}$ and $\mathbf{y} = \begin{pmatrix} 2 \\ 3 - i \\ 1 + 2i \end{pmatrix}$, show that

(a) $\mathbf{x}^T\mathbf{y} = \mathbf{y}^T\mathbf{x}$, (b) $(\mathbf{x}, \mathbf{y}) = \overline{(\mathbf{y}, \mathbf{x})}$.

In each of Problems 10 through 19 either compute the inverse of the given matrix, or else show that it is singular.

10. $\begin{pmatrix} 1 & 4 \\ -2 & 3 \end{pmatrix}$

11. $\begin{pmatrix} 3 & -1 \\ 6 & 2 \end{pmatrix}$

12. $\begin{pmatrix} 1 & 2 & 3 \\ 2 & 4 & 5 \\ 3 & 5 & 6 \end{pmatrix}$

13. $\begin{pmatrix} 1 & 1 & -1 \\ 2 & -1 & 1 \\ 1 & 1 & 2 \end{pmatrix}$

14. $\begin{pmatrix} 1 & 2 & 1 \\ -2 & 1 & 8 \\ 1 & -2 & -7 \end{pmatrix}$

15. $\begin{pmatrix} 2 & 1 & 0 \\ 0 & 2 & 1 \\ 0 & 0 & 2 \end{pmatrix}$

16. $\begin{pmatrix} 1 & -1 & -1 \\ 2 & 1 & 0 \\ 3 & -2 & 1 \end{pmatrix}$

17. $\begin{pmatrix} 2 & 3 & 1 \\ -1 & 2 & 1 \\ 4 & -1 & -1 \end{pmatrix}$

18. $\begin{pmatrix} 1 & 0 & 0 & -1 \\ 0 & -1 & 1 & 0 \\ -1 & 0 & 1 & 0 \\ 0 & 1 & -1 & 1 \end{pmatrix}$

19. $\begin{pmatrix} 1 & -1 & 2 & 0 \\ -1 & 2 & -4 & 2 \\ 1 & 0 & 1 & 3 \\ -2 & 2 & 0 & -1 \end{pmatrix}$

20. Prove that if A is nonsingular, then A^{-1} is uniquely determined; that is, show that there cannot be two different matrices B and C such that $AB = I$ and $AC = I$.

21. Prove that if **A** is nonsingular, then $\mathbf{AA}^{-1} = \mathbf{A}^{-1}\mathbf{A}$; that is, multiplication is commutative between any nonsingular matrix and its inverse.

22. If $\mathbf{A}(t) = \begin{pmatrix} e^t & 2e^{-t} & e^{2t} \\ 2e^t & e^{-t} & -e^{2t} \\ -e^t & 3e^{-t} & 2e^{2t} \end{pmatrix}$ and $\mathbf{B}(t) = \begin{pmatrix} 2e^t & e^{-t} & 3e^{2t} \\ -e^t & 2e^{-t} & e^{2t} \\ 3e^t & -e^{-t} & -e^{2t} \end{pmatrix}$,

find

(a) $\mathbf{A} + 3\mathbf{B}$,   (b) $\mathbf{AB}$,   (c) $d\mathbf{A}/dt$,   (d) $\int_0^1 \mathbf{A}(t)\, dt$.

In each of Problems 23 through 25 verify that the given vector satisfies the given differential equation.

23. $\mathbf{x}' = \begin{pmatrix} 3 & -2 \\ 2 & -2 \end{pmatrix}\mathbf{x}, \qquad \mathbf{x} = \begin{pmatrix} 4 \\ 2 \end{pmatrix}e^{2t}$

24. $\mathbf{x}' = \begin{pmatrix} 2 & -1 \\ 3 & -2 \end{pmatrix}\mathbf{x} + \begin{pmatrix} 1 \\ -1 \end{pmatrix}e^t, \qquad \mathbf{x} = \begin{pmatrix} 1 \\ 0 \end{pmatrix}e^t + 2\begin{pmatrix} 1 \\ 1 \end{pmatrix}te^t$

25. $\mathbf{x}' = \begin{pmatrix} 1 & 1 & 1 \\ 2 & 1 & -1 \\ 0 & -1 & 1 \end{pmatrix}\mathbf{x}, \qquad \mathbf{x} = \begin{pmatrix} 6 \\ -8 \\ -4 \end{pmatrix}e^{-t} + 2\begin{pmatrix} 0 \\ 1 \\ -1 \end{pmatrix}e^{2t}$

In each of Problems 26 and 27 verify that the given matrix satisfies the given differential equation.

26. $\mathbf{\Psi}' = \begin{pmatrix} 1 & 1 \\ 4 & -2 \end{pmatrix}\mathbf{\Psi}, \qquad \mathbf{\Psi}(t) = \begin{pmatrix} e^{-3t} & e^{2t} \\ -4e^{-3t} & e^{2t} \end{pmatrix}$

27. $\mathbf{\Psi}' = \begin{pmatrix} 1 & -1 & 4 \\ 3 & 2 & -1 \\ 2 & 1 & -1 \end{pmatrix}\mathbf{\Psi}, \qquad \mathbf{\Psi}(t) = \begin{pmatrix} e^t & e^{-2t} & e^{3t} \\ -4e^t & -e^{-2t} & 2e^{3t} \\ -e^t & -e^{-2t} & e^{3t} \end{pmatrix}$

7.4 SYSTEMS OF LINEAR ALGEBRAIC EQUATIONS; LINEAR INDEPENDENCE, EIGENVALUES, EIGENVECTORS

In this section we review some results from linear algebra that are important for the solution of systems of linear differential equations. Some of these results are easily proved and others are not; since we are interested simply in summarizing some useful information in compact form, we will give no indication of proofs in either case. All of the results in this section depend on some basic facts about the solution of systems of linear algebraic equations.

Systems of linear algebraic equations.   A set of n simultaneous linear algebraic equations in n variables

$$a_{11}x_1 + a_{12}x_2 + \cdots + a_{1n}x_n = b_1,$$

$$\cdot$$

$$\cdot \qquad (1)$$

$$\cdot$$

$$a_{n1}x_1 + a_{n2}x_2 + \cdots + a_{nn}x_n = b_n,$$

can be written as

$$\mathbf{Ax} = \mathbf{b}, \qquad (2)$$

where the $n \times n$ matrix $\mathbf{A}$ and the vector $\mathbf{b}$ are given, and the components of $\mathbf{x}$ are to be determined. If $\mathbf{b} = \mathbf{0}$, the system is said to be *homogeneous*; otherwise, it is *nonhomogeneous*. Two systems having precisely the same set of solutions are said to be *equivalent systems*.

If the coefficient matrix $\mathbf{A}$ is nonsingular, that is, if det $\mathbf{A}$ is not zero, then there is a unique solution of the system (2). Since $\mathbf{A}$ is nonsingular, $\mathbf{A}^{-1}$ exists, and the solution can be found by multiplying each side of Eq. (2) on the left by $\mathbf{A}^{-1}$; thus

$$\mathbf{x} = \mathbf{A}^{-1}\mathbf{b}. \qquad (3)$$

In particular, the homogeneous problem $\mathbf{Ax} = \mathbf{0}$, corresponding to $\mathbf{b} = \mathbf{0}$ in Eq. (2), has only the trivial solution $\mathbf{x} = \mathbf{0}$.

On the other hand, if $\mathbf{A}$ is singular, that is, if det $\mathbf{A}$ is zero, then solutions of Eq. (2) either do not exist, or do exist but are not unique. Since $\mathbf{A}$ is singular, $\mathbf{A}^{-1}$ does not exist, so Eq. (3) is no longer valid. The homogeneous system

$$\mathbf{Ax} = \mathbf{0} \qquad (4)$$

has (infinitely many) nonzero solutions in addition to the trivial solution. The situation for the nonhomogeneous system (2) is more complicated. This system has no solution unless the vector $\mathbf{b}$ satisfies a certain further condition that is by no means obviously necessary. This condition is that

$$(\mathbf{b}, \mathbf{y}) = 0, \qquad (5)$$

for all vectors $\mathbf{y}$ satisfying $\mathbf{A}^*\mathbf{y} = \mathbf{0}$, where $\mathbf{A}^*$ is the adjoint of $\mathbf{A}$. If condition (5) is met, then the system (2) has (infinitely many) solutions. Each of these solutions has the form

$$\mathbf{x} = \mathbf{x}^{(0)} + \boldsymbol{\xi}, \qquad (6)$$

where $\mathbf{x}^{(0)}$ is a particular solution of Eq. (2), and $\boldsymbol{\xi}$ is any solution of the homogeneous system (4). Note the resemblance between Eq. (6) and the solution of a nonhomogeneous linear differential equation, Eq. (4) of Section 3.6; $\mathbf{x}^{(0)}$ and $\boldsymbol{\xi}$ correspond, respectively, to the particular solution y_p and the complementary solution y_c. The proofs of some of the above statements are outlined in Problems 21 through 25.

While Eq. (3) is a compact formula for the solution of Eq. (2), at least when $\mathbf{A}$ is nonsingular, it is often easier to solve Eq. (2) by using elementary row operations to obtain a much simpler equivalent system from which the solution can be easily written down. In particular, it may be desirable to transform the coefficient matrix into triangular form; that is, to obtain a matrix whose elements below the main diagonal are all zero. The following example illustrates this procedure.

Example 1.   Solve the system of equations

$$\begin{aligned}
x_1 - 2x_2 + 3x_3 &= 7, \\
-x_1 + x_2 - 2x_3 &= -5, \\
2x_1 - x_2 - x_3 &= 4.
\end{aligned} \qquad (7)$$

It is convenient to write Eqs. (7) in the form

$$\begin{pmatrix} 1 & -2 & 3 \\ -1 & 1 & -2 \\ 2 & -1 & -1 \end{pmatrix} \begin{pmatrix} x_1 \\ x_2 \\ x_3 \end{pmatrix} = \begin{pmatrix} 7 \\ -5 \\ 4 \end{pmatrix}. \qquad (8)$$

The crucial parts of Eq. (8) are the elements of the coefficient matrix and of the vector on the right-hand side, in particular, the value and position of each element. Thus we can dispense with the x's and use the more compact notation

$$\left(\begin{array}{ccc|c} 1 & -2 & 3 & 7 \\ -1 & 1 & -2 & -5 \\ 2 & -1 & -1 & 4 \end{array} \right). \qquad (9)$$

The matrix (9) is called the augmented matrix for the system (7). The dashed line replaces the equals sign and is said to partition the matrix. We now perform elementary row operations on the matrix (9) with a view to introducing zeros in the lower left part of the matrix. Observe that elementary row operations on the matrix (9) correspond to legitimate operations on the equations in the system (7). Each step is described below, and the result is recorded at the right.

(a) Add the first row to the second row and add (-2) times the first row to the third row.

$$\left(\begin{array}{ccc|c} 1 & -2 & 3 & 7 \\ 0 & -1 & 1 & 2 \\ 0 & 3 & -7 & -10 \end{array} \right)$$

(b) Multiply the second row by -1.

$$\left(\begin{array}{ccc|c} 1 & -2 & 3 & 7 \\ 0 & 1 & -1 & -2 \\ 0 & 3 & -7 & -10 \end{array} \right)$$

(c) Add (-3) times the second row to the third row.

$$\begin{pmatrix} 1 & -2 & 3 & | & 7 \\ 0 & 1 & -1 & | & -2 \\ 0 & 0 & -4 & | & -4 \end{pmatrix}$$

(d) Divide the third row by -4.

$$\begin{pmatrix} 1 & -2 & 3 & | & 7 \\ 0 & 1 & -1 & | & -2 \\ 0 & 0 & 1 & | & 1 \end{pmatrix}$$

The matrix obtained in this manner corresponds to the system of equations

$$\begin{align} x_1 - 2x_2 + 3x_3 &= 7, \\ x_2 - x_3 &= -2, \\ x_3 &= 1, \end{align} \tag{10}$$

which is equivalent to the original system (7). Note that the coefficients in Eqs. (10) form a triangular matrix. From the last of Eqs. (10) we have $x_3 = 1$, from the second equation $x_2 = -2 + x_3 = -1$, and from the first equation $x_1 = 7 + 2x_2 - 3x_3 = 2$. Thus we obtain

$$\mathbf{x} = \begin{pmatrix} 2 \\ -1 \\ 1 \end{pmatrix},$$

which is the solution of the given system (7) or (8).

This procedure is also useful in solving homogeneous systems, and systems in which the number of equations is different from the number of unknowns.

Linear Independence. A set of k vectors $\mathbf{x}^{(1)}, \ldots, \mathbf{x}^{(k)}$ is said to be *linearly dependent* if there exists a set of (complex) numbers $c_1, \ldots, c_k$, at least one of which is nonzero, such that

$$c_1 \mathbf{x}^{(1)} + \cdots + c_k \mathbf{x}^{(k)} = \mathbf{0}. \tag{11}$$

In other words, $\mathbf{x}^{(1)}, \ldots, \mathbf{x}^{(k)}$ are linearly dependent if there is a linear relation among them. On the other hand, if the only set $c_1, \ldots, c_k$ for which Eq. (11) is satisfied is $c_1 = c_2 = \cdots = c_k = 0$, then $\mathbf{x}^{(1)}, \ldots, \mathbf{x}^{(k)}$ are said to be *linearly independent*.

Consider now a set of n vectors, each of which has n components. Let $x_{ij} = x_i^{(j)}$ be the ith component of the vector $\mathbf{x}^{(j)}$, and let $\mathbf{X} = (x_{ij})$. Then Eq. (11) can be written as

$$\begin{pmatrix} x_1^{(1)}c_1 + \cdots + x_1^{(n)}c_n \\ \cdot \\ \cdot \\ \cdot \\ x_n^{(1)}c_1 + \cdots + x_n^{(n)}c_n \end{pmatrix} = \begin{pmatrix} x_{11}c_1 + \cdots + x_{1n}c_n \\ \cdot \\ \cdot \\ \cdot \\ x_{n1}c_1 + \cdots + x_{nn}c_n \end{pmatrix} = \mathbf{Xc} = \mathbf{0}. \tag{12}$$

If det $\mathbf{X} \neq 0$, then the only solution of Eq. (12) is $\mathbf{c} = \mathbf{0}$, but if det $\mathbf{X} = 0$, there are nonzero solutions. Thus the set of vectors $\mathbf{x}^{(1)}, \ldots, \mathbf{x}^{(n)}$ is linearly independent if and only if det $\mathbf{X} \neq 0$.

Example 2. Determine whether the vectors

$$\mathbf{x}^{(1)} = \begin{pmatrix} 1 \\ 2 \\ -1 \end{pmatrix}, \quad \mathbf{x}^{(2)} = \begin{pmatrix} 2 \\ 1 \\ 3 \end{pmatrix}, \quad \mathbf{x}^{(3)} = \begin{pmatrix} -4 \\ 1 \\ -11 \end{pmatrix} \tag{13}$$

are linearly independent or linearly dependent. If linearly dependent, find a linear relation among them.

To determine whether $\mathbf{x}^{(1)}$, $\mathbf{x}^{(2)}$, and $\mathbf{x}^{(3)}$ are linearly dependent we compute det (x_{ij}), whose columns are the components of $\mathbf{x}^{(1)}$, $\mathbf{x}^{(2)}$, and $\mathbf{x}^{(3)}$, respectively. Thus

$$\det (x_{ij}) = \begin{vmatrix} 1 & 2 & -4 \\ 2 & 1 & 1 \\ -1 & 3 & -11 \end{vmatrix},$$

and an elementary calculation shows that it is zero. Thus $\mathbf{x}^{(1)}$, $\mathbf{x}^{(2)}$, and $\mathbf{x}^{(3)}$ are linearly dependent, and there are constants c_1, c_2, and c_3 such that

$$c_1\mathbf{x}^{(1)} + c_2\mathbf{x}^{(2)} + c_3\mathbf{x}^{(3)} = \mathbf{0}. \tag{14}$$

Equation (14) can also be written in the form

$$\begin{pmatrix} 1 & 2 & -4 \\ 2 & 1 & 1 \\ -1 & 3 & -11 \end{pmatrix} \begin{pmatrix} c_1 \\ c_2 \\ c_3 \end{pmatrix} = \begin{pmatrix} 0 \\ 0 \\ 0 \end{pmatrix}, \tag{15}$$

and solved by means of elementary row operations starting from the matrix

$$\begin{pmatrix} 1 & 2 & -4 & \vert & 0 \\ 2 & 1 & 1 & \vert & 0 \\ -1 & 3 & -11 & \vert & 0 \end{pmatrix}. \tag{16}$$

We proceed as in Example 1.

(a) Add (-2) times the first row to the second row, and add the first row to the third row.

$$\begin{pmatrix} 1 & 2 & -4 & \vert & 0 \\ 0 & -3 & 9 & \vert & 0 \\ 0 & 5 & -15 & \vert & 0 \end{pmatrix}$$

(b) Divide the second row by -3; then add (-5) times the second row to the third row.

$$\begin{pmatrix} 1 & 2 & -4 & \vert & 0 \\ 0 & 1 & -3 & \vert & 0 \\ 0 & 0 & 0 & \vert & 0 \end{pmatrix}$$

Thus we obtain the equivalent system

$$c_1 + 2c_2 - 4c_3 = 0,$$
$$c_2 - 3c_3 = 0. \tag{17}$$

From the second of Eqs. (17) we have $c_2 = 3c_3$, and from the first we obtain $c_1 = 4c_3 - 2c_2 = -2c_3$. Thus we have solved for c_1 and c_2 in terms of c_3, with the latter remaining arbitrary. If we choose $c_3 = -1$ for convenience, then $c_1 = 2$ and $c_2 = -3$. In this case the desired relation (14) becomes

$$2\mathbf{x}^{(1)} - 3\mathbf{x}^{(2)} - \mathbf{x}^{(3)} = \mathbf{0}.$$

Frequently it is useful to think of the columns (or rows) of a matrix $\mathbf{A}$ as vectors. These column (or row) vectors are linearly independent if and only if $\det \mathbf{A} \neq 0$. Further, if $\mathbf{C} = \mathbf{AB}$, then it can be shown that $\det \mathbf{C} = (\det \mathbf{A})(\det \mathbf{B})$. Therefore, if the columns (or rows) of both $\mathbf{A}$ and $\mathbf{B}$ are linearly independent, then the columns (or rows) of $\mathbf{C}$ are also linearly independent.

Now let us extend the concepts of linear dependence and independence to a set of vector functions $\mathbf{x}^{(1)}(t), \ldots, \mathbf{x}^{(k)}(t)$ defined on an interval $\alpha < t < \beta$. The vectors $\mathbf{x}^{(1)}(t), \ldots, \mathbf{x}^{(k)}(t)$ are said to be linearly dependent on $\alpha < t < \beta$ if there exists a set of constants $c_1, \ldots, c_k$, not all of which are zero, such that $c_1\mathbf{x}^{(1)}(t) + \cdots + c_k\mathbf{x}^{(k)}(t) = \mathbf{0}$ for all t in the interval. Otherwise, $\mathbf{x}^{(1)}(t), \cdots, \mathbf{x}^{(k)}(t)$ are said to be linearly independent. Note that if $\mathbf{x}^{(1)}(t), \ldots, \mathbf{x}^{(k)}(t)$ are linearly dependent on an interval, they are linearly dependent at each point in the interval. However, if $\mathbf{x}^{(1)}(t), \ldots, \mathbf{x}^{(k)}(t)$ are linearly independent on an interval, they may or may not be linearly independent at each point; they may, in fact, be linearly dependent at each point, but with different sets of constants at different points. See Problem 9 for an example.

Eigenvalues and Eigenvectors.   The equation

$$\mathbf{Ax} = \mathbf{y} \tag{18}$$

can be viewed as a linear transformation that maps (or transforms) a given vector $\mathbf{x}$ into a new vector $\mathbf{y}$. Vectors that are transformed into multiples of themselves play an important role in many applications.* To find such vectors we set $\mathbf{y} = \lambda\mathbf{x}$, where λ is a scalar proportionality factor, and seek solutions of the equations

$$\mathbf{Ax} = \lambda\mathbf{x}, \tag{19}$$

or

$$(\mathbf{A} - \lambda\mathbf{I})\mathbf{x} = \mathbf{0}. \tag{20}$$

The latter equation has nonzero solutions if and only if λ is chosen so that

$$\Delta(\lambda) = \det(\mathbf{A} - \lambda\mathbf{I}) = 0. \tag{21}$$

* For example, this problem is encountered in finding the principal axes of stress or strain in an elastic body, and in finding the modes of free vibration in a conservative system with a finite number of degrees of freedom.

Values of λ that satisfy Eq. (21) are called *eigenvalues* of the matrix $\mathbf{A}$, and the solutions of Eq. (19) or (20) that are obtained by using such a value of λ are called the *eigenvectors* corresponding to that eigenvalue. Eigenvectors are determined only up to an arbitrary multiplicative constant; if this constant is specified in some way, then the eigenvectors are said to be *normalized*. Frequently it is convenient to normalize an eigenvector $\mathbf{x}$ by requiring that $(\mathbf{x}, \mathbf{x}) = 1$. Alternatively, we may wish to set one of the components equal to one.

Since Eq. (21) is a polynomial equation of degree n in λ, there are n eigenvalues $\lambda_1, \ldots, \lambda_n$, some of which may be repeated. If a given eigenvalue appears m times as a root of Eq. (21), then that eigenvalue is said to have *multiplicity m*. Each eigenvalue has at least one associated eigenvector, and an eigenvalue of multiplicity m may have q linearly independent eigenvectors, where

$$1 \leq q \leq m. \tag{22}$$

That q can be less than m is illustrated by Example 3 below. If all of the eigenvalues of a matrix $\mathbf{A}$ are *simple* (have multiplicity one), then it is possible to show that the n eigenvectors of $\mathbf{A}$, one for each eigenvalue, are linearly independent. On the other hand, if $\mathbf{A}$ has one or more repeated eigenvalues, then there may be fewer than n linearly independent eigenvectors associated with $\mathbf{A}$, since for a repeated eigenvalue we may have $q < m$. This fact leads to complications later on in the solution of systems of differential equations.

Example 3.   Find the eigenvalues and eigenvectors of the matrix

$$\mathbf{A} = \begin{pmatrix} 1 & -1 \\ 1 & 3 \end{pmatrix}. \tag{23}$$

The eigenvalues λ and eigenvectors $\mathbf{x}$ satisfy the equation $(\mathbf{A} - \lambda\mathbf{I})\mathbf{x} = \mathbf{0}$, or

$$\begin{pmatrix} 1 - \lambda & -1 \\ 1 & 3 - \lambda \end{pmatrix} \begin{pmatrix} x_1 \\ x_2 \end{pmatrix} = \begin{pmatrix} 0 \\ 0 \end{pmatrix}. \tag{24}$$

The eigenvalues are the roots of the equation

$$\det (\mathbf{A} - \lambda\mathbf{I}) = \begin{vmatrix} 1 - \lambda & -1 \\ 1 & 3 - \lambda \end{vmatrix}$$

$$= \lambda^2 - 4\lambda + 4 = 0. \tag{25}$$

Thus the two eigenvalues are $\lambda_1 = 2$ and $\lambda_2 = 2$; that is, the eigenvalue 2 has multiplicity two.

To determine the eigenvectors we must return to Eq. (24) and use for λ the value 2. This gives

$$\begin{pmatrix} -1 & -1 \\ 1 & 1 \end{pmatrix} \begin{pmatrix} x_1 \\ x_2 \end{pmatrix} = \begin{pmatrix} 0 \\ 0 \end{pmatrix}. \tag{26}$$

Hence we obtain the single condition $x_1 + x_2 = 0$, which determines x_2 in terms of x_1, or vice versa. If $x_1 = c$, then $x_2 = -c$ and the eigenvector $\mathbf{x}^{(1)}$ is

$$\mathbf{x}^{(1)} = c\begin{pmatrix} 1 \\ -1 \end{pmatrix}. \tag{27}$$

Usually we will drop the arbitrary constant c when finding eigenvectors; thus instead of Eq. (27) we write

$$\mathbf{x}^{(1)} = \begin{pmatrix} 1 \\ -1 \end{pmatrix}, \tag{28}$$

and remember that any multiple of this vector is also an eigenvector. Observe that there is only one linearly independent eigenvector associated with the double eigenvalue.

Example 4.   Find the eigenvalues and eigenvectors of the matrix

$$\mathbf{A} = \begin{pmatrix} 0 & 1 & 1 \\ 1 & 0 & 1 \\ 1 & 1 & 0 \end{pmatrix}. \tag{29}$$

The eigenvalues λ and eigenvectors $\mathbf{x}$ satisfy the equation $(\mathbf{A} - \lambda\mathbf{I})\mathbf{x} = \mathbf{0}$, or

$$\begin{pmatrix} -\lambda & 1 & 1 \\ 1 & -\lambda & 1 \\ 1 & 1 & -\lambda \end{pmatrix}\begin{pmatrix} x_1 \\ x_2 \\ x_3 \end{pmatrix} = \begin{pmatrix} 0 \\ 0 \\ 0 \end{pmatrix}. \tag{30}$$

The eigenvalues are the roots of the equation

$$\det(\mathbf{A} - \lambda\mathbf{I}) = \begin{vmatrix} -\lambda & 1 & 1 \\ 1 & -\lambda & 1 \\ 1 & 1 & -\lambda \end{vmatrix}$$

$$= -\lambda^3 + 3\lambda + 2 = 0. \tag{31}$$

Solving Eq. (31), perhaps by trial and error, we obtain three roots, namely, $\lambda_1 = 2$, $\lambda_2 = -1$, and $\lambda_3 = -1$. Thus 2 is a simple eigenvalue, and -1 is an eigenvalue of multiplicity two.

To find the eigenvector $\mathbf{x}^{(1)}$ corresponding to the eigenvalue λ_1 we substitute $\lambda = 2$ in Eq. (30); this gives the system

$$\begin{pmatrix} -2 & 1 & 1 \\ 1 & -2 & 1 \\ 1 & 1 & -2 \end{pmatrix}\begin{pmatrix} x_1 \\ x_2 \\ x_3 \end{pmatrix} = \begin{pmatrix} 0 \\ 0 \\ 0 \end{pmatrix}. \tag{32}$$

We can reduce this to the equivalent system

$$
\begin{pmatrix} 2 & -1 & -1 \\ 0 & 1 & -1 \\ 0 & 0 & 0 \end{pmatrix} \begin{pmatrix} x_1 \\ x_2 \\ x_3 \end{pmatrix} = \begin{pmatrix} 0 \\ 0 \\ 0 \end{pmatrix}
\tag{33}
$$

by elementary row operations. Solving this system we obtain the eigenvector

$$
\mathbf{x}^{(1)} = \begin{pmatrix} 1 \\ 1 \\ 1 \end{pmatrix}.
\tag{34}
$$

For $\lambda = -1$, Eqs. (30) reduce immediately to the single equation

$$
x_1 + x_2 + x_3 = 0.
\tag{35}
$$

Thus values for two of the quantities x_1, x_2, x_3 can be chosen arbitrarily and the third is determined from Eq. (35). For example, if $x_1 = 1$ and $x_2 = 0$, then $x_3 = -1$ and

$$
\mathbf{x}^{(2)} = \begin{pmatrix} 1 \\ 0 \\ -1 \end{pmatrix}
\tag{36}
$$

is an eigenvector. Any multiple of $\mathbf{x}^{(2)}$ is also an eigenvector, but a second independent eigenvector can be found by making another choice of x_1 and x_2; for instance, $x_1 = 0$ and $x_2 = 1$. Again $x_3 = -1$ and

$$
\mathbf{x}^{(3)} = \begin{pmatrix} 0 \\ 1 \\ -1 \end{pmatrix}
\tag{37}
$$

is an eigenvector linearly independent of $\mathbf{x}^{(2)}$. Therefore in this example two linearly independent eigenvectors are associated with the double eigenvalue.

An important special class of matrices, called *self-adjoint* or *Hermitian* matrices, are those for which $\mathbf{A}^* = \mathbf{A}$; that is, $\bar{a}_{ji} = a_{ij}$. Hermitian matrices include as a subclass real symmetric matrices; that is, matrices that have real elements and for which $\mathbf{A}^T = \mathbf{A}$. The eigenvalues and eigenvectors of Hermitian matrices always have the following useful properties.

1. All eigenvalues are real.

2. There always exists a full set of n linearly independent eigenvectors, regardless of the multiplicities of the eigenvalues.

3. If $\mathbf{x}^{(1)}$ and $\mathbf{x}^{(2)}$ are eigenvectors that correspond to different eigenvalues,

then $(\mathbf{x}^{(1)}, \mathbf{x}^{(2)}) = 0$. Thus, if all eigenvalues are simple, then the associated eigenvectors form an orthogonal set of vectors.

4. Corresponding to an eigenvalue of multiplicity m, it is possible to choose m eigenvectors that are mutually orthogonal. Thus the full set of n eigenvectors can always be chosen to be orthogonal as well as linearly independent.

Example 4 above involves a real symmetric matrix and illustrates properties 1, 2, and 3, but the choice we have made for $\mathbf{x}^{(2)}$ and $\mathbf{x}^{(3)}$ does not illustrate property 4. However, it is always possible to choose an $\mathbf{x}^{(2)}$ and an $\mathbf{x}^{(3)}$ so that $(\mathbf{x}^{(2)}, \mathbf{x}^{(3)}) = 0$. For example, we could have chosen

$$\mathbf{x}^{(2)} = \begin{pmatrix} 1 \\ 0 \\ -1 \end{pmatrix}, \qquad \mathbf{x}^{(3)} = \begin{pmatrix} 1 \\ -2 \\ 1 \end{pmatrix}$$

as the eigenvectors associated with the eigenvalue $\lambda = -1$ in Example 4. Example 3 shows that if the matrix is not Hermitian, then there may be a deficiency of eigenvectors when an eigenvalue is repeated. The proofs of statements 1 and 3 are outlined in Problems 27 through 29.

In the solution of systems of algebraic equations, as well as in other situations, it is sometimes useful to transform a given matrix into a diagonal matrix, that is, one having nonzero elements only on the diagonal. Eigenvectors are useful in accomplishing such a transformation. Suppose that $\mathbf{A}$ has a full set of n linearly independent eigenvectors (whether $\mathbf{A}$ is Hermitian or not). Letting $\mathbf{x}^{(1)}, \ldots, \mathbf{x}^{(n)}$ denote these eigenvectors and $\lambda_1, \ldots, \lambda_n$ the corresponding eigenvalues, form the matrix $\mathbf{T}$ whose columns are the eigenvectors $\mathbf{x}^{(1)}, \ldots, \mathbf{x}^{(n)}$. Since the columns of $\mathbf{T}$ are linearly independent vectors, $\det \mathbf{T} \neq 0$; hence $\mathbf{T}$ is nonsingular and $\mathbf{T}^{-1}$ exists. A straightforward calculation shows that the columns of the matrix $\mathbf{AT}$ are just the vectors $\mathbf{Ax}^{(1)}, \ldots, \mathbf{Ax}^{(n)}$. Since $\mathbf{Ax}^{(k)} = \lambda_k \mathbf{x}^{(k)}$, it follows that

$$\mathbf{AT} = \begin{pmatrix} \lambda_1 x_1^{(1)} & \cdots & \lambda_n x_1^{(n)} \\ & \cdot & \\ \cdot & & \cdot \\ & \cdot & \\ \lambda_1 x_n^{(1)} & \cdots & \lambda_n x_n^{(n)} \end{pmatrix}$$

$$= \mathbf{TD}, \tag{38}$$

where

$$\mathbf{D} = \begin{pmatrix} \lambda_1 & 0 & \cdots & 0 \\ 0 & \lambda_2 & \cdots & 0 \\ \cdot & \cdot & & \cdot \\ \cdot & \cdot & & \cdot \\ \cdot & \cdot & & \cdot \\ 0 & 0 & \cdots & \lambda_n \end{pmatrix} \tag{39}$$

is a diagonal matrix whose diagonal elements are the eigenvalues of $\mathbf{A}$. From Eq. (38) it follows that

$$\mathbf{T}^{-1}\mathbf{A}\mathbf{T} = \mathbf{D}. \tag{40}$$

Thus, if the eigenvalues and eigenvectors of $\mathbf{A}$ are known, $\mathbf{A}$ can be transformed into a diagonal matrix by the process shown in Eq. (40). This process is known as a *similarity transformation*, and Eq. (40) is summed up in words by saying that $\mathbf{A}$ is *similar* to the diagonal matrix $\mathbf{D}$. Alternatively, we may say that $\mathbf{A}$ is *diagonalizable*. The possibility of carrying out such a diagonalization will be important later in the chapter.

If $\mathbf{A}$ is Hermitian, then the determination of $\mathbf{T}^{-1}$ is very simple. We choose the eigenvectors $\mathbf{x}^{(1)}, \ldots, \mathbf{x}^{(n)}$ of $\mathbf{A}$ so that they are normalized by $(\mathbf{x}^{(i)}, \mathbf{x}^{(i)}) = 1$ for each i, as well as orthogonal. Then it is easy to verify that $\mathbf{T}^{-1} = \mathbf{T}^*$; in other words, the inverse of $\mathbf{T}$ is the same as its adjoint (the transpose of its complex conjugate).

Finally, we note that if $\mathbf{A}$ has fewer than n linearly independent eigenvectors, then there is no matrix $\mathbf{T}$ such that $\mathbf{T}^{-1}\mathbf{A}\mathbf{T} = \mathbf{D}$. In this case, $\mathbf{A}$ is not similar to a diagonal matrix, or is not diagonalizable.

PROBLEMS

In each of Problems 1 through 5 either solve the given set of equations, or else show that there is no solution.

1.
$$\begin{aligned} x_1 \quad\quad - x_3 &= 0 \\ 3x_1 + x_2 + x_3 &= 1 \\ -x_1 + x_2 + 2x_3 &= 2 \end{aligned}$$

2.
$$\begin{aligned} x_1 + 2x_2 - x_3 &= 1 \\ 2x_1 + x_2 + x_3 &= 1 \\ x_1 - x_2 + 2x_3 &= 1 \end{aligned}$$

3.
$$\begin{aligned} x_1 + 2x_2 - x_3 &= 2 \\ 2x_1 + x_2 + x_3 &= 1 \\ x_1 - x_2 + 2x_3 &= -1 \end{aligned}$$

4.
$$\begin{aligned} x_1 + 2x_2 - x_3 &= 0 \\ 2x_1 + x_2 + x_3 &= 0 \\ x_1 - x_2 + 2x_3 &= 0 \end{aligned}$$

5.
$$\begin{aligned} x_1 \quad\quad - x_3 &= 0 \\ 3x_1 + x_2 + x_3 &= 0 \\ -x_1 + x_2 + 2x_3 &= 0 \end{aligned}$$

6. Determine whether each of the following sets of vectors is linearly independent. If linearly dependent, find the linear relation among them. The vectors are written as row vectors to save space, but may be considered as column vectors; that is, the transposes of the given vectors may be used instead of the vectors themselves.

(a) $\mathbf{x}^{(1)} = (1, 1, 0)$,　$\mathbf{x}^{(2)} = (0, 1, 1)$,　$\mathbf{x}^{(3)} = (1, 0, 1)$

(b) $\mathbf{x}^{(1)} = (2, 1, 0)$,　$\mathbf{x}^{(2)} = (0, 1, 0)$,　$\mathbf{x}^{(3)} = (-1, 2, 0)$

(c) $\mathbf{x}^{(1)} = (1, 2, 2, 3)$,　$\mathbf{x}^{(2)} = (-1, 0, 3, 1)$,　$\mathbf{x}^{(3)} = (-2, -1, 1, 0)$
　　$\mathbf{x}^{(4)} = (-3, 0, -1, 3)$

(d) $\mathbf{x}^{(1)} = (1, 2, -1, 0)$,　$\mathbf{x}^{(2)} = (2, 3, 1, -1)$,　$\mathbf{x}^{(3)} = (-1, 0, 2, 2)$
　　$\mathbf{x}^{(4)} = (3, -1, 1, 3)$

(e) $\mathbf{x}^{(1)} = (1, 2, -2)$,　$\mathbf{x}^{(2)} = (3, 1, 0)$,　$\mathbf{x}^{(3)} = (2, -1, 1)$,　$\mathbf{x}^{(4)} = (4, 3, -2)$

7. Suppose that the vectors $x^{(1)}, \ldots, x^{(m)}$ each have n components where $n < m$. Show that $x^{(1)}, \ldots, x^{(m)}$ are linearly dependent.

8. Determine whether each of the following sets of vectors is linearly independent for $-\infty < t < \infty$. If linearly dependent, find the linear relation among them. As in Problem 6 the vectors are written as row vectors to save space.

(a) $x^{(1)}(t) = (e^{-t}, 2e^{-t})$, $x^{(2)}(t) = (e^{-t}, e^{-t})$, $x^{(3)}(t) = (3e^{-t}, 0)$

(b) $x^{(1)}(t) = (2 \sin t, \sin t)$, $x^{(2)}(t) = (\sin t, 2 \sin t)$

9. Let

$$x^{(1)}(t) = \begin{pmatrix} e^t \\ te^t \end{pmatrix}, \qquad x^{(2)}(t) = \begin{pmatrix} 1 \\ t \end{pmatrix}.$$

Show that $x^{(1)}(t)$ and $x^{(2)}(t)$ are linearly dependent at each point in the interval $0 \leq t \leq 1$. Nevertheless, show that $x^{(1)}(t)$ and $x^{(2)}(t)$ are linearly independent on $0 \leq t \leq 1$.

In each of Problems 10 through 19 find all eigenvalues and eigenvectors of the given matrix.

10. $\begin{pmatrix} 5 & -1 \\ 3 & 1 \end{pmatrix}$

11. $\begin{pmatrix} 3 & -2 \\ 4 & -1 \end{pmatrix}$

12. $\begin{pmatrix} -2 & 1 \\ 1 & -2 \end{pmatrix}$

13. $\begin{pmatrix} 1 & i \\ -i & 1 \end{pmatrix}$

14. $\begin{pmatrix} 1 & \sqrt{3} \\ \sqrt{3} & -1 \end{pmatrix}$

15. $\begin{pmatrix} 1 & -4 \\ 4 & -7 \end{pmatrix}$

16. $\begin{pmatrix} 1 & 0 & 0 \\ 2 & 1 & -2 \\ 3 & 2 & 1 \end{pmatrix}$

17. $\begin{pmatrix} 3 & 2 & 2 \\ 1 & 4 & 1 \\ -2 & -4 & -1 \end{pmatrix}$

18. $\begin{pmatrix} 1 & 1 & 1 \\ 2 & 1 & -1 \\ -3 & 2 & 4 \end{pmatrix}$

19. $\begin{pmatrix} 3 & 2 & 4 \\ 2 & 0 & 2 \\ 4 & 2 & 3 \end{pmatrix}$

20. For each given matrix A, find T such that $T^{-1}AT = D$ where D is a diagonal matrix.

(a) A is the matrix of Problem 10.

(b) A is the matrix of Problem 11.

(c) A is the matrix of Problem 14.

Problems 21 through 25 deal with the problem of solving $Ax = b$ when det $A = 0$.

21. Suppose that, for a given matrix A, there is a nonzero vector x such that $Ax = 0$. Show that there is also a nonzero vector y such that $A^*y = 0$.

22. Show that $(Ax, y) = (x, A^*y)$ for any vectors x and y.

23. Suppose that det $A = 0$ but that $Ax = b$ has solutions. Show that $(b, y) = 0$, where y is any solution of $A^*y = 0$. Verify that this statement is true for the set of equations in Problem 3.

Hint: Use the result of Problem 22.

24. Suppose that det $A = 0$ but that $x = x^{(0)}$ is a solution of $Ax = b$. Show that if ξ is a solution of $A\xi = 0$ and α is any constant, then $x = x^{(0)} + \alpha\xi$ is also a solution of $Ax = b$.

*25. Suppose that det $\mathbf{A} = 0$, and that $\mathbf{y}$ is a solution of $\mathbf{A}^*\mathbf{y} = \mathbf{0}$. Show that if $(\mathbf{b}, \mathbf{y}) = 0$ then $\mathbf{A}\mathbf{x} = \mathbf{b}$ has solutions. Note that this is the converse of Problem 23; the form of the solution is given by Problem 24.

26. Prove that $\lambda = 0$ is an eigenvalue of $\mathbf{A}$ if and only if $\mathbf{A}$ is singular.

27. Prove that if $\mathbf{A}$ is Hermitian, then $(\mathbf{A}\mathbf{x}, \mathbf{y}) = (\mathbf{x}, \mathbf{A}\mathbf{y})$, where $\mathbf{x}$ and $\mathbf{y}$ are any vectors.

28. In this problem we show that the eigenvalues of a Hermitian matrix $\mathbf{A}$ are real. Let $\mathbf{x}$ be an eigenvector corresponding to the eigenvalue λ.

(a) Show that $(\mathbf{A}\mathbf{x}, \mathbf{x}) = (\mathbf{x}, \mathbf{A}\mathbf{x})$. *Hint:* See Problem 27.

(b) Show that $\lambda(\mathbf{x}, \mathbf{x}) = \bar{\lambda}(\mathbf{x}, \mathbf{x})$. *Hint:* Recall that $\mathbf{A}\mathbf{x} = \lambda\mathbf{x}$.

(c) Show that $\lambda = \bar{\lambda}$; that is, the eigenvalue λ is real.

29. Show that if λ_1 and λ_2 are eigenvalues of a Hermitian matrix $\mathbf{A}$, and if $\lambda_1 \neq \lambda_2$, then the corresponding eigenvectors $\mathbf{x}^{(1)}$ and $\mathbf{x}^{(2)}$ are orthogonal.
Hint: Use the results of Problems 27 and 28 to show that $(\lambda_1 - \lambda_2)(\mathbf{x}^{(1)}, \mathbf{x}^{(2)}) = 0$.

7.5   BASIC THEORY OF SYSTEMS OF FIRST ORDER LINEAR EQUATIONS

The general theory of a system of n first order linear equations

$$x_1' = p_{11}(t)x_1 + \cdots + p_{1n}(t)x_n + g_1(t),$$

$$\vdots \tag{1}$$

$$x_n' = p_{n1}(t)x_1 + \cdots + p_{nn}(t)x_n + g_n(t),$$

closely parallels that of a single linear equation of nth order. The discussion in this section therefore follows the same general lines as that in Sections 3.2, 3.3, and 5.2. In order to discuss the system (1) most effectively, we will write it in matrix notation. That is, we will consider $x_1 = \phi_1(t), \ldots, x_n = \phi_n(t)$ to be components of a vector $\mathbf{x} = \boldsymbol{\phi}(t)$; similarly $g_1(t), \ldots, g_n(t)$ are components of a vector $\mathbf{g}(t)$, and $p_{11}(t), \ldots, p_{nn}(t)$ are elements of an $n \times n$ matrix $\mathbf{P}(t)$. Equation (1) then takes the form

$$\mathbf{x}' = \mathbf{P}(t)\mathbf{x} + \mathbf{g}(t). \tag{2}$$

The use of vectors and matrices not only saves a great deal of space and facilitates calculations but also emphasizes the similarity between systems of equations and single (scalar) equations.

A vector $\mathbf{x} = \boldsymbol{\phi}(t)$ is said to be a solution of Eq. (2) if its components satisfy the system of equations (1). Throughout this section we will assume that $\mathbf{P}$ and $\mathbf{g}$ are continuous on some interval $\alpha < t < \beta$; that is, each of the scalar functions $p_{11}, \ldots, p_{nn}, g_1, \ldots, g_n$ is continuous there. According to Theorem 7.2, this is sufficient to guarantee the existence of solutions of Eq. (2) on the interval $\alpha < t < \beta$.

It is convenient to consider first the homogeneous equation

$$\mathbf{x}' = \mathbf{P}(t)\mathbf{x} \tag{3}$$

obtained from Eq. (2) by setting $\mathbf{g}(t) = \mathbf{0}$. Once the homogeneous equation has been solved, the method of variation of parameters can be used to solve the nonhomogeneous equation (2); this is taken up in Section 7.10. We will use the notation

$$\mathbf{x}^{(1)}(t) = \begin{pmatrix} x_{11}(t) \\ x_{21}(t) \\ \cdot \\ \cdot \\ \cdot \\ x_{n1}(t) \end{pmatrix}, \ldots, \mathbf{x}^{(k)}(t) = \begin{pmatrix} x_{1k}(t) \\ x_{2k}(t) \\ \cdot \\ \cdot \\ \cdot \\ x_{nk}(t) \end{pmatrix}, \ldots \tag{4}$$

to designate specific solutions of the system (3). Note that $x_{ij}(t) = x_i^{(j)}(t)$ refers to the ith component of the jth solution $\mathbf{x}^{(j)}(t)$. The main facts about the structure of solutions of the system (3) are stated in Theorems 7.3 to 7.6. They closely resemble the corresponding theorems in Sections 3.2, 3.3, and 5.2; some of the proofs are left to the reader as exercises.

Theorem 7.3. *If the vector functions* $\mathbf{x}^{(1)}$ *and* $\mathbf{x}^{(2)}$ *are solutions of the system* (3), *then the linear combination* $c_1\mathbf{x}^{(1)} + c_2\mathbf{x}^{(2)}$ *is also a solution for any constants* c_1 *and* c_2.

This is the principle of superposition; it is proved simply by differentiating $c_1\mathbf{x}^{(1)} + c_2\mathbf{x}^{(2)}$ and using the fact that $\mathbf{x}^{(1)}$ and $\mathbf{x}^{(2)}$ satisfy Eq. (3). By repeated application of Theorem 7.3 we reach the conclusion that if $\mathbf{x}^{(1)}, \ldots, \mathbf{x}^{(k)}$ are solutions of Eq. (3) then

$$\mathbf{x} = c_1\mathbf{x}^{(1)}(t) + \cdots + c_k\mathbf{x}^{(k)}(t) \tag{5}$$

is also a solution for any constants $c_1, \ldots, c_k$. As an example, it can be verified that

$$\mathbf{x}^{(1)}(t) = \begin{pmatrix} e^{3t} \\ 2e^{3t} \end{pmatrix} = \begin{pmatrix} 1 \\ 2 \end{pmatrix} e^{3t}, \qquad \mathbf{x}^{(2)}(t) = \begin{pmatrix} e^{-t} \\ -2e^{-t} \end{pmatrix} = \begin{pmatrix} 1 \\ -2 \end{pmatrix} e^{-t} \tag{6}$$

satisfy the equation

$$\mathbf{x}' = \begin{pmatrix} 1 & 1 \\ 4 & 1 \end{pmatrix} \mathbf{x}. \tag{7}$$

According to Theorem 7.3

$$\mathbf{x} = c_1 \begin{pmatrix} 1 \\ 2 \end{pmatrix} e^{3t} + c_2 \begin{pmatrix} 1 \\ -2 \end{pmatrix} e^{-t}$$

$$= c_1\mathbf{x}^{(1)}(t) + c_2\mathbf{x}^{(2)}(t) \tag{8}$$

also satisfies Eq. (7). This solution of the system (7) was found in Section 7.2 by the method of elimination.

As we indicated above, by repeatedly applying Theorem 7.3 it follows that every finite linear combination of solutions of Eq. (3) is again a solution. The question now arises as to whether all solutions of Eq. (3) can be found in this way. By analogy with previous cases it is reasonable to expect that for a system of the form (3) of nth order it is sufficient to form linear combinations of n properly chosen solutions. Therefore let $\mathbf{x}^{(1)}, \ldots, \mathbf{x}^{(n)}$ be n solutions of the nth order system (3), and consider the matrix $\mathbf{X}(t)$ whose columns are the vectors $\mathbf{x}^{(1)}(t), \ldots, \mathbf{x}^{(n)}(t)$;

$$
\mathbf{X}(t) = \begin{pmatrix} x_{11}(t) & \cdots & x_{1n}(t) \\ \vdots & & \vdots \\ x_{n1}(t) & \cdots & x_{nn}(t) \end{pmatrix}. \tag{9}
$$

Recall from Section 7.4 that the columns of $\mathbf{X}(t)$ are linearly independent for a given value of t if and only if $\det \mathbf{X} \neq 0$ for that value of t. This determinant is called the Wronskian of the n solutions $\mathbf{x}^{(1)}, \ldots, \mathbf{x}^{(n)}$ and is also denoted by $W[\mathbf{x}^{(1)}, \ldots, \mathbf{x}^{(n)}]$; that is

$$
W[\mathbf{x}^{(1)}, \ldots, \mathbf{x}^{(n)}] = \det \mathbf{X}. \tag{10}
$$

The solutions $\mathbf{x}^{(1)}, \ldots, \mathbf{x}^{(n)}$ are then linearly independent at a point if and only if $W[\mathbf{x}^{(1)}, \ldots, \mathbf{x}^{(n)}]$ is not zero there.

Theorem 7.4. *If the vector functions $\mathbf{x}^{(1)}, \ldots, \mathbf{x}^{(n)}$ are linearly independent solutions of the system (3) for each point in the interval $\alpha < t < \beta$, then each solution $\mathbf{x} = \boldsymbol{\phi}(t)$ of the system (3) can be expressed as a linear combination of $\mathbf{x}^{(1)}, \ldots, \mathbf{x}^{(n)}$,*

$$
\boldsymbol{\phi}(t) = c_1 \mathbf{x}^{(1)}(t) + \cdots + c_n \mathbf{x}^{(n)}(t), \tag{11}
$$

in exactly one way.

Before proving Theorem 7.4, note that according to Theorem 7.3 all expressions of the form (11) are solutions of the system (3), while by Theorem 7.4 all solutions of Eq. (3) can be written in the form (11). If the constants $c_1, \ldots, c_n$ are thought of as arbitrary, then Eq. (11) includes all solutions of the system (3), and it is customary to call it the *general solution*. Any set of solutions $\mathbf{x}^{(1)}, \ldots, \mathbf{x}^{(n)}$ of Eq. (3), which is linearly independent at each point in the interval $\alpha < t < \beta$, is said to be a *fundamental set of solutions* for that interval.

To prove Theorem 7.4 we will show, given any solution $\boldsymbol{\phi}$ of Eq. (3), that $\boldsymbol{\phi}(t) = c_1 \mathbf{x}^{(1)}(t) + \cdots + c_n \mathbf{x}^{(n)}(t)$ for suitable values of $c_1, \ldots, c_n$. Let $t = t_0$ be some point in the interval $\alpha < t < \beta$ and let $\boldsymbol{\xi} = \boldsymbol{\phi}(t_0)$. We now wish to determine whether there is any solution of the form $\mathbf{x} = c_1 \mathbf{x}^{(1)}(t) + \cdots + c_n \mathbf{x}^{(n)}(t)$ that also satisfies the same initial condition $\mathbf{x}(t_0) = \boldsymbol{\xi}$. That is, we wish to know whether there are values of $c_1, \ldots, c_n$ such that

$$
c_1 \mathbf{x}^{(1)}(t_0) + \cdots + c_n \mathbf{x}^{(n)}(t_0) = \boldsymbol{\xi}, \tag{12}
$$

or in scalar form
$$c_1 x_{11}(t_0) + \cdots + c_n x_{1n}(t_0) = \xi_1,$$

. (13)

$$c_1 x_{n1}(t_0) + \cdots + c_n x_{nn}(t_0) = \xi_n.$$

The necessary and sufficient condition that Eqs. (13) possess a unique solution $c_1, \ldots, c_n$ is precisely the nonvanishing of the determinant of coefficients, which is the Wronskian $W[\mathbf{x}^{(1)}, \ldots, \mathbf{x}^{(n)}]$ evaluated at $t = t_0$. The hypothesis that $\mathbf{x}^{(1)}, \ldots, \mathbf{x}^{(n)}$ are linearly independent throughout $\alpha < t < \beta$ guarantees that $W[\mathbf{x}^{(1)}, \ldots, \mathbf{x}^{(n)}]$ is not zero at $t = t_0$, and therefore there is a (unique) solution of Eq. (3) of the form $\mathbf{x} = c_1 \mathbf{x}^{(1)}(t) + \cdots + c_n \mathbf{x}^{(n)}(t)$ that also satisfies the initial condition (12). By the uniqueness part of Theorem 7.2 this solution is identical to $\boldsymbol{\phi}(t)$, and hence $\boldsymbol{\phi}(t) = c_1 \mathbf{x}^{(1)}(t) + \cdots + c_n \mathbf{x}^{(n)}(t)$, as was to be proved.

Theorem 7.5. *If $\mathbf{x}^{(1)}, \ldots, \mathbf{x}^{(n)}$ are solutions of Eq. (3) on the interval $\alpha < t < \beta$, then in this interval $W[\mathbf{x}^{(1)}, \ldots, \mathbf{x}^{(n)}]$ either is identically zero or else never vanishes.*

The significance of Theorem 7.5 lies in the fact that it relieves us of the necessity of examining $W[\mathbf{x}^{(1)}, \ldots, \mathbf{x}^{(n)}]$ at all points in the interval of interest, and enables us to determine whether $\mathbf{x}^{(1)}, \ldots, \mathbf{x}^{(n)}$ form a fundamental set of solutions merely by evaluating their Wronskian at any convenient point in the interval.

Theorem 7.5 is proved by first establishing that the Wronskian of $\mathbf{x}^{(1)}, \ldots, \mathbf{x}^{(n)}$ satisfies the differential equation (see Problem 2)

$$\frac{dW}{dt} = (p_{11} + p_{22} + \cdots + p_{nn})W. \tag{14}$$

Hence W is an exponential function, and the conclusion of the theorem follows immediately. The expression for W obtained by solving Eq. (14) is known as Abel's formula; note the analogy with Eq. (13) of Section 3.2.

Alternatively, Theorem 7.5 can be established by showing that if n solutions $\mathbf{x}^{(1)}, \ldots, \mathbf{x}^{(n)}$ of Eq. (3) are linearly dependent at one point $t = t_0$, then they must be linearly dependent at each point in $\alpha < t < \beta$ (see Problem 8). Consequently, if $\mathbf{x}^{(1)}, \ldots, \mathbf{x}^{(n)}$ are linearly independent at one point, they must be linearly independent at each point in the interval.

The next theorem states that the system (3) always has at least one fundamental set of solutions.

Theorem 7.6. Let

$$\mathbf{e}^{(1)} = \begin{pmatrix} 1 \\ 0 \\ 0 \\ \cdot \\ \cdot \\ \cdot \\ 0 \end{pmatrix}, \quad \mathbf{e}^{(2)} = \begin{pmatrix} 0 \\ 1 \\ 0 \\ \cdot \\ \cdot \\ \cdot \\ 0 \end{pmatrix}, \ldots, \mathbf{e}^{(n)} = \begin{pmatrix} 0 \\ \cdot \\ \cdot \\ \cdot \\ 0 \\ 1 \end{pmatrix};$$

further let $\mathbf{x}^{(1)}, \ldots, \mathbf{x}^{(n)}$ *be the solutions of the system* (3) *satisfying the initial conditions*

$$\mathbf{x}^{(1)}(t_0) = \mathbf{e}^{(1)}, \ldots, \mathbf{x}^{(n)}(t_0) = \mathbf{e}^{(n)}, \tag{15}$$

respectively, where t_0 *is any point in* $\alpha < t < \beta$. *Then* $\mathbf{x}^{(1)}, \ldots, \mathbf{x}^{(n)}$ *form a fundamental set of solutions of the system* (3).

To prove this theorem, note that the existence and uniqueness of the solutions $\mathbf{x}^{(1)}, \ldots, \mathbf{x}^{(n)}$ mentioned in Theorem 7.6 is assured by Theorem 7.2. It is not hard to see that the Wronskian of these solutions is equal to one when $t = t_0$; therefore $\mathbf{x}^{(1)}, \ldots, \mathbf{x}^{(n)}$ are a fundamental set of solutions.

Once one fundamental set of solutions has been found, other sets can be generated by forming (independent) linear combinations of the first set. For theoretical purposes the set given by Theorem 7.6 is usually the simplest.

To summarize, any set of n linearly independent solutions of the system (3) constitutes a fundamental set of solutions. Under the conditions given in this section, such fundamental sets always exist, and every solution of the system (3) can be represented as a linear combination of any fundamental set of solutions.

PROBLEMS

1. Using matrix algebra, prove the statement following Theorem 7.3 for an arbitrary value of the integer k.

2. In this problem we outline a proof of Theorem 7.5 in the case $n = 2$. Let $\mathbf{x}^{(1)}$ and $\mathbf{x}^{(2)}$ be solutions of Eq. (3) for $\alpha < t < \beta$, and let W be the Wronskian of $\mathbf{x}^{(1)}$ and $\mathbf{x}^{(2)}$.
   (a) Show that

$$\frac{dW}{dt} = \begin{vmatrix} \dfrac{dx_1^{(1)}}{dt} & \dfrac{dx_1^{(2)}}{dt} \\ x_2^{(1)} & x_2^{(2)} \end{vmatrix} + \begin{vmatrix} x_1^{(1)} & x_1^{(2)} \\ \dfrac{dx_2^{(1)}}{dt} & \dfrac{dx_2^{(2)}}{dt} \end{vmatrix}.$$

   (b) Using Eq. (3), show that

$$\frac{dW}{dt} = (p_{11} + p_{22})W.$$

   (c) Find $W(t)$ by solving the differential equation obtained in part (b). Use this expression to obtain the conclusion stated in Theorem 7.5.
   *(d) Generalize this procedure so as to prove Theorem 7.5 for an arbitrary value of n.

3. Show that the Wronskians of two fundamental sets of solutions of the system (3) can differ at most by a multiplicative constant.
Hint: Use Eq. (14).

4. If $x_1 = y$ and $x_2 = y'$, then the second order equation

$$y'' + p(t)y' + q(t)y = 0 \tag{i}$$

corresponds to the system

$$x_1' = x_2, \tag{ii}$$
$$x_2' = -q(t)x_1 - p(t)x_2.$$

Show that if $\mathbf{x}^{(1)}$ and $\mathbf{x}^{(2)}$ are a fundamental set of solutions of Eqs. (ii), and if $y^{(1)}$ and $y^{(2)}$ are a fundamental set of solutions of Eq. (i), then $W[y^{(1)}, y^{(2)}] = cW[\mathbf{x}^{(1)}, \mathbf{x}^{(2)}]$, where c is a nonzero constant.

Hint: $y^{(1)}(t)$ and $y^{(2)}(t)$ must be linear combinations of $x_{11}(t)$ and $x_{12}(t)$.

5. Show that the general solution of $\mathbf{x}' = \mathbf{P}(t)\mathbf{x} + \mathbf{g}(t)$ is the sum of any particular solution $\mathbf{x}^{(p)}$ of this equation and the general solution $\mathbf{x}^{(c)}$ of the corresponding homogeneous equation.

6. Consider the vectors $\mathbf{x}^{(1)}(t) = \begin{pmatrix} t \\ 1 \end{pmatrix}$ and $\mathbf{x}^{(2)}(t) = \begin{pmatrix} t^2 \\ 2t \end{pmatrix}$.

(a) Compute the Wronskian of $\mathbf{x}^{(1)}$ and $\mathbf{x}^{(2)}$.

(b) In what intervals are $\mathbf{x}^{(1)}$ and $\mathbf{x}^{(2)}$ linearly independent?

(c) What conclusion can be drawn about the coefficients in the system of homogeneous differential equations satisfied by $\mathbf{x}^{(1)}$ and $\mathbf{x}^{(2)}$?

(d) Find this system of equations and verify the conclusions of part (b).

7. Consider the vectors $\mathbf{x}^{(1)}(t) = \begin{pmatrix} t^2 \\ 2t \end{pmatrix}$ and $\mathbf{x}^{(2)}(t) = \begin{pmatrix} e^t \\ e^t \end{pmatrix}$, and answer the same questions as in Problem 6.

The following two problems indicate an alternative derivation of Theorem 7.4.

8. Let $\mathbf{x}^{(1)}, \ldots, \mathbf{x}^{(m)}$ be solutions of $\mathbf{x}' = \mathbf{P}(t)\mathbf{x}$ on the interval $\alpha < t < \beta$. Assume that $\mathbf{P}$ is continuous and let t_0 be an arbitrary point in the given interval. Show that $\mathbf{x}^{(1)}, \ldots, \mathbf{x}^{(m)}$ are linearly dependent for $\alpha < t < \beta$ if (and only if) $\mathbf{x}^{(1)}(t_0), \ldots, \mathbf{x}^{(m)}(t_0)$ are linearly dependent. In other words, $\mathbf{x}^{(1)}, \ldots, \mathbf{x}^{(m)}$ are linearly dependent on the interval (α, β) if they are linearly dependent at any point in it.

Hint: There are constants $c_1, \ldots, c_m$ such that $c_1\mathbf{x}^{(1)}(t_0) + \cdots + c_m\mathbf{x}^{(m)}(t_0) = \mathbf{0}$. Let $\mathbf{z}(t) = c_1\mathbf{x}^{(1)}(t) + \cdots + c_m\mathbf{x}^{(m)}(t)$, and use the uniqueness theorem to show that $\mathbf{z}(t) = \mathbf{0}$ for each t in $\alpha < t < \beta$.

9. Let $\mathbf{x}^{(1)}, \ldots, \mathbf{x}^{(n)}$ be linearly independent solutions of $\mathbf{x}' = \mathbf{P}(t)\mathbf{x}$, where $\mathbf{P}$ is continuous on $\alpha < t < \beta$.

(a) Show that any solution $\mathbf{x} = \mathbf{z}(t)$ can be written in the form

$$\mathbf{z}(t) = c_1\mathbf{x}^{(1)}(t) + \cdots + c_n\mathbf{x}^{(n)}(t)$$

for suitable constants $c_1, \ldots, c_n$.

Hint: Use the result of Problem 7 of Section 7.4, and also Problem 8 above.

(b) Show that the expression for the solution $\mathbf{z}(t)$ in part (a) is unique; that is, if $\mathbf{z}(t) = k_1\mathbf{x}^{(1)}(t) + \cdots + k_n\mathbf{x}^{(n)}(t)$, then $k_1 = c_1, \ldots, k_n = c_n$.

Hint: Show that $(k_1 - c_1)\mathbf{x}^{(1)}(t) + \cdots + (k_n - c_n)\mathbf{x}^{(n)}(t) = \mathbf{0}$ for each t in $\alpha < t < \beta$ and use the linear independence of $\mathbf{x}^{(1)}, \ldots, \mathbf{x}^{(n)}$.

7.6 HOMOGENEOUS LINEAR SYSTEMS WITH CONSTANT COEFFICIENTS

In this section we will begin to show how to construct the general solution of a system of homogeneous linear equations with constant coefficients—that is, a system of the form

$$\mathbf{x}' = \mathbf{A}\mathbf{x}, \tag{1}$$

where $\mathbf{A}$ is a constant $n \times n$ matrix. By analogy with the treatment of second order linear equations in Section 3.5, we will seek solutions of Eq. (1) of the form

$$\mathbf{x} = \boldsymbol{\xi} e^{rt}, \tag{2}$$

where r and the constant vector $\boldsymbol{\xi}$ are to be determined. Substituting from Eq. (2) for $\mathbf{x}$ in the system (1) gives

$$r\boldsymbol{\xi} e^{rt} = \mathbf{A}\boldsymbol{\xi} e^{rt}.$$

Upon canceling the nonzero scalar factor e^{rt} we obtain $\mathbf{A}\boldsymbol{\xi} = r\boldsymbol{\xi}$, or

$$(\mathbf{A} - r\mathbf{I})\boldsymbol{\xi} = \mathbf{0}, \tag{3}$$

where $\mathbf{I}$ is the $n \times n$ identity matrix. Thus, in order to solve the system of differential equations (1) we must solve the system of algebraic equations (3). This latter problem is precisely the one that determines the eigenvalues and eigenvectors of the matrix $\mathbf{A}$. Therefore the vector $\mathbf{x}$ given by Eq. (2) is a solution of Eq. (1) provided r is an eigenvalue and $\boldsymbol{\xi}$ an associated eigenvector of the coefficient matrix $\mathbf{A}$.

Example 1. Find the general solution of the system

$$\mathbf{x}' = \begin{pmatrix} 1 & 1 \\ 4 & 1 \end{pmatrix} \mathbf{x}. \tag{4}$$

Assuming that $\mathbf{x} = \boldsymbol{\xi} e^{rt}$, and substituting for $\mathbf{x}$ in Eq. (4), we are led to the system of algebraic equations

$$\begin{pmatrix} 1 - r & 1 \\ 4 & 1 - r \end{pmatrix} \begin{pmatrix} \xi_1 \\ \xi_2 \end{pmatrix} = \begin{pmatrix} 0 \\ 0 \end{pmatrix}. \tag{5}$$

Equations (5) have a nontrivial solution if and only if the determinant of coefficients vanishes. Thus allowable values of r are found from the equation

$$\begin{vmatrix} 1 - r & 1 \\ 4 & 1 - r \end{vmatrix} = (1 - r)^2 - 4$$

$$= r^2 - 2r - 3 = 0. \tag{6}$$

Equation (6) has the roots $r_1 = 3$ and $r_2 = -1$; these are the eigenvalues of the coefficient matrix in Eq. (4). If $r = 3$, then the system (5) reduces to the single equation

$$-2\xi_1 + \xi_2 = 0. \tag{7}$$

Thus $\xi_2 = 2\xi_1$ and the eigenvector corresponding to $r_1 = 3$ can be taken as

$$\boldsymbol{\xi}^{(1)} = \begin{pmatrix} 1 \\ 2 \end{pmatrix}. \tag{8}$$

Similarly, corresponding to $r_2 = -1$, we find that $\xi_2 = -2\xi_1$, so the eigenvector is

$$\boldsymbol{\xi}^{(2)} = \begin{pmatrix} 1 \\ -2 \end{pmatrix}. \tag{9}$$

The corresponding solutions of the differential equation are

$$\mathbf{x}^{(1)}(t) = \begin{pmatrix} 1 \\ 2 \end{pmatrix} e^{3t}, \qquad \mathbf{x}^{(2)}(t) = \begin{pmatrix} 1 \\ -2 \end{pmatrix} e^{-t}. \tag{10}$$

The Wronskian of these solutions is

$$W[\mathbf{x}^{(1)}, \mathbf{x}^{(2)}](t) = \begin{vmatrix} e^{3t} & e^{-t} \\ 2e^{3t} & -2e^{-t} \end{vmatrix} = -4e^{2t}, \tag{11}$$

which is never zero. Hence the solutions $\mathbf{x}^{(1)}$ and $\mathbf{x}^{(2)}$ form a fundamental set, and the general solution of the system (4) is

$$\mathbf{x} = c_1 \mathbf{x}^{(1)}(t) + c_2 \mathbf{x}^{(2)}(t)$$

$$= c_1 \begin{pmatrix} 1 \\ 2 \end{pmatrix} e^{3t} + c_2 \begin{pmatrix} 1 \\ -2 \end{pmatrix} e^{-t} \tag{12}$$

where c_1 and c_2 are arbitrary constants. This is the same as the solution found in Section 7.2 by the method of elimination.

Returning to the general system (1) we proceed as in the example. To find solutions of the differential equation (1) we must find the eigenvalues and eigenvectors of $\mathbf{A}$ from the associated algebraic system (3). The eigenvalues $r_1, \ldots, r_n$ (which need not all be different) are roots of the polynomial equation

$$\det (\mathbf{A} - r\mathbf{I}) = 0. \tag{13}$$

The nature of the eigenvalues and the corresponding eigenvectors determines the nature of the general solution of the system (1).

Hermitian Systems.   The situation is simplest when $\mathbf{A}$ is a Hermitian matrix. As stated in Section 7.4, the eigenvalues $r_1, \ldots, r_n$ are all real in this case. Further, even if some of the eigenvalues are repeated, there is always a full set of n linearly independent eigenvectors $\boldsymbol{\xi}^{(1)}, \ldots, \boldsymbol{\xi}^{(n)}$. Hence the corresponding solutions of the differential system (1) are

$$\mathbf{x}^{(1)}(t) = \boldsymbol{\xi}^{(1)} e^{r_1 t}, \ldots, \mathbf{x}^{(n)}(t) = \boldsymbol{\xi}^{(n)} e^{r_n t}. \tag{14}$$

To show that these solutions form a fundamental set, we evaluate their Wronskian:

$$W[\mathbf{x}^{(1)}, \ldots, \mathbf{x}^{(n)}](t) = \begin{vmatrix} \xi_1^{(1)} e^{r_1 t} & \cdots & \xi_1^{(n)} e^{r_n t} \\ \vdots & & \vdots \\ \xi_n^{(1)} e^{r_1 t} & \cdots & \xi_n^{(n)} e^{r_n t} \end{vmatrix} = e^{(r_1 + \cdots + r_n)t} \begin{vmatrix} \xi_1^{(1)} & \cdots & \xi_1^{(n)} \\ \vdots & & \vdots \\ \xi_n^{(1)} & \cdots & \xi_n^{(n)} \end{vmatrix}. \tag{15}$$

First we observe that the exponential function is never zero. Next, since the eigenvectors $\xi^{(1)}, \ldots, \xi^{(n)}$ are linearly independent, the determinant in the last term of Eq. (15) is nonzero. As a consequence, the Wronskian $W[\mathbf{x}^{(1)}, \ldots, \mathbf{x}^{(n)}](t)$ is never zero, and hence $\mathbf{x}^{(1)}, \ldots, \mathbf{x}^{(n)}$ form a fundamental set of solutions. Thus when $\mathbf{A}$ is a Hermitian matrix, the general solution of Eq. (1) is

$$\mathbf{x} = c_1\xi^{(1)}e^{r_1 t} + \cdots + c_n\xi^{(n)}e^{r_n t}. \tag{16}$$

An important subclass of Hermitian matrices is the class of real, symmetric matrices. If $\mathbf{A}$ is real and symmetric, then the eigenvectors $\xi^{(1)}, \ldots, \xi^{(n)}$ as well as the eigenvalues $r_1, \ldots, r_n$ are all real. Hence the solutions given by Eq. (14) are real-valued. However, if the Hermitian matrix $\mathbf{A}$ is not real, then in general the eigenvectors have nonzero imaginary parts, and the solutions (14) are complex-valued.

Example 2. Find the general solution of

$$\mathbf{x}' = \begin{pmatrix} 1 & \sqrt{3} \\ \sqrt{3} & -1 \end{pmatrix}\mathbf{x}. \tag{17}$$

The coefficient matrix in Eq. (17) is real and symmetric, so we can proceed in the manner just described. Assuming that $\mathbf{x} = \xi e^{rt}$ leads to the algebraic system

$$\begin{pmatrix} 1 - r & \sqrt{3} \\ \sqrt{3} & -1 - r \end{pmatrix}\begin{pmatrix} \xi_1 \\ \xi_2 \end{pmatrix} = \begin{pmatrix} 0 \\ 0 \end{pmatrix}. \tag{18}$$

The eigenvalues satisfy

$$\begin{vmatrix} 1 - r & \sqrt{3} \\ \sqrt{3} & -1 - r \end{vmatrix} = (1 - r)(-1 - r) - 3$$

$$= r^2 - 4 = 0. \tag{19}$$

Thus $r_1 = 2$ and $r_2 = -2$. For $r = 2$, Eq. (18) becomes

$$\begin{pmatrix} -1 & \sqrt{3} \\ \sqrt{3} & -3 \end{pmatrix}\begin{pmatrix} \xi_1 \\ \xi_2 \end{pmatrix} = \begin{pmatrix} 0 \\ 0 \end{pmatrix}. \tag{20}$$

Hence $\xi_1 = \sqrt{3}\xi_2$ and the eigenvector $\xi^{(1)}$ corresponding to the eigenvalue $r_1 = 2$ can be taken as

$$\xi^{(1)} = \begin{pmatrix} \sqrt{3} \\ 1 \end{pmatrix}. \tag{21}$$

Similarly, corresponding to the eigenvalue $r_2 = -2$, we have $\xi_2 = -\sqrt{3}\xi_1$, so the eigenvector is

$$\xi^{(2)} = \begin{pmatrix} 1 \\ -\sqrt{3} \end{pmatrix}. \tag{22}$$

Thus the general solution of Eq. (17) is

$$\mathbf{x} = c_1 \begin{pmatrix} \sqrt{3} \\ 1 \end{pmatrix} e^{2t} + c_2 \begin{pmatrix} 1 \\ -\sqrt{3} \end{pmatrix} e^{-2t}. \tag{23}$$

Example 3. Find the general solution of

$$\mathbf{x}' = \begin{pmatrix} 0 & 1 & 1 \\ 1 & 0 & 1 \\ 1 & 1 & 0 \end{pmatrix} \mathbf{x}. \tag{24}$$

Again, we observe that the coefficient matrix is real and symmetric. The eigenvalues and eigenvectors of this matrix were found in Example 4 of Section 7.4, namely,

$$r_1 = 2; \quad \boldsymbol{\xi}^{(1)} = \begin{pmatrix} 1 \\ 1 \\ 1 \end{pmatrix} \tag{25}$$

$$r_2 = -1, \quad r_3 = -1; \quad \boldsymbol{\xi}^{(2)} = \begin{pmatrix} 1 \\ 0 \\ -1 \end{pmatrix}, \quad \boldsymbol{\xi}^{(3)} = \begin{pmatrix} 0 \\ 1 \\ -1 \end{pmatrix}. \tag{26}$$

Hence a fundamental set of solutions of Eq. (24) is

$$\mathbf{x}^{(1)}(t) = \begin{pmatrix} 1 \\ 1 \\ 1 \end{pmatrix} e^{2t}, \quad \mathbf{x}^{(2)}(t) = \begin{pmatrix} 1 \\ 0 \\ -1 \end{pmatrix} e^{-t}, \quad \mathbf{x}^{(3)}(t) = \begin{pmatrix} 0 \\ 1 \\ -1 \end{pmatrix} e^{-t} \tag{27}$$

and the general solution is

$$\mathbf{x} = c_1 \begin{pmatrix} 1 \\ 1 \\ 1 \end{pmatrix} e^{2t} + c_2 \begin{pmatrix} 1 \\ 0 \\ -1 \end{pmatrix} e^{-t} + c_3 \begin{pmatrix} 0 \\ 1 \\ -1 \end{pmatrix} e^{-t}. \tag{28}$$

This example illustrates the fact that even though an eigenvalue ($r = -1$) is repeated twice, it may still be possible to find two linearly independent eigenvectors $\boldsymbol{\xi}^{(2)}$ and $\boldsymbol{\xi}^{(3)}$ and, as a consequence, to construct the general solution (28).

Nonhermitian Systems. If the coefficient matrix $\mathbf{A}$ in the system (1)

$$\mathbf{x}' = \mathbf{A}\mathbf{x}$$

is not Hermitian, then the situation regarding the solution is more complicated.

Let us suppose first that $\mathbf{A}$ is real. Then there are three possibilities for the eigenvalues of $\mathbf{A}$:

1. All eigenvalues are real and distinct.
2. Some eigenvalues occur in complex conjugate pairs.
3. Some eigenvalues are repeated.

The first case leads to no difficulties. There is a single linearly independent real eigenvector for each eigenvalue, and consequently there are n linearly independent solutions of the form (14). Hence the general solution is still given by Eq. (16), where it is now understood that $r_1, \ldots, r_n$ are all different. This case is illustrated by Example 1 above.

If some of the eigenvalues occur in complex conjugate pairs, then there are still n linearly independent solutions of the form (14), provided all of the eigenvalues are different. Of course, the solutions arising from complex eigenvalues are complex-valued. However, as in Section 3.5.1, it is possible to obtain a full set of real-valued solutions. This is discussed in Section 7.7.

More serious difficulties can occur if an eigenvalue is repeated. In this event the number of corresponding linearly independent eigenvectors may be smaller than the multiplicity of the eigenvalue, as in Example 3 of Section 7.4. If so, the number of linearly independent solutions of the form $\boldsymbol{\xi}e^{rt}$ will be smaller than n. To construct a fundamental set of solutions it is then necessary to seek additional solutions of another form. The situation is somewhat analogous to that for an nth order linear equation with constant coefficients; a repeated root of the characteristic equation gave rise to solutions of the form e^{rx}, xe^{rx}, x^2e^{rx}, ... The case of repeated eigenvalues is treated in Section 7.8.

Finally, if $\mathbf{A}$ is complex, but not Hermitian, then complex eigenvalues need not occur in conjugate pairs, and the eigenvectors are normally complex-valued even though the associated eigenvalue may be real. The solutions of the differential equation (1) are still of the form (14), provided the eigenvalues are distinct, but in general all of the solutions are complex-valued.

PROBLEMS

In each of Problems 1 through 10 find the general solution of the given system of equations.

1. $\mathbf{x}' = \begin{pmatrix} 3 & -2 \\ 2 & -2 \end{pmatrix}\mathbf{x}$ 2. $\mathbf{x}' = \begin{pmatrix} 4 & -3 \\ 8 & -6 \end{pmatrix}\mathbf{x}$ 3. $\mathbf{x}' = \begin{pmatrix} 2 & -1 \\ 3 & -2 \end{pmatrix}\mathbf{x}$

4. $\mathbf{x}' = \begin{pmatrix} 1 & 1 \\ 4 & -2 \end{pmatrix}\mathbf{x}$ 5. $\mathbf{x}' = \begin{pmatrix} -2 & 1 \\ 1 & -2 \end{pmatrix}\mathbf{x}$ 6. $\mathbf{x}' = \begin{pmatrix} 1 & i \\ -i & 1 \end{pmatrix}\mathbf{x}$

7. $\mathbf{x}' = \begin{pmatrix} 1 & 1 & 2 \\ 1 & 2 & 1 \\ 2 & 1 & 1 \end{pmatrix}\mathbf{x}$ 8. $\mathbf{x}' = \begin{pmatrix} 3 & 2 & 4 \\ 2 & 0 & 2 \\ 4 & 2 & 3 \end{pmatrix}\mathbf{x}$

9. $\mathbf{x}' = \begin{pmatrix} 1 & 1 & 1 \\ 2 & 1 & -1 \\ -8 & -5 & -3 \end{pmatrix} \mathbf{x}$ 10. $\mathbf{x}' = \begin{pmatrix} 1 & -1 & 4 \\ 3 & 2 & -1 \\ 2 & 1 & -1 \end{pmatrix} \mathbf{x}$

In each of Problems 11 through 14 solve the given initial value problem.

11. $\mathbf{x}' = \begin{pmatrix} 5 & -1 \\ 3 & 1 \end{pmatrix} \mathbf{x}, \qquad \mathbf{x}(0) = \begin{pmatrix} 2 \\ -1 \end{pmatrix}$

12. $\mathbf{x}' = \begin{pmatrix} -2 & 1 \\ -5 & 4 \end{pmatrix} \mathbf{x}, \qquad \mathbf{x}(0) = \begin{pmatrix} 1 \\ 3 \end{pmatrix}$

13. $\mathbf{x}' = \begin{pmatrix} 1 & 1 & 2 \\ 0 & 2 & 2 \\ -1 & 1 & 3 \end{pmatrix} \mathbf{x}, \qquad \mathbf{x}(0) = \begin{pmatrix} 2 \\ 0 \\ 1 \end{pmatrix}$

14. $\mathbf{x}' = \begin{pmatrix} 0 & 0 & -1 \\ 2 & 0 & 0 \\ -1 & 2 & 4 \end{pmatrix} \mathbf{x}, \qquad \mathbf{x}(0) = \begin{pmatrix} 7 \\ 5 \\ 5 \end{pmatrix}$

15. The system $t\mathbf{x}' = \mathbf{Ax}$ is analogous to the second order Euler equation (Section 4.4). Assuming that $\mathbf{x} = \boldsymbol{\xi} t^r$, where $\boldsymbol{\xi}$ is a constant vector, show that $\boldsymbol{\xi}$ and r must satisfy $(\mathbf{A} - r\mathbf{I})\boldsymbol{\xi} = \mathbf{0}$; also show that to obtain nontrivial solutions of the given differential equation, r must be a root of the characteristic equation $\det (\mathbf{A} - r\mathbf{I}) = 0$.

Referring to Problem 15, solve the given system of equations in each of Problems 16 through 19. Assume that $t > 0$.

16. $t\mathbf{x}' = \begin{pmatrix} 2 & -1 \\ 3 & -2 \end{pmatrix} \mathbf{x}$ 17. $t\mathbf{x}' = \begin{pmatrix} 5 & -1 \\ 3 & 1 \end{pmatrix} \mathbf{x}$

18. $t\mathbf{x}' = \begin{pmatrix} 4 & -3 \\ 8 & -6 \end{pmatrix} \mathbf{x}$ 19. $t\mathbf{x}' = \begin{pmatrix} 3 & -2 \\ 2 & -2 \end{pmatrix} \mathbf{x}$

20. Consider a second order system $\mathbf{x}' = \mathbf{Ax}$. Assuming that $r_1 \neq r_2$, the general solution is $\mathbf{x} = c_1 \boldsymbol{\xi}^{(1)} e^{r_1 t} + c_2 \boldsymbol{\xi}^{(2)} e^{r_2 t}$, provided $\boldsymbol{\xi}^{(1)}$ and $\boldsymbol{\xi}^{(2)}$ are linearly independent. In this problem we will establish the linear independence of $\boldsymbol{\xi}^{(1)}$ and $\boldsymbol{\xi}^{(2)}$ by assuming that they are linearly dependent, and then showing that this leads to a contradiction.

(a) Note that $\boldsymbol{\xi}^{(1)}$ satisfies the matrix equation $(\mathbf{A} - r_1\mathbf{I})\boldsymbol{\xi}^{(1)} = \mathbf{0}$; similarly, note that $(\mathbf{A} - r_2\mathbf{I})\boldsymbol{\xi}^{(2)} = \mathbf{0}$.

(b) Show that $(\mathbf{A} - r_2\mathbf{I})\boldsymbol{\xi}^{(1)} = (r_1 - r_2)\boldsymbol{\xi}^{(1)}$.

(c) Suppose that $\boldsymbol{\xi}^{(1)}$ and $\boldsymbol{\xi}^{(2)}$ are linearly dependent. Then $c_1 \boldsymbol{\xi}^{(1)} + c_2 \boldsymbol{\xi}^{(2)} = \mathbf{0}$ and at least one of c_1 and c_2 is not zero; suppose $c_1 \neq 0$. Show that $(\mathbf{A} - r_2\mathbf{I})(c_1 \boldsymbol{\xi}^{(1)} + c_2 \boldsymbol{\xi}^{(2)}) = \mathbf{0}$, and also show that $(\mathbf{A} - r_2\mathbf{I})(c_1 \boldsymbol{\xi}^{(1)} + c_2 \boldsymbol{\xi}^{(2)}) = c_1(r_1 - r_2)\boldsymbol{\xi}^{(1)}$. Hence $c_1 = 0$, which is a contradiction. Therefore $\boldsymbol{\xi}^{(1)}$ and $\boldsymbol{\xi}^{(2)}$ are linearly independent.

(d) Modify the argument of part (c) in case c_1 is zero but c_2 is not.

(e) Carry out a similar argument for the case in which $n = 3$; note that the procedure can be extended to cover an arbitrary value of n.

21. Consider the equation

$$ay'' + by' + cy = 0 \tag{i}$$

where a, b, and c are constants. In Chapter 3 it was shown that the general solution depended upon the roots of the characteristic equation

$$ar^2 + br + c = 0. \tag{ii}$$

(a) Transform Eq. (i) into a system of first order equations by letting $x_1 = y$, $x_2 = y'$. Find the system of equations $\mathbf{x}' = \mathbf{A}\mathbf{x}$ satisfied by $\mathbf{x} = \begin{pmatrix} x_1 \\ x_2 \end{pmatrix}$.

(b) Find the equation that determines the eigenvalues of the coefficient matrix A in part (a). Note that this equation is just the characteristic equation (ii) of Eq. (i).

22. In this problem we show how the method of reduction of order can be used for systems of equations. It provides a method for dealing with systems that do not have a complete set of solutions of the form $\boldsymbol{\xi}e^{rt}$. Consider the system

$$\mathbf{x}' = \begin{pmatrix} 3 & -2 \\ 2 & -2 \end{pmatrix}\mathbf{x}. \tag{i}$$

(a) Verify that $\mathbf{x} = \begin{pmatrix} 2 \\ 1 \end{pmatrix}e^{2t}$ satisfies the given differential equation.

(b) Introduce a new dependent variable by the transformation

$$\mathbf{x} = \begin{pmatrix} 1 & 2e^{2t} \\ 0 & e^{2t} \end{pmatrix}\mathbf{y}. \tag{ii}$$

Observe that this transformation is obtained by replacing the second column of the identity matrix by the known solution. By substituting for $\mathbf{x}$ in Eq. (i), show that $\mathbf{y}$ satisfies the system of equations

$$\begin{pmatrix} 1 & 2e^{2t} \\ 0 & e^{2t} \end{pmatrix}\mathbf{y}' = \begin{pmatrix} 3 & 0 \\ 2 & 0 \end{pmatrix}\mathbf{y}. \tag{iii}$$

(c) Solve Eq. (iii) and show that

$$\mathbf{y} = c_1 \begin{pmatrix} \frac{1}{2}e^{-t} \\ -\frac{1}{3}e^{-3t} \end{pmatrix} + c_2 \begin{pmatrix} 0 \\ 1 \end{pmatrix} \tag{iv}$$

where c_1 and c_2 are arbitrary constants.

(d) Using Eq. (ii), show that

$$\mathbf{x} = -\frac{c_1}{6}\begin{pmatrix} 1 \\ 2 \end{pmatrix}e^{-t} + c_2\begin{pmatrix} 2 \\ 1 \end{pmatrix}e^{2t}; \tag{v}$$

the first term is a second independent solution of Eq. (i). This is the method of reduction of order as it applies to a second order system of equations.

In Problems 23 and 24 use the method of reduction of order (Problem 22) to solve the given system of equations.

23. $\mathbf{x}' = \begin{pmatrix} 1 & -4 \\ 4 & -7 \end{pmatrix} \mathbf{x}$ 24. $\mathbf{x}' = \begin{pmatrix} 3 & -4 \\ 1 & -1 \end{pmatrix} \mathbf{x}$

Electric Circuits. Problems 25 and 26 are concerned with the electric circuit described by the system of differential equations derived in Problem 9 of Section 7.1:

$$\frac{d}{dt}\begin{pmatrix} I \\ V \end{pmatrix} = \begin{pmatrix} -\dfrac{R_1}{L} & -\dfrac{1}{L} \\ \dfrac{1}{C} & -\dfrac{1}{CR_2} \end{pmatrix}\begin{pmatrix} I \\ V \end{pmatrix}. \tag{i}$$

25. (a) Find the general solution of Eq. (i) if $R_1 = 1$ ohm, $R_2 = \frac{3}{5}$ ohm, $L = 2$ henrys, and $C = \frac{2}{3}$ farad.

(b) Show that $I(t) \to 0$ and $V(t) \to 0$ as $t \to \infty$ regardless of the initial values $I(0)$ and $V(0)$.

26. Consider the system of differential equations (i) above.

(a) Find a condition on R_1, R_2, C, and L that must be satisfied if the eigenvalues of the coefficient matrix are to be real and different.

(b) If the condition found in part (a) is satisfied, show that both eigenvalues are negative. Then show that $I(t) \to 0$ and $V(t) \to 0$ as $t \to \infty$ regardless of the initial conditions.

*(c) If the condition found in part (a) is not satisfied, then the eigenvalues are either complex or repeated. Do you think that $I(t) \to 0$ and $V(t) \to 0$ as $t \to \infty$ in these cases as well?

Hint: In part (c) one approach is to change the system (i) into a single second order equation. We will also discuss complex and repeated eigenvalues in Sections 7.7 and 7.8.

7.7 COMPLEX EIGENVALUES

In this section we again consider a system of n linear homogeneous equations with constant coefficients

$$\mathbf{x}' = \mathbf{A}\mathbf{x}, \tag{1}$$

where we now assume that the coefficient matrix $\mathbf{A}$ is real-valued. If we seek solutions of the form $\mathbf{x} = \boldsymbol{\xi}e^{rt}$, then it follows as in Section 7.6 that r must be an eigenvalue and $\boldsymbol{\xi}$ a corresponding eigenvector of the coefficient matrix $\mathbf{A}$. Recall that the eigenvalues $r_1, \ldots, r_n$ of $\mathbf{A}$ are the roots of the equation

$$\det(\mathbf{A} - r\mathbf{I}) = 0, \tag{2}$$

and that the corresponding eigenvectors satisfy

$$(\mathbf{A} - r\mathbf{I})\boldsymbol{\xi} = \mathbf{0}. \tag{3}$$

If $\mathbf{A}$ is real, then the coefficients in the polynomial equation (2) for r are real, and any complex eigenvalues must occur in conjugate pairs. For example, if $r_1 = \lambda + i\mu$, where λ and μ are real, is an eigenvalue of $\mathbf{A}$, then so is $r_2 = \lambda - i\mu$.

Further, the corresponding eigenvectors $\boldsymbol{\xi}^{(1)}$ and $\boldsymbol{\xi}^{(2)}$ are also complex conjugates. To see that this is so, suppose that r_1 and $\boldsymbol{\xi}^{(1)}$ satisfy

$$(\mathbf{A} - r_1\mathbf{I})\boldsymbol{\xi}^{(1)} = \mathbf{0}. \tag{4}$$

On taking the complex conjugate of this equation, and noting that $\mathbf{A}$ and $\mathbf{I}$ are real-valued, we obtain

$$(\mathbf{A} - \bar{r}_1\mathbf{I})\bar{\boldsymbol{\xi}}^{(1)} = \mathbf{0}, \tag{5}$$

where $\bar{r}_1$ and $\bar{\boldsymbol{\xi}}^{(1)}$ are the complex conjugates of r_1 and $\boldsymbol{\xi}^{(1)}$, respectively. In other words, $r_2 = \bar{r}_1$ is also an eigenvalue, and $\boldsymbol{\xi}^{(2)} = \bar{\boldsymbol{\xi}}^{(1)}$ is the corresponding eigenvector. The corresponding solutions

$$\mathbf{x}^{(1)}(t) = \boldsymbol{\xi}^{(1)}e^{r_1 t}, \qquad \mathbf{x}^{(2)}(t) = \bar{\boldsymbol{\xi}}^{(1)}e^{\bar{r}_1 t} \tag{6}$$

of the differential equation (1) are then complex conjugates of each other. Therefore, by an extension of Theorem 3.9 to systems (which is a simple matter), we can find two real-valued solutions of Eq. (1) corresponding to the eigenvalues r_1 and r_2 by taking the real and imaginary parts of $\mathbf{x}^{(1)}(t)$ or $\mathbf{x}^{(2)}(t)$ given by Eq. (6).

Let us write $\boldsymbol{\xi}^{(1)} = \mathbf{a} + i\mathbf{b}$, where $\mathbf{a}$ and $\mathbf{b}$ are real; then we have

$$\begin{aligned} \mathbf{x}^{(1)}(t) &= (\mathbf{a} + i\mathbf{b})e^{(\lambda+i\mu)t} \\ &= (\mathbf{a} + i\mathbf{b})e^{\lambda t}(\cos \mu t + i \sin \mu t). \end{aligned} \tag{7}$$

Upon separating $\mathbf{x}^{(1)}(t)$ into its real and imaginary parts, we obtain

$$\mathbf{x}^{(1)}(t) = e^{\lambda t}(\mathbf{a} \cos \mu t - \mathbf{b} \sin \mu t) + ie^{\lambda t}(\mathbf{a} \sin \mu t + \mathbf{b} \cos \mu t). \tag{8}$$

If we write $\mathbf{x}^{(1)}(t) = \mathbf{u}(t) + i\mathbf{v}(t)$, then the vectors

$$\mathbf{u}(t) = e^{\lambda t}(\mathbf{a} \cos \mu t - \mathbf{b} \sin \mu t), \tag{9a}$$

$$\mathbf{v}(t) = e^{\lambda t}(\mathbf{a} \sin \mu t + \mathbf{b} \cos \mu t) \tag{9b}$$

are real-valued solutions of Eq. (1). It is possible to show that $\mathbf{u}$ and $\mathbf{v}$ are linearly independent solutions (see Problem 13).

For example, suppose that $r_1 = \lambda + i\mu$, $r_2 = \lambda - i\mu$, and that $r_3, \ldots, r_n$ are all real and distinct. Let the corresponding eigenvectors be $\boldsymbol{\xi}^{(1)} = \mathbf{a} + i\mathbf{b}$, $\boldsymbol{\xi}^{(2)} = \mathbf{a} - i\mathbf{b}$, $\boldsymbol{\xi}^{(3)}, \ldots, \boldsymbol{\xi}^{(n)}$. Then the general solution of Eq. (1) is

$$\mathbf{x} = c_1\mathbf{u}(t) + c_2\mathbf{v}(t) + c_3\boldsymbol{\xi}^{(3)}e^{r_3 t} + \ldots + c_n\boldsymbol{\xi}^{(n)}e^{r_n t}, \tag{10}$$

where $\mathbf{u}(t)$ and $\mathbf{v}(t)$ are given by Eqs. (9). We emphasize that this analysis applies only if the coefficient matrix $\mathbf{A}$ in Eq. (1) is real, for it is only then that complex eigenvalues and eigenvectors occur in conjugate pairs.

Example 1. Find a fundamental set of real-valued solutions of the system

$$\mathbf{x}' = \begin{pmatrix} 1 & -1 \\ 5 & -3 \end{pmatrix}\mathbf{x}. \tag{11}$$

Assuming that

$$\mathbf{x} = \boldsymbol{\xi} e^{rt} \tag{12}$$

leads to the set of linear algebraic equations

$$\begin{pmatrix} 1-r & -1 \\ 5 & -3-r \end{pmatrix} \begin{pmatrix} \xi_1 \\ \xi_2 \end{pmatrix} = \begin{pmatrix} 0 \\ 0 \end{pmatrix}, \tag{13}$$

which determine the eigenvalues and eigenvectors of **A**. The characteristic equation is

$$\begin{vmatrix} 1-r & -1 \\ 5 & -3-r \end{vmatrix} = r^2 + 2r + 2 = 0; \tag{14}$$

therefore the eigenvalues are $r_1 = -1 + i$ and $r_2 = -1 - i$. From Eq. (13) a simple calculation shows that the corresponding eigenvectors are

$$\boldsymbol{\xi}^{(1)} = \begin{pmatrix} 1 \\ 2-i \end{pmatrix}, \qquad \boldsymbol{\xi}^{(2)} = \begin{pmatrix} 1 \\ 2+i \end{pmatrix}. \tag{15}$$

Hence a fundamental set of solutions of the system (11) is

$$\mathbf{x}^{(1)}(t) = \begin{pmatrix} 1 \\ 2-i \end{pmatrix} e^{(-1+i)t}, \qquad \mathbf{x}^{(2)}(t) = \begin{pmatrix} 1 \\ 2+i \end{pmatrix} e^{(-1-i)t}. \tag{16}$$

To obtain a set of real-valued solutions, we must find the real and imaginary parts of either $\mathbf{x}^{(1)}$ or $\mathbf{x}^{(2)}$. In fact,

$$\mathbf{x}^{(1)}(t) = \begin{pmatrix} 1 \\ 2-i \end{pmatrix} e^{-t}(\cos t + i \sin t)$$

$$= \begin{pmatrix} e^{-t} \cos t \\ 2e^{-t} \cos t + e^{-t} \sin t \end{pmatrix} + i \begin{pmatrix} e^{-t} \sin t \\ -e^{-t} \cos t + 2e^{-t} \sin t \end{pmatrix}. \tag{17}$$

Hence

$$\mathbf{u}(t) = e^{-t} \begin{pmatrix} \cos t \\ 2 \cos t + \sin t \end{pmatrix}, \qquad \mathbf{v}(t) = e^{-t} \begin{pmatrix} \sin t \\ -\cos t + 2 \sin t \end{pmatrix} \tag{18}$$

are a set of real-valued solutions of Eq. (11). To verify that $\mathbf{u}(t)$ and $\mathbf{v}(t)$ are linearly independent, we compute their Wronskian:

$$W(\mathbf{u}, \mathbf{v})(t) = \begin{vmatrix} e^{-t} \cos t & e^{-t} \sin t \\ e^{-t}(2 \cos t + \sin t) & e^{-t}(-\cos t + 2 \sin t) \end{vmatrix}$$

$$= e^{-2t}(-\cos^2 t + 2 \sin t \cos t - 2 \sin t \cos t - \sin^2 t)$$

$$= -e^{-2t}.$$

Since the Wronskian is never zero, it follows that $\mathbf{u}(t)$ and $\mathbf{v}(t)$ constitute a fundamental set of (real-valued) solutions of the system (11).

Example 2. The electric circuit shown in Figure 7.6 is described by the system of differential equations

$$\frac{d}{dt}\begin{pmatrix} I \\ V \end{pmatrix} = \begin{pmatrix} -1 & -1 \\ 2 & -1 \end{pmatrix}\begin{pmatrix} I \\ V \end{pmatrix} \tag{19}$$

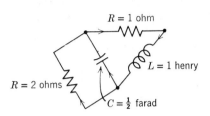

FIGURE 7.6

where I is the current through the inductance and V is the voltage drop across the capacitor. These equations were derived in Problem 8 of Section 7.1. Suppose that at time $t = 0$ the current is 1 ampere and the voltage drop is 2 volts. Find $I(t)$ and $V(t)$ at any time.

Assuming that

$$\begin{pmatrix} I \\ V \end{pmatrix} = \boldsymbol{\xi}e^{rt}, \tag{20}$$

we obtain the algebraic equations

$$\begin{pmatrix} -1 - r & -1 \\ 2 & -1 - r \end{pmatrix}\begin{pmatrix} \xi_1 \\ \xi_2 \end{pmatrix} = \begin{pmatrix} 0 \\ 0 \end{pmatrix}. \tag{21}$$

The eigenvalues are determined from the condition

$$\begin{vmatrix} -1 - r & -1 \\ 2 & -1 - r \end{vmatrix} = r^2 + 2r + 3 = 0; \tag{22}$$

thus $r_1 = -1 + \sqrt{2}i$ and $r_2 = -1 - \sqrt{2}i$. The corresponding eigenvectors are then found from Eq. (21), namely,

$$\boldsymbol{\xi}^{(1)} = \begin{pmatrix} 1 \\ -\sqrt{2}\,i \end{pmatrix}, \qquad \boldsymbol{\xi}^{(2)} = \begin{pmatrix} 1 \\ \sqrt{2}\,i \end{pmatrix}. \tag{23}$$

Thus the eigenvectors are of the form $\mathbf{a} \pm i\mathbf{b}$, where

$$\mathbf{a} = \begin{pmatrix} 1 \\ 0 \end{pmatrix}, \qquad \mathbf{b} = \begin{pmatrix} 0 \\ -\sqrt{2} \end{pmatrix}. \tag{24}$$

Proceeding as in Example 1, or by using Eqs. (9), we obtain the following linearly independent real-valued solutions of Eqs. (19):

$$\mathbf{u}(t) = e^{-t}\begin{pmatrix} \cos\sqrt{2}\,t \\ \sqrt{2}\sin\sqrt{2}\,t \end{pmatrix}, \qquad \mathbf{v}(t) = e^{-t}\begin{pmatrix} \sin\sqrt{2}\,t \\ -\sqrt{2}\cos\sqrt{2}\,t \end{pmatrix}. \tag{25}$$

Hence the general solution of Eqs. (19) is

$$\begin{pmatrix} I \\ V \end{pmatrix} = c_1 e^{-t} \begin{pmatrix} \cos \sqrt{2}\,t \\ \sqrt{2} \sin \sqrt{2}\,t \end{pmatrix} + c_2 e^{-t} \begin{pmatrix} \sin \sqrt{2}\,t \\ -\sqrt{2} \cos \sqrt{2}\,t \end{pmatrix}. \tag{26}$$

Upon imposing the initial conditions

$$\begin{pmatrix} I \\ V \end{pmatrix}(0) = \begin{pmatrix} 1 \\ 2 \end{pmatrix}, \tag{27}$$

we find that

$$c_1 \begin{pmatrix} 1 \\ 0 \end{pmatrix} + c_2 \begin{pmatrix} 0 \\ -\sqrt{2} \end{pmatrix} = \begin{pmatrix} 1 \\ 2 \end{pmatrix}. \tag{28}$$

Thus $c_1 = 1$ and $c_2 = -\sqrt{2}$. The solution of the stated problem is then given by Eq. (26) with these values of c_1 and c_2.

PROBLEMS

In each of Problems 1 through 6, express the general solution of the given system of equations in terms of real-valued functions.

1. $\mathbf{x}' = \begin{pmatrix} 3 & -2 \\ 4 & -1 \end{pmatrix} \mathbf{x}$

2. $\mathbf{x}' = \begin{pmatrix} -1 & -4 \\ 1 & -1 \end{pmatrix} \mathbf{x}$

3. $\mathbf{x}' = \begin{pmatrix} 2 & -5 \\ 1 & -2 \end{pmatrix} \mathbf{x}$

4. $\mathbf{x}' = \begin{pmatrix} 2 & -\frac{5}{2} \\ \frac{9}{5} & -1 \end{pmatrix} \mathbf{x}$

5. $\mathbf{x}' = \begin{pmatrix} 1 & 0 & 0 \\ 2 & 1 & -2 \\ 3 & 2 & 1 \end{pmatrix} \mathbf{x}$

6. $\mathbf{x}' = \begin{pmatrix} -3 & 0 & 2 \\ 1 & -1 & 0 \\ -2 & -1 & 0 \end{pmatrix} \mathbf{x}$

In each of Problems 7 and 8 find the solution of the given initial value problem.

7. $\mathbf{x}' = \begin{pmatrix} 1 & -5 \\ 1 & -3 \end{pmatrix} \mathbf{x}, \quad \mathbf{x}(0) = \begin{pmatrix} 1 \\ 1 \end{pmatrix}$

8. $\mathbf{x}' = \begin{pmatrix} -3 & 2 \\ -1 & -1 \end{pmatrix} \mathbf{x}, \quad \mathbf{x}(0) = \begin{pmatrix} 1 \\ -2 \end{pmatrix}$

In each of Problems 9 and 10 solve the given system of equations by the method of Problem 15 of Section 7.6. Assume that $t > 0$.

9. $t\mathbf{x}' = \begin{pmatrix} -1 & -1 \\ 2 & -1 \end{pmatrix} \mathbf{x}$

10. $t\mathbf{x}' = \begin{pmatrix} 2 & -5 \\ 1 & -2 \end{pmatrix} \mathbf{x}$

11. Consider the electric circuit shown in Figure 7.7. Suppose that $R_1 = R_2 = 4$ ohms, $C = \frac{1}{2}$ farad, and $L = 8$ henrys.

FIGURE 7.7

(a) Show that this circuit is described by the system of differential equations

$$\frac{d}{dt}\begin{pmatrix} I \\ V \end{pmatrix} = \begin{pmatrix} -\frac{1}{2} & -\frac{1}{8} \\ 2 & -\frac{1}{2} \end{pmatrix}\begin{pmatrix} I \\ V \end{pmatrix} \tag{i}$$

where I is the current through the inductance, and V is the voltage drop across the capacitor.

Hint: see Problem 7 of Section 7.1.

(b) Find the general solution of Eqs. (i) in terms of real-valued functions.

(c) Find $I(t)$ and $V(t)$ if $I(0) = 2$ amperes and $V(0) = 3$ volts.

(d) Determine the limiting values of $I(t)$ and $V(t)$ as $t \to \infty$. Do these limiting values depend on the initial conditions?

12. The electric circuit shown in Figure 7.8 is described by the system of differential equations

$$\frac{d}{dt}\begin{pmatrix} I \\ V \end{pmatrix} = \begin{pmatrix} 0 & \dfrac{1}{L} \\ -\dfrac{1}{C} & -\dfrac{1}{RC} \end{pmatrix}\begin{pmatrix} I \\ V \end{pmatrix} \tag{i}$$

FIGURE 7.8

where I is the current through the inductance and V is the voltage drop across the capacitor. These differential equations were derived in Problem 7 of Section 7.1.

(a) Show that the eigenvalues of the coefficient matrix are real and different if $L > 4R^2C$; show that they are complex conjugates if $L < 4R^2C$.

(b) Suppose that $R = 1$ ohm, $C = \frac{1}{2}$ farad, and $L = 1$ henry. Find the general solution of the system (i) in this case.

(c) Find $I(t)$ and $V(t)$ if $I(0) = 2$ amperes and $V(0) = 1$ volt.

(d) For the circuit of part (b) determine the limiting values of $I(t)$ and $V(t)$ as $t \to \infty$. Do these limiting values depend on the initial conditions?

13. In this problem we indicate how to show that $\mathbf{u}(t)$ and $\mathbf{v}(t)$, as given by Eqs. (9), are linearly independent.

(a) Let $r_1 = \lambda + i\mu$ and $\bar{r}_1 = \lambda - i\mu$ be a pair of conjugate eigenvalues of the coefficient matrix $\mathbf{A}$ of Eq. (1); let $\boldsymbol{\xi}^{(1)} = \mathbf{a} + i\mathbf{b}$ and $\bar{\boldsymbol{\xi}}^{(1)} = \mathbf{a} - i\mathbf{b}$ be the corresponding eigenvectors. Recall that it was stated in Section 7.4 that if $r_1 \neq \bar{r}_1$, then $\boldsymbol{\xi}^{(1)}$ and $\bar{\boldsymbol{\xi}}^{(1)}$ are linearly independent. By computing the Wronskian of $\boldsymbol{\xi}^{(1)}$ and $\bar{\boldsymbol{\xi}}^{(1)}$, show that $\mathbf{a}$ and $\mathbf{b}$ are also linearly independent.

(b) By computing the Wronskian of $\mathbf{u}(t)$ and $\mathbf{v}(t)$, and by using the fact that $\mathbf{a}$ and $\mathbf{b}$ are linearly independent, show that $\mathbf{u}(t)$ and $\mathbf{v}(t)$ are linearly independent.

7.8 REPEATED EIGENVALUES

We conclude our consideration of the linear homogeneous system with constant coefficients,

$$\mathbf{x}' = \mathbf{A}\mathbf{x}, \tag{1}$$

with a discussion of the case in which the matrix $\mathbf{A}$ has a repeated eigen value

Once again, the discussion in this section is valid whether A is real or complex. If $r = \rho$ is a k-fold root of the determinantal equation

$$\det (\mathbf{A} - r\mathbf{I}) = 0, \tag{2}$$

then ρ is said to be an eigenvalue of multiplicity k of the matrix $\mathbf{A}$. In this event, there are two possibilities: either there are k linearly independent eigenvectors corresponding to the eigenvalue ρ, or else there are fewer than k such eigenvectors. These possibilities were illustrated in Examples 3 and 4 of Section 7.4; in Example 4 a double eigenvalue was accompanied by two linearly independent eigenvectors, while in Example 3 a double eigenvalue was associated with only one linearly independent eigenvector.

In the first case, let $\boldsymbol{\xi}^{(1)}, \ldots, \boldsymbol{\xi}^{(k)}$ be k linearly independent eigenvectors associated with the eigenvalue ρ of multiplicity k. Then $\mathbf{x}^{(1)}(t) = \boldsymbol{\xi}^{(1)}e^{\rho t}, \ldots,$ $\mathbf{x}^{(k)}(t) = \boldsymbol{\xi}^{(k)}e^{\rho t}$ are k linearly independent solutions of Eq. (1). Thus in this case it makes no difference that the eigenvalue $r = \rho$ is repeated; there is still a fundamental set of solutions of Eq. (1) of the form $\boldsymbol{\xi}e^{rt}$. This case always occurs if the coefficient matrix $\mathbf{A}$ is Hermitian.

However, if the coefficient matrix is not Hermitian, then there may be fewer than k independent eigenvectors corresponding to an eigenvalue ρ of multiplicity k, and if so, there will be fewer than k solutions of Eq. (1) of the form $\boldsymbol{\xi}e^{rt}$ associated with this eigenvalue. Therefore, in order to construct the general solution of Eq. (1) it is necessary to find other solutions of a different form. By analogy with previous results for linear equations of order n, it is natural to seek additional solutions involving products of polynomials and exponential functions. We will first consider an example.

Example.   Find a fundamental set of solutions of the system

$$\mathbf{x}' = \mathbf{A}\mathbf{x} = \begin{pmatrix} 1 & -1 \\ 1 & 3 \end{pmatrix}\mathbf{x}. \tag{3}$$

Assuming that $\mathbf{x} = \boldsymbol{\xi}e^{rt}$ and substituting for $\mathbf{x}$ in Eq. (3) gives

$$\begin{pmatrix} 1 - r & -1 \\ 1 & 3 - r \end{pmatrix}\begin{pmatrix} \xi_1 \\ \xi_2 \end{pmatrix} = \begin{pmatrix} 0 \\ 0 \end{pmatrix}, \tag{4}$$

from which it follows that the eigenvalues of $\mathbf{A}$ are $r_1 = r_2 = 2$. For this value of r, Eqs. (4) imply that $\xi_1 + \xi_2 = 0$. Hence there is only the one eigenvector $\boldsymbol{\xi}$, given by $\boldsymbol{\xi}^T = (1, -1)$, corresponding to the double eigenvalue. Thus one solution of the system (3) is

$$\mathbf{x}^{(1)}(t) = \begin{pmatrix} 1 \\ -1 \end{pmatrix}e^{2t}, \tag{5}$$

but there is no second solution of the form $\mathbf{x} = \boldsymbol{\xi}e^{rt}$. It is natural to attempt to find a second solution of the system (3) of the form

$$\mathbf{x} = \boldsymbol{\eta}te^{2t}, \tag{6}$$

where $\boldsymbol{\eta}$ is a constant vector. Substituting for $\mathbf{x}$ in Eq. (3) gives

$$2\boldsymbol{\eta}te^{2t} + \boldsymbol{\eta}e^{2t} - \mathbf{A}\boldsymbol{\eta}te^{2t} = 0. \tag{7}$$

In order for Eq. (7) to be satisfied for all t it is necessary for the coefficients of te^{2t} and e^{2t} each to be zero. From the term in e^{2t} we find that

$$\boldsymbol{\eta} = 0. \tag{8}$$

Hence there is no nonzero solution of the system (3) of the form (6).

Since Eq. (7) contains terms in both te^{2t} and e^{2t}, it appears that in addition to $\boldsymbol{\eta}te^{2t}$ the second solution must contain a term of the form $\boldsymbol{\zeta}e^{2t}$; in other words, we need to assume that

$$\mathbf{x} = \boldsymbol{\eta}te^{2t} + \boldsymbol{\zeta}e^{2t}, \tag{9}$$

where $\boldsymbol{\eta}$ and $\boldsymbol{\zeta}$ are constant vectors. Upon substituting this expression for $\mathbf{x}$ in Eq. (3) we obtain

$$2\boldsymbol{\eta}te^{2t} + (\boldsymbol{\eta} + 2\boldsymbol{\zeta})e^{2t} = \mathbf{A}(\boldsymbol{\eta}te^{2t} + \boldsymbol{\zeta}e^{2t}). \tag{10}$$

Equating coefficients of te^{2t} and e^{2t} on each side of Eq. (10) gives the conditions

$$(\mathbf{A} - 2\mathbf{I})\boldsymbol{\eta} = 0, \tag{11}$$

and

$$(\mathbf{A} - 2\mathbf{I})\boldsymbol{\zeta} = \boldsymbol{\eta} \tag{12}$$

for the determination of $\boldsymbol{\eta}$ and $\boldsymbol{\zeta}$. Equation (11) is satisfied if $\boldsymbol{\eta}$ is the eigenvector of $\mathbf{A}$ corresponding to the eigenvalue $r = 2$, that is, $\boldsymbol{\eta}^T = (1, -1)$. Since $\det(\mathbf{A} - 2\mathbf{I})$ is zero, we might expect that Eq. (12) cannot be solved. However, this is not necessarily true, since for some vectors $\boldsymbol{\eta}$ it is possible to solve Eq. (12). The required condition on $\boldsymbol{\eta}$ is that $(\boldsymbol{\eta}, \mathbf{y}) = 0$ for each solution $\mathbf{y}$ of $(\mathbf{A}^* - 2\mathbf{I})\mathbf{y} = 0$. It is straightforward to verify that this condition is satisfied here; the details are outlined in Problem 10. The solution of Eq. (12) is not unique since an arbitrary multiple of $\boldsymbol{\eta}$ can be included. Upon solving Eq. (12), we find that

$$\boldsymbol{\zeta} = \begin{pmatrix} 0 \\ -1 \end{pmatrix} + k\begin{pmatrix} 1 \\ -1 \end{pmatrix}, \tag{13}$$

where k is an arbitrary constant. Substituting for $\boldsymbol{\eta}$ and $\boldsymbol{\zeta}$ in Eq. (9) gives

$$\mathbf{x} = \begin{pmatrix} 1 \\ -1 \end{pmatrix}te^{2t} + \begin{pmatrix} 0 \\ -1 \end{pmatrix}e^{2t} + k\begin{pmatrix} 1 \\ -1 \end{pmatrix}e^{2t}. \tag{14}$$

The last term in Eq. (14) is merely a multiple of the first solution $\mathbf{x}^{(1)}(t)$, and may be ignored, but the first two terms constitute a new solution

$$\mathbf{x}^{(2)}(t) = \begin{pmatrix} 1 \\ -1 \end{pmatrix}te^{2t} + \begin{pmatrix} 0 \\ -1 \end{pmatrix}e^{2t}. \tag{15}$$

An elementary calculation shows that $W[\mathbf{x}^{(1)}, \mathbf{x}^{(2)}](t) = -e^{4t}$, and therefore

$\mathbf{x}^{(1)}$ and $\mathbf{x}^{(2)}$ form a fundamental set of solutions of the system (3). The general solution is

$$\mathbf{x} = c_1 \mathbf{x}^{(1)}(t) + c_2 \mathbf{x}^{(2)}(t)$$

$$= c_1 \begin{pmatrix} 1 \\ -1 \end{pmatrix} e^{2t} + c_2 \left[\begin{pmatrix} 1 \\ -1 \end{pmatrix} t e^{2t} + \begin{pmatrix} 0 \\ -1 \end{pmatrix} e^{2t} \right]. \tag{16}$$

One difference between a system of two first order equations and a single second order equation is evident from the example. Recall that for a second order linear equation with a repeated root r_1 of the characteristic equation, a term $c e^{r_1 t}$ in the second solution is not required, since it is a multiple of the first solution. On the other hand, for a system of two first order equations, the term $\boldsymbol{\zeta} e^{r_1 t}$ of Eq. (9) with $r_1 = 2$ is not a multiple of the first solution $\boldsymbol{\xi} e^{r_1 t}$ unless $\boldsymbol{\zeta}$ is proportional to the eigenvector $\boldsymbol{\xi}$ associated with the eigenvalue r_1. Since this is not usually the case, the term $\boldsymbol{\zeta} e^{r_1 t}$ must be retained.

The same procedure can be applied in the general case. Consider again the system (1), and suppose that $r = \rho$ is a double eigenvalue of $\mathbf{A}$, but that there is only one corresponding eigenvector $\boldsymbol{\xi}$. Then one solution [similar to Eq. (5)] is

$$\mathbf{x}^{(1)}(t) = \boldsymbol{\xi} e^{\rho t}, \tag{17}$$

where $\boldsymbol{\xi}$ satisfies

$$(\mathbf{A} - \rho \mathbf{I}) \boldsymbol{\xi} = \mathbf{0}. \tag{18}$$

By proceeding as in the example we find that a second solution [similar to Eq. (15)] is

$$\mathbf{x}^{(2)}(t) = \boldsymbol{\xi} t e^{\rho t} + \boldsymbol{\eta} e^{\rho t}, \tag{19}$$

where $\boldsymbol{\xi}$ satisfies Eq. (18) and $\boldsymbol{\eta}$ is determined from

$$(\mathbf{A} - \rho \mathbf{I}) \boldsymbol{\eta} = \boldsymbol{\xi}. \tag{20}$$

Even though $\det (\mathbf{A} - \rho \mathbf{I}) = 0$, it can be shown that it is always possible to solve Eq. (20) for $\boldsymbol{\eta}$.

If $r = \rho$ is an eigenvalue of the matrix $\mathbf{A}$ of multiplicity greater than two, then there is a greater variety of possibilities. We can illustrate this by considering an eigenvalue of multiplicity three. Suppose first that the triple eigenvalue $r = \rho$ has three linearly independent eigenvectors $\boldsymbol{\xi}^{(1)}$, $\boldsymbol{\xi}^{(2)}$, and $\boldsymbol{\xi}^{(3)}$. In this case the corresponding set of independent solutions is simply

$$\mathbf{x}^{(1)}(t) = \boldsymbol{\xi}^{(1)} e^{\rho t}, \qquad \mathbf{x}^{(2)}(t) = \boldsymbol{\xi}^{(2)} e^{\rho t}, \qquad \mathbf{x}^{(3)}(t) = \boldsymbol{\xi}^{(3)} e^{\rho t}. \tag{21}$$

Now suppose that there is a single linearly independent eigenvector associated with the triple eigenvalue $r = \rho$. Then the first solution is given by Eq. (17), a second solution by Eq. (19), and a third solution is of the form

$$\mathbf{x}^{(3)}(t) = \boldsymbol{\xi} \frac{t^2}{2!} e^{\rho t} + \boldsymbol{\eta} t e^{\rho t} + \boldsymbol{\zeta} e^{\rho t}, \tag{22}$$

where $\boldsymbol{\xi}$ satisfies Eq. (18), $\boldsymbol{\eta}$ satisfies Eq. (20), and $\boldsymbol{\zeta}$ is determined from

$$(\mathbf{A} - \rho\mathbf{I})\boldsymbol{\zeta} = \boldsymbol{\eta}. \tag{23}$$

Again it is always possible to solve Eq. (23) for $\boldsymbol{\zeta}$.

The final possibility is that there are two linearly independent eigenvectors $\boldsymbol{\xi}^{(1)}$ and $\boldsymbol{\xi}^{(2)}$ corresponding to the eigenvalue $r = \rho$. Then two solutions of the system (1) are

$$\mathbf{x}^{(1)}(t) = \boldsymbol{\xi}^{(1)}e^{\rho t}, \qquad \mathbf{x}^{(2)}(t) = \boldsymbol{\xi}^{(2)}e^{\rho t}. \tag{24}$$

A third solution is of the form

$$\mathbf{x}^{(3)}(t) = \boldsymbol{\xi}te^{\rho t} + \boldsymbol{\eta}e^{\rho t}, \tag{25}$$

where $\boldsymbol{\xi}$ satisfies Eq. (18) and $\boldsymbol{\eta}$ is a solution of Eq. (20). Here it is necessary to choose $\boldsymbol{\xi}$ as a linear combination of the eigenvectors $\boldsymbol{\xi}^{(1)}$ and $\boldsymbol{\xi}^{(2)}$ in such a way that the equation

$$(\mathbf{A} - \rho\mathbf{I})\boldsymbol{\eta} = \boldsymbol{\xi} \tag{26}$$

is solvable. If we write

$$\boldsymbol{\xi} = c_1\boldsymbol{\xi}^{(1)} + c_2\boldsymbol{\xi}^{(2)} \tag{27}$$

then we must choose c_1 and c_2 so that $(\boldsymbol{\xi}, \mathbf{y}) = 0$ for every $\mathbf{y}$ that satisfies $(\mathbf{A}^* - \rho\mathbf{I})\mathbf{y} = \mathbf{0}$. For this choice of $\boldsymbol{\xi}$ it is then possible to solve Eq. (26) for $\boldsymbol{\eta}$. The solutions $\mathbf{x}^{(1)}(t)$, $\mathbf{x}^{(2)}(t)$, and $\mathbf{x}^{(3)}(t)$, given by Eqs. (24) and (25), form a set of independent solutions corresponding to the eigenvalue $r = \rho$ in this case. Problem 11 illustrates this situation.

It should be clear that if there is an eigenvalue of even higher multiplicity, then the situation may become even more complicated. For example, if $r = \rho$ is an eigenvalue of multiplicity four, then we must deal with different cases according to whether there are four, three, two, or one linearly independent eigenvectors. A general treatment of problems with repeated eigenvalues is possible, but it involves advanced topics from matrix theory and is beyond the scope of this book. We will concentrate on the computational techniques necessary to find a fundamental set of solutions when the multiplicities of the eigenvalues do not exceed three.

PROBLEMS

In each of Problems 1 through 4 find the general solution of the given system of equations.

1. $\mathbf{x}' = \begin{pmatrix} 3 & -4 \\ 1 & -1 \end{pmatrix}\mathbf{x}$

2. $\mathbf{x}' = \begin{pmatrix} 4 & -2 \\ 8 & -4 \end{pmatrix}\mathbf{x}$

3. $\mathbf{x}' = \begin{pmatrix} 1 & 1 & 1 \\ 2 & 1 & -1 \\ 0 & -1 & 1 \end{pmatrix}\mathbf{x}$

4. $\mathbf{x}' = \begin{pmatrix} 0 & 1 & 1 \\ 1 & 0 & 1 \\ 1 & 1 & 0 \end{pmatrix}\mathbf{x}$

In each of Problems 5 and 6 find the solution of the given initial value problem.

5. $\mathbf{x}' = \begin{pmatrix} 1 & -4 \\ 4 & -7 \end{pmatrix}\mathbf{x}, \qquad \mathbf{x}(0) = \begin{pmatrix} 3 \\ 2 \end{pmatrix}$

6. $\mathbf{x}' = \begin{pmatrix} 1 & 0 & 0 \\ -4 & 1 & 0 \\ 3 & 6 & 2 \end{pmatrix}\mathbf{x}, \qquad \mathbf{x}(0) = \begin{pmatrix} -1 \\ 2 \\ -30 \end{pmatrix}$

7. Show that $r = 2$ is a triple root of the characteristic equation for the system

$$\mathbf{x}' = \begin{pmatrix} 1 & 1 & 1 \\ 2 & 1 & -1 \\ -3 & 2 & 4 \end{pmatrix}\mathbf{x},$$

and find three linearly independent solutions of this system.

In each of Problems 8 and 9 solve the given system of equations by the method of Problem 15 of Section 7.6. Assume that $t > 0$.

8. $t\mathbf{x}' = \begin{pmatrix} 3 & -4 \\ 1 & -1 \end{pmatrix}\mathbf{x}$

9. $t\mathbf{x}' = \begin{pmatrix} 1 & -4 \\ 4 & -7 \end{pmatrix}\mathbf{x}$

10. Consider the example problem in the text,

$$\mathbf{x}' = \mathbf{A}\mathbf{x} = \begin{pmatrix} 1 & -1 \\ 1 & 3 \end{pmatrix}\mathbf{x}$$

with the eigenvector $\boldsymbol{\eta}$ given by $\boldsymbol{\eta}^T = (1, -1)$ corresponding to the double eigenvalue $r = 2$.

(a) Show that all solutions of $(\mathbf{A}^* - 2\mathbf{I})\mathbf{y} = \mathbf{0}$ are of the form $\mathbf{y}^T = c(1, 1)$, where c is an arbitrary constant.

(b) Show that $(\boldsymbol{\eta}, \mathbf{y}) = 0$ for all $\mathbf{y}$ found in part (a).

(c) Solve the system $(\mathbf{A} - 2\mathbf{I})\boldsymbol{\zeta} = \boldsymbol{\eta}$, showing that $\boldsymbol{\zeta}$ must be as given by Eq. (13) in the text.

11. In this problem we indicate how to proceed when there is a triple eigenvalue and only two associated eigenvectors. Consider the system

$$\mathbf{x}' = \mathbf{A}\mathbf{x} = \begin{pmatrix} 5 & -3 & -2 \\ 8 & -5 & -4 \\ -4 & 3 & 3 \end{pmatrix}\mathbf{x}. \qquad \text{(i)}$$

(a) Show that $r = 1$ is a triple eigenvalue of the coefficient matrix $\mathbf{A}$, and that there are only two linearly independent eigenvectors

$$\boldsymbol{\xi}^{(1)} = \begin{pmatrix} 1 \\ 0 \\ 2 \end{pmatrix}, \qquad \boldsymbol{\xi}^{(2)} = \begin{pmatrix} 0 \\ 2 \\ -3 \end{pmatrix}. \qquad \text{(ii)}$$

Find two linearly independent solutions $\mathbf{x}^{(1)}(t)$ and $\mathbf{x}^{(2)}(t)$ of Eq. (i).

(b) To find a third solution assume that

$$\mathbf{x}^{(3)}(t) = \boldsymbol{\xi} t e^t + \boldsymbol{\eta} e^t; \tag{iii}$$

then show that $\boldsymbol{\xi}$ and $\boldsymbol{\eta}$ must satisfy

$$(\mathbf{A} - \mathbf{I})\boldsymbol{\xi} = \mathbf{0}, \tag{iv}$$

$$(\mathbf{A} - \mathbf{I})\boldsymbol{\eta} = \boldsymbol{\xi}. \tag{v}$$

(c) Show that $\boldsymbol{\xi} = c_1 \boldsymbol{\xi}^{(1)} + c_2 \boldsymbol{\xi}^{(2)}$, where c_1 and c_2 are arbitrary constants, is the most general solution of Eq. (iv). In order to solve Eq. (v) it is necessary to choose c_1 and c_2 so that $\boldsymbol{\xi}$ is orthogonal to all solutions of the adjoint system $(\mathbf{A}^* - \mathbf{I})\mathbf{y} = \mathbf{0}$. Determine $\mathbf{A}^*$ and show that the general solution of the adjoint system is

$$\mathbf{y} = \alpha \begin{pmatrix} -2 \\ 1 \\ 0 \end{pmatrix} + \beta \begin{pmatrix} 1 \\ 0 \\ 1 \end{pmatrix}, \tag{vi}$$

where α and β are arbitrary constants.

(d) Calculate $(\boldsymbol{\xi}, \mathbf{y})$ and verify that this quantity is zero for all α and β only if $c_1 = c_2$. It is convenient to choose $c_1 = c_2 = 2$. Then

$$\boldsymbol{\xi} = \begin{pmatrix} 2 \\ 4 \\ -2 \end{pmatrix}. \tag{vii}$$

(e) Solve Eq. (v), using $\boldsymbol{\xi}$ as given by Eq. (vii), and show that

$$\boldsymbol{\eta} = \begin{pmatrix} 0 \\ 0 \\ -1 \end{pmatrix} + k_1 \begin{pmatrix} 1 \\ 0 \\ 2 \end{pmatrix} + k_2 \begin{pmatrix} 0 \\ 2 \\ -3 \end{pmatrix}, \tag{viii}$$

where k_1 and k_2 are arbitrary constants. Use the results given in Eqs. (vii) and (viii) to find a third linearly independent solution $\mathbf{x}^{(3)}(t)$ of Eq. (i).

12. Show that all solutions of the system

$$\mathbf{x}' = \begin{pmatrix} a & b \\ c & d \end{pmatrix} \mathbf{x}$$

approach zero as $t \to \infty$ if and only if $a + d < 0$ and $ad - bc > 0$. Compare this result with that of Problem 10 in Section 3.5.1.

13. Consider again the electric circuit in Problem 12 of Section 7.7. This circuit is described by the system of differential equations

$$\frac{d}{dt} \begin{pmatrix} I \\ V \end{pmatrix} = \begin{pmatrix} 0 & \dfrac{1}{L} \\ -\dfrac{1}{C} & -\dfrac{1}{RC} \end{pmatrix} \begin{pmatrix} I \\ V \end{pmatrix}.$$

(a) Show that the eigenvalues are real and equal if $L = 4R^2 C$.

(b) Suppose that $R = 1$ ohm, $C = 1$ farad, and $L = 4$ henrys. Suppose also that $I(0) = 1$ ampere and $V(0) = 2$ volts. Find $I(t)$ and $V(t)$.

7.9 FUNDAMENTAL MATRICES

The theory of systems of linear differential equations can be further illuminated by introducing the idea of a fundamental matrix. This concept will be particularly useful to us in the next section where we extend the method of variation of parameters to systems of first order linear equations. Suppose that $\mathbf{x}^{(1)}, \ldots, \mathbf{x}^{(n)}$ form a fundamental set of solutions for the equation

$$\mathbf{x}' = \mathbf{P}(t)\mathbf{x} \tag{1}$$

on some interval $\alpha < t < \beta$. Then the matrix

$$\mathbf{\Psi}(t) = \begin{pmatrix} x_1^{(1)}(t) \cdots x_1^{(n)}(t) \\ \cdot \qquad \cdot \\ \cdot \qquad \cdot \\ \cdot \qquad \cdot \\ x_n^{(1)}(t) \cdots x_n^{(n)}(t) \end{pmatrix}, \tag{2}$$

whose columns are the vectors $\mathbf{x}^{(1)}, \ldots, \mathbf{x}^{(n)}$, is said to be a fundamental matrix for the system (1). Note that any fundamental matrix is nonsingular since its columns are linearly independent vectors.

Example 1. Find a fundamental matrix for the system

$$\mathbf{x}' = \begin{pmatrix} 1 & 1 \\ 4 & 1 \end{pmatrix}\mathbf{x}. \tag{3}$$

In Section 7.6 we found that

$$\mathbf{x}^{(1)}(t) = \begin{pmatrix} e^{3t} \\ 2e^{3t} \end{pmatrix}, \qquad \mathbf{x}^{(2)}(t) = \begin{pmatrix} e^{-t} \\ -2e^{-t} \end{pmatrix} \tag{4}$$

are linearly independent solutions of Eq. (3). Thus a fundamental matrix for the system (3) is

$$\mathbf{\Psi}(t) = \begin{pmatrix} e^{3t} & e^{-t} \\ 2e^{3t} & -2e^{-t} \end{pmatrix}. \tag{5}$$

The solution of an initial value problem can be written very compactly in terms of a fundamental matrix. The general solution of Eq. (1) is

$$\mathbf{x} = c_1\mathbf{x}^{(1)}(t) + \ldots + c_n\mathbf{x}^{(n)}(t) \tag{6}$$

or, in terms of $\mathbf{\Psi}(t)$,

$$\mathbf{x} = \mathbf{\Psi}(t)\mathbf{c}, \tag{7}$$

where $\mathbf{c}$ is a constant vector with arbitrary components $c_1, \ldots, c_n$. For an initial value problem consisting of the differential equation (1) and the initial condition

$$\mathbf{x}(t_0) = \mathbf{x}^0, \tag{8}$$

where t_0 is a given point in $\alpha < t < \beta$ and $\mathbf{x}^0$ is a given initial vector, it is only necessary to choose the vector $\mathbf{c}$ in Eq. (7) so as to satisfy the initial condition (8). Hence $\mathbf{c}$ must satisfy

$$\mathbf{\Psi}(t_0)\mathbf{c} = \mathbf{x}^0. \tag{9}$$

Therefore, since $\mathbf{\Psi}(t)$ is nonsingular,

$$\mathbf{c} = \mathbf{\Psi}^{-1}(t_0)\mathbf{x}^0, \tag{10}$$

and

$$\mathbf{x} = \mathbf{\Psi}(t)\mathbf{\Psi}^{-1}(t_0)\mathbf{x}^0 \tag{11}$$

is the solution of the given initial value problem. We emphasize, however, that to solve a given initial value problem one would ordinarily solve Eq. (9) by row reduction and then substitute for $\mathbf{c}$ in Eq. (7), rather than compute $\mathbf{\Psi}^{-1}(t_0)$ and use Eq. (11).

Sometimes it is convenient to make use of the special fundamental matrix, denoted by $\mathbf{\Phi}(t)$, whose columns are the vectors $\mathbf{x}^{(1)}, \ldots, \mathbf{x}^{(n)}$ designated in Theorem 7.6 of Section 7.5. Besides the differential equation (1) these vectors satisfy the initial conditions

$$\mathbf{x}^{(j)}(t_0) = \mathbf{e}^{(j)}, \tag{12}$$

where $\mathbf{e}^{(j)}$ is the unit vector, defined in Theorem 7.6, with a one in the jth position and zeros elsewhere. Thus $\mathbf{\Phi}(t)$ has the property that

$$\mathbf{\Phi}(t_0) = \begin{pmatrix} 1 & 0 & \cdots & 0 \\ 0 & 1 & \cdots & 0 \\ & \cdot & & \cdot \\ & \cdot & & \cdot \\ & \cdot & & \cdot \\ 0 & 0 & \cdots & 1 \end{pmatrix} = \mathbf{I}. \tag{13}$$

We will always reserve the symbol $\mathbf{\Phi}$ to denote the fundamental matrix satisfying the initial condition (13), and use $\mathbf{\Psi}$ when an arbitrary fundamental matrix is intended. In terms of $\mathbf{\Phi}(t)$ the solution of the initial value problem (1) and (8) is even simpler in appearance; since $\mathbf{\Phi}^{-1}(t_0) = \mathbf{I}$, it follows from Eq. (11) that

$$\mathbf{x} = \mathbf{\Phi}(t)\mathbf{x}^0. \tag{14}$$

While the fundamental matrix $\mathbf{\Phi}(t)$ is often more complicated than $\mathbf{\Psi}(t)$, it is especially helpful if the same system of differential equations is to be solved repeatedly subject to many different initial conditions. This corresponds to a given physical system that can be started from many different initial states. If the fundamental matrix $\boldsymbol{\Phi}(t)$ has been determined, then the solution for each set of initial conditions can be found simply by matrix multiplication, as indicated by Eq. (14). The matrix $\boldsymbol{\Phi}(t)$ thus represents a transformation of the initial conditions $\mathbf{x}^0$ into the solution $\mathbf{x}(t)$ at an arbitrary time t. By comparing Eqs. (11) and (14) it is clear that $\boldsymbol{\Phi}(t) = \boldsymbol{\Psi}(t)\boldsymbol{\Psi}^{-1}(t_0)$.

Example 2. For the system (3)

$$\mathbf{x}' = \begin{pmatrix} 1 & 1 \\ 4 & 1 \end{pmatrix}\mathbf{x},$$

just discussed in Example 1, find the fundamental matrix $\boldsymbol{\Phi}$ such that $\boldsymbol{\Phi}(0) = \mathbf{I}$. The columns of $\boldsymbol{\Phi}$ are solutions of Eq. (3) which satisfy the initial conditions

$$\mathbf{x}^{(1)}(0) = \begin{pmatrix} 1 \\ 0 \end{pmatrix}, \qquad \mathbf{x}^{(2)}(0) = \begin{pmatrix} 0 \\ 1 \end{pmatrix}. \tag{15}$$

Since the general solution of Eq. (3) is

$$\mathbf{x} = c_1 \begin{pmatrix} 1 \\ 2 \end{pmatrix} e^{3t} + c_2 \begin{pmatrix} 1 \\ -2 \end{pmatrix} e^{-t},$$

we can find the solution satisfying the first set of these initial conditions by choosing $c_1 = c_2 = \frac{1}{2}$; similarly we obtain the solution satisfying the second set of initial conditions by choosing $c_1 = \frac{1}{4}$ and $c_2 = -\frac{1}{4}$. Hence

$$\boldsymbol{\Phi}(t) = \begin{pmatrix} \frac{1}{2}e^{3t} + \frac{1}{2}e^{-t} & \frac{1}{4}e^{3t} - \frac{1}{4}e^{-t} \\ e^{3t} - e^{-t} & \frac{1}{2}e^{3t} + \frac{1}{2}e^{-t} \end{pmatrix}. \tag{16}$$

Note that the elements of $\boldsymbol{\Phi}(t)$ are more complicated than those of the fundamental matrix $\boldsymbol{\Psi}(t)$ given by Eq. (5); however, it is now easy to determine the solution corresponding to any set of initial conditions.

Now let us turn again to the system

$$\mathbf{x}' = \mathbf{A}\mathbf{x}, \tag{17}$$

where $\mathbf{A}$ is a constant matrix, not necessarily real. We have already described two methods of solving such systems: in Section 7.2 by the method of elimination, and in Section 7.6 by assuming that $\mathbf{x} = \boldsymbol{\xi}e^{rt}$. Here we wish to provide still another viewpoint. The basic reason why a system of equations presents some difficulty is that the equations are usually *coupled*; in other words, some or all of the equations involve more than one, perhaps all, of the dependent variables. This occurs whenever the coefficient matrix $\mathbf{A}$ is not a diagonal matrix. Hence the equations in the system must be solved *simultaneously*, rather than *consecutively*. This observation suggests that one way to solve a system of equations might be by transforming it into an equivalent *uncoupled* system in which each equation contains only one dependent variable. This corresponds to transforming the coefficient matrix **A** into a *diagonal* matrix.

According to the results quoted in Section 7.4, it is possible to do this whenever **A** has a full set of n linearly independent eigenvectors. Let $\boldsymbol{\xi}^{(1)}, \ldots, \boldsymbol{\xi}^{(n)}$ be eigenvectors of **A** corresponding to the eigenvalues $r_1, \ldots, r_n$, and form the transformation matrix **T** whose columns are $\boldsymbol{\xi}^{(1)}, \ldots, \boldsymbol{\xi}^{(n)}$. Then

$$
\mathbf{T} = \begin{pmatrix} \xi_1^{(1)} & \cdots & \xi_1^{(n)} \\ & & \\ \cdot & & \cdot \\ \cdot & & \cdot \\ \cdot & & \cdot \\ \xi_n^{(1)} & \cdots & \xi_n^{(n)} \end{pmatrix}. \tag{18}
$$

Defining a new dependent variable **y** by the relation

$$
\mathbf{x} = \mathbf{T}\mathbf{y}, \tag{19}
$$

we have from Eq. (17) that

$$
\mathbf{T}\mathbf{y}' = \mathbf{A}\mathbf{T}\mathbf{y}, \tag{20}
$$

or

$$
\mathbf{y}' = (\mathbf{T}^{-1}\mathbf{A}\mathbf{T})\mathbf{y}. \tag{21}
$$

From Eq. (40) of Section 7.4,

$$
\mathbf{T}^{-1}\mathbf{A}\mathbf{T} = \mathbf{D} = \begin{pmatrix} r_1 & 0 & \cdots & 0 \\ 0 & r_2 & \cdots & 0 \\ \cdot & \cdot & & \cdot \\ \cdot & \cdot & & \cdot \\ \cdot & \cdot & & \cdot \\ 0 & 0 & \cdots & r_n \end{pmatrix} \tag{22}
$$

is the diagonal matrix whose diagonal elements are the eigenvalues of **A**. Thus **y** satisfies the uncoupled system of equations

$$
\mathbf{y}' = \mathbf{D}\mathbf{y}, \tag{23}
$$

for which a fundamental matrix is the diagonal matrix

$$
\mathbf{Q}(t) = \begin{pmatrix} e^{r_1 t} & 0 & \cdots & 0 \\ 0 & e^{r_2 t} & \cdots & 0 \\ \cdot & \cdot & & \cdot \\ \cdot & \cdot & & \cdot \\ \cdot & \cdot & & \cdot \\ 0 & 0 & \cdots & e^{r_n t} \end{pmatrix}. \tag{24}
$$

A fundamental matrix Ψ for the system (17) is then found from Q by the transformation (19),

$$\Psi = TQ; \tag{25}$$

that is,

$$\Psi(t) = \begin{pmatrix} \xi_1^{(1)} e^{r_1 t} & \cdots & \xi_1^{(n)} e^{r_n t} \\ \cdot & & \cdot \\ \cdot & & \cdot \\ \cdot & & \cdot \\ \xi_n^{(1)} e^{r_1 t} & \cdots & \xi_n^{(n)} e^{r_n t} \end{pmatrix}. \tag{26}$$

Equation (26) is the same result that was obtained earlier. This diagonalization procedure does not afford any computational advantage over the method of Section 7.6 since in either case it is necessary to calculate the eigenvalues and eigenvectors of the coefficient matrix in the system of differential equations. Nevertheless, it is noteworthy that the problem of solving a system of differential equations and the problem of diagonalizing a matrix are mathematically one and the same.

PROBLEMS

In each of Problems 1 through 10 find a fundamental matrix for the given system of equations. In each case also find the fundamental matrix $\Phi(t)$ satisfying $\Phi(0) = I$.

1. $x' = \begin{pmatrix} 3 & -2 \\ 2 & -2 \end{pmatrix} x$

2. $x' = \begin{pmatrix} 4 & -3 \\ 8 & -6 \end{pmatrix} x$

3. $x' = \begin{pmatrix} 2 & -1 \\ 3 & -2 \end{pmatrix} x$

4. $x' = \begin{pmatrix} 1 & 1 \\ 4 & -2 \end{pmatrix} x$

5. $x' = \begin{pmatrix} 2 & -5 \\ 1 & -2 \end{pmatrix} x$

6. $x' = \begin{pmatrix} -1 & -4 \\ 1 & -1 \end{pmatrix} x$

7. $x' = \begin{pmatrix} 5 & -1 \\ 3 & 1 \end{pmatrix} x$

8. $x' = \begin{pmatrix} 3 & -4 \\ 1 & -1 \end{pmatrix} x$

9. $x' = \begin{pmatrix} 1 & 1 & 1 \\ 2 & 1 & -1 \\ -8 & -5 & -3 \end{pmatrix} x$

10. $x' = \begin{pmatrix} 1 & -1 & 4 \\ 3 & 2 & -1 \\ 2 & 1 & -1 \end{pmatrix} x$

11. Show that $\Phi(t) = \Psi(t)\Psi^{-1}(t_0)$, where $\Phi(t)$ and $\Psi(t)$ are as defined in this section.

*12. The method of successive approximations (see Section 2.11) can also be applied to systems of equations. In this problem we indicate formally how it works for linear systems with constant coefficients. Consider the initial value problem

$$x' = Ax, \qquad x(0) = \eta, \tag{i}$$

where A is a constant matrix, and η a prescribed vector.

(a) Assuming that a solution $\mathbf{x} = \boldsymbol{\phi}(t)$ of Eq. (i) exists, show that it must satisfy the integral equation

$$\boldsymbol{\phi}(t) = \boldsymbol{\eta} + \int_0^t \mathbf{A}\boldsymbol{\phi}(s) \, ds. \tag{ii}$$

(b) Start with the initial approximation $\boldsymbol{\phi}^{(0)}(t) = \boldsymbol{\eta}$. Substitute this expression for $\boldsymbol{\phi}(s)$ in the right side of Eq. (ii), and obtain a new approximation $\boldsymbol{\phi}^{(1)}(t)$. Show that

$$\boldsymbol{\phi}^{(1)}(t) = (\mathbf{I} + \mathbf{A}t)\boldsymbol{\eta}. \tag{iii}$$

(c) Repeat this process by substituting the expression in Eq. (iii) for $\boldsymbol{\phi}(s)$ in the right side of Eq. (ii) to obtain a new approximation $\boldsymbol{\phi}^{(2)}(t)$. Show that

$$\boldsymbol{\phi}^{(2)}(t) = \left(\mathbf{I} + \mathbf{A}t + \mathbf{A}^2\frac{t^2}{2}\right)\boldsymbol{\eta}. \tag{iv}$$

(d) By continuing this process obtain a sequence of approximations $\boldsymbol{\phi}^{(0)}$, $\boldsymbol{\phi}^{(1)}$, $\boldsymbol{\phi}^{(2)}, \ldots, \boldsymbol{\phi}^{(n)}, \ldots$. Show by induction that

$$\boldsymbol{\phi}^{(n)}(t) = \left(\mathbf{I} + \mathbf{A}t + \mathbf{A}^2\frac{t^2}{2!} + \cdots + \mathbf{A}^n\frac{t^n}{n!}\right)\boldsymbol{\eta}. \tag{v}$$

(e) The sum appearing in Eq. (v) is a sum of matrices. It is possible to show that each component of this sum converges as $n \to \infty$, and therefore a new matrix $\mathbf{M}(t)$ can be defined as

$$\mathbf{M}(t) = \lim_{n \to \infty} \left(\mathbf{I} + \mathbf{A}t + \mathbf{A}^2\frac{t^2}{2!} + \cdots + \mathbf{A}^n\frac{t^n}{n!}\right) = \mathbf{I} + \sum_{k=1}^{\infty} \mathbf{A}^k\frac{t^k}{k!}. \tag{vi}$$

Assuming that term-by-term differentiation is permissible, show that $\mathbf{M}(t)$ satisfies $\mathbf{M}'(t) = \mathbf{A}\mathbf{M}(t)$, $\mathbf{M}(0) = \mathbf{I}$. Hence show that $\mathbf{M}(t) = \boldsymbol{\Phi}(t)$.

(f) The series in Eq. (vi) resembles the exponential series. Thus it is natural to adopt the notation $e^{\mathbf{A}t}$ for this sum:

$$e^{\mathbf{A}t} = \mathbf{I} + \sum_{k=1}^{\infty} \mathbf{A}^k\frac{t^k}{k!}. \tag{vii}$$

Show that the solution of the initial value problem (i) can then be written as $\mathbf{x} = e^{\mathbf{A}t}\boldsymbol{\eta}$.

*13. Let $\boldsymbol{\Phi}(t)$ denote the fundamental matrix satisfying $\boldsymbol{\Phi}' = \mathbf{A}\boldsymbol{\Phi}$, $\boldsymbol{\Phi}(0) = \mathbf{I}$. In Problem 12 we indicated that it was also reasonable to denote this matrix by $e^{\mathbf{A}t}$. In this problem we show that $\boldsymbol{\Phi}$ does indeed have the principal algebraic properties associated with the exponential function.

(a) Show that $\boldsymbol{\Phi}(t)\boldsymbol{\Phi}(s) = \boldsymbol{\Phi}(t + s)$; that is, $e^{\mathbf{A}t}e^{\mathbf{A}s} = e^{\mathbf{A}(t+s)}$.
Hint: Show that if s is fixed and t is variable, then both $\boldsymbol{\Phi}(t)\boldsymbol{\Phi}(s)$ and $\boldsymbol{\Phi}(t + s)$ satisfy the initial value problem $\mathbf{Z}' = \mathbf{A}\mathbf{Z}$, $\mathbf{Z}(0) = \boldsymbol{\Phi}(s)$.

(b) Show that $\boldsymbol{\Phi}(t)\boldsymbol{\Phi}(-t) = \mathbf{I}$, or $e^{\mathbf{A}t}e^{\mathbf{A}(-t)} = \mathbf{I}$, and hence that $\boldsymbol{\Phi}(-t) = \boldsymbol{\Phi}^{-1}(t)$.

(c) Show that $\boldsymbol{\Phi}(t - s) = \boldsymbol{\Phi}(t)\boldsymbol{\Phi}^{-1}(s)$.

7.10 NONHOMOGENEOUS LINEAR SYSTEMS

In this section we will briefly discuss ways of solving nonhomogeneous linear systems. We begin with systems of the form

$$\mathbf{x}' = \mathbf{A}\mathbf{x} + \mathbf{g}(t), \tag{1}$$

where $\mathbf{A}$ is an $n \times n$ diagonalizable constant matrix and $\mathbf{g}(t)$ is continuous for $\alpha < t < \beta$. We will proceed by the method indicated at the end of Section 7.9. Let $\mathbf{T}$ be the matrix whose columns are the eigenvectors $\xi^{(1)}, \ldots, \xi^{(n)}$ of $\mathbf{A}$, and define a new dependent variable $\mathbf{y}$ by

$$\mathbf{x} = \mathbf{T}\mathbf{y}. \tag{2}$$

Then substituting for $\mathbf{x}$ in Eq. (1), we obtain

$$\mathbf{T}\mathbf{y}' = \mathbf{A}\mathbf{T}\mathbf{y} + \mathbf{g}(t).$$

By multiplying by $\mathbf{T}^{-1}$ it follows that

$$\begin{aligned} \mathbf{y}' &= (\mathbf{T}^{-1}\mathbf{A}\mathbf{T})\mathbf{y} + \mathbf{T}^{-1}\mathbf{g}(t) \\ &= \mathbf{D}\mathbf{y} + \mathbf{h}(t). \end{aligned} \tag{3}$$

where $\mathbf{h}(t) = \mathbf{T}^{-1}\mathbf{g}(t)$, and where $\mathbf{D}$ is the diagonal matrix whose diagonal entries are the eigenvalues $r_1, \ldots, r_n$ of $\mathbf{A}$, arranged in the same order as the corresponding eigenvectors $\xi^{(1)}, \ldots, \xi^{(n)}$ appear as columns of $\mathbf{T}$. Equation (3) is a system of n uncoupled equations for $y_1(t), \ldots, y_n(t)$; as a consequence the equations can be solved separately. In scalar form Eq. (3) has the form

$$y_j'(t) = r_j y_j(t) + h_j(t), \qquad j = 1, \ldots, n, \tag{4}$$

where $h_j(t)$ is a certain linear combination of $g_1(t), \ldots, g_n(t)$. Equation (4) is a first order linear equation and can be solved by the methods of Section 2.1. In fact, we have

$$y_j(t) = e^{r_j t} \int_{t_0}^t e^{-r_j s} h_j(s) \, ds + c_j e^{r_j t}, \qquad j = 1, \ldots, n, \tag{5}$$

where the c_j are arbitrary constants. Finally, the solution $\mathbf{x}$ of Eq. (1) is obtained from Eq. (2).

Example. Find the general solution of the system

$$\mathbf{x}' = \begin{pmatrix} -4 & 2 \\ 2 & -1 \end{pmatrix} \mathbf{x} + \begin{pmatrix} t^{-1} \\ 2t^{-1} + 4 \end{pmatrix} = \mathbf{A}\mathbf{x} + \mathbf{g}(t) \tag{6}$$

on the interval $t > 0$.

Proceeding as in Section 7.6, we find that the eigenvalues of the coefficient matrix are $r_1 = 0$ and $r_2 = -5$, and the corresponding eigenvectors are

$$\xi^{(1)} = \begin{pmatrix} 1 \\ 2 \end{pmatrix}, \qquad \xi^{(2)} = \begin{pmatrix} -2 \\ 1 \end{pmatrix}. \tag{7}$$

Before writing down the matrix $\mathbf{T}$ of eigenvectors we recall that eventually we must find $\mathbf{T}^{-1}$. The coefficient matrix $\mathbf{A}$ is real and symmetric, so we can use the result stated at the end of Section 7.4: $\mathbf{T}^{-1}$ is simply the adjoint or (since $\mathbf{T}$ is real) the transpose of $\mathbf{T}$, provided the eigenvectors of $\mathbf{A}$ are normalized so that $(\boldsymbol{\xi}, \boldsymbol{\xi}) = 1$. Thus, on normalizing $\boldsymbol{\xi}^{(1)}$ and $\boldsymbol{\xi}^{(2)}$ we have

$$\mathbf{T} = \frac{1}{\sqrt{5}}\begin{pmatrix} 1 & -2 \\ 2 & 1 \end{pmatrix}, \qquad \mathbf{T}^{-1} = \frac{1}{\sqrt{5}}\begin{pmatrix} 1 & 2 \\ -2 & 1 \end{pmatrix}. \tag{8}$$

Letting $\mathbf{x} = \mathbf{Ty}$ and substituting for $\mathbf{x}$ in Eq. (6), we obtain the following system of equations for the new dependent variable $\mathbf{y}$:

$$\mathbf{y}' = \mathbf{Dy} + \mathbf{T}^{-1}\mathbf{g}(t) = \begin{pmatrix} 0 & 0 \\ 0 & -5 \end{pmatrix}\mathbf{y} + \frac{1}{\sqrt{5}}\begin{pmatrix} 5t^{-1} + 8 \\ 4 \end{pmatrix}. \tag{9}$$

Thus

$$y_1' = \frac{1}{\sqrt{5}}\left(\frac{5}{t} + 8\right), \qquad y_2' = -5y_2 + \frac{4}{\sqrt{5}} \tag{10}$$

and

$$\mathbf{y} = \frac{1}{\sqrt{5}}\begin{pmatrix} 5 \ln t + 8t + c_1 \\ \dfrac{4}{5} + c_2 e^{-5t} \end{pmatrix}. \tag{11}$$

Finally

$$\mathbf{x} = \mathbf{Ty} = \begin{pmatrix} 1 \\ 2 \end{pmatrix}\ln t + \frac{8}{5}\begin{pmatrix} 1 \\ 2 \end{pmatrix}t + \frac{4}{25}\begin{pmatrix} -2 \\ 1 \end{pmatrix} + \hat{c}_1\begin{pmatrix} 1 \\ 2 \end{pmatrix} + \hat{c}_2\begin{pmatrix} -2 \\ 1 \end{pmatrix}e^{-5t}. \tag{12}$$

The last two terms of the solution (12) of Eq. (6) comprise the general solution of the corresponding homogeneous system. The first three terms are a particular solution of the given nonhomogeneous system. Thus, the solution obtained here has the same basic structure as that obtained in Chapters 3 and 5 for linear scalar equations.

Now let us turn to more general problems in which the coefficient matrix is not constant or not diagonalizable. Let

$$\mathbf{x}' = \mathbf{P}(t)\mathbf{x} + \mathbf{g}(t) \tag{13}$$

where $\mathbf{P}(t)$ and $\mathbf{g}(t)$ are continuous on $\alpha < t < \beta$. Assume that a fundamental matrix $\boldsymbol{\Psi}(t)$ for the corresponding homogeneous system

$$\mathbf{x}' = \mathbf{P}(t)\mathbf{x} \tag{14}$$

has been found. We will use the method of variation of parameters to construct a particular solution, and hence the general solution, of the nonhomogeneous system (13).

Since the general solution of the homogeneous system (14) is $\mathbf{\Psi}(t)\mathbf{c}$, it is natural to proceed as in Section 3.6.2, and to seek a solution of the nonhomogeneous system (13) by replacing the constant vector $\mathbf{c}$ by a vector function $\mathbf{u}(t)$. Thus, we assume that

$$\mathbf{x} = \mathbf{\Psi}(t)\mathbf{u}(t), \tag{15}$$

where $\mathbf{u}(t)$ is a vector to be found. Upon differentiating $\mathbf{x}$ as given by Eq. (15) and requiring that Eq. (13) be satisfied, we obtain

$$\mathbf{\Psi}'(t)\mathbf{u}(t) + \mathbf{\Psi}(t)\mathbf{u}'(t) = \mathbf{P}(t)\mathbf{\Psi}(t)\mathbf{u}(t) + \mathbf{g}(t). \tag{16}$$

Since $\mathbf{\Psi}(t)$ is a fundamental matrix, $\mathbf{\Psi}'(t) = \mathbf{P}(t)\mathbf{\Psi}(t)$, and hence Eq. (16) reduces to

$$\mathbf{\Psi}(t)\mathbf{u}'(t) = \mathbf{g}(t). \tag{17}$$

Recall that $\mathbf{\Psi}(t)$ is nonsingular on any interval where $\mathbf{P}$ is continuous. Hence $\mathbf{\Psi}^{-1}(t)$ exists, and therefore

$$\mathbf{u}'(t) = \mathbf{\Psi}^{-1}(t)\mathbf{g}(t). \tag{18}$$

Thus for $\mathbf{u}(t)$ we can select any vector from the class of vectors that satisfy Eq. (18); these vectors are determined only up to an arbitrary additive constant (vector); therefore we denote $\mathbf{u}(t)$ by

$$\mathbf{u}(t) = \int^{t} \mathbf{\Psi}^{-1}(s)\mathbf{g}(s)\,ds + \mathbf{c}, \tag{19}$$

where the constant vector $\mathbf{c}$ is arbitrary. Finally, substituting for $\mathbf{u}(t)$ in Eq. (15) gives the solution $\mathbf{x}$ of the system (13):

$$\mathbf{x} = \mathbf{\Psi}(t)\mathbf{c} + \mathbf{\Psi}(t)\int^{t} \mathbf{\Psi}^{-1}(s)\mathbf{g}(s)\,ds. \tag{20}$$

Since $\mathbf{c}$ is arbitrary, any initial condition at a point $t = t_0$ can be satisfied by an appropriate choice of $\mathbf{c}$. Thus, every solution of the system (13) is contained in the expression given by Eq. (20); it is therefore the general solution of Eq. (13). Note that the first term on the right side of Eq. (20) is the general solution of the corresponding homogeneous system (14), and the second term is a particular solution of Eq. (13) itself.

Now let us consider the initial value problem consisting of the differential equation (13) and the initial condition

$$\mathbf{x}(t_0) = \mathbf{x}^0. \tag{21}$$

We can write the solution of this problem most conveniently if we choose for the particular solution in Eq. (20) the specific one that is zero when $t = t_0$. This can be done by using t_0 as the lower limit of integration in Eq. (20), so that the general solution of the differential equation takes the form

$$\mathbf{x} = \mathbf{\Psi}(t)\mathbf{c} + \mathbf{\Psi}(t)\int_{t_0}^{t} \mathbf{\Psi}^{-1}(s)\mathbf{g}(s)\,ds. \tag{22}$$

The initial condition (21) can also be satisfied provided that

$$\mathbf{c} = \Psi^{-1}(t_0)\mathbf{x}^0. \tag{23}$$

Therefore

$$\mathbf{x} = \Psi(t)\Psi^{-1}(t_0)\mathbf{x}^0 + \Psi(t) \int_{t_0}^t \Psi^{-1}(s)\mathbf{g}(s)\, ds \tag{24}$$

is the solution of the given initial value problem. Again, while it is helpful to use Ψ^{-1} in order to write the solutions (20) and (24), it is usually better in particular cases to solve the necessary equations by row reduction rather than to calculate Ψ^{-1} and substitute into Eqs. (20) and (24).

The solution (24) takes a slightly simpler form if we use the fundamental matrix $\Phi(t)$ satisfying $\Phi(t_0) = \mathbf{I}$. In this case we have

$$\mathbf{x} = \Phi(t)\mathbf{x}^0 + \Phi(t) \int_{t_0}^t \Phi^{-1}(s)\mathbf{g}(s)\, ds. \tag{25}$$

Equation (25) can be simplified further if the coefficient matrix $\mathbf{P}(t)$ is a constant matrix (see Problem 13).

In some cases a more convenient way to find a particular solution of the nonhomogeneous system (13) is the method of undetermined coefficients. To make use of this method, one assumes the form of the solution with some or all of the coefficients unspecified, and then seeks to determine these coefficients so as to satisfy the differential equation. As a practical matter, this method is applicable only if the coefficient matrix $\mathbf{P}$ is a constant matrix, and if the components of $\mathbf{g}$ are polynomial, exponential, or sinusoidal functions, or sums of products of these. Only in these cases can the correct form of the solution be predicted in a simple and systematic manner. The procedure for choosing the form of the solution is substantially the same as that given in Section 3.6.1 for linear second order equations. The main difference is illustrated by the case of a nonhomogeneous term of the form $\mathbf{v}e^{\lambda t}$, where λ is a simple root of the characteristic equation. In this situation, rather than assuming a solution of the form $\mathbf{a}te^{\lambda t}$, it is necessary to use $\mathbf{a}te^{\lambda t} + \mathbf{b}e^{\lambda t}$, where $\mathbf{a}$ and $\mathbf{b}$ are determined by substituting into the differential equation.

PROBLEMS

In each of Problems 1 through 7 solve the given system of equations.

1. $\mathbf{x}' = \begin{pmatrix} 2 & -1 \\ 3 & -2 \end{pmatrix}\mathbf{x} + \begin{pmatrix} e^t \\ t \end{pmatrix}$

2. $\mathbf{x}' = \begin{pmatrix} 1 & \sqrt{3} \\ \sqrt{3} & -1 \end{pmatrix}\mathbf{x} + \begin{pmatrix} e^t \\ \sqrt{3}e^{-t} \end{pmatrix}$

3. $\mathbf{x}' = \begin{pmatrix} 2 & -5 \\ 1 & -2 \end{pmatrix}\mathbf{x} + \begin{pmatrix} -\cos t \\ \sin t \end{pmatrix}$

4. $\mathbf{x}' = \begin{pmatrix} 1 & 1 \\ 4 & -2 \end{pmatrix}\mathbf{x} + \begin{pmatrix} e^{-2t} \\ -2e^t \end{pmatrix}$

5. $\mathbf{x}' = \begin{pmatrix} 4 & -2 \\ 8 & -4 \end{pmatrix}\mathbf{x} + \begin{pmatrix} t^{-3} \\ -t^{-2} \end{pmatrix}$, $t > 0$

6. $\mathbf{x'} = \begin{pmatrix} 2 & -5 \\ 1 & -2 \end{pmatrix} \mathbf{x} + \begin{pmatrix} 0 \\ \cos t \end{pmatrix}$, $0 < t < \pi$

7. $\mathbf{x'} = \begin{pmatrix} 2 & -5 \\ 1 & -2 \end{pmatrix} \mathbf{x} + \begin{pmatrix} \csc t \\ \sec t \end{pmatrix}$, $\dfrac{\pi}{2} < t < \pi$

8. The electric circuit shown in Figure 7.9 is described by the system of differential equations

$$\frac{d\mathbf{x}}{dt} = \begin{pmatrix} -\frac{1}{2} & -\frac{1}{8} \\ 2 & -\frac{1}{2} \end{pmatrix} \mathbf{x} + \begin{pmatrix} \frac{1}{2} \\ 0 \end{pmatrix} I(t), \qquad \text{(i)}$$

where x_1 is the current through the inductance, x_2 is the voltage drop across the capacitance, and $I(t)$ is the current supplied by the external source.

$L = 8$ henrys

$I(t)$

$R = 4$ ohms

$C = \frac{1}{2}$ farad

$R = 4$ ohms

FIGURE 7.9

(a) Determine a fundamental matrix $\mathbf{\Psi}(t)$ for the homogeneous system corresponding to Eqs. (i). Refer to Problem 11 of Section 7.7.

(b) If $I(t) = e^{-t/2}$, determine the solution of the system (i) that also satisfies the initial conditions $\mathbf{x}(0) = \mathbf{0}$.

9. The electric circuit shown in Figure 7.10 is described by the system of differential equations

$C = 1$ farad

$V(t)$

$R = \frac{1}{3}$ ohm

$L = \frac{1}{2}$ henry

FIGURE 7.10

$$\frac{d\mathbf{x}}{dt} = \begin{pmatrix} 0 & 2 \\ -1 & -3 \end{pmatrix} \mathbf{x} + \begin{pmatrix} 1 \\ 3 \end{pmatrix} V(t), \qquad \text{(i)}$$

where x_1 is the current through the inductance, x_2 is the voltage drop across the capacitor, and $V(t)$ is the voltage supplied by the external source.

(a) Determine a fundamental matrix $\mathbf{\Psi}(t)$ for the homogeneous system corresponding to Eqs. (i).

(b) Let

$$V(t) = \begin{cases} 1, & 0 \le t \le 1 \\ 0, & t > 1 \end{cases}$$

and determine the solution of the system (i) that also satisfies the initial conditions $\mathbf{x}(0) = \mathbf{0}$.

In each of Problems 10 and 11 verify that the given vector is the general solution of the corresponding homogeneous system, and then solve the nonhomogeneous system. Assume that $t > 0$.

10. $t\mathbf{x'} = \begin{pmatrix} 2 & -1 \\ 3 & -2 \end{pmatrix} \mathbf{x} + \begin{pmatrix} 1 - t^2 \\ 2t \end{pmatrix}$, $\mathbf{x}^{(c)} = c_1 \begin{pmatrix} 1 \\ 1 \end{pmatrix} t + c_2 \begin{pmatrix} 1 \\ 3 \end{pmatrix} t^{-1}$

11. $t\mathbf{x'} = \begin{pmatrix} 3 & -2 \\ 2 & -2 \end{pmatrix} \mathbf{x} + \begin{pmatrix} -2t \\ t^4 - 1 \end{pmatrix}$, $\mathbf{x}^{(c)} = c_1 \begin{pmatrix} 1 \\ 2 \end{pmatrix} t^{-1} + c_2 \begin{pmatrix} 2 \\ 1 \end{pmatrix} t^2$

12. Show that the general solution of the system $\mathbf{x}' = \mathbf{P}(t)\mathbf{x} + \mathbf{g}(t)$ may be expressed as $\mathbf{x} = \mathbf{x}^{(c)}(t) + \mathbf{x}^{(p)}(t)$, where $\mathbf{x}^{(c)}$ is the general solution of the corresponding homogeneous system, and $\mathbf{x}^{(p)}$ is any solution of the given nonhomogeneous system.

*13. Consider the initial value problem

$$\mathbf{x}' = \mathbf{A}\mathbf{x} + \mathbf{g}(t), \qquad \mathbf{x}(0) = \mathbf{x}^0.$$

(a) By referring to Problem 13(c) in Section 7.9, show that

$$\mathbf{x} = \mathbf{\Phi}(t)\mathbf{x}^0 + \int_0^t \mathbf{\Phi}(t - s)\mathbf{g}(s) \, ds.$$

(b) Show also that

$$\mathbf{x} = e^{\mathbf{A}t}\mathbf{x}^0 + \int_0^t e^{\mathbf{A}(t-s)} \mathbf{g}(s) \, ds.$$

14. Use the method of undetermined coefficients to find the general solution of each of the following systems.

(a) $\mathbf{x}' = \begin{pmatrix} 1 & 1 \\ 4 & 1 \end{pmatrix} \mathbf{x} + \begin{pmatrix} 2 \\ -1 \end{pmatrix} e^t$ (b) $\mathbf{x}' = \begin{pmatrix} 2 & -1 \\ 3 & -2 \end{pmatrix} \mathbf{x} + \begin{pmatrix} 1 \\ -1 \end{pmatrix} e^t$

REFERENCES

The books listed below are representative of many recent introductory books on matrices and linear algebra:

Campbell, H. G., *Linear Algebra with Applications*, Appleton-Century-Crofts, New York, 1971.

Kolman, B., *Elementary Linear Algebra*, Macmillan, New York, 1970.

Noble, B., *Applied Linear Algebra*, Prentice-Hall, Englewood Cliffs, N.J., 1969.

Schneider, H. and Barker, G. P., *Matrices and Linear Algebra*, Holt, New York, 1968.

Stewart, F. M., *Introduction to Linear Algebra*, Van Nostrand, Princeton, 1963.

Strang, G., *Linear Algebra and its Applications*, Academic Press, New York, 1976.

Williams, G., *Computational Linear Algebra with Models*, Allyn and Bacon, Boston, 1975.

There are many books on electrical networks, of which the following is a sample:

Bose, A. G. and Stevens, K. N., *Introductory Network Theory*, Harper and Row, New York, 1965.

CHAPTER 8

Numerical Methods

8.1 INTRODUCTION

In this chapter we will consider numerical methods by which we can construct, in tabular form, the solution of a given differential equation and initial conditions. For example, a table of values of e^x, which is the solution of the initial value problem

$$y' - y = 0, \qquad y(0) = 1,$$

would represent a numerical solution of this problem. We will primarily consider the first order differential equation

$$y' = f(x, y) \tag{1}$$

with the initial condition

$$y(x_0) = y_0. \tag{2}$$

We will also assume that the functions f and f_y are continuous in some rectangle in the xy plane including the point (x_0, y_0) so that, according to Theorem 2.2, there exists a unique solution $y = \phi(x)$ of the given problem in some interval about x_0. As we emphasized in Section 2.3, if Eq. (1) is nonlinear the interval about x_0 in which the solution ϕ exists may be difficult to determine, and may have no simple relationship to the function f. In all of our discussion of Eq. (1) we will assume that a unique solution satisfying the initial condition (2) exists on the interval of interest.

In some cases $f(x, y)$ is so simple that Eq. (1) can be integrated directly. However, in many engineering and scientific problems this is not the case, and it is natural to consider numerical procedures for obtaining an approximate solution of the problem. Even if Eq. (1) can be integrated directly, it may be more difficult to evaluate the analytic solution than to solve the original initial value problem numerically. For example, the analytic solution may be in the form of a complicated implicit relation $F(x, y) = 0$. Nevertheless, the importance of analytic methods should not be underestimated. They provide valuable information about the existence, uniqueness, and qualitative behavior of solutions.

By a numerical procedure for solving the initial value problem given by Eqs. (1) and (2) is meant a procedure for constructing approximate values $y_0, y_1, y_2, \ldots, y_n, \ldots$ of the solution ϕ at the points $x_0 < x_1 < x_2 < \cdots < x_n < \cdots$ (see Figure 8.1). More precisely, such a numerical procedure is referred to

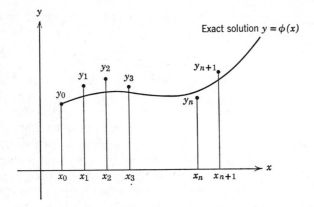

FIGURE 8.1

as a *discrete variable method* since we are replacing a problem involving continuous variables by one involving discrete variables.

Our first concern is to determine y_1, knowing y_0 from the initial condition and $\phi'(x_0) = f(x_0, y_0)$ from the differential equation (1). Then knowing y_1 we determine y_2, and so on. In determining y_2 we can either use the same method as in going from x_0 to x_1, or we can use a different method that takes account of the fact that we know both y_0 at x_0 and y_1 at x_1.

Methods that require only a knowledge of y_n to determine y_{n+1} are known as *one-step*, or *stepwise*, or *starting methods*. Methods that make use of data at more than the previous point, say y_n, y_{n-1}, y_{n-2} to determine y_{n+1}, are known as *multistep* or *continuing methods*. Clearly, if a continuing method involving y_n and y_{n-1} is used, we must use a starting method (hence the name starting method) at the first step to determine y_1; the continuing method can then be used to determine y_2, and so on. On the other hand, a stepwise or starting method may be used for all of the computations. In Sections 8.2 through 8.7 we will discuss several stepwise methods, and in Section 8.8 we will discuss a typical multistep method. In Section 8.9 we will briefly consider systems of equations.

With the advent of high-speed electronic digital computers that can do tens or hundreds of thousands of computations per second, the use of numerical methods to solve initial value problems has become commonplace. Many problems that are hopeless to solve analytically, and that would require years to solve by hand computation, can now be solved in a matter of minutes* using high-speed computers. On the other hand, not all questions about solving differential equations can be answered by using numerical methods. For example, it may be difficult to answer such questions as how a solution depends on y_0 (which is often of interest)

---

* That is, the actual computation may require only a few minutes, but much more time may be required to set up the problem for computation—perhaps weeks or even months. The preparation time depends strongly on whether this is the *first* such problem to be solved, or whether it falls into a class of problems for which a computational program is already available. Accurate programs for solving general classes of initial value programs are widely available.

or on other parameters in the problem. Such questions would be considerably simpler if the analytic solution were known.

Moreover, numerical methods raise serious questions of their own. In the first place, there is the question of convergence. That is, as the spacing between the points $x_0, x_1, x_2, \ldots, x_n, \ldots$ approaches zero, do the values of the numerical solution $y_1, y_2, \ldots, y_n, \ldots$ approach the values of the exact solution? There is also the question of estimating the error made in computing the values $y_1, y_2, \ldots, y_n$. This error arises from two sources: first, the formula used in the numerical method is only an approximate one, which causes a *formula error*, or *truncation error*, or *discretization error;* second, it is possible to carry only a limited number of digits in any computation, which causes a *round-off error*. In Sections 8.3 and 8.7 we will discuss briefly the significance of formula and round-off errors. Another important question arises when multistep methods are used. It is possible that the numerical formula can have solutions, usually called parasitic solutions, that do not correspond to solutions of the original differential equation. If these extraneous solutions grow with increasing x they can cause serious difficulties. This is discussed in Section 8.8.

The notation that will be used throughout this chapter is as follows. The exact solution of the initial value problem given by Eqs. (1) and (2) will be denoted by ϕ (or $y = \phi(x)$); then the value of the exact solution at x_n is $\phi(x_n)$. For a given numerical procedure the symbols y_n and $y'_n = f(x_n, y_n)$ will denote the approximate values of the exact solution and its derivative at the point x_n. Clearly $\phi(x_0) = y_0$, but in general $\phi(x_n) \neq y_n$ for $n \geq 1$. Similarly $\phi'(x_0) = y'_0$, but in general $\phi'(x_n) = f[x_n, \phi(x_n)]$ is not equal to $y'_n = f(x_n, y_n)$ for $n \geq 1$. Further, in all of the discussion we will use a uniform spacing or step size h on the x axis. Thus $x_1 = x_0 + h, x_2 = x_1 + h = x_0 + 2h$, and in general $x_n = x_0 + nh$.

Finally we would like to call the reader's attention to the following point. The first problem at the end of Sections 8.2 through 8.6 and 8.8 deals with the same initial value problem, and so do the second and third problems. By using one of these problems for practice work, it is possible to compare the accuracy and amount of labor required by the various numerical procedures.

8.2   THE EULER OR TANGENT LINE METHOD

One of the simplest stepwise methods of solving the initial value problem

$$y' = f(x, y), \qquad y(x_0) = y_0 \tag{1}$$

numerically is the tangent line method, which was first used by Euler. Since x_0 and y_0 are known, the slope of the tangent line to the solution at $x = x_0$, namely $\phi'(x_0) = f(x_0, y_0)$, is also known. Hence we can construct the tangent line to the solution at x_0, and then obtain an approximate value y_1 of $\phi(x_1)$ by moving along the tangent line to x_1 (see Figure 8.2). Thus

$$y_1 = y_0 + \phi'(x_0)(x_1 - x_0)$$

$$= y_0 + f(x_0, y_0)(x_1 - x_0). \tag{2}$$

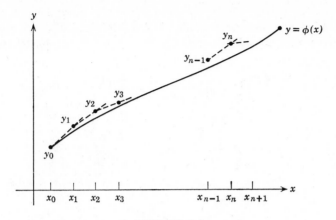

FIGURE 8.2

Once y_1 is determined we can compute $y_1' = f(x_1, y_1)$, which is an approximate value of $\phi'(x_1) = f[x_1, \phi(x_1)]$, the slope of the tangent line to the actual solution at x_1. Using this slope we obtain

$$y_2 = y_1 + y_1'(x_2 - x_1)$$
$$= y_1 + f(x_1, y_1)(x_2 - x_1),$$

and in general

$$y_{n+1} = y_n + f(x_n, y_n)(x_{n+1} - x_n).$$

If we assume that there is a uniform step size h between the points $x_0, x_1, x_2, \ldots$, then $x_{n+1} = x_n + h$ and we obtain the Euler formula

$$y_{n+1} = y_n + hf(x_n, y_n)$$
$$= y_n + hy_n', \qquad n = 0, 1, 2, \ldots. \tag{3}$$

The Euler formula is very simple to use, and in discussing the application of the formula it is possible to illustrate many of the important ideas in numerical methods for solving initial value problems. On the other hand, it is not an accurate formula. Before discussing the Euler formula and methods of improving it, let us first illustrate how this result can be used to solve the initial value problem

$$y' = 1 - x + 4y, \tag{4}$$
$$y(0) = 1. \tag{5}$$

This example will be used throughout the rest of this chapter to illustrate and compare different numerical methods. While it is a very simple problem, it still illustrates the improvement that can be gained by using more accurate numerical procedures, and smaller step sizes h in the independent variable. Equation (4) is a linear first order equation, and it is easily verified that the solution satisfying the initial condition (5) is

$$y = \phi(x) = \tfrac{1}{4}x - \tfrac{3}{16} + \tfrac{19}{16}e^{4x}. \tag{6}$$

Example.   Using the Euler formula (3) and a step size $h = 0.1$, determine an approximate value of the solution $y = \phi(x)$ at $x = 0.2$ for the illustrative example $y' = 1 - x + 4y$, $y(0) = 1$.

First we compute $y'_0 = f(0, 1) = 5$; hence

$$y_1 = y_0 + hf(0, 1)$$
$$= 1 + (0.1)5 = 1.5.$$

Next

$$y_2 = y_1 + hf(x_1, y_1)$$
$$= 1.5 + (0.1)f(0.1, 1.5)$$
$$= 1.5 + (0.1)(1 - 0.1 + 6)$$
$$= 2.19.$$

This result should be compared with the exact value, which is $\phi(0.2) = 2.5053299$ correct through eight digits. The error is approximately $2.51 - 2.19 = 0.32$.

Normally, an error this large (a percentage error of 12 percent) is not acceptable.* A better result can be obtained by using a smaller step size. Thus using $h = 0.05$ and taking four steps to reach $x = 0.2$ gives the approximate value 2.3249 for $\phi(0.2)$ with a percentage error of 8 percent. A step size $h = 0.025$ leads to a value of 2.4080117 with a percentage error of 4 percent. Alternatively we can use more accurate procedures that will give satisfactory results without using a very small step size. For example, if the Runge-Kutta method, which is discussed in Section 8.6, is used with $h = 0.1$, the approximate value 2.5050062 for $\phi(0.2)$ is obtained. This result agrees with the exact solution through the first four figures. These points are discussed in more detail in the following sections. In this section we are primarily interested in developing a familiarity with the technique of using a numerical procedure to solve an initial value problem.

Today, when solving an initial value problem numerically one would normally use a high-speed computer to carry out the numerical calculations. The example calculations to illustrate different numerical procedures and difficulties associated with numerical methods have been done on a high-speed computer. However, the practice problems in this chapter are sufficiently simple, although still appropriately illustrative, that the numerical calculations can be done with a pocket calculator or even by hand. In carrying out such a computation it is generally wise to construct a table (essentially a program for a high-speed computer) in which to record the necessary results in a systematic way. Carrying out the computations in a systematic manner reduces the danger of making errors, and if errors are made this makes it easier to locate and correct them.

---

* Lest the student be discouraged we emphasize again that we have purposely chosen an example to illustrate the improvement to be gained by using more accurate procedures and/or a smaller step size h. On the other hand, the present example is extremely simple compared to what may be encountered in practice. It follows that, in general, more accurate procedures are used in constructing numerical solutions of initial value problems in engineering, science, and other fields of applications.

For the present problem we might set up Table 8.1, where we work from left to right along the rows. In Table 8.1 the last two columns contain the values of the exact solution ϕ correct through eight digits and the differences $\phi(x_n) - y_n$. It should be evident from these results that for the present problem the Euler method with a step size of $h = 0.1$ is unsatisfactory.

TABLE 8.1   Numerical Solution of $y' = 1 - x + 4y$, $y(0) = 1$ Using the Euler Method with $h = 0.1$

x_n	y_n	$y_{n+1} = y_n + hf(x_n, y_n)$ $= y_n + h(1 - x_n + 4y_n)$	$\phi(x_{n+1})$	$\phi(x_{n+1}) - y_{n+1}$
0	1.0	1.5	1.6090418	0.1090418
0.1	1.5	2.19	2.5053299	0.3153299
0.2	2.19	3.146	3.8301388	0.6841388
0.3	3.146	4.4744	5.7942260	1.3192260
0.4	4.4744	6.32416	8.7120041	2.3878441
0.5	6.32416		13.052522	
0.6			19.515518	
0.7				

A flow chart for the Euler method for solving the initial value problem (1) for $x_0 \le x \le x_0 + Nh$ is shown in Figure 8.3. Notice that in the flow chart the second box represents the calculation of $f(x_n, y_n)$ for given x_n and y_n. The first time this box is entered we have $x_n = x_0$ and $y_n = y_0$. In succeeding calculations the values of x_n and y_n are determined in the third box for entry into the second box until $x_n = x_0 + Nh$. When this happens the calculation stops.

In concluding this section we point out two alternate methods for deriving the Euler formula (3), which will suggest ways of obtaining better formulas and will also assist in our study of the error in using formula (3).

First, since $y = \phi(x)$ is a solution of the initial value problem (1), on integrating from x_n to x_{n+1} we obtain

$$\int_{x_n}^{x_{n+1}} \phi'(x)\, dx = \int_{x_n}^{x_{n+1}} f[x, \phi(x)]\, dx$$

or

$$\phi(x_{n+1}) = \phi(x_n) + \int_{x_n}^{x_{n+1}} f[x, \phi(x)]\, dx. \qquad (7)$$

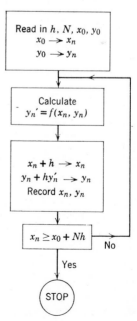

FIGURE 8.3   Flow chart for the Euler method.

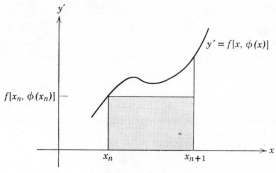

FIGURE 8.4

The integral in Eq. (7) is represented geometrically as the area under the curve in Figure 8.4 between $x = x_n$ and $x = x_{n+1}$. If we approximate the integral by replacing $f[x, \phi(x)]$ by its value $f[x_n, \phi(x_n)]$ at $x = x_n$, then we are approximating the actual area by the area of the shaded rectangle. In this way we obtain

$$\phi(x_{n+1}) \cong \phi(x_n) + f[x_n, \phi(x_n)](x_{n+1} - x_n)$$
$$= \phi(x_n) + hf[x_n, \phi(x_n)]. \tag{8}$$

Finally, to obtain an approximation y_{n+1} for $\phi(x_{n+1})$ we make a second approximation by replacing $\phi(x_n)$ by its approximate value y_n in Eq. (8). This gives the Euler formula $y_{n+1} = y_n + hf(x_n, y_n)$. A more accurate formula can be obtained by approximating the integral more accurately. This is discussed in Section 8.4.

Second, suppose we assume that the solution $y = \phi(x)$ has a Taylor series about the point x_n. Then

$$\phi(x_n + h) = \phi(x_n) + \phi'(x_n)h + \phi''(x_n)\frac{h^2}{2!} + \cdots$$

or

$$\phi(x_{n+1}) = \phi(x_n) + f[x_n, \phi(x_n)]h + \phi''(x_n)\frac{h^2}{2!} + \cdots. \tag{9}$$

If the series is terminated after the first two terms, and $\phi(x_{n+1})$ and $\phi(x_n)$ are replaced by their approximate values y_{n+1} and y_n, we again obtain the Euler formula (3). If more terms in the series are retained, a more accurate formula is obtained. This is discussed in Section 8.5. Further, by using a Taylor series with a remainder it is possible to estimate the magnitude of the error in the formula. This is discussed in the next section.

PROBLEMS

In all computations, unless otherwise specified, retain four digits starting from the first nonzero digit. Thus, for example, the number 1.6846 is to be rounded to 1.685 before further computation. If the fifth digit is 5 or more, round up.

1. Consider the initial value problem

$$y' = 2y - 1, \qquad y(0) = 1.$$

(a) Determine the solution $y = \phi(x)$ and evaluate $\phi(x)$ at $x = 0.1, 0.2, 0.3,$ and 0.4.

(b) Set up a table for determining y_n using the Euler formula (3). Take $h = 0.1$ and determine approximate values of $\phi(0.1), \phi(0.2), \phi(0.3),$ and $\phi(0.4)$. Compare the results with the exact values.

(c) Take $h = 0.05$ and determine approximate values of $\phi(0.05), \phi(0.1), \phi(0.15),$ and $\phi(0.2)$. Compare the results with the results of part (b) and the exact values. The differences in the results for $h = 0.1$ and $h = 0.05$ tend to indicate whether $h = 0.1$ is satisfactory, or whether a smaller step size must be used for this range of x.

2. Carry out parts (a), (b), and (c) of Problem 1 for the initial value problem

$$y' = \tfrac{1}{2} - x + 2y, \qquad y(0) = 1.$$

3. Consider the initial value problem

$$y' = x^2 + y^2, \qquad y(0) = 1.$$

The solution $y = \phi(x)$ of this initial value problem cannot be expressed in terms of elementary functions; hence the use of numerical procedures is more important in this case. However, this equation is a special case of the Riccati equation discussed in Problems 35 and 36 of Section 2.8.

(a) Set up a table for determining y_n using the Euler formula (3). Take $h = 0.1$, and determine approximate values of $\phi(0.1), \phi(0.2), \phi(0.3),$ and $\phi(0.4)$.

(b) Take $h = 0.05$, and determine approximate values of $\phi(0.05), \phi(0.1), \phi(0.15),$ and $\phi(0.2)$. The differences in the results for $h = 0.1$ and $h = 0.05$ tend to indicate whether $h = 0.1$ is satisfactory, or whether a smaller step size should be used for this range of x.

4. Complete the rows corresponding to $n = 5$ and $n = 6$ in Table 8.1.

5. Using three terms in the Taylor series given in Eq. (9), and taking $h = 0.1$, determine approximate values of the solution of the illustrative example $y' = 1 - x + 4y$, $y(0) = 1$ at $x = 0.1$ and 0.2. Compare the results with those using the Euler method and the exact values.

Hint: If $y' = f(x, y)$ what is y''? See Section 8.5.

*6. It can be shown that under suitable conditions on f the numerical solution generated by the Euler method for the initial value problem $y' = f(x, y), y(x_0) = y_0$ converges to the exact solution as the step size h decreases. This is illustrated by the following example. Consider the initial value problem

$$y' = 1 - x + y, \qquad y(x_0) = y_0.$$

(a) Show that the exact solution is $y = \phi(x) = (y_0 - x_0)e^{(x-x_0)} + x$.

(b) Show, using the Euler formula, that

$$y_k = (1 + h)y_{k-1} + h - hx_{k-1}, \qquad k = 1, 2, \ldots.$$

(c) Noting that $y_1 = (1 + h)(y_0 - x_0) + x_1$, show that

$$y_n = (1 + h)^n(y_0 - x_0) + x_n.$$

(d) Consider a fixed point $x > x_0$, and for a given n choose $h = (x - x_0)/n$. Then for any n, $x_n = x$. Note also that $h \to 0$ as $n \to \infty$. Substituting for h in the preceding formula and letting $n \to \infty$ gives the desired result.

Hint: $\lim\limits_{n \to \infty} (1 + a/n)^n = e^a$.

*7. Using the technique discussed in Problem 6, show that the approximate solution obtained by the Euler method converges to the exact solution at any fixed point as $h \to 0$ for each of the following problems.

(a) $y' = y$, $y(0) = 1$

(b) $y' = 2y - 1$, $y(0) = 1$ Hint: $y_1 = (1 + 2h)/2 + 1/2$

(c) $y' = \frac{1}{2} - x + 2y$, $y(0) = 1$ Hint: $y_1 = (1 + 2h) + x_1/2$

8. An alternative procedure for constructing the solution $y = \phi(x)$ of the initial value problem $y' = f(x, y)$, $y(x_0) = y_0$ is to use the method of *iteration*. Integrating the differential equation from x_0 to x gives

$$\phi(x) = y_0 + \int_{x_0}^{x} f[t, \phi(t)] \, dt. \tag{i}$$

If, on the right-hand side of Eq. (i), $\phi(t)$ is replaced by a particular function, the integral can be evaluated and a new function ϕ obtained. This suggests the iteration procedure

$$\phi_{n+1}(x) = y_0 + \int_{x_0}^{x} f[t, \phi_n(t)] \, dt. \tag{ii}$$

With an initial choice for $\phi_0(x)$, Eq. (ii) can be used to generate a sequence of functions $\phi_n(x)$ that approximates the exact solution and, under suitable conditions on $f(x, y)$, actually converges to $\phi(x)$ as $n \to \infty$. Indeed, this iteration procedure was used to establish the existence of a solution of the initial value problem in Section 2.11. For actually computing $\phi(x)$ this procedure is generally unwieldy since it may be difficult, if not impossible, to evaluate the integral of $f[t, \phi_n(t)]$, $n = 0, 1, 2, \ldots$.

Taking $\phi_0(x) = 1$, determine $\phi_3(x)$ for each of the following initial value problems. Also, compute $\phi_2(0.4)$ and $\phi_3(0.4)$ and compare with the result obtained using the Euler method.

(a) $y' = 2y - 1$, $y(0) = 1$ (b) $y' = \frac{1}{2} - x + 2y$, $y(0) = 1$

(c) $y' = x^2 + y^2$, $y(0) = 1$;   compute only $\phi_2(x)$

8.3   THE ERROR

As we mentioned in the introduction to this chapter there are two fundamental sources of error in solving the initial value problem

$$y' = f(x, y), \qquad y(x_0) = y_0 \tag{1}$$

by a numerical procedure such as the Euler method:

$$y_{n+1} = y_n + hf(x_n, y_n), \qquad x_n = x_0 + nh. \tag{2}$$

Let us first assume that our computer is such that we can carry out all computations with complete accuracy; that is, we can retain an infinite number of decimal places. The difference between the exact solution $y = \phi(x)$ and the approximate solution of the initial value problem (1),

$$E_n = \phi(x_n) - y_n, \tag{3}$$

is known as the *formula error*, or the *accumulated formula error*. It arises from two causes: (i) at each step we use an approximate formula to determine y_{n+1}; (ii) the input data at each step do not agree with the exact solution, since in general $\phi(x_n)$ is not equal to y_n. If we assume that the input data are correct, the only error in going one step is that due to the use of an approximate formula. This error is known as the *local formula error e_n*.

In practice, due to the limitations of all computers, it is in general impossible to compute y_{n+1} exactly from the given formula. Thus we have a round-off error due to a lack of computational accuracy. For example, if a computer that can carry only eight digits is used, and y_0 is 1.017325842, the last two digits must be rounded off, which immediately introduces an error in the computation of y_1. Alternatively, if $f(x, y)$ involves functions such as the logarithm or exponential we have a round-off error in carrying out these operations. Just as for the formula error, it is possible to speak of the local round-off error and the accumulated round-off error. The *accumulated round-off error R_n* is defined as

$$R_n = y_n - Y_n, \tag{4}$$

where Y_n is the value *actually computed* by the given numerical procedure—for example, the Euler formula (2).

The absolute value of the total error in computing $\phi(x_n)$ is given by

$$|\phi(x_n) - Y_n| = |\phi(x_n) - y_n + y_n - Y_n|. \tag{5}$$

Making use of the triangle inequality, $|a + b| \le |a| + |b|$, we obtain from Eq. (5)

$$|\phi(x_n) - Y_n| \le |\phi(x_n) - y_n| + |y_n - Y_n|$$

$$\le |E_n| + |R_n|. \tag{6}$$

Thus the total error is bounded by the sum of the absolute values of the formula and round-off errors. For the numerical procedures discussed in this book it is possible to obtain useful estimates of the formula error. However, we shall limit our discussion primarily to the local formula error, which is somewhat simpler. The round-off error is clearly more random in nature. It depends on the type of computer used, the sequence in which the computations are carried out, the method of rounding off, etc. While an analysis of round-off error is beyond the scope of this introductory text, it is possible to say more about it than one would at first expect (see, for example, Henrici). Some of the dangers due to round-off error are discussed in Problems 10, 11, 12, and 13 and in Section 8.7.

Local Formula Error for the Euler Method. Let us assume that the solution $y = \phi(x)$ of the initial value problem (1) has a continuous second derivative in the interval of interest. This assumption is equivalent to assuming that f, f_x, and f_y are continuous in the region of interest. To show this, observe that if f has these properties and if ϕ is a solution of the initial value problem (1), then

$$\phi'(x) = f[x, \phi(x)],$$

and by the chain rule

$$\phi''(x) = f_x[x, \phi(x)] + f_y[x, \phi(x)]\phi'(x)$$
$$= f_x[x, \phi(x)] + f_y[x, \phi(x)]f[x, \phi(x)]. \tag{7}$$

Since the right-hand side of this equation is continuous, ϕ'' is also continuous.

Then making use of a Taylor series with a remainder to expand ϕ about x_n, we obtain

$$\phi(x_n + h) = \phi(x_n) + \phi'(x_n)h + \tfrac{1}{2}\phi''(\bar{x}_n)h^2, \tag{8}$$

where $\bar{x}_n$ is some point in the interval $x_n < \bar{x}_n < x_n + h$. Subtracting Eq. (2) from Eq. (8), and noting that $\phi(x_n + h) = \phi(x_{n+1})$ and $\phi'(x_n) = f[x_n, \phi(x_n)]$, gives

$$\phi(x_{n+1}) - y_{n+1} = [\phi(x_n) - y_n] + h\{f[x_n, \phi(x_n)] - f(x_n, y_n)\} + \tfrac{1}{2}\phi''(\bar{x}_n)h^2. \tag{9}$$

To compute the local formula error we assume that the data at the nth step are correct, that is, $y_n = \phi(x_n)$. Then we immediately obtain from Eq. (9) that the local formula error e_{n+1} is

$$e_{n+1} = \phi(x_{n+1}) - y_{n+1} = \tfrac{1}{2}\phi''(\bar{x}_n)h^2. \tag{10}$$

Thus the local formula error for the Euler method is proportional to the square of the step size h and to the second derivative of ϕ. For the fixed interval $a = x_0 \le x \le b$, the absolute value of the local formula error for any step is bounded by $Mh^2/2$ where M is the maximum of $|\phi''(x)|$ on $a \le x \le b$. The primary difficulty in estimating the local formula error is that of obtaining an accurate estimate of M. However, notice that M is independent of h; hence, reducing h by a factor of $\tfrac{1}{2}$ reduces the error bound by a factor of $\tfrac{1}{4}$, and a reduction by a factor of $\tfrac{1}{10}$ in h reduces the error bound by a factor of $\tfrac{1}{100}$.

More important than the local formula error is the accumulated formula error E_n. Nevertheless, an estimate of the local formula error does give a better understanding of the numerical procedure and a means for comparing the accuracy of different numerical procedures. The analysis for estimating E_n is more difficult than that for e_n. However, knowing the local formula error we can make an *intuitive* estimate of the accumulated formula error at a fixed $\bar{x} > x_0$ as follows. Suppose we take n steps in going from x_0 to $\bar{x} = x_0 + nh$. In each step the error is at most $Mh^2/2$; thus the error in n steps is at most $nMh^2/2$. Noting that $n = (\bar{x} - x_0)/h$, we find that the accumulated formula error for the Euler method in going from x_0 to $\bar{x}$ is bounded by

$$n\frac{Mh^2}{2} = (\bar{x} - x_0)\frac{Mh}{2}. \tag{11}$$

While this argument is not correct since it does not take account of the effect that an error at one step will have in succeeding steps, it is shown in Problem 5 that on any finite interval the accumulated formula error using the Euler method is no greater than a constant times h. Thus in going from x_0 to a fixed point $\bar{x}$, the accumulated formula error can be reduced by making the step size h smaller.

Unfortunately, this is not the end of the story. If h is made too small, that is, too many steps are used in going from x_0 to $\bar{x}$, the accumulated round-off error may become more important than the accumulated formula error. In practice both sources of error must be considered and an "optimum" choice of h made so that neither is too large. This is discussed further in Section 8.7. A practical procedure for improving the accuracy of the computation, which requires computation with two different step sizes, is discussed in Problems 6 and 8.

As an example of how we can use the result (10) if we have a priori information about the solution of the given initial value problem consider the illustrative example

$$y' = 1 - x + 4y, \qquad y(0) = 1 \tag{12}$$

on the interval $0 \leq x \leq 1$. Let $y = \phi(x)$ be the solution of the initial value problem (12). Then $\phi'(x) = 1 - x + 4\phi(x)$, and

$$\begin{aligned}
\phi''(x) &= -1 + 4\phi'(x) \\
&= -1 + 4[1 - x + 4\phi(x)] \\
&= 3 - 4x + 16\phi(x).
\end{aligned}$$

From Eq. (10)

$$e_{n+1} = \frac{3 - 4\bar{x}_n + 16\phi(\bar{x}_n)}{2} h^2, \qquad x_n < \bar{x}_n < x_n + h. \tag{13}$$

Noting that $|3 - 4\bar{x}_n| \leq 3$ on $0 \leq x \leq 1$, a rough bound for e_{n+1} is given by

$$e_{n+1} \leq \left[\tfrac{3}{2} + 8 \max_{0 \leq x \leq 1} |\phi(x)| \right] h^2. \tag{14}$$

Since the right-hand side of Eq. (14) does not depend on n, this provides a uniform, although not necessarily an accurate, bound for the error at any step if a bound for $|\phi(x)|$ on $0 \leq x \leq 1$ is known. For the present problem, $\phi(x) = (4x - 3 + 19e^{4x})/16$. Substituting for $\phi(\bar{x}_n)$ in Eq. (13) gives

$$e_{n+1} = 19e^{4\bar{x}_n}h^2/2, \qquad x_n < \bar{x}_n < x_n + h. \tag{15}$$

The appearance of the factor 19 and the rapid growth of e^{4x} (at $x = 1, e^{4x} = 54.598$) explain why the results in the previous section with $h = 0.1$ were not very accurate. For example, the error in the first step is

$$e_1 = \phi(x_1) - y_1 = \frac{19e^{4\bar{x}_0}(0.01)}{2}, \qquad 0 < \bar{x}_0 < 0.1.$$

It is clear that e_1 is positive and, since $e^{4\bar{x}_0} < e^{0.4}$, we have

$$e_1 \leq \frac{19e^{0.4}(0.01)}{2} \cong 0.142. \tag{16}$$

Note also that $e^{4\bar{x}_0} > 1$; hence $e_1 \geq 19(0.01)/2 = 0.095$. The actual error is 0.1090418. It follows from Eq. (15) that the error becomes progressively worse with increasing x; this is clearly shown by the results in Table 8.1 of the last section. A similar computation for a bound for the local formula error in going from 0.4 to 0.5 gives

$$0.47 \cong \frac{19e^{1\cdot6}(0.01)}{2} \le e_5 \le \frac{19e^2(0.01)}{2} \cong 0.7.$$

In practice, of course, we will not know the solution ϕ of the given initial value problem. However, if we have bounds on the functions $f, f_x,$ and f_y we can obtain a bound on $|\phi''(x)|$, although not necessarily an accurate one, from Eq. (7). Then we can use Eq. (10) to estimate the local formula error. Even if we cannot find a useful bound for $|\phi''(x)|$, we still know that for the Euler method the local formula error and the accumulated formula error are bounded by a constant times h^2 and a constant times h, respectively.

We conclude this section with two practical comments based on experience.

1. One way of estimating whether the formula error is sufficiently small is, after completing the computations with a given h, to repeat the computations using a step size $h/2$ (see Problems 6 and 8). If the changes are greater than we are willing to accept, then it is necessary to use a smaller step size or a more accurate numerical procedure or possibly both. More accurate procedures are discussed in the following sections. The results* for the illustrative example using the Euler method with several different step sizes are given in Table 8.2. These results show that even with $h = 0.01$ it is not possible to obtain the first three digits correctly at $x = 0.10$. This could be expected without knowing the exact solution, since the results at $x = 0.1$ for $h = 0.025$ and $h = 0.01$ differ in the third digit.

TABLE 8.2   A Comparison of Results for the Numerical Solution of $y' = 1 - x + 4y$, $y(0) = 1$ Using the Euler Method for Different Step Sizes h

x	$h = 0.1$	$h = 0.05$	$h = 0.025$	$h = 0.01$	Exact
0	1.0000000	1.0000000	1.0000000	1.0000000	1.0000000
0.1	1.5000000	1.5475000	1.5761188	1.5952901	1.6090418
0.2	2.1900000	2.3249000	2.4080117	2.4644587	2.5053299
0.3	3.1460000	3.4333560	3.6143837	3.7390345	3.8301388
0.4	4.4744000	5.0185326	5.3690304	5.6137120	5.7942260
0.5	6.3241600	7.2901870	7.9264062	8.3766865	8.7120041
0.6	8.9038240	10.550369	11.659058	12.454558	13.052522
0.7	12.505354	15.234032	17.112430	18.478797	19.515518
0.8	17.537495	21.967506	25.085110	27.384136	29.144880
0.9	24.572493	31.652708	36.746308	40.554208	43.497903
1	34.411490	45.588400	53.807866	60.037126	64.897803

* In Table 8.2 as well as all others in this chapter except those in Section 8.7, all computations were carried out on a computer using twelve digits, and the answers were rounded to eight digits.

2. It is possible to estimate the effect of the round-off error after completing the computations using a certain number of digits by repeating the computations carrying one or two more digits. Again, if the changes are more than we are willing to accept, it is necessary to retain more digits in the computations. Some examples of the difficulties that arise due to round-off errors are discussed in Problems 10 through 13.

PROBLEMS

In Problems 1 and 2 estimate the local formula error in terms of the exact solution $y = \phi(x)$ if the Euler method is used. Obtain a bound for e_{n+1} in terms of x and $\phi(x)$ that is valid on the interval $0 \leq x \leq 1$. By using the exact solution obtain a more accurate error bound for e_{n+1}. For $h = 0.1$ compute a bound for e_1 and compare with the actual error at $x = 0.1$. Also compute a bound for the error e_4 in the fourth step.

1. $y' = 2y - 1$, $y(0) = 1$ 2. $y' = \frac{1}{2} - x + 2y$, $y(0) = 1$

3. Obtain a formula for the local formula error in terms of x and the exact solution ϕ if the Euler method is used for the initial value problem

$$y' = x^2 + y^2, \qquad y(0) = 1.$$

4. Consider the initial value problem

$$y' = \cos 5\pi x, \qquad y(0) = 1.$$

(a) Determine the exact solution $y = \phi(x)$, and draw a graph of $y = \phi(x) - 1$ for $0 \leq x \leq 1$. Use a scale for the ordinate so that $1/5\pi$ is about 1 in. or 2–3 cm.

(b) Determine approximate values of $\phi(x)$ at $x = 0.2$, 0.4, and 0.6 using the Euler method with $h = 0.2$. Draw a broken line graph for the approximate solution, and compare with the graph of the exact solution.

(c) Repeat the computations of part (b) for $0 \leq x \leq 0.4$, but take $h = 0.1$.

(d) Show by computing the local formula error that neither of these step sizes is sufficiently small. Determine a value of h to insure that the local formula error is less than 0.05 throughout the interval $0 \leq x \leq 1$. That such a small value of h is required results from the fact that max $|\phi''(x)|$ is large or, put in rough terms, that the solution is highly oscillatory.

*5. In this problem we discuss the accumulated formula error associated with the Euler method for the initial value problem $y' = f(x, y)$, $y(x_0) = y_0$. Assuming that the functions f and f_y are continuous in a region R of the xy plane that includes the point (x_0, y_0), it can be shown that there exists a constant L such that $|f(x, y) - f(x, \tilde{y})| < L |y - \tilde{y}|$ where (x, y) and $(x, \tilde{y})$ are any two points in R with the same x coordinate (see Problem 3 of Section 2.11). Further, we assume that f_x is continuous so that the solution ϕ has a continuous second derivative.

(a) Using Eq. (9) show that

$$|E_{n+1}| \leq |E_n| + h|f[x_n, \phi(x_n)] - f(x_n, y_n)| + \tfrac{1}{2}h^2 |\phi''(\bar{x}_n)| \leq \alpha |E_n| + \beta h^2, \qquad \text{(i)}$$

where $\alpha = 1 + hL$ and $\beta = \max |\phi''(x)|/2$ on $x_0 \leq x \leq x_n$.

(b) Accepting without proof that if $E_0 = 0$, and if $|E_n|$ satisfies Eq (i), then $|E_n| \leq \beta h^2(\alpha^n - 1)/(\alpha - 1)$ for $\alpha \neq 1$, show that

$$|E_n| \leq \frac{(1 + hL)^n - 1}{L} \beta h. \tag{ii}$$

Equation (ii) gives a bound for $|E_n|$ in terms of h, L, n, and β. Notice that for a fixed h, this error bound increases with increasing n; that is, the error bound increases with distance from the starting point x_0.

(c) Show that $(1 + hL)^n \leq e^{nhL}$, and hence that

$$|E_n| \leq \frac{e^{nhL} - 1}{L} \beta h$$

$$\leq \frac{e^{(x_n - x_0)L} - 1}{L} \beta h.$$

For a fixed point $\bar{x} = x_0 + nh$ (that is, nh is constant and $h = (\bar{x} - x_0)/n$) this error bound is of the form constant times h and approaches zero as $h \to 0$. Also note that for $nhL = (\bar{x} - x_0)L$ small the right side of the above equation is approximately $nh^2\beta = (\bar{x} - x_0)\beta h$ which was obtained in Eq. (11) by an intuitive argument.

6. **Richardson's Extrapolation Method.** Let $y = \phi(x)$ be the solution of the initial value problem $y' = f(x, y)$, $y(x_0) = y_0$. As was discussed in the text (also see part (c) of Problem (5)), the accumulated formula error using the Euler method to go from x_0 to a fixed point $\bar{x} = x_0 + nh$ is bounded by a constant times h. More precisely, it can be shown that $\phi(\bar{x}) = y_n(h) + Ch + M(h)$ where C is a constant that depends on the differential equation and the length of the interval but not on h, and $M(h)$ is a function that is proportional to h^2. The dependence of y_n on h is indicated by writing $y_n(h)$. The error is $Ch + M(h)$. Now suppose that the computation is repeated using a step size $h/2$. Since $2n$ steps are now required to reach $\bar{x}$, we have $\phi(\bar{x}) = y_{2n}(h/2) + Ch/2 + M(h/2)$. By solving the first equation for C and substituting in the second equation, show that

$$\phi(\bar{x}) \cong y_{2n}(h/2) + [y_{2n}(h/2) - y_n(h)]$$

where terms proportional to h^2 have been neglected. Thus, this new approximation to $\phi(\bar{x})$ has an error proportional to h^2 as compared to an error proportional to h for a single calculation with the Euler method. This procedure for computing an improved approximation to $\phi(\bar{x})$ in terms of $y_n(h)$ and $y_{2n}(h/2)$ is called Richardson's deferred approach to the limit or Richardson's extrapolation method. It was first used by L. F. Richardson in 1927.

7. Using the technique discussed in Problem 6 determine new estimates for the value of the exact solution $y = \phi(x)$ at $x = 0.2$, by making use of the results for $h = 0.2$ and $h = 0.1$ with the Euler method, for each of the following initial value problems. When possible compare the results with the values of the exact solution at $x = 0.2$.

(a) $y' = 2y$, $\quad y(0) = 1$ $\qquad$ (b) $y' = 2y - 1$, $\quad y(0) = 1$

(c) $y' = \frac{1}{2} - x + 2y$, $\quad y(0) = 1$ $\qquad$ (d) $y' = x^2 + y^2$, $\quad y(0) = 1$

8. A numerical method is said to have rth order accuracy if the error at $\bar{x} = x_0 + nh$ is proportional to h^r; or more precisely, if $\phi(\bar{x}) = y_n(h) + Ch^r + M(h)$, where the constant C depends on the differential equation and $\bar{x}$ but is independent of h, and $M(h)$ is proportional to h^{r+1}. Thus the Euler method has first order accuracy. Following the procedure of Problem 6, show that an improved estimate of $\phi(\bar{x})$ is given by $\phi(\bar{x}) = y_{2n}(h/2) + [y_{2n}(h/2) - y_n(h)]/(2^r - 1)$.

9. Consider the example problem $y' = 1 - x + 4y$, $y(0) = 1$. Using Richardson's extrapolation method (Problems 6 and 8), and using the approximate values of $\phi(1)$ for $h = 0.1, 0.05,$ and 0.025 given in Table 8.2, obtain new estimates for $\phi(1)$ for
    (a) $h = 0.1$ and $h = 0.05$ (b) $h = 0.05$ and $h = 0.025$
Compare the results with the value of the exact solution.

10. Using a step size $h = 0.05$ and the Euler method, but retaining only three digits throughout the computations, determine approximate values of the exact solution at $x = 0.05, 0.1, 0.15,$ and 0.2 for each of the following initial value problems.

    (a) $y' = 2y - 1,$ $y(0) = 1$ (b) $y' = \frac{1}{2} - x + 2y,$ $y(0) = 1$
    (c) $y' = x^2 + y^2,$ $y(0) = 1$

Compare the results with those using four digits that were obtained in the previous section. The small differences between some of those results rounded to three digits and the present results are due to round-off error. The round-off error would become important if the computation required many steps.

11. The following problem illustrates a danger that occurs because of round-off error when nearly equal numbers are subtracted, and the difference then multiplied by a large number. Evaluate the quantity

$$1000 \cdot \begin{vmatrix} 6.010 & 18.04 \\ 2.004 & 6.000 \end{vmatrix}$$

as follows.
    (a) First round each entry in the determinant to two digits.
    (b) First round each entry in the determinant to three digits.
    (c) Retain all four digits. Compare the exact value with the results in (a) and (b).

12. The distributive law $a(b - c) = ab - ac$ does not hold, in general, if the products are rounded off to a smaller number of digits. To show this in a specific case take $a = 0.22$, $b = 3.19$, and $c = 2.17$. After each multiplication round off the last digit.

13. To obtain some idea of the possible dangers of small errors in the initial conditions, such as those due to round-off, consider the initial value problem

$$y' = x + y - 3, \qquad y(0) = 2.$$

    (a) Show that the exact solution is $y = \phi_1(x) = 2 - x$.
    (b) Suppose that in the initial condition a mistake is made and 2.001 is used instead of 2. Determine the solution $y = \phi_2(x)$ in this case, and compare the difference $\phi_2(x) - \phi_1(x)$ at $x = 1$ and as $x \to \infty$.

8.4   AN IMPROVED EULER METHOD

Consider the initial value problem

$$y' = f(x, y), \qquad y(x_0) = y_0. \tag{1}$$

Letting ϕ be the exact solution of the initial value problem (1), and integrating from x_n to x_{n+1}, as we did in Eq. (7) of Section 8.2, we obtain

$$\phi(x_{n+1}) = \phi(x_n) + \int_{x_n}^{x_{n+1}} f[x, \phi(x)] \, dx. \tag{2}$$

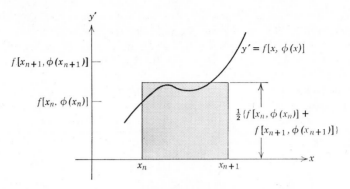

FIGURE 8.5

The Euler formula

$$y_{n+1} = y_n + hf(x_n, y_n) \tag{3}$$

was obtained by replacing $f[x, \phi(x)]$ in Eq. (2) by its approximate value $f(x_n, y_n)$ at the left-hand end point.

A better formula can be obtained if the integrand in Eq. (2) is approximated more accurately. One way to do this is to replace the integrand by the average of its values at the two end points, namely $\{f[x_n, \phi(x_n)] + f[x_{n+1}, \phi(x_{n+1})]\}/2$. Thus the area under the curve in Figure 8.5 between $x = x_n$ and $x = x_{n+1}$ is approximated by the area of the shaded rectangle. Further, we replace $\phi(x_n)$ and $\phi(x_{n+1})$ by their approximate values y_n and y_{n+1}. Then we obtain from Eq. (2)

$$y_{n+1} = y_n + \frac{f(x_n, y_n) + f(x_{n+1}, y_{n+1})}{2} h. \tag{4}$$

Since the unknown y_{n+1} appears as one of the arguments of f on the right-hand side of Eq. (4), it will in general be fairly difficult to solve this equation for y_{n+1}. This difficulty can be overcome by replacing y_{n+1} on the right-hand side of Eq. (4) by the value obtained using the simple Euler formula (3). Thus

$$y_{n+1} = y_n + \frac{f(x_n, y_n) + f[x_n + h, y_n + hf(x_n, y_n)]}{2} h$$

$$= y_n + \frac{y_n' + f[x_n + h, y_n + hy_n']}{2} h, \tag{5}$$

where x_{n+1} has been replaced by $x_n + h$.

Equation (5) gives a formula for computing y_{n+1}, the approximate value of $\phi(x_{n+1})$, in terms of the data at x_n. This formula is known as the *improved Euler formula* or the *Heun formula*. That Eq. (5) does represent an improvement over the Euler formula (3) rests on the fact that the local formula error in using Eq. (5) is proportional to h^3 while that for the Euler method is proportional to h^2. The error estimate for the improved Euler formula is established in Problem 4. It can also be shown that for a finite interval the accumulated formula error for the

TABLE 8.3 Numerical Solution of $y' = 1 - x + 4y$, $y(0) = 1$ Using the Improved Euler Method with $h = 0.1$

Table Instructions	A	B	$C = 1 - A + 4B$	$D = A + 0.1$	$E = B + 0.1C$	$F = 1 - D + 4E$	$G = C + F$	$H = (0.05)G$	$I = B + H$	
n	x_n	y_n	$y'_n = f(x_n, y_n)$	$x_n + h$	$y_n + hy'_n$	$f[x_n + h, y_n + hy'_n]$			y_{n+1}	$\phi(x_{n+1})$
0	0	1	5	0.1	1.5	6.9	11.9	1.595	1.595	1.6090418
1	0.1	1.595	7.280	0.2	2.323	10.092	17.372	0.8686	2.4636	2.5053299
2	0.2	2.4636	10.6544	0.3	3.52904	14.81616	25.47056	1.273528	3.737128	3.8301388
3	0.3									5.7942260
4	0.4									8.7120041
5										
6										

improved Euler formula is bounded by a constant times h^2. Note that this greater accuracy is achieved at the expense of more computational work, since it is now necessary to evaluate $f(x, y)$ twice in order to go from x_n to x_{n+1}.

Example.   To illustrate the use of the improved Euler formula (5) again consider the initial value problem

$$y' = 1 - x + 4y, \qquad y(0) = 1. \tag{6}$$

For the present problem $f(x, y) = 1 - x + 4y$; hence

$$y'_n = 1 - x_n + 4y_n,$$

and

$$f[x_n + h, y_n + hy'_n] = 1 - (x_n + h) + 4(y_n + hy'_n).$$

A possible format for carrying out the calculations is shown in Table 8.3. Actually, the number of entries can be reduced; for example, it is not necessary to record G in computing H.

A flow chart for the improved Euler method is similar to the one for the Euler method in Section 8.2. The main difference is that in the second and third boxes y'_n is replaced by $F(x_n, y_n)$, where

$$F(x_n, y_n) = \tfrac{1}{2}\{f(x_n, y_n) + f[x_n + h, y_n + hf(x_n, y_n)]\}.$$

The results using the improved Euler method with $h = 0.1$ are not only better than those obtained with the Euler method for $h = 0.1$, but also better than those obtained with the Euler method for $h = 0.05$. This is clearly shown by the results tabulated in Table 8.4. Thus even though more computations are required at

TABLE 8.4   A Comparison of Results Using the Euler and Improved Euler Methods for $h = 0.1$ and $h = 0.05$ for the Initial Value Problem $y' = 1 - x + 4y$, $y(0) = 1$

	Euler		Improved Euler		
x	$h = 0.1$	$h = 0.05$	$h = 0.1$	$h = 0.05$	Exact
0	1.0000000	1.0000000	1.0000000	1.0000000	1.0000000
0.1	1.5000000	1.5475000	1.5950000	1.6049750	1.6090418
0.2	2.1900000	2.3249000	2.4636000	2.4932098	2.5053299
0.3	3.1460000	3.4333560	3.7371280	3.8030484	3.8301388
0.4	4.4744000	5.0185326	5.6099494	5.7404023	5.7942260
0.5	6.3241600	7.2901870	8.3697252	8.6117498	8.7120041
0.6	8.9038240	10.550369	12.442193	12.873253	13.052522
0.7	12.505354	15.234032	18.457446	19.203865	19.515518
0.8	17.537495	21.967506	27.348020	28.614138	29.144880
0.9	24.572493	31.652708	40.494070	42.608178	43.497903
1.0	34.411490	45.588400	59.938223	63.424698	64.897803

each step to use the improved Euler formula, substantially better results can be obtained with only half as many steps. The results using the improved Euler formula with $h = 0.05$ are also tabulated in Table 8.4. They are in fairly close agreement with the values of the exact solution, the percentage errors at $x = 0.5$ and $x = 1.0$ being 1.15 and 2.3 percent, respectively.

PROBLEMS

In Problems 1 through 3 set up a table for use with the improved Euler method and determine approximate values of the exact solution at $x = 0.1, 0.2$, and 0.3. Take $h = 0.1$. Compare the results with those obtained using the Euler method and the exact results (if available). Carry four digits in the computations.

1. $y' = 2y - 1$, $y(0) = 1$ 2. $y' = \frac{1}{2} - x + 2y$, $y(0) = 1$
3. $y' = x^2 + y^2$, $y(0) = 1$

4. In this problem we will establish that the local formula error for the improved Euler formula is proportional to h^3. Assuming that the solution ϕ of the initial value problem $y' = f(x, y)$, $y(x_0) = y_0$ has derivatives that are continuous through the third order (f has continuous second partial derivatives), it follows that

$$\phi(x_n + h) = \phi(x_n) + \phi'(x_n)h + \frac{\phi''(x_n)h^2}{2!} + \frac{\phi'''(\bar{x}_n)h^3}{3!},$$

where $x_n < \bar{x}_n < x_n + h$. Assume that $y_n = \phi(x_n)$.

(a) Show that for y_{n+1} as given by Eq. (5)

$$e_{n+1} = \phi(x_{n+1}) - y_{n+1}$$

$$= \frac{\phi''(x_n)h - \{f[x_n + h, y_n + hf(x_n, y_n)] - f(x_n, y_n)\}}{2!} h + \frac{\phi'''(\bar{x}_n)h^3}{3!}. \qquad \text{(i)}$$

(b) Making use of the fact that $\phi''(x) = f_x[x, \phi(x)] + f_y[x, \phi(x)]\phi'(x)$, and that the Taylor series with a remainder for a function of two variables (which we accept without proof) is

$$F(a + h, b + k) = F(a, b) + F_x(a, b)h + F_y(a, b)k$$

$$+ \frac{1}{2!}(h^2 F_{xx} + 2hk F_{xy} + k^2 F_{yy})\big|_{x=\xi, y=\eta},$$

where ξ lies between a and $a + h$ and η lies between b and $b + k$, show that the first term on the right-hand side of Eq. (i) is proportional to h^3 plus higher order terms. This is the desired result.

(c) Show that if $f(x, y)$ is linear in x and y, then $e_{n+1} = \phi'''(\bar{x}_n)h^3/6$, where $x_n < \bar{x}_n < x_{n+1}$.

Hint: What are f_{xx}, f_{xy}, and f_{yy}?

5. Consider the improved Euler method for solving the initial value problem $y' = 1 - x + 4y$, $y(0) = 1$. Using the result of part (c) of Problem 4 and the exact solution of the initial value problem, determine e_{n+1} and a bound for the error at any step on $0 \leq x \leq 1$. Compare this error with the one obtained in the previous section, Eq. (15), using

the Euler method. Also obtain a bound for e_1 for $h = 0.1$ and compare with Eq. (16) of the previous section.

6. Making use of the exact solution, determine e_{n+1} and a bound for e_{n+1} at any step on $0 \leq x \leq 1$ for the improved Euler method for each of the following initial value problems. Also, obtain a bound for e_1 for $h = 0.1$ and compare it with the similar estimate for the Euler method and with the actual error using the improved Euler method.

(a) $y' = 2y - 1$, $y(0) = 1$ (b) $y' = \frac{1}{2} - x + 2y$, $y(0) = 1$

7. The *modified Euler formula* for the initial value problem $y' = f(x, y)$, $y(x_0) = y_0$ is given by

$$y_{n+1} = y_n + hf[x_n + \tfrac{1}{2}h, y_n + \tfrac{1}{2}hf(x_n, y_n)].$$

Following the procedure outlined in Problem 4 show that the local formula error in the modified Euler formula is proportional to h^3.

8. Using the modified Euler formula with $h = 0.1$ compute approximate values of the exact solution at $x = 0.1, 0.2$, and 0.3 for each of the following initial value problems.

(a) $y' = 2y - 1$, $y(0) = 1$ (b) $y' = \frac{1}{2} - x + 2y$, $y(0) = 1$
(c) $y' = x^2 + y^2$, $y(0) = 1$

Where appropriate, compare the results with those obtained earlier.

8.5 THE THREE-TERM TAYLOR SERIES METHOD

We saw in Section 8.2 that the Euler formula $y_{n+1} = y_n + hf(x_n, y_n)$ for solving the initial value problem

$$y' = f(x, y), \qquad y(x_0) = y_0 \tag{1}$$

can be derived by retaining the first two terms in the Taylor series for the solution $y = \phi(x)$ about the point $y = x_n$. A more accurate formula can be obtained by using the first three terms. Assuming that ϕ has at least three continuous derivatives in the interval of interest (f has continuous second partial derivatives) and using the Taylor series with a remainder, we obtain

$$\phi(x_n + h) = \phi(x_n) + \phi'(x_n)h + \phi''(x_n)\frac{h^2}{2!} + \phi'''(\bar{x}_n)\frac{h^3}{3!}, \tag{2}$$

where $\bar{x}_n$ is some point in the interval $x_n < \bar{x}_n < x_n + h$.
From Eq. (1)

$$\phi'(x_n) = f[x_n, \phi(x_n)]. \tag{3}$$

Further, $\phi''(x)$ can be computed from Eq. (1):

$$\phi''(x) = f_x[x, \phi(x)] + f_y[x, \phi(x)]\phi'(x). \tag{4}$$

Hence

$$\phi''(x_n) = f_x[x_n, \phi(x_n)] + f_y[x_n, \phi(x_n)]\phi'(x_n). \tag{5}$$

The three-term Taylor series formula is obtained by replacing $\phi(x_n)$ by its approximate value y_n in the formulas for $\phi'(x)$ and $\phi''(x)$ and then neglecting the term $\phi'''(\bar{x}_n)h^3/3!$ in Eq. (2). Thus we obtain

$$y_{n+1} = y_n + hy'_n + \frac{h^2}{2} y''_n, \tag{6}$$

where

$$y'_n = f(x_n, y_n), \tag{7}$$

and

$$y''_n = f_x(x_n, y_n) + f_y(x_n, y_n)y'_n. \tag{8}$$

It is trivial to show, assuming $\phi(x_n) = y_n$, that the local formula error e_{n+1} associated with formula (6) is

$$e_{n+1} = \phi(x_{n+1}) - y_{n+1} = \tfrac{1}{6}\phi'''(\bar{x}_n)h^3, \tag{9}$$

where $x_n < \bar{x}_n < x_n + h$. Thus the local formula error for the three-term Taylor series formula is proportional to h^3, just as for the improved Euler formula discussed in the previous section. Again, it can also be shown that for a finite interval the accumulated formula error is no greater than a constant times h^2.

The three-term Taylor series formula requires the computation of $f_x(x, y)$ and $f_y(x, y)$, and then the evaluation of these functions as well as $f(x, y)$ at (x_n, y_n). In some problems it may be difficult, or fairly lengthy, to compute f_x and f_y. If this is the case, it is probably better to use a formula with comparable accuracy, such as the improved Euler formula, that does not require the partial derivatives f_x and f_y. In principle, four-term, or even higher, Taylor series formulas can be developed. However, such formulas involve even higher partial derivatives of f and are in general rather awkward to use.

Example.   For the illustrative example

$$y' = 1 - x + 4y, \qquad y(0) = 1$$

the three-term Taylor series method is fairly convenient. We have

$$f(x, y) = 1 - x + 4y,$$
$$f_x(x, y) = -1, \qquad f_y(x, y) = 4.$$

Hence

$$y'_n = 1 - x_n + 4y_n, \qquad y''_n = -1 + 4y'_n. \tag{10}$$

It follows from Eqs. (10) that $y'_0 = 5$ and $y''_0 = 19$. Taking h equal to 0.1 and substituting in Eq. (6) gives

$$y_1 = 1 + (0.1)(5) + (0.005)(19)$$
$$= 1.595.$$

For the example problem the results using the three-term Taylor series method are identical, except for small differences due to round-off error, to those obtained using the improved Euler method. The reason for this is that when $f(x, y)$ is linear in x and y the two methods are identical. This can be shown by expanding the improved Euler formula about the point (x_n, y_n) and comparing with the three-term Taylor series formula. (See Problem 4(b) of Section 8.4. Note that the functions f_{xx}, f_{xy}, and f_{yy} are each zero.)

Thus the results given in Table 8.4 for the improved Euler method for $h = 0.1$ and $h = 0.05$ also apply for the three-term Taylor series method. On the other hand, if $f(x, y)$ is not linear in x and y the two methods will give slightly different results. For example, the results are different for the third of the exercise problems $y' = x^2 + y^2$, $y(0) = 1$.

PROBLEMS

In Problems 1 through 3 determine the three-term Taylor series formula for y_{n+1}. Set up an appropriate table and determine approximate values of the exact solution at $x = 0.1, 0.2$, and 0.3. Take $h = 0.1$. Compare your results with those obtained using the Euler method, the improved Euler method, and the exact results (if available). Carry four digits in your computations.

1. $y' = 2y - 1$, $\quad y(0) = 1$ $\qquad$ 2. $y' = \frac{1}{2} - x + 2y$, $\quad y(0) = 1$

3. $y' = x^2 + y^2$, $\quad y(0) = 1$

4. For the illustrative problem $y' = 1 - x + 4y$, $y(0) = 1$, determine approximate values of the exact solution at $x = 0.2$ and 0.3, using the three-term Taylor series method. Take $h = 0.1$. Retain four digits. Check your results with those given in Table 8.4.

Problems 5 through 7 deal with the initial value problem $y' = f(x, y)$, $y(x_0) = y_0$. Let $y = \phi(x)$ be the exact solution.

5. Show that the local formula error e_{n+1} using the three-term Taylor series formula is given by $\phi'''(\bar{x})h^3/6$, $x_n < \bar{x}_n < x_n + h$. Remember that in computing e_{n+1} it is assumed that $y_n = \phi(x_n)$.

6. Derive a four-term Taylor series formula for y_{n+1} in terms of $f(x, y)$ and its partial derivatives evaluated at (x_n, y_n). What is the local formula error?

7. Following the outline of Richardson's extrapolation method (Problems 6 and 8 of Section 8.3) show that an improved estimate for $\phi(\bar{x})$, $\bar{x} = x_0 + nh$, in terms of $y_n(h)$ and $y_{2n}(h/2)$ computed using the three-term Taylor series method or the improved Euler method, is given by $y_{2n}(h/2) + [y_{2n}(h/2) - y_n(h)]/(2^2 - 1)$. Consider the initial value problem $y' = 1 - x + 4y$, $y(0) = 1$. Using the approximate results for $\phi(1)$ obtained by the improved Euler method for $h = 0.1$ and $h = 0.05$ (see Table 8.4), determine a new estimate of $\phi(1)$. Compare this result with the value of the exact solution.

8.6 THE RUNGE-KUTTA METHOD

Again consider the initial value problem

$$y' = f(x, y), \qquad y(x_0) = y_0. \tag{1}$$

In the previous section we developed a three-term Taylor series formula that involved the functions f_x and f_y. Unfortunately, higher order Taylor series formulas rapidly become unwieldy because of the higher partial derivatives of f that must be computed.

However, it is possible to develop formulas equivalent to third, fourth, fifth, or even higher order Taylor series formulas that do not involve partial derivatives of f. By equivalent numerical procedure we mean that the local formula errors are each proportional to the same power of h plus (possibly different) higher order terms. The development of these formulas started with the work of Carl Runge (1856–1927) in 1895 and was continued by M. W. Kutta (1867–1944) in 1901. The classical Runge-Kutta formula is equivalent to a five-term Taylor formula*

$$y_{n+1} = y_n + hy_n' + \frac{h^2}{2!} y_n'' + \frac{h^3}{3!} y_n''' + \frac{h^4}{4!} y_n^{\mathrm{iv}}. \tag{2}$$

The Runge-Kutta formula involves a weighted average of values of $f(x, y)$ taken at different points in the interval $x_n \leq x \leq x_{n+1}$. It is given by

$$y_{n+1} = y_n + \frac{h}{6}(k_{n1} + 2k_{n2} + 2k_{n3} + k_{n4}), \tag{3}$$

where

$$k_{n1} = f(x_n, y_n), \tag{4a}$$

$$k_{n2} = f(x_n + \tfrac{1}{2}h, y_n + \tfrac{1}{2}hk_{n1}), \tag{4b}$$

$$k_{n3} = f(x_n + \tfrac{1}{2}h, y_n + \tfrac{1}{2}hk_{n2}), \tag{4c}$$

$$k_{n4} = f(x_n + h, y_n + hk_{n3}). \tag{4d}$$

The sum $(k_{n1} + 2k_{n2} + 2k_{n3} + k_{n4})/6$ can be interpreted as an average slope. Note that k_{n1} is the slope at the left-hand end of the interval, k_{n2} is the slope at the midpoint using the Euler formula to go from x_n to $x_n + h/2$, k_{n3} is a second approximation to the slope at the midpoint, and finally k_{n4} is the slope at $x_n + h$ using the Euler formula and the slope k_{n3} to go from x_n to $x_n + h$.

While in principle it is not too difficult to show that Eqs. (2) and (3) differ by terms that are proportional to h^5, the algebra is extremely lengthy. Thus we will accept the fact that the local formula error in using Eq. (3) is proportional to h^5 and that for a finite interval the accumulated formula error is at most a constant times h^4. A derivation of a Runge-Kutta formula that is equivalent to the three-term Taylor series formula is given in Problem 5.

Clearly the Runge-Kutta formula, Eqs. (3) and (4), is more complicated than any of the formulas discussed previously. On the other hand, it should be remembered that this is a very accurate formula (halving the step size reduces the local formula error by the factor $\frac{1}{32}$); further, it is not necessary to compute any partial derivatives of f. Moreover, the calculations described here are routine with a high-speed computer and can be done in seconds for any except extremely complicated and lengthy functions f.

* The numbers y_n'', y_n''', and y_n^{iv} are obtained by successively differentiating Eq. (1), and then evaluating at (x_n, y_n). See Eq. (8) of Section 8.5.

We also note that if f does not depend on y, then

$$k_{n1} = f(x_n), \qquad k_{n2} = k_{n3} = f(x_n + \tfrac{1}{2}h), \qquad k_{n4} = f(x_n + h),$$

and Eq. (3) is identical with that obtained by using Simpson's (1710–1761) rule (see Problem 6) to evaluate the integral of $y' = f(x)$:

$$\int_{y_n}^{y_{n+1}} dy = \int_{x_n}^{x_n+h} f(x)\,dx$$

or

$$y_{n+1} - y_n = \frac{h}{6}\,[f(x_n) + 4f(x_n + \tfrac{1}{2}h) + f(x_n + h)].$$

The fact that Simpson's rule has an error proportional to h^5 is in agreement with the earlier comment concerning the error in the Runge-Kutta formula.

The Runge-Kutta formula, Eqs. (3) and (4), is one of the most widely used and most successful of all one-step formulas.

Example. Using the Runge-Kutta method compute an approximate value of the exact solution $y = \phi(x)$ at $x = 0.2$ for the illustrative initial value problem

$$y' = 1 - x + 4y, \qquad y(0) = 1.$$

Taking $h = 0.2$ we have

$$k_{01} = f(0, 1) = 5; \qquad\qquad\qquad hk_{01} = 1.0$$
$$k_{02} = f(0 + 0.1, 1 + 0.5) = 6.9; \qquad hk_{02} = 1.38$$
$$k_{03} = f(0 + 0.1, 1 + 0.69) = 7.66; \qquad hk_{03} = 1.532$$
$$k_{04} = f(0 + 0.2, 1 + 1.532) = 10.928.$$

Thus

$$y_1 = 1 + \frac{0.2}{6}\,[5 + 2(6.9) + 2(7.66) + 10.928]$$

$$= 1 + 1.5016 = 2.5016.$$

The exact value of $\phi(0.2)$ is 2.5053299. Hence the Runge-Kutta method with a step size of $h = 0.2$ gives a better result than the Euler method (2.19) with $h = 0.1$, or the improved Euler method (2.4636) with $h = 0.1$.

A tabulation of the results for the illustrative example for the different methods is given in Table 8.5. The results for the three-term Taylor method are omitted since they are identical to those for the improved Euler method. The accuracy of the Runge-Kutta method for the present problem can be seen by comparing the approximate values of $\phi(1)$ using different methods and different step sizes. This is done in Table 8.6. Note that the results for the Runge-Kutta method with

TABLE 8.5 Comparison of Results for the Numerical Solution of the Initial Value Problem $y' = 1 - x + 4y$, $y(0) = 1$

x	Euler $h = 0.1$	Improved Euler $h = 0.1$	Runge-Kutta $h = 0.2$	Runge-Kutta $h = 0.1$	Exact
0	1.0000000	1.0000000	1.0000000	1.0000000	1.0000000
0.1	1.5000000	1.5950000		1.6089333	1.6090418
0.2	2.1900000	2.4636000	2.5016000	2.5050062	2.5053299
0.3	3.1460000	3.7371280		3.8294145	3.8301388
0.4	4.4774000	5.6099494	5.7776358	5.7927853	5.7942260
0.5	6.3241600	8.3697252		8.7093175	8.7120041
0.6	8.9038240	12.442193	12.997178	13.047713	13.052522
0.7	12.505354	18.457446		19.507148	19.515518
0.8	17.537495	27.348020	28.980768	29.130609	29.144880
0.9	24.572493	40.494070		43.473954	43.497903
1.0	34.411490	59.938223	64.441579	64.858107	64.897803

TABLE 8.6 Comparison of Results Using Different Numerical Procedures and Different Step Sizes for the Value at $x = 1$ of the Solution of the Initial Value Problem $y' = 1 - x + 4y$, $y(0) = 1$

h	Euler	Improved Euler	Runge-Kutta	Exact
0.2			64.441579	64.897803
0.1	34.411490	59.938223	64.858107	64.897803
0.05	45.588400	63.424698	64.894875	64.897803
0.025	53.807866	64.497931	64.897604	64.897803
0.01	60.037126	64.830722	64.897798	64.897803

$h = 0.1$ are better than the results with either of the other methods with $h = 0.01$, a hundred steps compared to ten steps—actually a hundred evaluations of $f(x, y)$ compared to forty!

PROBLEMS

In Problems 1 through 3 set up a table for use with the Runge-Kutta method and determine approximate values of the exact solution at $x = 0.2$ and 0.4. Take $h = 0.2$. Compare your results with those of other methods and the exact results (if available).

Carry four digits in your computations.

1. $y' = 2y - 1$, $y(0) = 1$ 2. $y' = \frac{1}{2} - x + 2y$, $y(0) = 1$
3. $y' = x^2 + y^2$, $y(0) = 1$

4. Compute an approximate value of the exact solution $y = \phi(x)$ at $x = 0.4$ for the illustrative example $y' = 1 - x + 4y$, $y(0) = 1$. Use $h = 0.2$ and the approximate value of $\phi(0.2)$ given in the example in the text. Compare the result with the value given in Table 8.5.

*5. Consider the initial value problem $y' = f(x, y)$, $y(x_0) = y_0$. The corresponding three-term Taylor series formula is

$$y_{n+1} = y_n + hy'_n + \frac{h^2}{2}\, y''_n. \tag{i}$$

For the moment consider the formula

$$y_{n+1} = y_n + h\{af(x_n, y_n) + bf[x_n + \alpha h, y_n + \beta hf(x_n, y_n)]\}, \tag{ii}$$

where a, b, α, and β are arbitrary.

(a) Show by expanding the right-hand side of formula (ii) about the point (x_n, y_n) that for $a + b = 1$, $b\alpha = \frac{1}{2}$, and $b\beta = \frac{1}{2}$ the difference between formulas (i) and (ii) is proportional to h^3.
Hint: Use the Taylor series expansion for a function of two variables given in Problem 4(b) of Section 8.4.

(b) Show that the equations for a, b, α, and β have the infinity of solutions $a = 1 - \lambda$, $b = \lambda$, and $\alpha = \beta = 1/2\lambda$, $\lambda \neq 0$.

(c) For $\lambda = \frac{1}{2}$, show that Eq. (ii) reduces to the improved Euler formula given in Section 8.4.

(d) For $\lambda = 1$, show that Eq. (ii) reduces to the modified Euler formula given in Problem 7 of Section 8.4.

6. To derive Simpson's rule for the approximate value of the integral of $f(x)$ from $x = 0$ to $x = h$, first show that

$$\int_0^h (Ax^2 + Bx + C)\, dx = \frac{Ah^3}{3} + \frac{Bh^2}{2} + Ch.$$

Next choose the constants A, B, and C so that the parabola $Ax^2 + Bx + C$ passes through the points $[0, f(0)]$, $[h/2, f(h/2)]$, and $[h, f(h)]$. Using this polynomial to represent $f(x)$ approximately on the interval $0 \leq x \leq h$, substitute for A, B, and C in the above formula to obtain

$$\int_0^h f(x)\, dx \cong \frac{h}{6}\left[f(0) + 4f\left(\frac{h}{2}\right) + f(h)\right].$$

It can be shown that the error in using this formula is proportional to h^5.

7. Following the outline of Richardson's extrapolation method (Problems 6 and 8 of Section 8.3), show that for the initial value problem $y' = f(x, y)$, $y(x_0) = y_0$ an improved estimate for $\phi(\bar{x})$, $\bar{x} = x_0 + nh$, in terms of $y_n(h)$ and $y_{2n}(h/2)$ for the Runge-Kutta method, is given by $y_{2n}(h/2) + [y_{2n}(h/2) - y_n(h)]/(2^4 - 1)$. Consider the illustrative initial value problem $y' = 1 - x + 4y$, $y(0) = 1$. Using the approximate values of $\phi(1)$ obtained by the Runge-Kutta method with $h = 0.2$ and $h = 0.1$ (Table 8.5) and the above formula, obtain a new estimate for $\phi(1)$. Compare the result with the exact solution.

8.7   SOME DIFFICULTIES WITH NUMERICAL METHODS

In Section 8.3 we discussed some ideas related to the errors that can occur in a numerical solution of the initial value problem

$$y' = f(x, y), \qquad y(x_0) = y_0. \tag{1}$$

In this section we will continue that discussion and also point out several more subtle difficulties that can arise. The points that we wish to make are fairly difficult to treat in detail at this level, and we will be content to illustrate them by means of examples.

First, recall that for the Euler method we showed that the local formula error is proportional to h^2 and that for a finite interval the accumulated formula error is at most a constant times h. Although we will not prove it, it is true in general that if the local formula error is proportional to h^p then for a finite interval the accumulated formula error is bounded by a constant times h^{p-1}. To achieve high accuracy we normally use a numerical procedure for which p is large (for the Runge-Kutta method $p = 5$). As p increases, the formula used in computing y_{n+1} normally becomes more complicated, and hence more calculations are required at each step; however, this is usually not a serious problem when using a computer unless $f(x, y)$ is very complicated or if the calculation must be repeated many times (for example, for many different values of y_0 and/or different values of parameters that may appear in f).

If the step size h is decreased, the accumulated formula error is decreased by the same factor raised to the power $p - 1$. However, as we mentioned in Section 8.3, if h is too small, too many steps will be required to cover a fixed interval, and the accumulated round-off error may be larger than the accumulated formula error. This is illustrated by solving our example problem

$$y' = 1 - x + 4y, \qquad y(0) = 1 \tag{2}$$

on the interval $0 \leq x \leq 1$ using the Runge-Kutta method* with different step sizes h. In Table 8.7 the difference between the value of the exact solution and the computed value at $x = 1$ is shown for several values of h. Notice that initially the error decreases as h is decreased to about 0.025 to 0.02 (N, the number of steps is increased to about 40 to 50), but then begins to increase with decreasing h. This is the result of the accumulated round-off error. It is not possible to name an optimum value of N, since for values of N from about 40 to 55 both the accumulated formula error and the accumulated round-off error are probably affecting the last two significant digits and the small error is oscillatory.

As a second example of the types of difficulties that can arise when we use numerical procedures carelessly, consider the problem of determining the solution $y = \phi(x)$ of

$$y' = x^2 + y^2, \qquad y(0) = 1. \tag{3}$$

---

* The calculations in this section were carried out on an I.B.M. 360-50 computer with seven-digit precision. Only a few seconds of computing is required to obtain all of the results given in Table 8.7 using a modern computer.

TABLE 8.7 A Comparison of the Error at $x = 1$ Using Different Step Sizes and the Runge-Kutta Method for the Initial Value Problem $y' = 1 - x + 4y$, $y(0) = 1$

$N = 1/h$	h	$(-1) \times$ Error at $x = 1$
10	0.1	0.04008
20	0.05	0.00336
30	0.03333333	0.00114
40	0.025	0.00078
50	0.02	0.00089
60	0.01666667	0.00101
80	0.0125	0.00111
160	0.00625	0.00212
320	0.003125	0.00357
640	0.0015625	0.00731
1280	0.00078125	0.01567

This is the third practice problem in the Problems for the previous sections. Since the differential equation is nonlinear, the existence and uniqueness Theorem 2.2 guarantees only that there is a solution in *some* interval about $x = 0$. Suppose we try to compute a solution of the initial value problem on the interval $0 \leq x \leq 1$ using different numerical procedures.

If we use the Euler method with $h = 0.1, 0.05$, and 0.01, we find the following approximate values at $x = 1$: 7.189500, 12.32054, and 90.68743, respectively. The large differences among the computed values are convincing evidence that we should use a more accurate numerical procedure—the Runge-Kutta method, for example. Using the Runge-Kutta method with $h = 0.1$ we find the approximate value 735.0004 at $x = 1$, which is quite different from those obtained using the Euler method. If we were naive and stopped now, we would make a serious mistake. Repeating the calculations using step sizes of $h = 0.05$ and $h = 0.01$ we obtain the interesting information listed in Table 8.8.

While the values at $x = 0.90$ are reasonable and we might well believe that the exact solution has a value of about 14.3 at $x = 0.90$, it is clear that something strange is happening between $x = 0.9$ and $x = 1.0$. To help determine what is

TABLE 8.8 Calculation of the Solution of the Initial Value Problem $y' = x^2 + y^2$, $y(0) = 1$ Using the Runge-Kutta Method

h	$x = 0.90$	$x = 1.0$
0.1	14.02158	735.0004
0.05	14.27042	1.755613×10^4
0.01	14.30200	$> 10^{15}$

happening let us turn to some analytical approximations to the solution of the initial value problem (3). Note that on $0 \le x \le 1$,

$$y^2 \le x^2 + y^2 \le 1 + y^2. \tag{4}$$

Thus the solution $y = \phi_1(x)$ of

$$y' = 1 + y^2, \qquad y(0) = 1 \tag{5}$$

and the solution $y = \phi_2(x)$ of

$$y' = y^2, \qquad y(0) = 1 \tag{6}$$

grow more rapidly and less rapidly, respectively, than the solution $y = \phi(x)$ of the original problem. Since all of these solutions pass through the same initial point, $\phi_2(x) \le \phi(x) \le \phi_1(x)$. The important thing to note is that we *can solve* Eqs. (5) and (6) for ϕ_1 and ϕ_2 by separation of variables. We find that

$$\phi_1(x) = \tan\left(x + \frac{\pi}{4}\right), \qquad \phi_2(x) = \frac{1}{1 - x}. \tag{7}$$

Thus $\phi_2(x) \to \infty$ as $x \to 1$ and $\phi_1(x) \to \infty$ as $x \to \pi/4 \simeq 0.785$. These calculations show that the solution of the original initial value problem must become unbounded somewhere between $x \simeq 0.785$ and $x = 1$. We now know that the problem (3) has no solution on the entire interval $0 \le x \le 1$.

However, our numerical calculations suggest that we can go beyond $x = 0.785$ and probably beyond $x = 0.9$. Assuming that the solution of the initial value problem exists at $x = 0.9$ and has the value 14.3, we can obtain a more accurate appraisal of what happens for larger x by considering the initial value problems (5) and (6) with $y(0) = 1$ replaced by $y(0.9) = 14.3$. Then we obtain

$$\phi_1(x) = \tan(x + 0.6010), \qquad \phi_2(x) = \frac{1}{0.9699 - x}, \tag{8}$$

where only four decimal places have been kept. Thus $\phi_1(x) \to \infty$ as $x \to \pi/2 - 0.6010 \simeq 0.9698$ and $\phi_2(x) \to \infty$ as $x \to 0.9699$. We conclude that the solution of the initial value problem (3) becomes unbounded near $x = 0.97$. We cannot be more precise than this because the initial condition $y(0.9) = 14.3$ is only approximate. This example illustrates the sort of information that can be obtained by a judicious combination of analytical and numerical work.

As a final example, consider the problem of determining two linearly independent solutions of the second order linear equation

$$y'' - 100y = 0 \tag{9}$$

for $x > 0$. The generalization of numerical techniques for first order equations to higher order equations or to systems of equations is discussed in Section 8.9, but that is not needed for the present discussion. Two linearly independent solutions of Eq. (9) are $\phi_1(x) = \cosh 10x$ and $\phi_2(x) = \sinh 10x$. The first solution $\phi_1(x) = \cosh 10x$ is generated by the initial conditions $\phi_1(0) = 1$, $\phi_1'(0) = 0$; the second solution $\phi_2(x) = \sinh 10x$ is generated by the initial conditions $\phi_2(0) = 0$, $\phi_2'(0) = 10$. While analytically we can tell the difference between $\cosh 10x$ and $\sinh 10x$, for large x we have $\cosh 10x \sim e^{10x}/2$ and $\sinh 10x \sim e^{10x}/2$; numerically these two functions look exactly the same if only a fixed number of digits are retained. For example, correct to eight significant figures,

$$\sinh 10 = \cosh 10 = 11{,}013.233.$$

If the calculations are carried out on a machine that carries only seven digits the two solutions ϕ_1 and ϕ_2 are identical at $x = 1$ and indeed for all $x > 1$! Thus, even though the solutions are linearly independent, their numerical tabulation would show that they are the same because we can retain only a finite number of digits. We call this phenomenon *numerical dependence*. It is one of the important and difficult problems always encountered when we are solving a second or higher order equation that has at least one solution that grows very rapidly.

For the present problem we can partially circumvent this difficulty by computing instead of $\sinh 10x$ and $\cosh 10x$ the linearly independent solutions $\phi_3(x) = e^{10x}$ and $\phi_4(x) = e^{-10x}$ corresponding to the initial conditions $\phi_3(0) = 1$, $\phi_3'(0) = 10$ and $\phi_4(0) = 1$, $\phi_4'(0) = -10$, respectively. The solution ϕ_3 grows exponentially while ϕ_4 decays exponentially. Even so, we will encounter difficulty in calculating ϕ_4 correctly on a large interval. The reason is that at each step of the calculation for ϕ_4 we will introduce formula and round-off errors. Thus at any point x_n the data to be used in going to the next point are not precisely the values of $\phi_4(x_n)$ and $\phi_4'(x_n)$. The solution of the initial value problem with these data at x_n will involve not only e^{-10x} but also e^{10x}. Because the error in the data at x_n is small, the latter function will appear with a very small coefficient. Nevertheless, since e^{-10x} tends to zero and e^{10x} grows very rapidly, the latter will eventually dominate, and the calculated solution will be simply a multiple of $e^{10x} = \phi_3(x)$.

Another problem that must be investigated for each numerical procedure is the convergence of the approximate solution to the exact solution as h tends to zero. While it does not occur for the one-step methods discussed in the previous sections, there is the following danger when using multistep procedures. It can happen that the numerical formula for y_n, which will involve y_{n-1} and y_{n-2}, etc., may have extraneous solutions that do not correspond to solutions of the original differential equation. If these extraneous solutions grow rather than decay they will cause difficulty. This is illustrated by a specific example in the next section.

Other practical problems, which the numerical analyst must often face, are how to compute solutions of a differential equation in the neighborhood of a regular singular point and how to calculate numerically the solution of a differential equation on an *unbounded* interval. Such problems generally require a combination of analytical and numerical work.

8.8 A MULTISTEP METHOD

In the previous sections we considered one-step or starting methods for solving numerically the initial value problem

$$y' = f(x, y), \qquad y(x_0) = y_0. \tag{1}$$

Once approximate values of the exact solution $y = \phi(x)$ have been obtained at a few points beyond x_0, it is natural to ask whether we can make use of some of this information, rather than just the value at the last point, in order to obtain the value of ϕ at the next point. Specifically, if y_1 at x_1, y_2 at x_2, ... , y_n at x_n are known, how can we use this information to determine y_{n+1} at x_{n+1}?

One of the best of the multistep methods is the Adams-Moulton predictor-corrector method that makes use of information at the points x_n, x_{n-1}, x_{n-2}, x_{n-3}. Before the method can be used, it is necessary to compute y_1, y_2, and y_3 by some starting method such as the Runge-Kutta method—the starting method should be as accurate as the multistep method. We will briefly discuss the development of the Adams-Moulton predictor-corrector method.

Suppose that we integrate $\phi'(x)$ from x_n to x_{n+1}; then

$$\phi(x_{n+1}) - \phi(x_n) = \int_{x_n}^{x_{n+1}} \phi'(x)\, dx. \tag{2}$$

Assuming that $\phi(x_{n-3})$, $\phi(x_{n-2})$, $\phi(x_{n-1})$, and $\phi(x_n)$ are known, then $\phi'(x_{n-3})$, ... , $\phi'(x_n)$ can be computed from Eq. (1). Next we approximate $\phi'(x)$ by a polynomial of degree three that passes through the four points $[x_{n-3}, \phi'(x_{n-3})]$, $[x_{n-2}, \phi'(x_{n-2})]$, $[x_{n-1}, \phi'(x_{n-1})]$, and $[x_n, \phi'(x_n)]$. It can be proved that this can always be done, and further that the polynomial is unique. This polynomial is called an *interpolation polynomial* since it provides us with an approximate formula* for $\phi'(x)$ at values of x other than x_{n-3}, x_{n-2}, x_{n-1}, and x_n. Substituting this interpolation polynomial for $\phi'(x)$ in Eq. (2), evaluating the integral, and replacing $\phi(x_{n-3})$, ... , $\phi(x_{n+1})$ by their approximate values gives the multistep formula

$$y_{n+1} = y_n + \frac{h}{24} (55y'_n - 59y'_{n-1} + 37y'_{n-2} - 9y'_{n-3}). \tag{3}$$

This formula is known as the Adams-Bashforth predictor formula; it predicts the approximate value y_{n+1} for $\phi(x)$ at x_{n+1}. It has a local formula error proportional to h^5.

The details of the derivation of Eq. (3) that have been omitted are the determination of the polynomial of degree three that passes through the points (x_{n-3}, y'_{n-3}), (x_{n-2}, y'_{n-2}), (x_{n-1}, y'_{n-1}), and (x_n, y'_n), and the integration of this polynomial from x_n to x_{n+1}. The method of constructing interpolation polynomials dates back to Newton and is discussed in books on numerical methods;

* Recall, for example, that we often use linear interpolation in determining the values of the trigonometric functions for arguments lying between those given in a table.

see, for example, Henrici (Chapter 5) or Conte and deBoor (Chapter 4). What is important at the present time is not these algebraic details, but the underlying general principle of multistep methods.

1. A polynomial of degree p is fitted to the $p + 1$ points $(x_{n-p}, y'_{n-p}), \ldots,$ (x_n, y'_n).

2. This interpolation polynomial is integrated over an interval in x terminating at x_{n+1}, which gives an equation for y_{n+1}.

Also note that different formulas can be obtained by integrating the same polynomial over different intervals. If the polynomial used in obtaining Eq. (3) is integrated from x_{n-3} to x_{n+1}, we obtain the Milne (1890–1971) predictor formula:

$$y_{n+1} = y_{n-3} + \frac{4h}{3}(2y'_n - y'_{n-1} + 2y'_{n-2}). \tag{4}$$

This formula also has a local formula error proportional to h^5.

The Adams-Moulton predictor-corrector method involves, in addition to the formula (3), a *corrector formula* for improving the value of y_{n+1} predicted by Eq. (3). The corrector formula is derived by integrating $\phi'(x)$ from x_n to x_{n+1} using an interpolation polynomial that passes through the points $[x_{n-2}, \phi'(x_{n-2})]$, $[x_{n-1}, \phi'(x_{n-1})]$, $[x_n, \phi'(x_n)]$, and $[x_{n+1}, \phi'(x_{n+1})]$. In the actual calculation the approximate values y'_{n-2}, y'_{n-1}, y'_n, and y'_{n+1} are used for $\phi'(x)$ at x_{n-2}, x_{n-1}, x_n, and x_{n+1}, and $y'_{n+1} = f(x_{n+1}, y_{n+1})$ is evaluated using the value of y_{n+1} determined by Eq. (3). The Adams-Moulton corrector formula is

$$y_{n+1} = y_n + \frac{h}{24}(9y'_{n+1} + 19y'_n - 5y'_{n-1} + y'_{n-2}). \tag{5}$$

This formula also has a local formula error proportional to h^5. However, the proportionality constant in the error bound for the corrector formula (5) is about $\frac{1}{15}$ of the proportionality constant in the error bound for the predictor formula (3).

Once $y_{n-3}, y_{n-2}, y_{n-1}$, and y_n are known we can compute $y'_{n-3}, y'_{n-2}, y'_{n-1}$ and y'_n, and then use the predictor formula (3) to obtain a first value for y^-_{n+1}. Then we compute y'_{n+1} and use the corrector formula (5) to obtain an improved value of y_{n+1}. We can, of course, continue to use the corrector formula (5) if the change in y_{n+1} is too large. As a general rule, however, if it is necessary to use the corrector formula more than once or twice it can be expected that the step size h is too large and should be made smaller. Note also that Eq. (5) serves as a check on the arithmetic (programming) in determining y_{n+1} from Eq. (3)—any substantial difference between the values of y_{n+1} from Eq. (3) and from Eq. (5) would indicate a definite possibility of an arithmetic error.

To summarize, we first compute y_1, y_2, and y_3 by a starting formula with a local formula error no greater than h^5. We then switch to the predictor formula (3) to compute y_{n+1}, $n \geq 3$ and the corrector formula (5) to correct y_{n+1}. We continue to correct y_{n+1} until the relative change is less than a prescribed allowable error. A more detailed discussion of the Adams-Moulton predictor-corrector method and other multistep methods can be found in Conte and deBoor (Chapter 6) and Henrici (Chapter 5).

Example. Using the Adams-Moulton predictor-corrector method with $h = 0.1$ determine an approximate value of the exact solution $y = \phi(x)$ at $x = 0.4$ for the initial value problem

$$y' = 1 - x + 4y, \qquad y(0) = 1. \tag{6}$$

For starting data we use the values of y_1, y_2, and y_3 determined using the Runge-Kutta method. These are tabulated in Table 8.5. Then, using Eq. (6), we obtain

$$
\begin{aligned}
y_0 &= 1 & y_0' &= 5 \\
y_1 &= 1.6089333 & y_1' &= 7.3357332 \\
y_2 &= 2.5050062 & y_2' &= 10.820025 \\
y_3 &= 3.8294145 & y_3' &= 16.017658.
\end{aligned}
$$

From Eq. (3) we find that the predicted value of y_4 is

$$y_4 = 3.8294145 + \frac{0.1(469.011853)}{24} = 5.783631.$$

Next we use the corrector formula (5) to correct y_4. Corresponding to the predicted value of y_4 we find from Eq. (6) that $y_4' = 23.734524$. Hence, from Eq. (5), the corrected value of y_4 is

$$y_4 = 3.8294145 + \frac{0.1(471.181828)}{24} = 5.792673.$$

The exact value, correct through eight digits, is 5.7942260. Note that using the correction formula once reduces the error in y_4 to approximately one seventh of the error before its use.

A natural question to ask is why use a predictor-corrector method with a a local formula error proportional to h^5 when the Runge-Kutta method has an error of the same order of magnitude. One reason is that for many problems the most time-consuming part of the calculation, and hence the most expensive, is the evaluation of the function $f(x, y)$ at successive steps. For the Runge-Kutta method four evaluations of f are required in taking one step. On the other hand for the predictor-corrector method (3) and (5), the predictor requires only the evaluation of $f(x, y)$ at each x_n, and the use of the corrector requires an additional evaluation of $f(x, y)$ at x_{n+1} for each iteration performed. Thus the number of function evaluations can be reduced significantly. This saving can be very important in writing a general program for solving initial value problems. One drawback of predictor-corrector methods is the inconvenience in changing step size as the calculation proceeds; for example, one might want to use a step size h for

$n = 1, 2, \ldots, m$ and then a step size $h/2$ for $n > m$. In order to change the step size it is necessary to obtain new data at the new points $x_{m-1}, x_{m-2}, x_{m-3}$ consistent with the new step size. Such details are automatically handled in many general programs available at a large computer center.

Stability. We mentioned earlier in this chapter that numerical difficulties can arise when multistep methods are used. We will illustrate what can happen by considering the initial value problem

$$y' + 2y = 0, \qquad y(0) = 1, \tag{7}$$

and the multistep method

$$y_{n+1} = y_{n-1} + 2hy_n'. \tag{8}$$

The multistep method (8) is derived in Problems 4 and 5; it has a local formula error proportional to h^3. For later reference we note that the exact solution of the initial value problem (7) is $y = e^{-2x}$.

For this problem $y_n' = -2y_n$, so Eq. (8) becomes

$$y_{n+1} + 4hy_n - y_{n-1} = 0. \tag{9}$$

Equation (9) is a *linear difference equation*. It can be solved explicitly following a theory similar to the theory for linear constant coefficient differential equations. We look for a solution of Eq. (9) of the form $y_n = \lambda^n$ where λ is to be determined. We have

$$\lambda^{n+1} + 4h\lambda^n - \lambda^{n-1} = 0,$$

or

$$\lambda^{n-1}(\lambda^2 + 4h\lambda - 1) = 0. \tag{10}$$

The nonzero solutions of Eq. (10) are

$$\lambda_1 = -2h + \sqrt{1 + 4h^2} \quad \text{and} \quad \lambda_2 = -(2h + \sqrt{1 + 4h^2}). \tag{11}$$

The corresponding solutions of Eq. (9) are

$$y_n = \lambda_1^n \quad \text{and} \quad y_n = \lambda_2^n. \tag{12}$$

By analogy with the theory of linear second order differential equations it is reasonable to expect (and it can be shown) that the general solution of the difference equation (9) is

$$y_n = c_1\lambda_1^n + c_2\lambda_2^n, \tag{13}$$

where c_1 and c_2 are arbitrary constants.

What is important is that the difference equation (9) has two solutions, λ_1^n and λ_2^n, while the original differential equation has only the one solution e^{-2x}. This may cause trouble. We will now show that as $h \to 0$ one of the solutions of the difference equation will approach the true solution e^{-2x} of the differential equation, but that the other will oscillate between $\pm e^{2x}$. We suppose that h is small and

(i) approximate $(1 + 4h^2)^{1/2}$ by $1 + 2h^2$, (ii) approximate λ_1 by $1 - 2h$ and λ_2 by $-(1 + 2h)$, (iii) use the formula $\lim_{\alpha \to 0}(1 + \alpha)^{1/\alpha} = e$, and (iv) note that $nh = x_n$. We have

$$\lim_{h \to 0} \lambda_1{}^n = \lim_{h \to 0} (1 - 2h)^n = \lim_{h \to 0} (1 - 2h)^{x_n/h}$$

$$= \lim_{h \to 0} [(1 - 2h)^{-1/(2h)}]^{-2x_n} = e^{-2x_n}. \tag{14}$$

A similar calculation gives

$$\lim_{h \to 0} \lambda_2{}^n = (-1)^n e^{2x_n}. \tag{15}$$

Thus

$$y_n \to c_1 e^{-2x_n} + c_2(-1)^n e^{2x_n} \qquad \text{as} \qquad h \to 0. \tag{16}$$

The first term of Eq. (16) correctly approximates the general solution $c_1 e^{-2x}$ of the original differential equation (7). The second term is an extraneous solution that arises from the numerical procedure.

In using the multistep formula (8) or (9) we must know in addition to the initial value $y_0 = 1$ at $x_0 = 0$ the value of y_1 at $x_1 = h$. In theory, if the correct values of y_0 and y_1 are used, then the constants c_1 and c_2 of Eq. (16) would be determined as $c_1 = 1$ and $c_2 = 0$, and the extraneous solution $(-1)^n e^{2x_n}$ would not appear. However, as we discussed in the previous section, there will always be small errors, and as a consequence the term $c_2(-1)^n e^{2x_n}$ will be present in the solution of Eq. (9). Even though c_2 may be very small corresponding to the very small error in the calculation, the term $c_2(-1)^n e^{2x_n}$ will soon dominate the full solution as x_n increases. This is vividly shown by the graph in Figure 8.6 of the numerical solution of the initial value problem (7) using the multistep method (8) with $h = 7/224 \simeq 0.03$. Clearly, beyond $x = 3$ the extraneous solution $(-1)^n e^{2x_n}$ dominates.

This phenomenon is known as *numerical instability*. Roughly speaking, a numerical procedure is unstable if errors introduced into the calculations grow at an exponential rate as the computation proceeds. While it is beyond the scope of this book to give a detailed account of stability of numerical methods, we can make several observations. The question of stability (or instability) of a multistep method depends not only on the method but also on the particular differential equation; the method may be stable for one problem but not for another. Also, multistep methods may be stable for some values of h but not for all values of h. In order to assess the stability of a multistep method when applied to an initial value problem, one must examine the solutions of the resulting difference equation. The method is said to be *strongly stable* if the extraneous solutions λ^n of the difference equation are such that $|\lambda| < 1$ as $h \to 0$. If this is the case, then $\lambda^n \to 0$ as $n \to \infty$ for h sufficiently small, and errors in the computation will not grow as n increases. Note that for the example problem just discussed the extraneous solution $\lambda_2{}^n$ has $|\lambda_2| > 1$.

It is not possible, in general, to apply this criterion to the initial value problem $y' = f(x, y)$, $y(x_0) = y_0$. Usually, in discussing stability of a numerical method

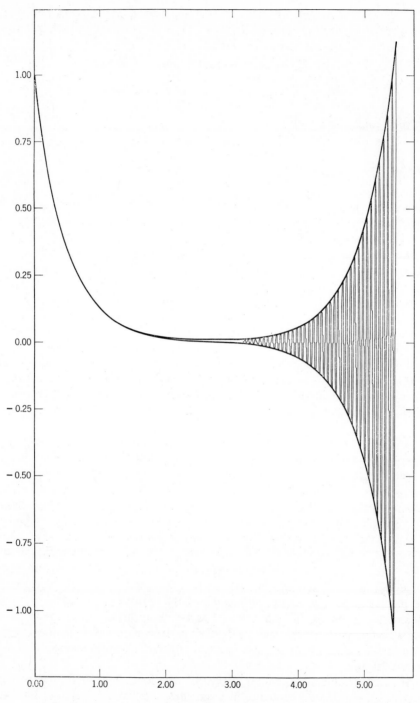

FIGURE 8.6 $y' + 2y = 0, y(0) = 1;$ $y_{n+1} = y_{n-1} + 2hy_n'.$

one is content to examine the *linear stability* of the method; is the procedure stable for the differential equation $y' = Ay$ where A is an arbitrary constant? It can be shown that the Adams-Moulton predictor-corrector method is strongly stable in the sense of linear stability for h sufficiently small. See Problems 7 and 8 for further discussion of these ideas.

PROBLEMS

In Problems 1 through 3 determine an approximate value of the exact solution at $x = 0.4$ using the Adams-Moulton predictor-corrector method. Take $h = 0.1$. The values for y_1, y_2, and y_3 are those computed by the Runge-Kutta method, rounded to six digits. Use the corrector formula to compute one correction.

1. $y' = 2y - 1$; $y_0 = 1$, $y_1 = 1.11070$, $y_2 = 1.24591$, $y_3 = 1.41105$

2. $y' = \frac{1}{2} - x + 2y$; $y_0 = 1$, $y_1 = 1.27140$, $y_2 = 1.59182$, $y_3 = 1.97211$

3. $y' = x^2 + y^2$; $y_0 = 1$, $y_1 = 1.11146$, $y_2 = 1.25302$, $y_3 = 1.43967$

4. Given that the values of $\phi'(x)$ at x_{n-1} and x_n are y'_{n-1} and y'_n, fit a first degree polynomial $ax + b$ to $\phi'(x)$ so that it passes through these two points. By integrating $\phi'(x)$ from x_{n-1} to x_{n+1} derive the multistep formula

$$y_{n+1} = y_{n-1} + 2hy'_n.$$

Hint: The calculations are not difficult if one sets $x_{n+1} = x_n + h$ and $x_{n-1} = x_n - h$.

5. Show that the multistep formula derived in Problem 4 has a local formula error proportional to h^3. Note that this formula is just as easy to use (after y_1 has been computed) as the Euler formula $y_{n+1} = y_n + h y'_n$ which has a local formula error proportional only to h^2. Unfortunately, however, as shown in the text this multistep formula is unstable.

Hint: Evaluate $\phi(x_n \pm h)$ by expanding $\phi(x)$ in a Taylor series about $x = x_n$.

6. This problem is concerned with the Milne predictor-corrector formula. The predictor formula was given in Eq. (4) of the text. The corrector formula is derived by evaluating the integral of $\phi'(x)$ from x_{n-1} to x_{n+1} by Simpson's rule. The formula is

$$y_{n+1} = y_{n-1} + \frac{h}{3}(y'_{n-1} + 4y'_n + y'_{n+1}).$$

Both the predictor and corrector formula have local formula errors proportional to h^5. The Milne method is often very effective, but the corrector can be unstable (see Problem 8). As a consequence, the Adams-Moulton predictor-corrector method is to be preferred. In each of the following problems determine an approximate value of the exact solution at $x = 0.4$ using the Milne predictor-corrector method. Use the corrector formula to compute one correction.

(a) Problem 1,   (b) Problem 2,   (c) Problem 3.

*7. In this problem we will show that the Adams-Moulton corrector formula (5) is stable for the linear differential equation $y' = Ay$.

(a) Show that the solution of the differential equation is $y = ce^{Ax}$.

(b) Show that the appropriate difference equation is

$$(1 - 9\alpha)y_{n+1} - (1 + 19\alpha)y_n + 5\alpha y_{n-1} - \alpha y_{n-2} = 0 \qquad \text{(i)}$$

where $\alpha = Ah/24$.

(c) Next show that $y_n = \lambda^n$ is a solution of Eq. (i) if λ is a root of

$$(1 - 9\alpha)\lambda^3 - (1 + 19\alpha)\lambda^2 + 5\alpha\lambda - \alpha = 0. \qquad \text{(ii)}$$

(d) In the limit as $h \to 0$, which implies $\alpha \to 0$, show that the roots of Eq. (ii) are $\lambda_1 = 1$, $\lambda_2 = 0$, $\lambda_3 = 0$.

(e) We examine the root λ_1 more closely for h small. Set $\lambda_1 = 1 + \lambda_{11}h$, substitute in Eq. (ii), and neglect terms proportional to h^2 or higher. Deduce that $\lambda_{11} = A$ and show that if $y_1 = \lambda_1{}^n \simeq (1 + Ah)^n$ then $y_1 \to e^{Ax_n}$ as $h \to 0$, on using $x_n = nh$.

Thus the solution corresponding to λ_1 approximates the true solution of the differential equation. For small h we have $|\lambda_2| < 1$ and $|\lambda_3| < 1$; thus the corresponding extraneous solutions will not grow. As a consequence, the Adams-Moulton corrector formula is strongly stable.

*8. In this problem we will show that the Milne corrector formula given in Problem 6 is unstable for the linear differential equation $y' = Ay$. The analysis parallels that of Problem 7.

(a) Show that the solution of the differential equation is $y = ce^{Ax}$.

(b) Show that the appropriate difference equation is

$$(1 - \alpha)y_{n+1} - 4\alpha y_n - (1 + \alpha)y_{n-1} = 0 \qquad \text{(i)}$$

where $\alpha = Ah/3$.

(c) Next show that $y_n = \lambda^n$ is a solution of Eq. (i) if λ is a root of

$$(1 - \alpha)\lambda^2 - 4\alpha\lambda - (1 + \alpha) = 0. \qquad \text{(ii)}$$

(d) In the limit $h \to 0$, which implies $\alpha \to 0$, show that the roots of Eq. (ii) are $\lambda_1 = 1$ and $\lambda_2 = -1$. Note that $|\lambda_1| = |\lambda_2| = 1$ and hence for small nonzero values of h one of the roots may have absolute value greater than one.

(e) Solve the quadratic equation (ii) and show that if h is small then $\lambda_1 \simeq 1 + Ah$ and $\lambda_2 \simeq -(1 - Ah/3)$. Next show that as $h \to 0$, with $x_n = nh$, the solution of the difference equation (i) is

$$y_n = c_1 e^{Ax_n} + c_2(-1)^n e^{-Ax_n/3}. \qquad \text{(iii)}$$

The first term in Eq. (iii) approximates the true solution of the differential equation. The second term, the extraneous solution corresponding to λ_2, decays if $A > 0$ but grows if $A < 0$. Thus the Milne corrector formula can be unstable. Methods whose stability (instability) depend on the sign of A are called *weakly stable*. Finally, we note that $|\lambda_2| > 1$ for $A < 0$.

8.9 SYSTEMS OF FIRST ORDER EQUATIONS

In the preceding sections we discussed numerical methods for solving initial value problems associated with first order differential equations. These methods can also be applied to a system of first order equations or to equations of higher order. Recalling that a higher order equation such as $d^2x/dt^2 = f(t, x, dx/dt)$ can be reduced to the system of first order equations $dx/dt = y$, $dy/dt = f(t, x, y)$, we will only consider systems of first order equations.* We will use t to denote the independent variable, and x, y, z, ... to denote the dependent variables. Primes will denote differentiation with respect to t. Further, for the purpose of simplicity, we will restrict our attention to a system of two first order equations

$$x' = f(t, x, y), \tag{1}$$
$$y' = g(t, x, y), \tag{2}$$

with the initial conditions

$$x(t_0) = x_0, \qquad y(t_0) = y_0. \tag{3}$$

The functions f and g are assumed to satisfy the conditions of Theorem 7.1 so that the initial value problem (1), (2), and (3) has a unique solution in some interval of the t axis containing the point t_0. We wish to determine approximate values x_1, x_2, ..., x_n, ... and y_1, y_2, ..., y_n, .. of the exact solution $x = \phi(t)$, $y = \psi(t)$ at the points $t_n = t_0 + nh$.

As an example of how the methods of the previous sections can be generalized consider the Euler method of Section 8.2. In this case we use the initial conditions to determine $\phi'(t_0)$ and $\psi'(t_0)$, and then move along the tangent lines to the curves $x = \phi(t)$ and $y = \psi(t)$. Thus

$$x_1 = x_0 + hf(t_0, x_0, y_0), \qquad y_1 = y_0 + hg(t_0, x_0, y_0).$$

In general

$$\begin{aligned} x_{n+1} &= x_n + hf(t_n, x_n, y_n) & y_{n+1} &= y_n + hg(t_n, x_n, y_n) \\ &= x_n + hx'_n & &= y_n + hy'_n. \end{aligned} \tag{4}$$

The other numerical methods can be generalized in a similar manner. The formulas are given in Problems 2, 4, and 5.

Example.   Determine approximate values of $x = \phi(t)$, $y = \psi(t)$ at the points $t = 0.1$ and $t = 0.2$ using the Euler formula (4) for the initial value problem

$$x' = x - 4y, \qquad y' = -x + y, \tag{5}$$

with

$$x(0) = 1, \qquad y(0) = 0. \tag{6}$$

Take $h = 0.1$.

For this problem $x'_n = x_n - 4y_n$ and $y'_n = -x_n + y_n$; hence

$$\begin{aligned} x'_0 &= 1 - 4(0) & y'_0 &= -1 + 0 \\ &= 1 & &= -1. \end{aligned}$$

Thus

$$\begin{aligned} x_1 &= 1 + 0.1(1) & y_1 &= 0 + 0.1(-1) \\ &= 1.1 & &= -0.1. \end{aligned}$$

---

* From the point of view of numerical accuracy there is some disagreement on whether a higher order equation should be reduced to a system of first order equations and solved by procedures of the type discussed in this section, or whether the higher order equation should be solved by procedures that are directly applicable. Such questions are well beyond the scope of an elementary text such as this, but are discussed in advanced books on numerical methods. See, for example, Henrici. In practice an nth order differential equation is usually written as a system of n first order differential equations.

Similarly

$$x'_1 = 1.1 - 4(-0.1) \qquad y'_1 = -1.1 + (-0.1)$$
$$= 1.5 \qquad\qquad = -1.2;$$

and

$$x_2 = 1.1 + 0.1(1.5) \qquad y_2 = -0.1 + 0.1(-1.2)$$
$$= 1.25 \qquad\qquad = -0.22.$$

It is easily verified that the exact solution of the initial value problem (5) and (6) is

$$\phi(t) = \frac{e^{-t} + e^{3t}}{2}, \qquad \psi(t) = \frac{e^{-t} - e^{3t}}{4}. \tag{7}$$

The values of the exact solution, rounded to eight digits, at $t = 0.2$ are $\phi(0.2) = 1.3204248$ and $\psi(0.2) = -0.25084701$.

PROBLEMS

Some of the problems for this section will deal with the following initial value problems. In all computations carry four digits unless otherwise specified.

(a) $x' = x + y + t$, $y' = 4x - 2y$; $x(0) = 1$, $y(0) = 0$
(b) $x' = 2x + ty$, $y' = xy$; $x(0) = 1$, $y(0) = 1$
(c) $x' = -tx - y - 1$, $y' = x$; $x(0) = 1$, $y(0) = 1$

1. For each of initial value problems (a), (b), and (c), determine approximate values of the exact solution at $t = 0.1$ and $t = 0.2$ using the Euler method with $h = 0.1$.

2. Show that the generalization of the three-term Taylor formula of Section 8.5 for the initial value problem $x' = f(t, x, y)$, $y' = g(t, x, y)$, with $x(t_0) = x_0$, $y(x_0) = y_0$ is

$$x_{n+1} = x_n + hx'_n + \tfrac{1}{2}h^2 x''_n, \qquad y_{n+1} = y_n + hy'_n + \tfrac{1}{2}h^2 y''_n$$

where

$$x''_n = f_t(t_n, x_n, y_n) + f_x(t_n, x_n, y_n)x'_n + f_y(t_n, x_n, y_n)y'_n,$$
$$y''_n = g_t(t_n, x_n, y_n) + g_x(t_n, x_n, y_n)x'_n + g_y(t_n, x_n, y_n)y'_n.$$

Determine an approximate value of the exact solution at $t = 0.2$ for the illustrative initial value problem in the text. Take $h = 0.2$ and compare the results with the exact values.

3. Using the three-term Taylor formula derived in Problem 2 determine an approximate value of the exact solution at $x = 0.2$ for each of the initial value problems (a), (b), and (c). Take $h = 0.2$.

4. The generalization of the Runge-Kutta method of Section 8.6 for the initial value problem $x' = f(t, x, y)$, $y' = g(t, x, y)$, with $x(t_0) = x_0$, $y(t_0) = y_0$ is

$$x_{n+1} = x_n + \tfrac{1}{6}h(k_{n1} + 2k_{n2} + 2k_{n3} + k_{n4}), \qquad y_{n+1} = y_n + \tfrac{1}{6}h(l_{n1} + 2l_{n2} + 2l_{n3} + l_{n4})$$

where

$$k_{n1} = f(t_n, x_n, y_n), \qquad\qquad l_{n1} = g(t_n, x_n, y_n),$$
$$k_{n2} = f(t_n + \tfrac{1}{2}h, x_n + \tfrac{1}{2}k_{n1}, y_n + \tfrac{1}{2}l_{n1}), \qquad l_{n2} = g(t_n + \tfrac{1}{2}h, x_n + \tfrac{1}{2}k_{n1}, y_n + \tfrac{1}{2}l_{n1}),$$
$$k_{n3} = f(t_n + \tfrac{1}{2}h, x_n + \tfrac{1}{2}k_{n2}, y_n + \tfrac{1}{2}l_{n2}), \qquad l_{n3} = g(t_n + \tfrac{1}{2}h, x_n + \tfrac{1}{2}k_{n2}, y_n + \tfrac{1}{2}l_{n2}),$$
$$k_{n4} = f(t_n + h, x_n + k_{n3}, y_n + l_{n3}), \qquad\qquad l_{n4} = g(t_n + h, x_n + k_{n3}, y_n + l_{n3}).$$

Determine an approximate value of the exact solution at $t = 0.2$ for the illustrative initial value problem in the text. Take $h = 0.2$ and compare the results with the exact values.

5. Consider the initial value problem $x' = f(t, x, y)$ and $y' = g(t, x, y)$ with $x(t_0) = x_0$ and $y(t_0) = y_0$. The generalization of the Adams-Moulton predictor-corrector method of Section 8.8 is

$$x_{n+1} = x_n + \tfrac{1}{24} h \left(55x'_n - 59x'_{n-1} + 37x'_{n-2} - 9x'_{n-3}\right),$$

$$y_{n+1} = y_n + \tfrac{1}{24} h \left(55y'_n - 59y'_{n-1} + 37y'_{n-2} - 9y'_{n-3}\right),$$

and

$$x_{n+1} = x_n + \tfrac{1}{24} h \left(9x'_{n+1} + 19x'_n - 5x'_{n-1} + x'_{n-2}\right),$$

$$y_{n+1} = y_n + \tfrac{1}{24} h \left(9y'_{n+1} + 19y'_n - 5y'_{n-1} + y'_{n-2}\right),$$

Determine an approximate value of the exact solution at $t = 0.4$ for the illustrative initial value problem $x' = x - 4y$, $y' = -x + y$ with $x(0) = 1$, $y(0) = 0$. Take $h = 0.1$. Correct the predicted value once. For the values of $x_1, \ldots, y_3$ use the values of the exact solution rounded to six digits: $x_1 = 1.12883$, $x_2 = 1.32042$, $x_3 = 1.60021$, $y_1 = -0.110527$, $y_2 = -0.250847$, and $y_3 = -0.429696$.

6. Consider the initial value problem

$$x'' + t^2 x' + 3x = t, \qquad x(0) = 1, \qquad x'(0) = 2.$$

Change this problem to a system of two equations and determine approximate values of the exact solution at $t = 0.1$ and $t = 0.2$ using the Euler method with a step size $h = 0.1$.

REFERENCES

There are many books of varying degrees of sophistication dealing with numerical analysis in general, programming, and the numerical solution of ordinary differential equations and partial differential equations. The following books were mentioned in the text:

Conte, S. D. and deBoor, Carl, *Elementary Numerical Analysis*, Second Edition, McGraw-Hill, New York, 1972.

Henrici, Peter, *Discrete Variable Methods in Ordinary Differential Equations*, Wiley, New York, 1962.

The text by Henrici provides a wealth of theoretical information about numerical methods for solving ordinary differential equations; however, it is fairly advanced. Two additional books on numerical methods for ordinary differential equations are:

Gear, C. William, *Numerical Initial Value Problems in Ordinary Differential Equations*, Prentice-Hall, Englewood, Cliffs, N.J., 1971.

Lambert, J. D., *Computational Methods in Ordinary Differential Equations*, Wiley, New York, 1973.

Several of the many elementary general texts with chapters on ordinary differential equations are:

Acton, Forman S., *Numerical Methods That Work*, Harper and Row, New York, 1970.

Dorn, William S. and McCracken, Daniel D., *Numerical Methods with Fortran IV Case Studies*, Wiley, New York, 1972.

Henrici, Peter, *Elements of Numerical Analysis*, Wiley, New York, 1964.

Hildebrand, F. B., *Introduction to Numerical Analysis*, Second Edition, McGraw-Hill, New York, 1974.

CHAPTER 9

Nonlinear Differential Equations and Stability

9.1 INTRODUCTION

Up to this point most of our discussion has dealt with methods for solving differential equations. Primarily, we have considered linear equations for which there is an elegant and extensive theory. However, some simple first order nonlinear equations were considered in Chapter 2, two special cases of second order nonlinear equations were discussed in the problems following Section 3.1, and the numerical methods of Chapter 8 apply equally well to nonlinear equations.

We must face the fact that it is usually very difficult, if not impossible, to find a solution of a given differential equation in a reasonably convenient and explicit form, especially if the equation is nonlinear. Therefore, it is important to consider what qualitative information can be obtained about the solutions of differential equations, particularly nonlinear ones, without actually solving the equations. The questions considered in this chapter are mainly associated with the idea of "stability" of a solution; this idea will be made more precise in Section 9.4. As an example, in applications such as automatic control theory, an important question is whether small changes in the initial conditions (input)* lead to small changes (stability) or to large changes (instability) in the solution (output). A second type of question arises when a nonlinear equation is approximated by a simpler linear one; for example, as mentioned in Section 1.1, when the nonlinear pendulum equation $d^2\theta/dt^2 + (g/l) \sin \theta = 0$ is replaced by $d^2\theta/dt^2 + (g/l)\theta = 0$ for θ small. If the solution of the linear equation is not a reasonable approximation to the solution of the original nonlinear equation, the linearization is of dubious value. Many of the arguments that we will present follow from geometric

---

* For purposes of discussion it may be helpful to think of the initial state as the input and the differential equation as an operation that starts once the input is given and gives an output, the solution.

considerations and were developed by the French mathematician H. Poincaré*
and the Russian mathematician A. M. Liapounov.†

To illustrate some of the ideas in their simplest form, consider the nonlinear
first order equation

$$\frac{dA}{dt} = \epsilon A - \sigma A^2, \tag{1}$$

with the initial condition

$$A(0) = A_0 \geq 0, \tag{2}$$

where ϵ and σ are given positive numbers. While negative values of A_0 are permissible mathematically, in physical applications A_0 will be nonnegative. Equation
(1) occurs in many different applications. Let us see how it arises in ecology
in an analysis of the growth of a single species having access to limited resources.

Let A be the size of a population at time t. In the simplest model of population growth the rate of growth is assumed to be directly proportional to the size
of the population: $dA/dt = \epsilon A$, where the constant $\epsilon > 0$ is known as the *growth
rate*. As we know, the solution of this differential equation satisfying the initial
condition (2) is

$$A(t) = A_0 e^{\epsilon t}, \tag{3}$$

which gives an exponentially growing population that becomes unbounded as
$t \to \infty$. However, the growth of a population must eventually be limited by the
availability of resources, which in turn implies that the growth rate should be a
function of the population:

$$dA/dt = Af(A). \tag{4}$$

When the population size A is small, there will be sufficient resources so the
growth rate will be independent of the resources and hence approximately proportional to the population, $f(A) \approx \epsilon$. However, as the population increases
the available resources per individual will decrease and produce an inhibitory
effect on the population growth. Thus $f(A)$ should decrease as A increases,
so eventually $df(A)/dA < 0$. The simplest function that meets these conditions is
$f(A) = \epsilon - \sigma A$ where $\sigma > 0$. This gives Eq. (1). In ecology Eq. (1) is known as
the *Verhulst*‡ equation or the *logistic* equation. An equation similar to Eq. (1)

* H. Poincaré (1854–1912) was generally regarded as the outstanding mathematician in the world
around 1900. He made many fundamental contributions in several different areas of mathematics.
He founded the subject of topological dynamics and was the father of modern topology, which is
one of the most active fields in mathematical research. In his work on celestial mechanics he
developed his theory of asymptotic expansions, which today is one of the most powerful tools of
the applied mathematician.
† A. M. Liapounov's (1857–1918) work on stability and nonlinear differential equations, which is
now realized to be of great importance, was not fully appreciated during his lifetime.
‡ P. F. Verhulst (1804–1849) was a Belgian mathematician who in 1839 introduced the solution of
Eq. (1) into the theory of population dynamics.

also arises in the study of the transition from laminar to turbulent flow in fluid dynamics; there it is often called the *Landau** equation.

As we have noted, if the nonlinear term σA^2 is neglected in Eq. (1) then the population grows exponentially with time. The situation is entirely different if the nonlinear term is retained. Note that if $A = 0$ or if $A = \epsilon/\sigma$ the right-hand side of Eq. (1) vanishes. Thus we have the two constant solutions $A = \phi_1(t) = 0$ and $A = \phi_2(t) = \epsilon/\sigma$. As we shall see, these two solutions are of particular importance. In the theory of differential equations they are called *critical points* or, since $dA/dt = 0$ at these points, *equilibrium points*. Provided that $A \neq 0$ and $A \neq \epsilon/\sigma$, we can write Eq. (1) in the form

$$\frac{dA}{A(\epsilon - \sigma A)} = dt. \tag{5}$$

Using partial fractions we have

$$\left[\frac{1}{A} + \frac{\sigma}{\epsilon - \sigma A}\right] dA = \epsilon \, dt. \tag{6}$$

Integrating,

$$\ln |A| - \ln |\epsilon - \sigma A| = \epsilon t + K, \tag{7}$$

where K is a constant of integration to be chosen so that the initial condition $A(0) = A_0$ is satisfied. The cases $0 < A_0 < \epsilon/\sigma$, and $A_0 > \epsilon/\sigma$ must be considered separately since the sign of $(\epsilon - \sigma A)$ is different in the intervals $0 < A < \epsilon/\sigma$ and $A > \epsilon/\sigma$. Nevertheless, the final result can be expressed in a single formula. Omitting several steps of algebra, we find that

$$A = \phi(t) = \frac{\epsilon}{\sigma + [(\epsilon - \sigma A_0)/A_0]e^{-\epsilon t}} \tag{8}$$

or $A_0 \neq 0$. While the formula (8) is not valid for $A_0 = 0$, it does contain the constant solution $A = \phi_2(t) = \epsilon/\sigma$ corresponding to $A_0 = \epsilon/\sigma$. Since $\epsilon > 0$, it follows that $\exp(-\epsilon t) \to 0$ as $t \to \infty$, and hence $A \to \epsilon/\sigma$ as $t \to \infty$. Several typical solutions for different values of A_0 as well as the solutions $A = \phi_1(t) = 0$ and $A = \phi_2(t) = \epsilon/\sigma$ are shown in Figure 9.1.

For A_0 sufficiently small (actually $A_0 < \epsilon/2\sigma$, see Problem 14) the curve is S-shaped or sigmoid. While there are numerous experimental data for the growth of protozoa and bacteria that fit a sigmoid curve, there are also populations for which the fit of the growth curve is not so good. In ecology the ratio ϵ/σ is called the *saturation level* or *total carrying capacity*; a population starting below this level cannot surpass it.

It is clear from Figure 9.1 or from Eq. (8) that any solution of Eq. (1) starting at an initial point $A_0 > 0$ ultimately approaches the solution $A = \phi_2(t) = \epsilon/\sigma$ as $t \to \infty$. This limiting behavior as $t \to \infty$ is an example of the type of qualitative

---

* L. D. Landau (1908–1968) was a Russian physicist who received the Nobel prize in 1962 for his contributions to the understanding of condensed states of matters, particularly liquid helium. He was also the coauthor, with E. M. Lifshitz, of a well-known series of physics textbooks.

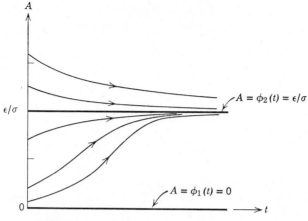

FIGURE 9.1 $dA/dt = \epsilon A - \sigma A^2$, $\epsilon > 0$, $\sigma > 0$.

information that is often of interest. Here we have obtained it by first solving the differential equation, but it is significant for more difficult problems that the limiting behavior can often be determined without a complete knowledge of the solutions. To show this for the present problem, let us plot dA/dt as a function of A as given by Eq. (1). The graph is a parabola with $dA/dt = 0$ at $A = 0$ and at $A = \epsilon/\sigma$. Furthermore, dA/dt is positive for A small; thus the graph must have the appearance indicated in Figure 9.2. If $0 < A < \epsilon/\sigma$, then $dA/dt > 0$ and A increases toward ϵ/σ. On the other hand, if $A > \epsilon/\sigma$, then $dA/dt < 0$ and A decreases toward ϵ/σ. As indicated, A always approaches the solution $A = \phi_2(t) = \epsilon/\sigma$ provided that the initial value $A_0 > 0$.

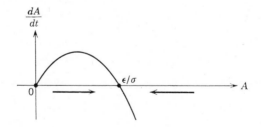

FIGURE 9.2 $dA/dt = \epsilon A - \sigma A^2$, $\epsilon > 0$, $\sigma > 0$.

Let us examine the solution $A = \phi_1(t) = 0$ in more detail. To emphasize the point we wish to make, suppose that we think of Eq. (1) as representing a physical system, or black box, as depicted in Figure 9.3. If $A_0 = 0$ is the input, then the output is $A = 0$. But suppose that a slight error is made, and a small but nonzero value of A_0 is the input. The question is whether the output remains close to $A = 0$. It is clear from Figure 9.1 that the answer is no. Instead, $A \to \epsilon/\sigma$ as

$$A_0 \longrightarrow \boxed{\dfrac{dA}{dt} = \epsilon A - \sigma A^2} \longrightarrow A = \phi(t; A_0)$$

FIGURE 9.3

$t \to \infty$. It is natural to say that the solution $A = \phi_1(t) = 0$ is an *unstable solution* of Eq. (1) or that $A = 0$ is an *unstable critical (equilibrium) point*.

Now consider the solution $A = \phi_2(t) = \epsilon/\sigma$. If the input is $A_0 = \epsilon/\sigma$, then the output of the black box is $A = \epsilon/\sigma$. If a small error is made in the input, the output will approach ϵ/σ asymptotically (in fact, exponentially) as $t \to \infty$. Hence we say that the solution $A = \phi_2(t) = \epsilon/\sigma$ is an *asymptotically stable solution* of Eq. (1) or that $A = \epsilon/\sigma$ is an *asymptotically stable critical (equilibrium) point*.

An ecological interpretation of these two cases is as follows. First, suppose we are trying to perform a bacteria-free experiment in an environment for which the growth of bacteria is given by Eq. (1). After considerable effort the bacteria population is reduced to an extremely low, but nonzero, level. Is it safe to carry out the experiment on the assumption that the bacteria population will stay very low? No; the initial level of bacteria (A_0) is greater than zero and the bacteria will grow toward ϵ/σ as t increases. Second, suppose that we are trying to perform an experiment with a level of bacteria of approximately ϵ/σ but that from time to time contaminants are introduced that kill a few bacteria. Is it possible that this will cause the bacteria population to continue to decrease and eventually become extinct? No; if the population is moved slightly below ϵ/σ it will increase toward ϵ/σ; if it is moved slightly above ϵ/σ it will decrease toward ϵ/σ.

As a second mathematical example again consider Eq. (1) with the initial condition (2), but suppose that $\epsilon < 0$ and $\sigma < 0$. Without computing the explicit solution of Eq. (1), let us consider what information can be obtained from the graph of dA/dt vs A shown in Figure 9.4. Since $dA/dt < 0$ for $0 < A < \epsilon/\sigma$, it is clear that if the initial value A_0 is in the interval $0 < A_0 < \epsilon/\sigma$, then A will approach the solution $A = \phi_1(t) = 0$ as $t \to \infty$. On the other hand, since $dA/dt > 0$ for $A > \epsilon/\sigma$ it follows that if $A_0 > \epsilon/\sigma$, then A will increase without bound as t

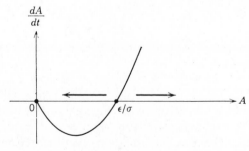

FIGURE 9.4 $dA/dt = \epsilon A - \sigma A^2$, $\epsilon < 0$, $\sigma < 0$.

increases. It also follows from Figure 9.4 that the solution $A = \phi_2(t) = \epsilon/\sigma$ is unstable to small changes in the initial condition from $A_0 = \epsilon/\sigma$; a small positive change causes A to increase without bound, and a small negative change causes A to approach zero with increasing t.

On the other hand, the solution $A = \phi_1(t) = 0$ is asymptotically stable to small changes in the initial condition $A_0 = 0$, in the sense that the system causes a small initial deviation from the zero solution to decay. However, the solution $A = \phi_1(t) = 0$ is not stable to all initial deviations from $A_0 = 0$. If $A_0 > \epsilon/\sigma$, then rather than dying out, A will grow without bound. A sketch of the solutions of Eq. (1) with $\epsilon < 0$ and $\sigma < 0$ for different values of A_0 is shown in Figure 9.5.

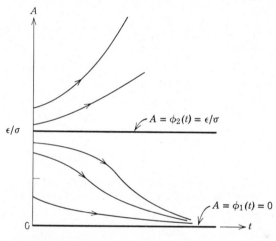

FIGURE 9.5 $dA/dt = \epsilon A - \sigma A^2$, $\epsilon < 0$, $\sigma < 0$.

We can interpret these results in the context of our black box of Figure 9.3. Usually in applications where a mechanism is to deliver a certain output when a certain input (or something very close to that input) is provided, we are interested only in stability with respect to small disturbances. It is, of course, conceivable that large disturbances can occur. For example, suppose an automatic control is set to keep a flap on an airplane wing at a certain inclination. In the normal motion of the plane the changing aerodynamic forces on the flap will cause it to move from its set position, but then the automatic control (the black box) will come into action to damp out the small deviation and return the flap to its set position. (Mathematically, for Eq. (1) with $\epsilon < 0$ and $\sigma < 0$ the set position is $A = 0$.) However, if the airplane is caught in a high gust of wind the flap may be deflected so much that the automatic control cannot bring it back to the set position (this would correspond to a deviation greater than ϵ/σ). Presumably the pilot would then take control!

In concluding this introductory section let us consider the question of neglecting the nonlinear term σA^2 in Eq. (1). In the first case ($\epsilon > 0$, $\sigma > 0$) the linear equation does give the correct behavior of A for a small time interval if A is initially small. However, in studying the solution for all time it *is not* permissible to neglect the term σA^2. The reason is that even if A is initially very small, so that the term σA^2 is initially small compared to the term ϵA, A grows exponentially with t, and the term σA^2 eventually becomes important. In the second case ($\epsilon < 0$, $\sigma < 0$), it *is* permissible to neglect the quadratic term σA^2 provided that A is small. The solution of the linear equation, $A_0 \exp(\epsilon t)$, approaches zero for any value of A_0. The same result is obtained from the nonlinear equation provided that $A_0 < \epsilon/\sigma$, and this is what is meant by "small" in this particular example. Thus, in the first case, the solution of the linearized problem is not a good approximation (at least not for all time) to that of the nonlinear problem; in the second case it is, provided only that $A_0 < \epsilon/\sigma$.

In the following sections we will develop the ideas suggested here, and others, in greater depth. While many books dealing with nonlinear differential equations have appeared in recent years, almost all are at an advanced level and assume considerable mathematical sophistication. However, several texts, listed in the references at the end of this chapter, have chapters dealing with theory or applications that should be readable after completing the present chapter.

PROBLEMS

In Problems 1 through 7 determine the limiting behavior of the solution of each equation as $t \to \infty$ for different initial conditions $A(0) = A_0$ by plotting dA/dt vs A.

1. $dA/dt = \epsilon A - \sigma A^2$, $\epsilon > 0$, $\sigma > 0$, $-\infty < A_0 < \infty$
2. $dA/dt = \epsilon A - \sigma A^2$, $\epsilon < 0$, $\sigma < 0$, $-\infty < A_0 < \infty$
3. $dA/dt = \epsilon A + \sigma A^2$, $\epsilon > 0$, $\sigma > 0$, $A_0 \geq 0$
4. $dA/dt = \epsilon A + \sigma A^2$, $\epsilon > 0$, $\sigma > 0$, $-\infty < A_0 < \infty$
5. $dA/dt = A(A - 1)(A - 2)$, $A_0 \geq 0$
6. $dA/dt = A(1 - A)(A - 2)$, $A_0 \geq 0$
7. $dA/dt = A^2(A^2 - 1)$, $-\infty < A_0 < \infty$
8. The population equation

$$dA/dt = \epsilon A^{1/2}(1 - A/K), \qquad \epsilon > 0, \qquad K > 0$$

appears to give results in good agreement with some experimental data. Determine the behavior of A as $t \to \infty$ for different values of $A(0) = A_0 \geq 0$.

9. Suppose that the number of trout $N(t)$ in a pond with a limited food supply is governed by the logistic equation $dN/dt = \alpha N(N_E - N)$ where α and N_E are positive constants.

(a) Show that $N = N_E$ is a stable critical point and that $N = 0$ is an unstable critical point.

(b) How is the logistic equation modified if a fisherman removes trout at a rate proportional to the number of trout in the pond? What happens to the stable critical point? Make a reasonable assumption about the constant of proportionality.

(c) How is the logistic equation modified if a fisherman removes trout at a uniform rate per unit time? Determine the number of trout in the pond for large time as a function of the initial number N_0. Make a reasonable assumption about the rate at which trout are removed.

10. In a simple model of economic growth the differential equation

$$dk/dt = sf(k) - gk$$

is obtained. Here k is the ratio of the supply of capital to the supply of labor, $f(k)$ is the productivity function (output per unit labor), g is the growth rate of the supply of labor, and s is a constant with $0 < s < 1$. It is known that $f'(k) > 0$, $f''(k) < 0$, and $f(k)$ is bounded. Show that if $0 < g/s < f'(0)$, then there exists a unique stable equilibrium solution (critical point) k given by the solution of the equation $sf(k) - gk = 0$.

11. Show that all solutions of the linear equation $dA/dt = -\epsilon A$ with $\epsilon > 0$ approach the solution $A = \phi(t) = 0$ as $t \to \infty$ independent of the choice of the initial condition $A(0) = A_0$, $-\infty < A_0 < \infty$. The solution $A = \phi(t) = 0$ is said to be *globally asymptotically stable* to emphasize that this solution is approached as $t \to \infty$ regardless of the initial state. This may be contrasted with the situation in Figure 9.4, where the solution $A = 0$ is approached as $t \to \infty$ only if $A_0 < \epsilon/\sigma$.

12. Derive Eq. (8) by choosing K in Eq. (7) so that the initial condition $A(0) = A_0$ is satisfied.

13. Determine the solution of the equation

$$dA/dt = \epsilon A - \sigma A^2, \qquad \epsilon < 0, \qquad \sigma < 0$$

with $A(0) = A_0$, $0 \le A_0 < \infty$, and verify that the sketch shown in Figure 9.5 is correct.

14. Show that the graph of the solution of the logistic equation $dA/dt = \epsilon A - \sigma A^2$, $\epsilon > 0$, $\sigma > 0$ is concave up if $0 < A < \epsilon/2\sigma$, concave down if $\epsilon/2\sigma < A < \epsilon/\sigma$, and concave up if $A > \epsilon/\sigma$.

Hint: Compute d^2A/dt^2 by differentiating the logistic equation and then eliminate dA/dt to express d^2A/dt^2 as a function of A.

15. **Linear Stability.** Consider the logistic equation $dA/dt = \epsilon A - \sigma A^2$, $\epsilon > 0$, $\sigma > 0$. We have seen that the constant solutions (critical points) $A = 0$ and $A = \epsilon/\sigma$ play a crucial role in the analysis of this differential equation. In this problem we will discuss an analytic method, rather than the geometric arguments of the text, for analyzing the stability of these solutions.

Let A_E be any constant solution (critical point) of the logistic equation. Suppose that this solution is very slightly perturbed, what happens? We write $A(t) = A_E + u(t)$ where $u(0)$ is very small and ask what happens to u as $t \to \infty$.

(a) Derive the differential equation satisfied by $u(t)$.

(b) If $u(0)$ is very small then, at least initially, $u^2 \ll u$. Neglecting quadratic terms in u, derive a linear first order differential equation for u. This is known as *linearization*.

(c) If $u(t) \to \infty$ as $t \to \infty$ the constant solution is said to be linearly unstable; on the other hand, if $u(t) \to 0$ as $t \to \infty$ the constant solution is said to be linearly stable (actually asymptotically stable). Show that the constant solution $A_E = 0$ is linearly unstable and that the constant solution $A_E = \epsilon/\sigma$ is linearly stable. Note that the stability characteristics may be modified by the inclusion of the nonlinear term involving u^2 which has been neglected. That is why the stability is referred to as *linear stability* or *linear instability*.

16. Following the outline of Problem 15 analyze the linear stability of the constant solutions of the equation $dA/dt = \epsilon A - \sigma A^2$, $\epsilon < 0$, $\sigma < 0$.

9.2 SOLUTIONS OF AUTONOMOUS SYSTEMS

In the remainder of this chapter we will be concerned with systems of two simultaneous differential equations of the form

$$\frac{dx}{dt} = F(x, y), \qquad \frac{dy}{dt} = G(x, y). \tag{1}$$

We assume that the functions F and G are continuous and have continuous partial derivatives in some domain D in the xy plane. Then, by Theorem 7.1, if (x_0, y_0) is a point in this domain, there exists a unique solution $x = \phi(t)$, $y = \psi(t)$ of the system (1) satisfying the initial conditions

$$x(t_0) = x_0, \qquad y(t_0) = y_0. \tag{2}$$

The solution is defined in some interval $\alpha < t < \beta$ which contains the point t_0.

Notice that the independent variable t does not appear explicitly in Eqs. (1). Such a system is said to be *autonomous*. Physically, an autonomous system is one in which the parameters of the system are not time-dependent. Autonomous systems occur frequently in practice; for example, the motion of an undamped pendulum of length l is governed by the differential equation

$$\frac{d^2\theta}{dt^2} + \frac{g}{l}\sin\theta = 0. \tag{3}$$

Letting $x = \theta$ and $y = d\theta/dt$, we can rewrite Eq. (3) as a nonlinear autonomous system of two equations:

$$\frac{dx}{dt} = y, \qquad \frac{dy}{dt} = -\left(\frac{g}{l}\right)\sin x. \tag{4}$$

The geometric reasoning introduced in the previous section can be effectively extended to systems of two autonomous equations. In order to initiate this discussion and to motivate some of the more mathematical details that will appear in the following sections we consider the ecological problem of two competing species.

Two Competing Species.   Suppose that we have two similar species competing for a limited food supply; for example, two species of fish in a pond that do not prey on each other but do compete for the available food.* Let x and y be the populations of the two species at time t. The discussion of the previous section

* Another example is that of two yeasts competing in an anaerobic (absence of oxygen) medium. Note further that yeasts produce alcohol, which in turn reduces their growth rate; this is consistent with Eqs. (5), (6), and (7).

suggests that in the absence of species y the growth of species x is governed by an equation of the form

$$dx/dt = x(\epsilon_1 - \sigma_1 x), \qquad (5)$$

and that in the absence of species x the growth of species y is governed by an equation of the form

$$dy/dt = y(\epsilon_2 - \sigma_2 y). \qquad (6)$$

However, when both species are present each will impinge on the available food supply for the other. In effect, they reduce the growth rates and saturation populations of each other. The simplest expression for reducing the growth rate of species x due to the presence of species y is to replace the growth rate factor $\epsilon_1 - \sigma_1 x$ of Eq. (5) by $\epsilon_1 - \sigma_1 x - \alpha_1 y$ where α_1 is a measure of the degree to which species y interferes with species x. Similarly, in Eq. (6) we replace $\epsilon_2 - \sigma_2 y$ by $\epsilon_2 - \sigma_2 y - \alpha_2 x$. Thus we have the system of equations

$$dx/dt = x(\epsilon_1 - \sigma_1 x - \alpha_1 y), \qquad (7a)$$

$$dy/dt = y(\epsilon_2 - \sigma_2 y - \alpha_2 x). \qquad (7b)$$

The actual values of the positive constants ϵ_1, σ_1, α_1, ϵ_2, σ_2, α_2 depend on the physical problem under consideration.

By analogy with our discussion of the logistic equation in Section 9.1, we determine the constant solutions of Eqs. (7). This is done by setting the right-hand sides of Eqs. (7) equal to zero

$$x(\epsilon_1 - \sigma_1 x - \alpha_1 y) = 0,$$

$$y(\epsilon_2 - \sigma_2 y - \alpha_2 x) = 0. \qquad (8)$$

The solutions corresponding to either $x = 0$ or $y = 0$ are readily seen to be $x = 0$, $y = 0$; $x = 0$, $y = \epsilon_2/\sigma_2$; $x = \epsilon_1/\sigma_1$, $y = 0$. In addition, there is a constant solution corresponding to the intersection of the lines $\epsilon_1 - \sigma_1 x - \alpha_1 y = 0$ and $\epsilon_2 - \sigma_2 y - \alpha_2 x = 0$ if these lines intersect. There are no other constant solutions of Eqs. (7). Geometrically these solutions can be represented as points in the xy plane; they are called critical points or equilibrium points. Moreover, in the same xy plane it is very helpful to visualize a solution of the system (7) as a point (x, y) moving as a function of time. At time $t = 0$ the initial populations of the two species provide an initial point (x_0, y_0) in the plane; then we follow the motion of the point (x, y) representing the populations of the two species at time t as it traces a curve in the plane. We can obtain considerable information about the behavior of solutions of Eqs. (7) without actually solving the problem.

First, from Eq. (7a) we observe that x increases or decreases as $\epsilon_1 - \sigma_1 x - \alpha_1 y > 0$ or $\epsilon_1 - \sigma_1 x - \alpha_1 y < 0$; similarly from Eq. (7b) we see that y increases or decreases as $\epsilon_2 - \sigma_2 y - \alpha_2 x > 0$ or $\epsilon_2 - \sigma_2 y - \alpha_2 x < 0$. This situation is depicted geometrically in Figure 9.6. In order to see what is happening to the two populations simultaneously we must superimpose the diagrams of Figure 9.6.

FIGURE 9.6

FIGURE 9.7

There are four possibilities as shown in Figure 9.7. The critical points are indicated by heavy dots. We will examine only cases (a) and (d) in more detail; cases (b) and (c) are similar and are left as problems. Also, for convenience we will assume that the initial populations x_0 and y_0 are each nonzero.

Consider case (a). If the initial populations are in region I then both x and y will increase; if the point moves into region II then species y will continue to increase but species x will start to decrease. Similarly, if the initial point is in region III, then both x and y will decrease; if the point moves into region II then x will continue to decrease while y now starts to increase. This suggests, for populations initially reasonably close to $(0, \epsilon_2/\sigma_2)$ that the point (x, y) representing the populations at time t approaches the critical point $(0, \epsilon_2/\sigma_2)$ as $t \to \infty$. This is shown in Figure 9.8a for several different initial states. This situation corresponds to the extinction of population x with population y reaching an equilibrium state of size ϵ_2/σ_2.

One might ask whether the point $(0, \epsilon_1/\alpha_1)$ is also a possible limiting state, since superficially populations that start near this point may seem to approach it as $t \to \infty$. The answer is no. In region I the point (x, y) moves away from the y axis while moving upward, and in region II, while moving toward the y axis the point (x, y) still moves upward. Moreover, note that $(0, \epsilon_1/\alpha_1)$ is not a critical point; that is $x = 0$, $y = \epsilon_1/\alpha_1$ is not a solution of Eqs. (7). The other critical points in Figure 9.7a are $(0, 0)$ and $(\epsilon_1/\sigma_1, 0)$. However, an inspection of Figure 9.8a readily shows that a solution (x, y) starting from nonzero values (x_0, y_0) cannot approach either of these points as $t \to \infty$.

Consider case (d). An examination of Figure 9.7d suggests that the population point (x, y) will move toward the intersection of the two straight dividing lines as t increases. This is shown schematically in Figure 9.8b for several different initial states. In this case both species can coexist with equilibrium populations given by the coordinates of the critical point B.

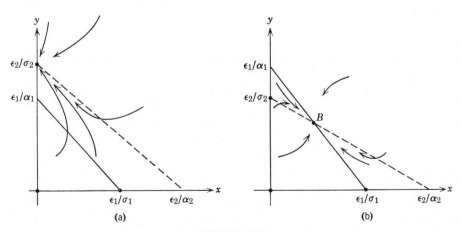

FIGURE 9.8

We turn now from this specific example to a more general discussion in order to develop the necessary mathematical ideas so as to make our geometric analysis simpler and more precise. With this knowledge we will return to the problem of two competing species in Section 9.5.

General Systems. Consider the general autonomous system (1). We have seen that it is very helpful to interpret a solution $x = \phi(t)$, $y = \psi(t)$ as a point tracing a curve in the xy plane. The equations $x = \phi(t)$, $y = \psi(t)$ are a parametric representation of this curve. The xy plane is referred to as the *phase plane*. A curve described parametrically by a solution of the system (1) is usually referred to as a *trajectory*, *orbit*, or *path* of the system.

Such a geometric interpretation is helpful mainly when the functions F and G do not depend on t; that is, when the system is autonomous. To illustrate the differences between autonomous and nonautonomous systems, we will consider two simple examples. First, suppose that particles are being continuously emitted at the point $x = 1$, $y = 2$, and then move in the xy plane according to the law

$$\frac{dx}{dt} = x, \qquad \frac{dy}{dt} = y. \tag{9}$$

The particle emitted at the time $t = s$ is specified by the initial conditions $x(s) = 1$, $y(s) = 2$. What is the path of the particle for $t \geq s$? The solution of Eqs. (9) satisfying these initial conditions is

$$x = \phi(t;\ s) = e^{t-s}, \qquad y = \psi(t;\ s) = 2e^{t-s},\ t \geq s. \tag{10}$$

The paths followed by the particles emitted at different times s can be most readily visualized by eliminating t from Eqs. (10). This gives

$$y = 2x, \tag{11}$$

where it is clear from Eqs. (10) that $x \geq 1$, $y \geq 2$. The portion of the straight line $y = 2x$ corresponding to Eqs. (10), with the direction of motion shown by an arrow, is sketched in Figure 9.9a. What is important is that the initial time s

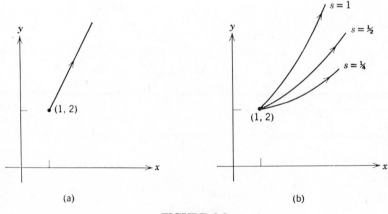

(a)   (b)

FIGURE 9.9

does not appear in Eq. (11); therefore, no matter when a particle is emitted from the point $(1, 2)$, it always moves along the same curve.

Now consider a similar problem in which the functions F and/or G depend on t; for example,

$$\frac{dx}{dt} = \frac{x}{t}, \qquad \frac{dy}{dt} = y. \tag{12}$$

The solution of Eqs. (12) satisfying the initial conditions $x(s) = 1$, $y(s) = 2$ is

$$x = \phi(t; s) = t/s, \qquad y = \psi(t; s) = 2e^{t-s}, \qquad t \geq s. \tag{13}$$

Solving the first of Eqs. (13) gives $t = sx$. Then substituting in the second of Eqs. (13) we obtain

$$y = 2e^{s(x-1)}, \tag{14}$$

where again $x \geq 1$. Since s appears in Eq. (14), the path that a particle follows depends on the time at which it is emitted; that is, the time at which the initial condition is applied. The paths followed by particles emitted at several different times are sketched in Figure 9.9b.

In the first example, when we eliminated the parameter t we automatically eliminated s, since t and s appeared only in the combination $t - s$. In the second example this was not the case, and s remains in the final result (14). These examples illustrate the special property of autonomous systems, not shared by nonautonomous systems, that all particles passing through a given point follow the same curve in the phase plane. An equivalent, and the more usual, statement is the following. Suppose that $x = \phi(t)$, $y = \psi(t)$ is a solution of Eqs. (1) for $-\infty < t < \infty$. If s is any constant then $x = \phi(t - s)$, $y = \psi(t - s)$ is also a solution of Eqs. (1) for $-\infty < t < \infty$. This is readily verified by direct substitution. Thus the same curve (trajectory) in the phase plane is represented parametrically by many different solutions differing from one another by a translation of the parameter t. This is illustrated in Figure 9.9a where every solution of the system (9) passing through the point $(1, 2)$ lies on the straight line $y = 2x$. Also consider the following example.

Example. Sketch the trajectories of the linearized pendulum equation, obtained by replacing $\sin x$ by x in Eqs. (4):

$$\frac{dx}{dt} = y, \qquad \frac{dy}{dt} = -\left(\frac{g}{l}\right)x. \tag{15}$$

It can be verified by direct substitution that for $-\infty < t < \infty$,

$$x = \phi(t; s) = C \sin\left(\sqrt{\frac{g}{l}}\, t - s\right),$$

$$y = \psi(t; s) = C\sqrt{\frac{g}{l}} \cos\left(\sqrt{\frac{g}{l}}\, t - s\right), \tag{16}$$

where C and s are arbitrary constants, is the general solution of Eqs. (15). Eliminating t by dividing the equation for y by $\sqrt{g/l}$ and then squaring and adding the equations for x and y shows that the trajectories of the system (15) lie on the ellipses

$$x^2 + \frac{y^2}{g/l} = C^2.$$

Several trajectories are sketched in Figure 9.10. The direction of motion with increasing t can be determined from the differential equations (15). For points in the first quadrant ($x > 0$, $y > 0$), Eqs. (15) imply that $dx/dt > 0$ and $dy/dt < 0$; therefore the motion is clockwise. The value of C in Eqs. (16) singles out one particular ellipse, and the value of s simply represents a displacement along the ellipse. For example, the solution $x = \sqrt{l/g} \sin \sqrt{g/l}\, t$, $y = \cos \sqrt{g/l}\, t$, which satisfies the initial conditions $x = 0$, $y = 1$ at $t = 0$, describes the trajectory $(g/l)x^2 + y^2 = 1$. Similarly, the solution $x = \sqrt{l/g} \sin (\sqrt{g/l}\, t + \pi/2)$, $y = \cos (\sqrt{g/l}\, t + \pi/2)$, which satisfies the initial conditions $x = \sqrt{l/g}$, $y = 0$ at $t = 0$, also lies on the same trajectory. Clearly, infinitely many solutions follow the same trajectory.

FIGURE 9.10

For this example, it is clear from Eqs. (16) that every solution of the linearized pendulum equations (15) is periodic with period $2\pi/\sqrt{g/l}$. More generally, a solution $x = \phi(t)$, $y = \psi(t)$ of Eqs. (1) is said to be *periodic* with period T if it exists for all t, and if $\phi(t + T) = \phi(t)$, $\psi(t + T) = \psi(t)$ for all t. The trajectory of a periodic solution is a closed curve, as illustrated by the elliptical trajectories in Figure 9.10.

In determining the trajectories of the system (1), it is often convenient to eliminate the parameter t from the solution $x = \phi(t)$, $y = \psi(t)$, obtaining a relation between x and y, as was just illustrated for Eqs. (16). Alternatively, in a region where $F(x, y) \neq 0$, we have from Eqs. (1)

$$\frac{dy}{dx} = \frac{dy/dt}{dx/dt} = \frac{G(x, y)}{F(x, y)}, \tag{17}$$

which is a first order differential equation* for $y = f(x)$. The one parameter family of solutions of Eq. (17) is the set of trajectories of the system (1). Equation (17) is particularly convenient for determining the slope at a point on a trajectory. Near a point where $F(x, y) = 0$, but $G(x, y) \neq 0$ one can consider in place of Eq. (17) the equation $dx/dy = F(x, y)/G(x, y)$.

However, at a point (x_0, y_0) at which both F and G vanish we cannot solve for dy/dx or dx/dy. Thus we again see that such a point, a *critical point*, has a special significance. If the point (x_0, y_0) is a critical point of the system (1), then $x = x_0, y = y_0$ is a solution of the system (1). Indeed, it follows from the existence and uniqueness theorem that the only solution of the system (1) passing through the point (x_0, y_0) is the constant solution itself. The trajectory of this solution is, of course, simply the single point (x_0, y_0). The particle at (x_0, y_0) is then often said to be at rest or in equilibrium. A trajectory, represented by the solution $x = \phi(t)$, $y = \psi(t), t \geq \alpha$, is said to approach the critical point (x_0, y_0) as $t \to \infty$ if $\phi(t) \to x_0$ and $\psi(t) \to y_0$ as $t \to \infty$.

We have already seen the important role that critical points played in our preliminary analysis of the problem of two competing species. The nature of a critical point of an autonomous system can be further illustrated by an appeal to mechanics. The equation

$$m \frac{d^2x}{dt^2} = f\left(x, \frac{dx}{dt}\right) \tag{18}$$

can be interpreted as describing the motion of a particle (position x at time t) of a mass m acted on by a force that depends on the position x and velocity dx/dt. If we let $y = dx/dt$, then Eq. (18) can be rewritten as

$$\frac{dx}{dt} = y, \qquad \frac{dy}{dt} = \frac{1}{m} f(x, y). \tag{19}$$

At a critical point of this system, the velocity $y = dx/dt$ and acceleration dy/dt vanish simultaneously. Thus, when the mass is at a critical point it is at rest (zero velocity) with zero force acting on it; in mechanics such a state (point) is called an *equilibrium state* (*point*).

A question of interest is whether a particle slightly displaced from an equilibrium point moves back toward the equilibrium point, away from it, or perhaps just moves in the neighborhood. Imagine the possibilities of a small displacement of a pendulum bob from: (a) a vertical downward position with air resistance, (b) a vertical upward position with or without air resistance, and (c) a vertical downward position without air resistance (see Figure 9.11). In case (a) we say that the critical (equilibrium) point is asymptotically stable, in case (b) it is unstable, and in case (c) it is stable but not asymptotically stable.

* The condition $F(x, y) \neq 0$ is needed so that we can divide by $F(x, y)$ to form Eq. (17). Moreover, the condition $F(x, y) \neq 0$ implies that $dx/dt \neq 0$; as a consequence we are assured that the solution $x = \phi(t)$ can be solved for t as a function of x and, in turn, y becomes a function of x.

(a) Air resistance   (b)   (c) No air resistance

FIGURE 9.11

For the problem of competing species, the pattern of trajectories in Figure 9.8a suggest that the critical point $(0, \epsilon_2/\sigma_2)$ is an asymptotically stable critical point. If the populations are slightly disturbed from these values they will return to $(0, \epsilon_2/\sigma_2)$. The critical points $(0, 0)$ and $(\epsilon_1/\sigma_1, 0)$ are unstable. Similarly, when coexistence is possible (Figure 9.8b) it would appear that the coexistence critical point is asymptotically stable, and that the critical points $(0, 0)$, $(\epsilon_1/\sigma_1, 0)$, and $(0, \epsilon_2/\sigma_2)$ are unstable. The intuitive understanding of stability is clear for these specific physical examples; precise mathematical definitions will be given in Section 9.4.

The significance of critical points in the analysis of the autonomous system (1) is due to the fact that they are constant solutions of the system, and that the qualitative behavior of all trajectories in the phase plane is determined to a very large extent by the location of the critical points and the behavior of trajectories near them.

The following statements can be made about the trajectories of autonomous systems. The proofs follow from the existence and uniqueness theorem and are outlined in the problems.

1. Through any point (x_0, y_0) in the phase plane there is at most one trajectory of the system (1) (see Problem 8).

2. A particle starting at a point that is not a critical point cannot reach a critical point in a finite time. Thus if (x_0, y_0) is a critical point, and a trajectory corresponding to a solution approaches (x_0, y_0), then necessarily $t \to \infty$ (see Problem 9).

3. A trajectory passing through at least one point that is not a critical point cannot cross itself unless it is a closed curve. In this case the trajectory corresponds to a periodic solution of the system (1) (see Problem 10).

The substance of these three results is the following. If a particle (solution) starts at a point that is not a critical point, then it moves on the same trajectory no matter at what time it starts, it can never come back to its initial point unless the motion is periodic, it can never cross another trajectory, and it can only "reach" a critical point in the limit as $t \to \infty$. This suggests that such a particle (solution) either approaches a critical point (x_0, y_0), moves on a closed trajectory or approaches a closed trajectory as $t \to \infty$, or else goes off to infinity. Thus for autonomous systems a study of the critical points and periodic solutions is of fundamental importance. These facts will be developed in greater detail in the next two sections.

In conclusion, we note that many results for autonomous systems can be generalized to nonautonomous systems though the analysis becomes more complicated. Formally, a nonautonomous system of n first order equations for $x_1(t), x_2(t), \ldots, x_n(t)$ can be written as an autonomous system of $n + 1$ equations by letting $t = x_{n+1}$ wherever t appears explicitly and by appending the equation $dx_{n+1}/dt = 1$. For $n = 2$, this would require an analysis of the trajectories in three-dimensional space, which is considerably more difficult than in two dimensions.

PROBLEMS

1. Sketch the trajectory corresponding to the solution satisfying the specified initial conditions, and indicate the direction of motion for increasing t.

(a) $dx/dt = -x$, $\quad dy/dt = -2y$; $\quad x(0) = 4$, $\quad y(0) = 2$

(b) $dx/dt = -x$, $\quad dy/dt = 2y$;

$\qquad x(0) = 4$, $\quad y(0) = 2$ $\quad$ and $\quad x(0) = 4$, $\quad y(0) = 0$

(c) $dx/dt = -y$, $\quad dy/dt = x$;

$\qquad x(0) = 4$, $\quad y(0) = 0$ $\quad$ and $\quad x(0) = 0$, $\quad y(0) = 4$

2. Determine the critical points for each of the following systems.

(a) $dx/dt = 2x - 3y$, $\quad dy/dt = 2x - 2y$

(b) $dx/dt = x - xy$, $\quad dy/dt = y + 2xy$

(c) $dx/dt = x - x^2 - xy$, $\quad dy/dt = \frac{1}{2}y - \frac{1}{4}y^2 - \frac{3}{4}xy$

(d) $dx/dt = y$, $\quad dy/dt = -(g/l)x - (c/ml)y$; $\quad g, l, c, m > 0$

(e) $dx/dt = y$, $\quad dy/dt = -(g/l)\sin x - (c/lm)y$; $\quad g, l, c, m > 0$

(f) The van der Pol equation: $dx/dt = y$, $\quad dy/dt = \mu(1 - x^2)y - x$, $\quad \mu > 0$

3. Consider the linear system

$$dx/dt = ax + by, \qquad dy/dt = cx + dy,$$

where a, b, c, and d are real constants.

(a) Show that if $ad - bc \neq 0$, then the only critical point is $(0, 0)$.

(b) Show that if $ad - bc = 0$, then in addition to the critical point $(0,0)$ there is a line through the origin for which every point is a critical point of the system. Thus in the latter case the critical point $(0, 0)$ is not "isolated" from the other critical points of the system.

4. Sketch the trajectories of the following system either by solving the system and eliminating t to obtain a one-parameter family of curves, or by integrating the differential equation [Eq. (17)], which gives the slope of the tangent of a trajectory. Indicate the direction of motion with increasing t.

(a) $dx/dt = -x$, $\quad dy/dt = -2y$ $\qquad$ (b) $dx/dt = -x$, $\quad dy/dt = 2y$

(c) $dx/dt = x$, $\quad dy/dt = -2y$ $\qquad$ (d) $dx/dt = -x$, $\quad dy/dt = 3x + 2y$

5. Given that $x = \phi(t)$, $y = \psi(t)$ is a solution of the autonomous system

$$dx/dt = F(x, y), \qquad dy/dt = G(x, y)$$

for $\alpha < t < \beta$, show that $x = \Phi(t) = \phi(t - s)$, $y = \Psi(t) = \psi(t - s)$ is a solution for $\alpha + s < t < \beta + s$.

6. Show by direct integration that even though the right-hand side of the system

$$\frac{dx}{dt} = \frac{x}{1 + t}, \qquad \frac{dy}{dt} = \frac{y}{1 + t}$$

depends on t, the paths followed by particles emitted at (x_0, y_0) at $t = s$ are the same regardless of the value of s. Why is this so?
Hint: Consider Eq. (17) for this case.

7. Consider the system

$$dx/dt = F(x, y, t), \qquad dy/dt = G(x, y, t).$$

Show that if the functions $F/(F^2 + G^2)^{1/2}$ and $G/(F^2 + G^2)^{1/2}$ are independent of t, then the solutions corresponding to the initial conditions $x(s) = x_0$, $y(s) = y_0$ give the same trajectories regardless of the value of s.
Hint: Use Eq. (17).

8. Prove that for the system

$$dx/dt = F(x, y), \qquad dy/dt = G(x, y)$$

there is at most one trajectory passing through a given point (x_0, y_0).
Hint: Let C_0 be the trajectory generated by the solution $x = \phi_0(t)$, $y = \psi_0(t)$, with $\phi_0(t_0) = x_0$, $\psi_0(t_0) = y_0$, and let C_1 be the trajectory generated by the solution $x = \phi_1(t)$, $y = \psi_1(t)$, with $\phi_1(t_1) = x_0$, $\psi_1(t_1) = y_0$. Use the fact that the system is autonomous and the existence and uniqueness theorem to show that C_0 and C_1 are the same.

9. Prove that if a trajectory starts at a noncritical point of the system

$$dx/dt = F(x, y), \qquad dy/dt = G(x, y)$$

then it cannot reach a critical point (x_0, y_0) in a finite length of time.
Hint: Assume the contrary; that is, assume that the solution $x = \phi(t)$, $y = \psi(t)$ satisfies $\phi(a) = x_0$, $\psi(a) = y_0$. Then use the fact that $x = x_0$, $y = y_0$ is a solution of the given system satisfying the initial condition $x = x_0$, $y = y_0$, at $t = a$.

10. Assuming that the trajectory corresponding to a solution $x = \phi(t)$, $y = \psi(t)$, $-\infty < t < \infty$, of an autonomous system is closed, show that the solution is periodic.
Hint: Since the trajectory is closed there exists at least one point (x_0, y_0) such that $\phi(t_0) = x_0$, $\psi(t_0) = y_0$ and a number $T > 0$ such that $\phi(t_0 + T) = x_0$, $\psi(t_0 + T) = y_0$. Show that $x = \Phi(t) = \phi(t + T)$ and $y = \Psi(t) = \psi(t + T)$ is a solution and then use the existence and uniqueness theorem to show that $\Phi(t) = \phi(t)$ and $\Psi(t) = \psi(t)$ for all t.*

Problems 11 through 13 deal with two competing species as described by Eqs. (7) of the text.

11. For the cases in which coexistence is possible (Figures 9.7c and 9.7d), determine the coordinates of this critical point.

---

* If it is originally known only that the functions ϕ and ψ are defined on an interval containing t_0 and $t_0 + T$, the proof is more difficult. However, it can be shown that the interval of definition must be $-\infty < t < \infty$ and that the solution is periodic with period T.

12. Draw several typical trajectories for the case shown in Figure 9.7b and determine what happens as $t \to \infty$.

13. Draw several typical trajectories for the case shown in Figure 9.7c and determine what happens as $t \to \infty$.

9.3   THE PHASE PLANE;   LINEAR SYSTEMS

In this section we will continue our discussion of the autonomous system

$$\frac{dx}{dt} = F(x, y), \qquad \frac{dy}{dt} = G(x, y). \tag{1}$$

In particular we will be concerned with the behavior of the trajectories of the system (1) in the neighborhood of a critical point (x_0, y_0) and the significance of this behavior. It is convenient to choose the critical point to be at the origin of the phase plane: $x_0 = 0$, $y_0 = 0$. This involves no loss of generality, since if $x_0 \neq 0$, $y_0 \neq 0$, it is always possible to make the substitution $x = x_0 + u$, $y = y_0 + v$ in Eqs. (1), so that u and v will satisfy an autonomous system of equations with a critical point at the origin.

We will further assume that the origin is an *isolated critical point* of the system (1); that is, we assume that there is some circle about the critical point, inside which there are no other critical points. Finally, we assume that in the neighborhood of $(0, 0)$, the functions F and G have the form

$$F(x, y) = ax + by + F_1(x, y),$$

$$G(x, y) = cx + dy + G_1(x, y), \tag{2}$$

where $ad - bc \neq 0$. Note that there are no constant terms in Eqs. (2) since $F(0, 0) = 0$ and $G(0, 0) = 0$. Also, we require that the functions F_1 and G_1 be continuous, have continuous first partial derivatives, and be small in the sense* that

$$\frac{F_1(x, y)}{r} \to 0, \qquad \frac{G_1(x, y)}{r} \to 0 \qquad \text{as} \qquad r \to 0, \tag{3}$$

where $r = (x^2 + y^2)^{\frac{1}{2}}$. Such a system is often referred to as an *almost linear system* in the neighborhood of the critical point $(0, 0)$. Essentially, Eqs. (2) and (3) say that for (x, y) near $(0, 0)$ the functions F and G are well approximated by the linear functions $ax + by$ and $cx + dy$, respectively.

The conditions on F and G are satisfied by many functions of two variables. For example, any polynomial in x and y with zero constant term is an almost linear function in the neighborhood of $(0, 0)$.

---

* Actually, if the functions F and G are continuous and have continuous first partial derivatives, as we assumed in Section 9.2, then it can be shown that F and G have the form (2) in the neighborhood of the critical point $(0, 0)$ and that condition (3) is satisfied. The reason for the restriction $ad - bc \neq 0$ will be made clear after Eq. (8) of this section; also see Problem 3 of Section 9.2.

As another example, consider the autonomous system for the undamped motion of a pendulum given in Eqs. (4) of Section 9.2: $dx/dt = y$, $dy/dt = -(g/l)\sin x$. The point $(0, 0)$ is a critical point of this system. In order to compare these equations with those of Eqs. (2), we write them in the form

$$\frac{dx}{dt} = y,$$

$$\frac{dy}{dt} = -\frac{g}{l} x - \frac{g}{l} (\sin x - x). \tag{4}$$

Thus $a = 0$, $b = 1$, $c = -g/l$, $d = 0$, $F_1(x, y) = 0$, and

$$G_1(x, y) = -\frac{g}{l}(\sin x - x) = -\frac{g}{l}\left(-\frac{x^3}{3!} + \frac{x^5}{5!} + \cdots\right). \tag{5}$$

From Eq. (5), $\sin x - x$ is similar to $-x^3/3! = -(r^3 \cos^3 \theta)/3!$ for x very small, and hence $(\sin x - x)/r \to 0$ as $r \to 0$. Consequently, the autonomous system (4) is almost linear in the neighborhood of the critical point $(0, 0)$.

A consequence of the assumption that $F_1(x, y)$ and $G_1(x, y)$ are small compared to the linear terms $ax + by$ and $cx + dy$ near the origin is that in many cases (but not all) the trajectories of the linear autonomous system

$$\frac{dx}{dt} = ax + by,$$

$$\frac{dy}{dt} = cx + dy, \tag{6}$$

are good approximations to those of the almost linear system in the neighborhood of the critical point $(0, 0)$. Thus we first consider the linear system (6), and then in Section 9.4 we will relate our conclusions to the almost linear system.

The system of equations (6) can be readily solved by elimination (Section 7.2) or by the matrix methods of Chapter 7. For our purposes it is not necessary to introduce the vector-matrix notation of that chapter. Of course, for the student who has read Chapter 7 it may be more convenient to rephrase some of the following results in vector-matrix notation.

We look for solutions of Eqs. (6) of the form $x = A \exp{(rt)}$, $y = B \exp{(rt)}$. Substituting for x and y in Eqs. (6), we obtain

$$(a - r)A + bB = 0$$

$$cA + (d - r)B = 0. \tag{7}$$

In order for these two linear homogeneous equations to have a nontrivial solution, it is necessary for the determinant of the coefficients to vanish. Thus r must be a root of the characteristic equation

$$r^2 - (a + d)r + (ad - bc) = 0. \tag{8}$$

Equation (8) is also obtained if one eliminates either x or y in system (6) and then looks for a solution proportional to exp (rt). The student who has read Section 7.4 will note that the roots r_1 and r_2 of Eq. (8) are the eigenvalues of the coefficient matrix of the system (6). When r is a root of Eq. (8), then Eqs. (7) are not independent; hence either one of them can be used to calculate the relation between A and B. For example, for $r = r_1$, we have from the first of Eqs. (7) that $B = -(a - r_1)A/b$, assuming that $b \neq 0$.

Notice that $r = 0$ cannot be a root of Eq. (8). If it were, then necessarily $ad - bc = 0$. As a consequence, the system of equations $ax + by = 0$, $cx + dy = 0$ would have nontrivial solutions since the determinant of the coefficients would be zero. These nontrivial solutions would be determined only up to a multiplicative constant and would lie on a line passing through the origin. Thus other critical points would be located arbitrarily close to the origin, violating the assumption that the origin is an isolated critical point.

A number of different cases must be considered depending on whether the roots of Eq. (8) are both positive, both negative, one positive and one negative, complex with positive real parts, complex with negative real parts, or pure imaginary. We will consider each case in turn, illustrating the behavior of the trajectories by a specific example. We emphasize that the general situation for each case is similar to that for the corresponding example, since a suitable change of variables can be made to transform a given system into a system having the form of the example for that case. Once the problem is solved in the new coordinate system the solution can be transformed back to the original coordinate system and constitutes a solution of the original system. For example, a circle in the transformed coordinate system may be an ellipse in the original coordinate system with its major axis not necessarily parallel to one of the coordinate axes. Also for each case we will show a more "general" picture of a typical set of trajectories. It is important that the student become familiar with the type of behavior that the trajectories have for each case, because these are the basic building blocks of the qualitative theory of differential equations.

Case 1. Real Unequal Roots of the Same Sign. The general solution of Eqs. (6) is

$$x = A_1 e^{r_1 t} + A_2 e^{r_2 t}, \qquad y = B_1 e^{r_1 t} + B_2 e^{r_2 t}, \tag{9}$$

where only one each of the constants A_1, B_1 and A_2, B_2, respectively, are independent. Suppose first that r_1 and r_2 are negative; then both x and y approach zero and the point (x, y) approaches the critical point $(0, 0)$ as $t \to \infty$ independent of the choice of the constants A_1, A_2, B_1, and B_2. Every solution of the system (6) approaches the solution $x = 0$, $y = 0$ exponentially as $t \to \infty$.

To illustrate this case consider the system

$$\frac{dx}{dt} = -x, \qquad \frac{dy}{dt} = -2y, \tag{10}$$

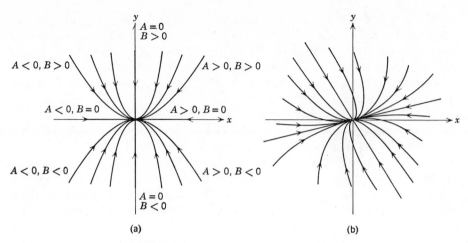

FIGURE 9.12   Improper node, $r_1 \neq r_2$.

which has the general solution

$$x = Ae^{-t}, \qquad y = Be^{-2t}, \tag{11}$$

where A and B are arbitrary. A few trajectories for this example are sketched in Figure 9.12a. If $A > 0$, $B = 0$, then $y = 0$ and $x \to 0$ through positive values as $t \to \infty$. The other halves of the coordinate axes correspond to the cases $A < 0$, $B = 0$; $A = 0$, $B > 0$; and $A = 0$, $B < 0$. If $A \neq 0$, $B \neq 0$, then $y = (B/A^2)x^2$, and we obtain the portions of the parabolas in the first, second, third, and fourth quadrants, as indicated in Figure 9.12a.

This type of critical point is called a *node*, or sometimes an *improper node*, to distinguish it from another type of node to be mentioned later. The distinguishing feature is that all trajectories except one pair approach the critical point tangent to the same line, while the exceptional pair of trajectories approaches tangent to a different line. In the example all trajectories approach the origin tangent to the x axis except one pair, which approaches along the y axis. A more general sketch of trajectories near an improper node is shown in Figure 9.12b. If the roots are real, unequal, and positive the situation is similar except that the direction of motion on the trajectories is away from the critical point $(0, 0)$. In this case every solution, except the solution $x = 0$, $y = 0$, even a solution started very near $(0, 0)$, recedes from the origin as $t \to \infty$.

Case 2. Real Roots of Opposite Sign. The general solution of Eqs. (6) is still given by Eqs. (9), but with $r_1 > 0$ and $r_2 < 0$. It is now possible for the direction of motion to be toward the critical point on some trajectories and away from the critical point on other trajectories. Typical of this case is the system

$$\frac{dx}{dt} = -x, \qquad \frac{dy}{dt} = 2y, \tag{12}$$

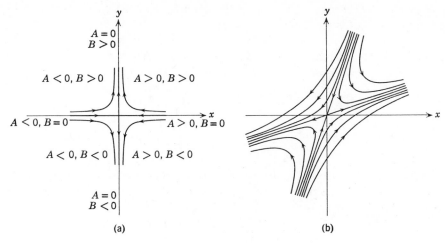

FIGURE 9.13   Saddle point, $r_1 r_2 < 0$.

which has the general solution

$$x = Ae^{-t}, \qquad y = Be^{2t}, \tag{13}$$

where A and B are arbitrary. For any initial point not on the x axis, $x \to 0$, $y \to \pm\infty$, depending on the sign of B, as $t \to \infty$; while for an initial point on the x axis, y stays zero and $x \to 0$ as $t \to \infty$. The trajectories are sketched in Figure 9.13a where use has been made of the fact that for $A \neq 0$ and $B \neq 0$, the solution lies on a portion of the curve $y = BA^2/x^2$. In this case the critical point is called a *saddle point*.

It is important to note that only the trajectories on the x axis approach the critical point; all others approach infinity as $t \to \infty$ because of the presence of $y = B \exp(2t)$, $B \neq 0$. The situation is always similar for a saddle point. To see this, note that the general solution of the system (6) is

$$\begin{pmatrix} x \\ y \end{pmatrix} = A \begin{pmatrix} k_{11} \\ k_{21} \end{pmatrix} e^{r_1 t} + B \begin{pmatrix} k_{12} \\ k_{22} \end{pmatrix} e^{r_2 t},$$

where A and B are arbitrary constants, k_{11}, k_{21}, k_{12}, and k_{22} are known, and r_1 and r_2 are real but of opposite sign ($r_1 < 0$, $r_2 > 0$). Hence only the trajectories corresponding to solutions for which $B = 0$ will approach the critical point. Taking $A > 0$ gives one trajectory, $A < 0$ gives a second. A more general sketch of trajectories near a saddle point is shown in Figure 9.13b.

Case 3.   Equal Roots.   The general solution of Eqs. (6) is of the form

$$x = (A_1 + A_2 t)e^{rt}, \qquad y = (B_1 + B_2 t)e^{rt}, \tag{14}$$

where only two of the four constants A_1, A_2, B_1, and B_2 are independent. Regardless of the values of A_1, A_2, B_1, and B_2 it is clear that if $r < 0$, the direction of motion on all trajectories is toward the critical point and, if $r > 0$, it is away from the critical point. We will consider the case $r < 0$.

Depending on whether the term te^{rt} is present, the trajectories for this case are of two quite different types. The simpler case is when this term is not present; that is, when $A_2 = B_2 = 0$. Illustrative of this situation is the system

$$\frac{dx}{dt} = -x, \qquad \frac{dy}{dt} = -y, \tag{15}$$

which has the general solution

$$x = Ae^{-t}, \qquad y = Be^{-t}, \tag{16}$$

where A and B are arbitrary. For $A \neq 0$, $B \neq 0$ the trajectories are the straight lines $y = (B/A)x$. The trajectories are sketched in Figure 9.14. Every trajectory has a different slope as it approaches (or recedes from, if r is greater than zero) the origin, and the critical point is called a *proper node*.

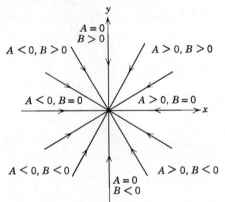

FIGURE 9.14 Proper node, $r_1 = r_2$.

Typical of the situation when the term te^{rt} is present is the system

$$\frac{dx}{dt} = -2x, \qquad \frac{dy}{dt} = x - 2y, \tag{17}$$

which has the general solution

$$x = Ae^{-2t}, \qquad y = Be^{-2t} + Ate^{-2t}, \tag{18}$$

where A and B are arbitrary. A detailed sketch of the trajectories represented by the solutions (18) is fairly complicated since, if t is eliminated between the equations for x and y, an expression is obtained for y that involves both x and $\ln x$. Assuming that $A > 0$, it then follows from the first of Eqs. (18) that x is positive and $e^{-2t} = x/A$ so $t = -(\frac{1}{2}) \ln (x/A)$. Substituting this expression for t in the second of Eqs. (18) yields

$$y = \frac{Bx}{A} - \frac{x}{2} \ln \frac{x}{A}. \tag{19}$$

For assigned values of $A > 0$ and B, the trajectories can be obtained from Eq. (19). For $A < 0$ a similar equation can be derived.

It is clear from Eq. (18) that all of the trajectories approach the origin as $t \to \infty$. The slope at any point can be obtained from Eq. (19) for $A > 0$ or from a similar equation for $A < 0$, or by recalling that

$$\frac{dy}{dx} = \frac{dy/dt}{dx/dt} = \frac{2Be^{-2t} - 2Ate^{-2t} + Ae^{-2t}}{-2Ae^{-2t}} = \frac{(-2B + A) - 2At}{-2A}. \quad (20)$$

For $t \to \infty$, $dy/dx \to \infty$; thus all of the trajectories enter the origin along the y axis. A qualitative picture of the trajectories is shown in Figure 9.15a. Again, the critical point is called an *improper node*. A more general sketch is shown in Figure 9.15b.

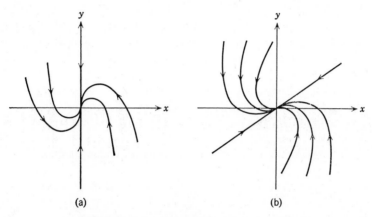

(a)   (b)

FIGURE 9.15 Improper node, $r_1 = r_2$.

Case 4. Complex Roots.   In this case the general solution of Eqs. (6) is

$$x = e^{\lambda t}(A_1 \cos \mu t + A_2 \sin \mu t), \qquad y = e^{\lambda t}(B_1 \cos \mu t + B_2 \sin \mu t), \quad (21)$$

where the roots of Eq. (8) are $r = \lambda \pm i\mu$, and only two of the constants A_1, A_2, B_1, and B_2 are independent. If $\lambda < 0$ the motion on every trajectory is toward the critical point and, if $\lambda > 0$, it is away from the critical point.

Consider the specific system

$$\frac{dx}{dt} = -x + 2y, \qquad \frac{dy}{dt} = -2x - y, \quad (22)$$

which has the general solution

$$x = e^{-t}(A \cos 2t + B \sin 2t), \qquad y = e^{-t}(B \cos 2t - A \sin 2t), \quad (23)$$

where A and B are arbitrary. The trajectories associated with the solution (23) can best be visualized by introducing polar coordinates. If we let

$$x = r \cos \theta, \qquad y = r \sin \theta,$$

and   (24)

$$R = (A^2 + B^2)^{\frac{1}{2}}, \qquad R \cos \alpha = A, \qquad R \sin \alpha = B,$$

then Eqs. (23) take the form

$$r \cos \theta = Re^{-t} \cos (2t - \alpha), \qquad r \sin \theta = -Re^{-t} \sin (2t - \alpha). \tag{25}$$

Thus

$$r = Re^{-t}, \qquad \theta = -(2t - \alpha), \tag{26}$$

Finally, eliminating t gives

$$r = Re^{(\theta - \alpha)/2}, \tag{27}$$

which represents a family of spirals, one of which is sketched in Figure 9.16a. Since $\lambda = -1 < 0$, the direction of motion is toward the critical point $(0, 0)$; since θ decreases with increasing t, the motion is clockwise. Whether the motion is clockwise or counterclockwise can also readily be obtained directly from the original differential equations (22). It is convenient to choose a point on one of the coordinate axes and to determine the direction of motion there. For example, if we choose a point on the positive x axis, then it follows from the second of Eqs. (22) that $dy/dt < 0$. Hence the point (x, y) on the trajectory is moving downward when it crosses the positive x axis, so the direction of motion on the trajectory is clockwise.

The critical point is called a *spiral point*. A more general sketch of trajectories near a spiral point is shown in Figure 9.16b. If λ were positive the spirals would be similar, but the direction of motion would be away from $(0, 0)$.

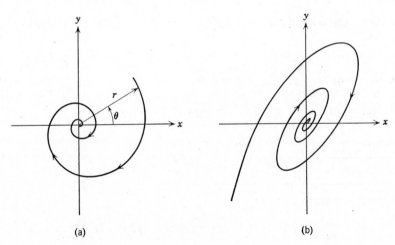

(a)   (b)

FIGURE 9.16 Spiral point, $r_1 = \lambda + i\mu$, $r_2 = \lambda - i\mu$.

Case 5. Pure Imaginary Roots.   While this is a special case of the previous one given by Eqs. (21) with $\lambda = 0$, it must be treated separately, since the trajectories are no longer spirals. In this case the motion is periodic in time and the trajectories are closed curves. The direction of motion around the closed curves can be either clockwise or counterclockwise, depending on the equations. Typical of this case are the linearized undamped pendulum equations discussed as an example in Section 9.2. The trajectories, as shown in Figure 9.10 of the previous section, are ellipses centered at the origin. The origin is called a *center*. A typical case in which the major axes of the ellipses are inclined at an angle to the coordinate axes is shown in Figure 9.17. The trajectories neither approach nor recede from the critical point.

FIGURE 9.17   Center, $r_1 = i\mu$, $r_2 = -i\mu$.

In each of the cases we have discussed, the trajectories of the system exhibited one of the following three types of behavior.

1. All of the trajectories approach the critical point as $t \to \infty$. This will be the case if the roots of the characteristic equation (8) are real and negative or complex with negative real parts.

2. The trajectories neither approach the critical point nor tend to infinity as $t \to \infty$. This will be the case if the roots of the characteristic equation (8) are pure imaginary.

3. At least one (possibly all) of the trajectories tends to infinity as $t \to \infty$. This will be the case if at least one of the roots of the characteristic equation (8) is positive or if the roots have positive real parts.

These three possibilities illustrate the concepts of asymptotic stability, stability, and instability, respectively, of the critical point at the origin of the system (6). The precise definitions of these terms will be given in Section 9.4, but their basic meaning should be clear from the geometrical discussion above and the discussion following Eq. (19) of Section 9.2. Our analysis of the trajectories of the linear system (6) is summarized in Table 9.1. Also see Problems 16 and 17 and Figure 9.20, which accompanies those problems.

TABLE 9.1

$\begin{aligned} dx/dt &= ax + by \\ dy/dt &= cx + dy \end{aligned}$		$\begin{aligned} r^2 - (a+d)r + (ad-bc) &= 0 \\ ad - bc &\neq 0 \end{aligned}$
Roots of the Characteristic Equation	Type of Critical Point	Stability
$r_1 > r_2 > 0$	Improper node	Unstable
$r_1 < r_2 < 0$		Asymptotically stable
$r_2 < 0 < r_1$	Saddle point	Unstable
$r_1 = r_2 > 0$	Proper or improper node	Unstable
$r_1 = r_2 < 0$		Asymptotically stable
$r_1, r_2 = \lambda \pm i\mu$	Spiral point	
$\quad \lambda > 0$		Unstable
$\quad \lambda < 0$		Asymptotically stable
$r_1 = i\mu, r_2 = -i\mu$	Center	Stable

It is clear from our discussion and Table 9.1 that the roots r_1, r_2 of Eq. (8) determine the type of critical point and its stability characteristics. In turn, the values of r_1, r_2 depend on the coefficients a, b, c, d in the system of differential equations (6). When such a system (6) arises in some applied field, the coefficients a, b, c, and d usually result from the measurements of certain physical quantities. Such measurements are often subject to small uncertainties, so it is of interest to investigate whether small changes (perturbations) in the coefficients can affect the stability or instability of a critical point and/or significantly alter the pattern of trajectories.

It is possible to show from Eq. (8) that *small* perturbations in some or all of the coefficients a, b, c, and d are reflected in *small* perturbations in the roots r_1 and r_2. The most sensitive situation occurs for $r_1 = i\mu$ and $r_2 = -i\mu$; that is, when the critical point is a center and the trajectories are closed curves. If a slight change is made in the coefficients, then the roots r_1 and r_2 will take on new values, $r_1' = \lambda' + i\mu'$ and $r_2' = \lambda' - i\mu'$, where λ' is small in magnitude and $\mu' \cong \mu$ (see Figure 9.18). If $\lambda' \neq 0$, which almost always occurs, then the trajectories of the

FIGURE 9.18 Schematic perturbation of $r_1 = i\mu$, $r_2 = -i\mu$.

perturbed system are spirals, rather than closed curves. The system is asymptotically stable if $\lambda' < 0$, but unstable if $\lambda' > 0$. Thus in the case of a center, small perturbations in the coefficients may well change a stable system into an unstable one, and in any case may be expected to alter radically the pattern of trajectories in the phase plane (see Problem 14).

Another slightly less sensitive case occurs if the roots r_1 and r_2 are equal; in this case the critical point is a node. Small perturbations in the coefficients will normally cause the two equal roots to separate (bifurcate). If the separated roots are real, then the critical point of the perturbed system remains a node, but if the separated roots are complex conjugates, then the critical point becomes a spiral point. These two possibilities are shown schematically in Figure 9.19. In this case the stability or instability of the system is not affected by small perturbations in the coefficients, but the trajectories may be altered considerably (see Problem 15).

FIGURE 9.19   Schematic perturbation of $r_1 = r_2$.

In all other cases the stability or instability of the system is not changed, nor is the type of critical point altered, by sufficiently small perturbations in the coefficients of the system. For example, if r_1 and r_2 are real, negative, and unequal, then a *small* change in the coefficients will not change the sign of r_1 and r_2 nor allow them to coalesce. Thus the critical point remains an asymptotically stable improper node.

In Section 9.4 we will discuss how the results of this section may be used to investigate almost linear systems.

PROBLEMS

In each of Problems 1 through 8 classify the critical point $(0, 0)$, and determine whether it is stable, asymptotically stable, or unstable.

1. $dx/dt = 3x - 2y$
   $dy/dt = 2x - 2y$

2. $dx/dt = 5x - y$
   $dy/dt = 3x + y$

3. $dx/dt = 2x - y$
   $dy/dt = 3x - 2y$

4. $dx/dt = x - 4y$
   $dy/dt = 4x - 7y$

5. $dx/dt = x - 5y$
   $dy/dt = x - 3y$

6. $dx/dt = 2x - 5y$
   $dy/dt = x - 2y$

7. $dx/dt = 3x - 2y$
   $dy/dt = 4x - y$

8. $dx/dt = -x - y$
   $dy/dt = \quad - \frac{1}{4}y$

In each of Problems 9 through 12 determine the critical point (x_0, y_0), and then classify its type and examine its stability by making the transformation $x = x_0 + u$, $y = y_0 + v$.

9. $dx/dt = x + y - 2$
   $dy/dt = x - y$

10. $dx/dt = -2x + y - 2$
    $dy/dt = x - 2y + 1$

11. $dx/dt = -x - y - 1$
    $dy/dt = 2x - y + 5$

12. $dx/dt = \alpha - \beta y$
    $dy/dt = -\gamma + \delta x,$ $\alpha, \beta, \gamma, \delta > 0$

13. The equation of motion of a spring-mass system with damping (see Section 3.7) is

$$m \frac{d^2u}{dt^2} + c \frac{du}{dt} + ku = 0,$$

where m, c, and k are positive. Write this second order equation as a system of two first order equations for $x = u$, $y = du/dt$. Show that $x = 0$, $y = 0$ is a critical point, and analyze the nature and stability of the critical point as a function of the parameters m, c, and k. A similar analysis can be applied to the electric circuit equation (see Section 3.8)

$$L \frac{d^2I}{dt^2} + R \frac{dI}{dt} + \frac{1}{C} I = 0.$$

14. In this problem we show how small changes in the coefficients of a system of linear equations can affect a critical point that is a center. Consider the system

$$dx/dt = 0x + y, \qquad dy/dt = -x + 0y.$$

Show that the roots of the characteristic equation are $\pm i$ so that $(0, 0)$ is a center. Now consider the system

$$dx/dt = \epsilon x + y, \qquad dy/dt = -x + \epsilon y,$$

where $|\epsilon|$ is arbitrarily small. Show that the characteristic equation has roots $\epsilon \pm i$. Thus no matter how small $|\epsilon| \neq 0$ is, the center is shifted to a spiral point. If $\epsilon < 0$ then the spiral point is asymptotically stable; if $\epsilon > 0$ the spiral point is unstable.

15. In this problem we show how small changes in the coefficients of a system of linear equations can affect the nature of a critical point when the roots of the characteristic equation are equal. Consider the system

$$dx/dt = -x + y, \qquad dy/dt = 0x - y.$$

Show that the roots of the characteristic equation are $r_1 = -1, r_2 = -1$ so that the critical point $(0, 0)$ is an asymptotically stable node. Now consider the system

$$dx/dt = -x + y, \qquad dy/dt = -\epsilon x - y,$$

where $|\epsilon|$ is arbitrarily small. Show that the characteristic equation has roots $-1 \pm i\sqrt{\epsilon}$. Thus if $\epsilon > 0$ then the asymptotically stable node is shifted to an asymptotically stable spiral point. If $\epsilon < 0$ then the roots are $-1 \pm \sqrt{|\epsilon|}$, and the critical point remains an asymptotically stable node.

16. Consider the linear autonomous system

$$dx/dt = ax + by, \qquad dy/dt = cx + dy,$$

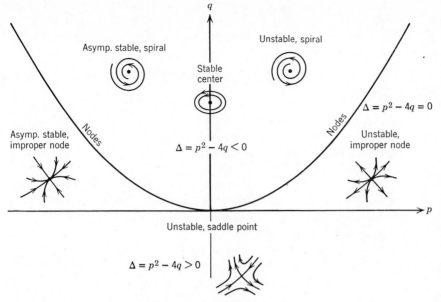

FIGURE 9.20   Stability diagram.

where a, b, c, and d are real constants. Let $p = a + d$, $q = ad - bc$, and $\Delta = p^2 - 4q$. Show that the critical point $(0, 0)$ is a

    (a) node if $q > 0$ and $\Delta \geq 0$;   (b) saddle point if $q < 0$;

    (c) spiral point if $p \neq 0$ and $\Delta < 0$;   (d) center if $p = 0$ and $q > 0$.

Hint: These conclusions can be obtained by studying the roots r_1 and r_2 of the characteristic equation. It may also be helpful to show, and then to use, the relations $r_1 r_2 = q$ and $r_1 + r_2 = p$.

    17. Continuing Problem 16, show that the critical point $(0, 0)$ is

    (a) asymptotically stable if $q > 0$ and $p < 0$;

    (b) stable if $q > 0$ and $p = 0$;

    (c) unstable if $q < 0$ or $p > 0$.

Notice that the results (a), (b), and (c) together with the fact that $q \neq 0$ show that the critical point is asymptotically stable if, and only if, $q > 0$ and $p < 0$

    The results of Problems 16 and 17 are conveniently summarized in Figure 9.20.

9.4   STABILITY; ALMOST LINEAR SYSTEMS

In the previous sections of this chapter, we have on several occasions referred to the concepts of stability, asymptotic stability, and instability of a solution of the autonomous system

$$\frac{dx}{dt} = F(x, y), \qquad \frac{dy}{dt} = G(x, y). \tag{1}$$

In this section we will finally give a precise mathematical meaning to these concepts, discuss an important theorem dealing with the stability of an almost linear system, and explore its consequences by considering an illustrative example.

A critical point $x = x_0, y = y_0$ (an equilibrium solution $x = x_0, y = y_0$) of the autonomous system (1) is said to be a *stable critical point* if, given any $\epsilon > 0$, it is possible to find a δ such that every solution $x = \phi(t), y = \psi(t)$ of the system (1), which at $t = 0$ satisfies

$$\{[\phi(0) - x_0]^2 + [\psi(0) - y_0]^2\}^{\frac{1}{2}} < \delta, \tag{2}$$

exists and satisfies

$$\{[\phi(t) - x_0]^2 + [\psi(t) - y_0]^2\}^{\frac{1}{2}} < \epsilon \tag{3}$$

for all $t \geq 0$. This is illustrated geometrically in Figures 9.21a and 9.21b. These mathematical statements say that all solutions that start "sufficiently close" to (x_0, y_0) stay "close" to (x_0, y_0). Note that in Figure 9.21a the trajectory is within the circle $(x - x_0)^2 + (y - y_0)^2 = \delta^2$ at $t = 0$ and, while it soon passes outside of this circle, it remains within the circle $(x - x_0)^2 + (y - y_0)^2 = \epsilon^2$ for all $t \geq 0$. However, the trajectory of the solution does not have to approach the critical point (x_0, y_0), as is illustrated in Figure 9.21b.

A critical point (x_0, y_0) is said to be *asymptotically stable* if it is stable and if there exists a $\delta_0, 0 < \delta_0 < \delta$, such that if a solution $x = \phi(t), y = \psi(t)$ satisfies

$$\{[\phi(0) - x_0]^2 + [\psi(0) - y_0]^2\}^{\frac{1}{2}} < \delta_0, \tag{4}$$

then

$$\lim_{t \to \infty} \phi(t) = x_0, \qquad \lim_{t \to \infty} \psi(t) = y_0. \tag{5}$$

Trajectories that start "sufficiently close" to (x_0, y_0) must not only stay "close" but must eventually approach (x_0, y_0) as t approaches infinity. This is the case for the trajectory in Figure 9.21a but not for the one in Figure 9.21b. Note that asymptotic stability is a stronger requirement than stability, since a critical point must be stable before we can even talk about whether it is asymptotically stable. On the other hand, the limit condition (5), which is an essential feature of asymptotic stability, does not by itself imply even ordinary stability. Indeed, examples can be constructed in which all of the trajectories approach (x_0, y_0) as $t \to \infty$, but for which (x_0, y_0) is not a stable critical point. Geometrically, all that is needed

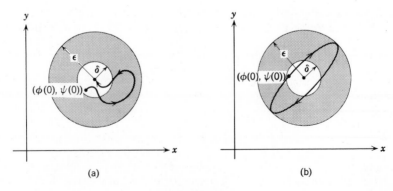

(a)   (b)

FIGURE 9.21

is a family of trajectories having members that start arbitrarily close to (x_0, y_0), then recede an arbitrarily large distance before eventually approaching (x_0, y_0) as t approaches infinity.* A critical point that is not stable is said to be *unstable*.

For the linear system

$$\frac{dx}{dt} = ax + by, \qquad \frac{dy}{dt} = cx + dy. \tag{6}$$

with $ad - bc \neq 0$, the type and stability of the critical point $(0, 0)$ as a function of the roots $r_1 \neq 0$ and $r_2 \neq 0$ of the characteristic equation

$$r^2 - (a + d)r + ad - bc = 0 \tag{7}$$

were listed in Table 9.1 of the preceding section. The stability characteristics are summarized in the following theorem.

Theorem 9.1. *The critical point* $(0, 0)$ *of the linear system* (6) *is*
(i) *asymptotically stable if the roots* r_1, r_2 *of the characteristic equation* (7) *are real and negative or have negative real parts;*
(ii) *stable, but not asymptotically stable, if* r_1 *and* r_2 *are pure imaginary;*
(iii) *unstable if* r_1 *and* r_2 *are real and either is positive, or if they have positive real parts.*

This theorem was established by intuitive arguments in the preceding section. For a rigorous proof, it is necessary to show how to compute δ for a given ϵ so that Eq. (3) is satisfied and, for asymptotic stability, how to compute δ_0 so that Eq. (5) is true. We will not take up this type of detailed analysis.

Notice that if a critical point of the linear system (6) is asymptotically stable, then not only do trajectories that start close to the critical point approach the critical point, but, in fact, since every solution is a linear combination of $e^{r_1 t}$ and $e^{r_2 t}$, *every* trajectory approaches the critical point. In this case the critical point is said to be *globally asymptotically stable*. This property of linear systems is not, in general, true for nonlinear systems. This was illustrated by the example $dA/dt = \epsilon A - \sigma A^2$ with $\epsilon < 0$ and $\sigma < 0$, whose solutions are sketched in Figure 9.5 in Section 9.1. The critical point $A = 0$ is asymptotically stable but not globally asymptotically stable.

Often an important practical problem in considering an asymptotically stable critical point of a nonlinear system is to estimate the set of initial conditions for which the critical point is asymptotically stable. This set of initial points is called the *region of asymptotic stability* for the critical point. Alternatively, we may wish to determine whether the critical point is asymptotically stable for a given set of initial conditions. Again, for the example sketched in Figure 9.5, the critical point $A = 0$ is asymptotically stable for all initial conditions $A(0)$ such that $A(0) < \epsilon/\sigma$.

---

* Such examples are fairly complicated and not often encountered in practice (see Cesari, page 96).

We now want to relate the results for the linear system (6) to the nonlinear system

$$\frac{dx}{dt} = ax + by + F_1(x, y),$$

$$\frac{dy}{dt} = cx + dy + G_1(x, y),$$

(8)

mentioned at the beginning of Section 9.3. We assume that $(0, 0)$ is a critical point of the system (8) and that $ad - bc \neq 0$. Also we assume that F_1 and G_1 have continuous first partial derivatives and are small near the origin in the sense that $F_1(x, y)/r \to 0$ and $G_1(x, y)/r \to 0$ as $r \to 0$, where $r = (x^2 + y^2)^{1/2}$. Recall that such a system is said to be almost linear in the neighborhood of the origin. In our discussion we will usually not mention the phrase "near the origin," since it will be clear that we are talking about the neighborhood of the critical point $(0, 0)$.

As an example, the system

$$\frac{dx}{dt} = x - x^2 - xy,$$

$$\frac{dy}{dt} = \tfrac{1}{2}y - \tfrac{1}{4}y^2 - \tfrac{3}{4}xy,$$

(9)

satisfies the stated conditions. Here $a = 1$, $b = 0$, $c = 0$, $d = \tfrac{1}{2}$, $F_1(x, y) = -x^2 - xy$, and $G_1(x, y) = -\tfrac{1}{4}y^2 - \tfrac{3}{4}xy$. To show that $F_1(x, y)/r \to 0$ as $r \to 0$, let $x = r \cos \theta$, $y = r \sin \theta$. Then

$$\frac{F_1(x, y)}{r} = \frac{-r^2 \cos^2 \theta - r^2 \sin \theta \cos \theta}{r} = -r(\cos^2 \theta + \cos \theta \sin \theta) \to 0 \quad (10)$$

as $r \to 0$. The argument that $G_1(x, y)/r \to 0$ as $r \to 0$ is similar.

The type and stability of a critical point of the almost linear system (8) are closely related in most cases to the type and stability of the critical point of the corresponding linear system (6).

Theorem 9.2.   *Let r_1 and r_2 be the roots of the characteristic equation (7) of the linear system (6) corresponding to the almost linear system (8). Then the type and stability of the critical point $(0, 0)$ of the linear system (6) and the almost linear system (8) are as shown in the table on page 413.*

At this stage, the proof of this theorem is too difficult to give, and we will accept the results of Theorem 9.2 without proof. The results for asymptotic stability and for instability follow as a consequence of a result discussed in Section 9.6, and a proof is sketched in Problems 5 to 7 of Section 9.6. Essentially, Theorem 9.2 says that for x and y near zero the nonlinear terms $F_1(x, y)$ and $G_1(x, y)$ are small and do not affect the stability and type of critical point as determined by the linear terms except in two sensitive cases: r_1 and r_2 pure imaginary, and r_1 and r_2

r_1, r_2	Linear System		Almost Linear System	
	Type	*Stability*	*Type*	*Stability*
$r_1 > r_2 > 0$	IN	Unstable	IN	Unstable
$r_1 < r_2 < 0$	IN	Asymptotically stable	IN	Asymptotically stable
$r_2 < 0 < r_1$	SP	Unstable	SP	Unstable
$r_1 = r_2 > 0$	PN or 'N	Unstable	PN, IN, or SpP	Unstable
$r_1 = r_2 < 0$	PN or IN	Asymptotically stable	PN, IN or SpP	Asymptotically stable
$r_1 = r_2 = \lambda \pm i\mu$				
$\lambda > 0$	SpP	Unstable	SpP	Unstable
$\lambda < 0$	SpP	Asymptotically stable	SpP	Asymptotically stable
$r_1 = i\mu, r_2 = -i\mu$	C	Stable	C or SpP	Indeterminate

IN = Improper node; PN = Proper node; SP = Saddle point; SpP = Spiral point; C = Center

real and equal. Recall that at the end of Section 9.3 we stated that small perturbations in the coefficients of the linear system (6), and hence in the roots r_1 and r_2 can alter the type and stability of the critical point only in these two sensitive cases. When r_1 and r_2 are pure imaginary, a small perturbation can change the stable center into either an asymptotically stable or an unstable spiral point or even leave it as a center. When $r_1 = r_2$ a small perturbation does not affect the stability of the critical point, but may change the node into a spiral point. It is reasonable to expect that the small nonlinear terms in Eqs. (8) might have a similarly substantial effect, at least in these two sensitive cases. This is so, but the main significance of Theorem 9.2 is that in *all other cases* the small nonlinear terms do not alter the type or stability of the critical point. Thus, except in the two sensitive cases, the type and stability of the critical point of the nonlinear system (8) can be determined from a study of the much simpler linear system (6).

Even if the critical point is of the same type as that of the linear system, the trajectories of the almost linear system may be considerably different in appearance from those of the corresponding linear system. This is illustrated in Problems 12 and 13. However, it can be shown that the slope at which trajectories "enter" or "leave" the critical point is given correctly by the linear equations.

We will illustrate some of these ideas by considering the motion of a damped pendulum and several problems in ecology. The pendulum problem is discussed here; the ecological problems are discussed in Section 9.5.

Damped Pendulum.   We consider the motion of a damped pendulum for which the damping is proportional to the speed (see Figure 9.22). It is not difficult to

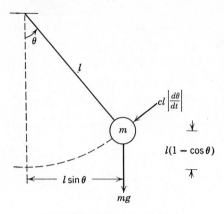

FIGURE 9.22   Pendulum.

show, by reasoning similar to that used in the derivation of the spring-mass equation (Section 3.7), that the governing equation is

$$ml^2 \frac{d^2\theta}{dt^2} + cl \frac{d\theta}{dt} + mgl \sin \theta = 0, \tag{11}$$

where the damping constant $c > 0$. Letting $x = \theta$ and $y = d\theta/dt$ gives the system

$$\frac{dx}{dt} = y, \qquad \frac{dy}{dt} = -\frac{g}{l} \sin x - \frac{c}{ml} y. \tag{12}$$

The point $x = 0$, $y = 0$ is a critical point of the system (12). Because of the damping mechanism, we expect any small motion about $\theta = 0$ to decay in amplitude. Thus intuitively the equilibrium point $(0, 0)$ should be asymptotically stable. To show this we rewrite the system (12) as

$$\frac{dx}{dt} = y,$$

$$\frac{dy}{dt} = -\frac{g}{l} x - \frac{c}{ml} y - \frac{g}{l} (\sin x - x). \tag{13}$$

It was shown in Section 9.3 [see Eq. (5)] that $(\sin x - x)/r \to 0$ as $r \to 0$, so the system (13) is an almost linear system; hence Theorem 9.2 is applicable. The roots of the characteristic equation associated with the corresponding linear system

$$\frac{dx}{dt} = y, \qquad \frac{dy}{dt} = -\frac{g}{l} x - \frac{c}{ml} y \tag{14}$$

are

$$r_1, r_2 = \frac{-c/ml \pm \sqrt{(c/ml)^2 - 4g/l}}{2}. \tag{15}$$

1. If $(c/ml)^2 - 4g/l > 0$ the roots are real, unequal, and negative. The critical point $(0, 0)$ is an asymptotically stable improper node of the linear system (14) and of the almost linear system (13).

2. If $(c/ml)^2 - 4g/l = 0$ the roots are real, equal, and negative. The critical point $(0, 0)$ is an asymptotically stable node of the linear system (14). It may be either an asymptotically stable node or an asymptotically stable spiral point of the almost linear system (13).

3. If $(c/ml)^2 - 4g/l < 0$ the roots are complex with negative real parts. The critical point $(0, 0)$ is an asymptotically stable spiral point of the linear system (14) and of the almost linear system (13).

In addition to the critical point $(0, 0)$, the almost linear system (12) has the critical points $x = n\pi$, $y = 0$, $n = \pm1$, ±2, ±3, ... corresponding to $\theta = \pm\pi, \pm2\pi, \pm3\pi, \ldots, d\theta/dt = 0$. Again we expect (from Figure 9.22) that the points corresponding to $\theta = \pm2\pi, \pm4\pi, \ldots$ are asymptotically stable spiral points, and the points corresponding to $\theta = \pm\pi, \pm3\pi, \ldots$ are unstable saddle points. Consider the critical point $x = \pi$, $y = 0$. To examine the stability of this point, let

$$x = \pi + u, \qquad y = 0 + v. \tag{16}$$

Substituting for x and y in Eqs. (12), and making use of the fact that $\sin(\pi + u) = -\sin u$, we obtain

$$\frac{du}{dt} = v,$$

$$\frac{dv}{dt} = -\frac{c}{ml}v + \frac{g}{l}\sin u. \tag{17}$$

We are interested in studying the critical point $u = v = 0$ of the system (17). We can rewrite the second of Eqs. (17) as

$$\frac{dv}{dt} = \frac{g}{l}u - \frac{c}{ml}v + \frac{g}{l}(\sin u - u). \tag{18}$$

It is clear that the first of Eqs. (17) and Eq. (18) are the same as the system (13), except that $-g/l$ is replaced by g/l. Thus the system is almost linear, and the roots of the characteristic equation of the corresponding linear system are given by

$$r_1, r_2 = \frac{-c/ml \pm \sqrt{(c/ml)^2 + 4g/l}}{2}. \tag{19}$$

One root (r_1) is positive and the other (r_2) is negative. Therefore, the critical point $x = \pi$, $y = 0$ is an unstable saddle point of both the linear system and the almost linear system, as expected.

Let us consider the case $(c/ml)^2 - 4g/l < 0$ in more detail. We will try to make a schematic sketch of the trajectories in the xy (phase) plane. The critical points are at $y = 0$, $x = n\pi$, $n = 0, \pm1, \pm2, \ldots$.

We have seen that the critical point $(0, 0)$ is an asymptotically stable spiral point. This point corresponds to the pendulum hanging vertically downward with zero velocity. It is not difficult to show that the critical points $(2n\pi, 0)$, $n = \pm 1, \pm 2, \dots$, which also correspond to the pendulum hanging vertically downward with zero velocity, are also asymptotically stable spiral points. (Let $x = 2n\pi + u$, $y = 0 + v$ and proceed as above.) The direction of motion on the spirals near $(0, 0)$ can be obtained from Eqs. (12). Consider the point at which a spiral intersects the positive y axis $(x = 0, y > 0)$. At such a point it follows from Eqs. (12) that $dx/dt > 0$. Thus the point (x, y) on the trajectory is moving to the right so the direction of motion on the spirals is clockwise. The situation is the same for the spirals near the critical points $(2n\pi, 0)$, $n = \pm 1, \pm 2, \dots$. The spirals are sketched in Figure 9.23.

FIGURE 9.23   Asymptotically stable spiral points for the damped pendulum.

We have shown above that the critical point $(\pi, 0)$ is an unstable saddle point. This point corresponds to the pendulum in a vertically upward position with zero velocity. Again, we can establish that the critical points $((2n + 1)\pi, 0)$, $n = 0$, $\pm 1, \pm 2, \dots$, are unstable saddle points. Thus in the phase plane the asymptotically stable spiral points are separated by unstable saddle points. Recall from Section 9.3 that only one pair of trajectories "enters" a saddle point. In order to determine the direction of the entering trajectories at the saddle point $(\pi, 0)$, we consider the linear system corresponding to Eqs. (17):

$$\frac{du}{dt} = v,$$

$$\frac{dv}{dt} = \frac{g}{l} u - \frac{c}{ml} v. \tag{20}$$

On making use of Eq. (19) and the first of Eqs. (20), we can write the general solution of Eq. (20) in the form

$$\binom{u}{v} = C_1 \binom{1}{r_1} e^{r_1 t} + C_2 \binom{1}{r_2} e^{r_2 t},$$

where C_1 and C_2 are arbitrary constants. Since $r_1 > 0$ and $r_2 < 0$, it follows that the solution that approaches zero as $t \to \infty$ corresponds to $C_1 = 0$. For this solution $v/u = r_2$ so the slope of the entering trajectories is negative; one lies in the second quadrant ($C_2 < 0$) and the other lies in the fourth quadrant ($C_2 > 0$). For $C_2 = 0$ we obtain the pair of trajectories "exiting" from the saddle point. These trajectories have slope $r_1 > 0$; one lies in the first quadrant ($C_1 > 0$) and the other lies in the third quadrant ($C_1 < 0$). The situation is the same for the other unstable saddle points (($2n + 1)\pi$, 0), $n = \pm 1, \pm 2, \ldots$. A possible picture of the trajectories in the neighborhood of the unstable saddle points is shown in Figure 9.24.

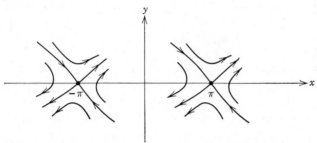

FIGURE 9.24   Unstable saddle points for the damped pendulum.

If we put Figures 9.23 and 9.24 together, we obtain the schematic picture of the trajectories in the phase plane shown in Figure 9.25. Note that the trajectories entering the saddle points separate the phase plane into sectors. Such a trajectory is called a *separatrix*. The initial conditions on θ and $d\theta/dt$ determine the position of an initial point (x, y) in the phase plane. The subsequent motion of the pendulum is represented by the trajectory passing through the initial point as it spirals towards the asymptotically stable point of that sector. Note that it is mathematically possible (but physically unrealizable) to choose initial conditions on a separatrix

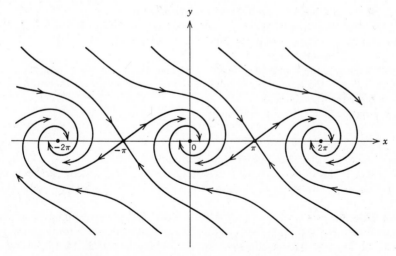

FIGURE 9.25   Phase plane for the damped pendulum.

so that the resulting motion leads to a balanced pendulum in a vertically upward position of unstable equilibrium.

PROBLEMS

In each of Problems 1 through 10, verify that $(0, 0)$ is a critical point, show that the system is almost linear, and discuss the type and stability of the critical point $(0, 0)$.

1. $dx/dt = x - y + xy$
   $dy/dt = 3x - 2y - xy$

2. $dx/dt = x + x^2 + y^2$
   $dy/dt = y - xy$

3. $dx/dt = -2x - y - x(x^2 + y^2)$
   $dy/dt = x - y + y(x^2 + y^2)$

4. $dx/dt = y + x(1 - x^2 - y^2)$
   $dy/dt = -x + y(1 - x^2 - y^2)$

5. $dx/dt = 2x + y + xy^3$
   $dy/dt = x - 2y - xy$

6. $dx/dt = x + 2x^2 - y^2$
   $dy/dt = x - 2y + x^3$

7. $dx/dt = y$
   $dy/dt = -x + \mu y(1 - x^2), \ \mu > 0$

8. $dx/dt = 1 + y - e^{-x}$
   $dy/dt = y - \sin x$

9. $dx/dt = (1 + x) \sin y$
   $dy/dt = 1 - x - \cos y$

10. $dx/dt = e^{-x+y} - \cos x$
    $dy/dt = \sin (x - 3y)$

11. Determine all real critical points of each of the following systems of equations and discuss their type and stability.

(a) $dx/dt = x + y^2$
    $dy/dt = x + y$

(b) $dx/dt = 1 - xy$
    $dy/dt = x - y^3$

(c) $dx/dt = x - x^2 - xy$
    $dy/dt = 3y - xy - 2y^2$

(d) $dx/dt = 1 - y$
    $dy/dt = x^2 - y^2$

12. Consider the autonomous system

$$dx/dt = y, \qquad dy/dt = x + 2x^3.$$

(a) Show that the critical point $(0, 0)$ is a saddle point.
(b) Sketch the trajectories for the corresponding linear system by integrating the equation for dy/dx. Show from the parametric form of the solution that the only trajectory on which $x \to 0$, $y \to 0$ as $t \to \infty$ is $y = -x$.
(c) Determine the trajectories for the nonlinear system by integrating the equation for dy/dx. Sketch the trajectories for the nonlinear system that correspond to $y = -x$ and $y = x$ for the linear system.

13. Consider the autonomous system

$$dx/dt = x, \qquad dy/dt = -2y + x^3.$$

(a) Show that the critical point $(0, 0)$ is a saddle point.
(b) Sketch the trajectories for the corresponding linear system [Problem 4(c) of Section 9.2], and show that the trajectory for which $x \to 0$, $y \to 0$ as $t \to \infty$ is given by $x = 0$.
(c) Determine the trajectories for the nonlinear system for $x \neq 0$ by integrating the equation for dy/dx. Show that the trajectory corresponding to $x = 0$ for the linear system is unaltered, but that the one corresponding to $y = 0$ is $y = x^5/5$. Sketch several of the trajectories for the nonlinear system.

14. Theorem 9.2 provides no information about the stability of a critical point of an almost linear system if that point is a center of the corresponding linear system. That this must be the case is illustrated by the following two systems

$$\text{(i)} \quad \begin{cases} dx/dt = y + x(x^2 + y^2) \\ dy/dt = -x + y(x^2 + y^2) \end{cases} \qquad \text{(ii)} \quad \begin{cases} dx/dt = y - x(x^2 + y^2) \\ dy/dt = -x - y(x^2 + y^2). \end{cases}$$

(a) Show that $(0, 0)$ is a critical point of each system and, furthermore, is a center of the corresponding linear system.

(b) Show that each system is almost linear.

(c) Let $r^2 = x^2 + y^2$, and note that $x\, dx/dt + y\, dy/dt = r\, dr/dt$. For system (ii) show that $dr/dt < 0$ and that $r \to 0$ as $t \to \infty$, and hence the critical point is asymptotically stable. For system (i) show that the solution of the initial value problem for r with $r = r_0$ at $t = 0$ becomes unbounded as $t \to 1/2r_0^2$, and hence the critical point is unstable.

15. The equation of motion of an undamped pendulum is $d^2\theta/dt^2 + (g/l) \sin \theta = 0$. Let $x = \theta$, $y = d\theta/dt$, and $k^2 = g/l$ to obtain the system of equations

$$dx/dt = y, \qquad dy/dt = -k^2 \sin x.$$

(a) Show that the critical points are $(\pm n\pi, 0)$, $n = 0, 1, 2, \ldots$, and that the system is almost linear in the neighborhood of each critical point.

(b) Show that the critical point $(0, 0)$ is a (stable) center of the corresponding linear system. Using Theorem 9.2 what can be said about the almost linear problem? The situation is similar at the critical points $(\pm 2n\pi, 0)$, $n = 1, 2, 3, \ldots$. What is the physical interpretation of these critical points?

(c) Show that the critical point $(\pi, 0)$ is an (unstable) saddle point. The situation is similar at the critical points $[\pm(2n - 1)\pi, 0]$, $n = 1, 2, 3, \ldots$. What is the physical interpretation of these critical points?

16. In this problem we give some of the details of the analysis for sketching the trajectories of the undamped pendulum of Problem 15. Show by eliminating t that the equation of the trajectories can be written as

$$\tfrac{1}{2}y^2 + k^2(1 - \cos x) = E.$$

To discover the significance of the constant E, observe that $\tfrac{1}{2}y^2 = \tfrac{1}{2}(dx/dt)^2 = \tfrac{1}{2}(d\theta/dt)^2$ is proportional to the kinetic energy of the pendulum. Also, $k^2(1 - \cos x) = \int_0^x k^2 \sin x\, dx$ is proportional to the potential energy of the pendulum due to the gravity force. Thus the constant E is the "energy" of the motion. It is constant along a trajectory (during the course of the motion), and is determined by the initial values of x and y.

In sketching the trajectories it is only necessary to consider the interval $-\pi \leq x \leq \pi$ since the equation is periodic in x with period 2π. For $E = 2k^2$ show that $y = \pm 2k \cos x/2$ and sketch these trajectories. Observe that these trajectories enter or leave the unstable saddle points at $(\pm \pi, 0)$. Determine the direction of motion on each trajectory by using the differential equations given in Problem 15.

It can be shown that the trajectories are closed curves for $E < 2k^2$ and are not closed curves for $E > 2k^2$. We will not pursue these details here, but a schematic sketch of the trajectories for an undamped pendulum is given in Figure 9.26. The trajectories for $E < 2k^2$ correspond to periodic motions about a center; the trajectories for $E > 2k^2$ correspond to whirling motions.

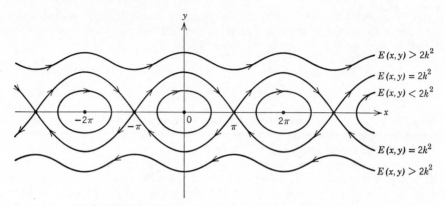

FIGURE 9.26   Phase plane for the undamped pendulum.

*17. In this problem we derive a formula for the natural period of an undamped non-linear pendulum [$c = 0$ in Eq. (11)]. Suppose that the bob is pulled through a positive angle α and then released with zero velocity. Assume that the relation θ as a function of t can be solved for t as a function of θ so that $d\theta/dt$ can be considered a function of θ. Derive the following sequence of equations:

$$\tfrac{1}{2}ml^2 \frac{d}{d\theta}\left[\left(\frac{d\theta}{dt}\right)^2\right] = -mgl \sin\theta,$$

$$\tfrac{1}{2}m\left(l\frac{d\theta}{dt}\right)^2 = mgl(\cos\theta - \cos\alpha),$$

$$dt = -\sqrt{\frac{l}{2g}}\frac{d\theta}{\sqrt{\cos\theta - \cos\alpha}}.$$

Why was the negative square root chosen in the last equation?

If T is the natural period of oscillation, derive the formula

$$\frac{T}{4} = -\sqrt{\frac{l}{2g}}\int_\alpha^0 \frac{d\theta}{\sqrt{\cos\theta - \cos\alpha}}$$

By making the change of variables $\cos\theta = 1 - 2\sin^2\theta/2$, $\cos\alpha = 1 - 2\sin^2\alpha/2$ followed by $\sin\theta/2 = k\sin\phi$ with $k = \sin\alpha/2$, show that

$$T = 4\sqrt{\frac{l}{g}}\int_0^{\pi/2}\frac{d\phi}{\sqrt{1 - k^2\sin^2\phi}} = 4\sqrt{\frac{l}{g}}F(k, \pi/2).$$

The function F is called the *elliptic integral of the first kind*. Note that the period depends on the ratio l/g and also the initial displacement α through $k = \sin\alpha/2$. The corresponding period for the linearized pendulum is $2\pi(l/g)^{1/2}$ and is independent of the initial displacement. To obtain this special result from the general formula we must consider the limiting case of small α (small angular displacement) in which case k is small. In the limit $k \to 0$ the above formula gives $T = 4(l/g)^{1/2} F(0, \pi/2) = 2\pi(l/g)^{1/2}$.

18. A generalization of the damped pendulum equation discussed in the text, or a damped spring-mass system, is the Liénard equation

$$\frac{d^2x}{dt^2} + c(x)\frac{dx}{dt} + g(x) = 0.$$

If $c(x)$ is a constant and $g(x) = kx$ then this equation has the form of the linear pendulum equation [replace $\sin\theta$ with θ in Eq. (11)]; otherwise the damping force $c(x)\,dx/dt$ and restoring force $g(x)$ are nonlinear. Assume that c is continuously differentiable, g is twice continuously differentiable, and $g(0) = 0$.

(a) Write the Liénard equation as a system of two first order equations by introducing the variable $y = dx/dt$.

(b) Show that $(0, 0)$ is a critical point and that the system is almost linear in the neighborhood of $(0, 0)$.

(c) Show that if $c(0) > 0$ and $g'(0) > 0$ then the critical point is asymptotically stable, and that if $c(0) < 0$ or $g'(0) < 0$ then the critical point is unstable.
Hint: Use Taylor series to approximate f and g in the neighborhood of $x = 0$.

19. Consider the equations for two competing species derived in Section 9.2:

$$dx/dt = x(\epsilon_1 - \sigma_1 x - \alpha_1 y), \qquad dy/dt = y(\epsilon_2 - \sigma_2 y - \alpha_2 x).$$

Suppose that $\epsilon_1/\sigma_1 < \epsilon_2/\alpha_2$ and $\epsilon_2/\sigma_2 < \epsilon_1/\alpha_1$.

(a) Find the critical point (x_E, y_E) for which both species can coexist.

(b) By making the change of variables $x = x_E + u, y = y_E + v$ transform the system of equations to one with a critical point at $u = 0$, $v = 0$. Observe that the system is almost linear.

(c) Classify the critical point as to type and stability.

20. Carry out the calculations of Problem 19 for the case $\epsilon_1/\sigma_1 > \epsilon_2/\alpha_2$ and $\epsilon_2/\sigma_2 > \epsilon_1/\alpha_1$.

9.5 COMPETING SPECIES AND PREDATOR-PREY PROBLEMS

In this section we consider two problems in ecology: competing species and predator-prey.

Competing Species. In Section 9.2 we showed that a model for the competition between two species with population densities x and y leads to the differential equations

$$dx/dt = x(\epsilon_1 - \sigma_1 x - \alpha_1 y), \tag{1a}$$

$$dy/dt = y(\epsilon_2 - \sigma_2 y - \alpha_2 x), \tag{1b}$$

where the parameters $\epsilon_1, \sigma_1, \ldots, \alpha_2$ are positive. As we saw then, we can analyze these equations by dividing the phase plane into regions according to the sign of dx/dt and dy/dt and then drawing typical trajectories. Let us now see how we can use the theory of almost linear systems to obtain a more precise understanding of what happens.

We start by considering the following specific example:

$$dx/dt = x(1 - x - y), \tag{2a}$$

$$dy/dt = y(\tfrac{1}{2} - \tfrac{1}{4}y - \tfrac{3}{4}x). \tag{2b}$$

For convenience we think of x and y as the population densities of two species of bacteria competing with each other for the same supply of food. We ask whether there are equilibrium states that might be reached, or whether a periodic growth and decay will be observed, and how such possibilities depend on the initial state of the two cultures.

The critical points of the system (2) are the solutions of the nonlinear algebraic equations

$$x(1 - x - y) = 0,$$
$$y(\tfrac{1}{2} - \tfrac{1}{4}y - \tfrac{3}{4}x) = 0. \tag{3}$$

Clearly one solution is $x = y = 0$; a second solution is $x = 1$, $y = 0$; a third is $x = 0$, $y = 2$. Finally, if $x \neq 0$, $y \neq 0$, we obtain from Eqs. (3) the system

$$x + y = 1,$$
$$3x + y = 2,$$

which has the solution $x = \tfrac{1}{2}$, $y = \tfrac{1}{2}$. These four points in the xy plane are the only critical points of the system (2). We will consider each separately.

$x = 0, y = 0$. This corresponds to a state in which both bacteria die as a result of their competition. For this case the corresponding linear system, from Eqs. (2), is

$$dx/dt = x, \quad dy/dt = \tfrac{1}{2}y, \tag{4}$$

and the roots of the characteristic equation are 1 and $\tfrac{1}{2}$. Thus the origin is an unstable improper node: the solution $x = 0$, $y = 0$ of the interaction problem will not occur in practice.

$x = 1, y = 0$. Clearly this corresponds to a state in which bacteria x survives the competition but bacteria y does not. To examine this critical point, let $x = 1 + u$, $y = 0 + v$. Substituting for x and y in Eqs. (2) and simplifying, we obtain

$$du/dt = -u - v - u^2 - uv,$$
$$dv/dt = -\tfrac{1}{4}v - \tfrac{1}{4}v^2 - \tfrac{3}{4}uv. \tag{5}$$

It is not difficult to show that the system (5) is an almost linear system. The corresponding linear system is $du/dt = -u - v$, $dv/dt = -v/4$ and he roots of the characteristic equation are $-\tfrac{1}{4}$ and -1. The general solution is

$$u = -\tfrac{4}{3} Ae^{-t/4} + Be^{-t}, \quad v = Ae^{-t/4}.$$

where A and B are arbitrary. Thus $x = 1$, $y = 0$ is an asymptotically stable improper node. If the initial values of x and y are sufficiently close to $x = 1$, $y = 0$ the interaction will lead finally to that state.

For this critical point we will indicate how the trajectories of the linear system behave in the neighborhood of $x = 1$, $y = 0$. It is clear that $u \to 0$, $v \to 0$ as $t \to \infty$ so that all of the trajectories enter the critical point $(1, 0)$ as $t \to \infty$. For $A = 0$ we have $x = 1 + u = 1 + Be^{-t}$ and $y = v = 0$ so one pair ($B < 0$ and

$B > 0$) of trajectories enters along the x axis. For $A \neq 0$ we can compute the slope at any point on a trajectory by noting that

$$\frac{dy}{dx} = \frac{dv/dt}{d(1+u)/dt} = \frac{-\frac{1}{4}Ae^{-t/4}}{\frac{1}{3}Ae^{-t/4} - Be^{-t}} = \frac{-\frac{1}{4}A}{\frac{1}{3}A - Be^{-3t/4}} \; .$$

In particular as we approach $(t \to \infty)$ the critical point along any trajectory with $A \neq 0$ we see that $dy/dx \to -\frac{3}{4}$. Thus all the trajectories except one pair enter the critical point along a line with slope $-\frac{3}{4}$.

x = 0, y = 2.   The analysis is exactly similar to that for the critical point $x = 1$, $y = 0$. The critical point $x = 0$, $y = 2$ is also an asymptotically stable improper node. In this case bacteria y survives, but bacteria x does not.

x = 1/2, y = 1/2.   This critical point corresponds to a mixed equilibrium state or coexistence; a standoff, so to speak, in the competition between the two bacteria cultures. To examine the nature of this critical point we let $x = \frac{1}{2} + u$, $y = \frac{1}{2} + v$. Substituting for x and y in Eqs. (2) we obtain

$$du/dt = -\tfrac{1}{2}u - \tfrac{1}{2}v - u^2 - uv,$$

$$dv/dt = -\tfrac{3}{8}u - \tfrac{1}{8}v - \tfrac{1}{4}v^2 - \tfrac{3}{4}uv. \tag{6}$$

The system (6) is an almost linear system, and the roots of the characteristic equation of the corresponding linear system are $(-5 \pm \sqrt{57})/16$. Since these roots are real and of opposite sign, the critical point $(\frac{1}{2}, \frac{1}{2})$ is an unstable saddle point. One pair of trajectories enters the critical point; the others recede from it. It can be shown by considering the general solution of the corresponding linear system that the slope of the pair of entering trajectories as $(x, y) \to (\frac{1}{2}, \frac{1}{2})$ is $(\sqrt{57} - 3)/8 \simeq 0.57$ (see Problem 1).

A schematic sketch of what the trajectories might look like in the neighborhood of each critical point is shown in Figure 9.27a. With a little detective work it is possible to extend the local pictures and obtain a global picture of the trajectories in the phase plane. First, we are only interested in x and y positive. Since trajectories cannot cross other trajectories and since the x and y axes are trajectories it follows that a trajectory that starts in the first quadrant must stay in the first quadrant, and a trajectory that starts in any other quadrant cannot enter the first quadrant. Second, we accept without proof two facts that follow from advanced theory: (i) the system (2) does not have any periodic solutions, that is, trajectories that are closed curves (see Theorem 9.7 of Section 9.7); and (ii) a trajectory that is not a closed curve must either enter a critical point or go off to infinity as $t \to \infty$. But consider what is happening for x and y large. The nonlinear terms $-(x^2 + xy)$ and $-\frac{1}{4}(y^2 + 3xy)$ in the first and second of Eqs. (2), respectively, outweigh the linear terms. Since they are negative, dx/dt and dy/dt are negative for x and y large. Thus for x and y large the direction of motion on any trajectory is inward. The trajectories cannot escape to infinity! Eventually they must head toward one or the other of the two stable nodes. The *schematic*

FIGURE 9.27a

FIGURE 9.27b

sketch shown in Figure 9.27b is not an unreasonable representation* of what must be happening in the first quadrant. If the initial values of x and y are in region I of Figure 9.27b, then x wins the competition; if the initial values are in region II, then y wins. "Peaceful coexistence" is not possible unless the initial point lies exactly on the dividing trajectory (separatrix). Of particular interest would be the determination of the dividing trajectories that enter the saddle point $(\frac{1}{2}, \frac{1}{2})$ which separate regions I and II.

Now let us return to the general system (1). Recall from Section 9.2 that four cases must be considered depending on the relative orientation of the lines

$$\epsilon_1 - \sigma_1 x - \alpha_1 y = 0 \quad \text{and} \quad \epsilon_2 - \sigma_2 y - \alpha_2 x = 0, \tag{7}$$

as shown in Figure 9.28. Let (X_E, Y_E) denote any critical point in any one of the four cases. To study the system (1) in the neighborhood of this critical point we let

$$x = X_E + u, \quad y = Y_E + v, \tag{8}$$

FIGURE 9.28

* The dotted trajectories that appear to go, for example, from $(0, 0)$ to $(1, 0)$ actually correspond to trajectories through particular initial points. As $t \to \infty$ they approach the stable node $(1, 0)$; as $t \to -\infty$ they approach the unstable node $(0, 0)$.

and substitute in Eqs. (1):

$$\frac{d}{dt}(X_E + u) = (X_E + u)[\epsilon_1 - \sigma_1(X_E + u) - \alpha_1(Y_E + v)],$$

$$\frac{d}{dt}(Y_E + v) = (Y_E + v)[\epsilon_2 - \sigma_2(Y_E + v) - \alpha_2(X_E + u)].$$
(9)

Since $dX_E/dt = dY_E/dt = 0$, we have

$$\frac{du}{dt} = (X_E + u)[(\epsilon_1 - \sigma_1 X_E - \alpha_1 Y_E) - \sigma_1 u - \alpha_1 v],$$
(10a)

$$\frac{dv}{dt} = (Y_E + v)[(\epsilon_2 - \sigma_2 Y_E - \alpha_2 X_E) - \sigma_2 v - \alpha_2 u].$$
(10b)

The right-hand side of Eq. (10a) has the form $X_E(\epsilon_1 - \sigma_1 X_E - \alpha_1 Y_E) + (\ )u + (\ )v + (\ )u^2 + (\ )uv$. The constant term is zero since either $X_E = 0$ or $\epsilon_1 - \sigma_1 X_E - \alpha_1 Y_E = 0$. Similarly, the right-hand side of Eq. (10b) reduces to the form $(\ )u + (\ )v + (\ )uv + (\ )v^2$. Next, it is clear that Eqs. (10) are almost linear, so we consider the corresponding linear system

$$\frac{du}{dt} = [(\epsilon_1 - \sigma_1 X_E - \alpha_1 Y_E) - \sigma_1 X_E]u - \alpha_1 X_E v,$$

$$\frac{dv}{dt} = -\alpha_2 Y_E u + [(\epsilon_2 - \sigma_2 Y_E - \alpha_2 X_E) - \sigma_2 Y_E]v.$$
(11)

Equations (11), along with Theorem 9.2 of Section 9.4, can be used to determine the type and stability of any critical point (X_E, Y_E) of the original system (1). These linear equations are referred to as the linearized equations for small perturbations in the neighborhood of the critical point (X_E, Y_E). The process of deriving them is called linearization.

We will use Eqs. (11) to determine whether the model given by the system (1) can ever lead to coexistence for the two species x and y, and if so, under what conditions on the parameters $\epsilon_1, \sigma_1, \ldots, \alpha_2$. The four possible situations are shown in Figure 9.28; coexistence is possible only in cases (c) and (d). The values of $X_E \neq 0$ and $Y_E \neq 0$ are obtained by solving the system of simultaneous linear algebraic equations (7). We readily obtain

$$X_E = \frac{\epsilon_1 \sigma_2 - \epsilon_2 \alpha_1}{\sigma_1 \sigma_2 - \alpha_1 \alpha_2}, \qquad Y_E = \frac{\epsilon_2 \sigma_1 - \epsilon_1 \alpha_2}{\sigma_1 \sigma_2 - \alpha_1 \alpha_2}.$$
(12)

Moreover, since $\epsilon_1 - \sigma_1 X_E - \alpha_1 Y_E = 0$ and $\epsilon_2 - \sigma_2 Y_E - \alpha_2 X_E = 0$, Eqs. (11) immediately reduce to

$$\frac{du}{dt} = -\sigma_1 X_E u - \alpha_1 X_E v,$$

$$\frac{dv}{dt} = -\alpha_2 Y_E u - \sigma_2 Y_E v.$$
(13)

The corresponding characteristic equation is

$$r^2 + (\sigma_1 X_E + \sigma_2 Y_E)r + (\sigma_1\sigma_2 X_E Y_E - \alpha_1\alpha_2 X_E Y_E) = 0. \tag{14}$$

Thus

$$r = \frac{-(\sigma_1 X_E + \sigma_2 Y_E) \pm \sqrt{(\sigma_1 X_E + \sigma_2 Y_E)^2 - 4(\sigma_1\sigma_2 - \alpha_1\alpha_2)X_E Y_E}}{2}. \tag{15}$$

If $\sigma_1\sigma_2 - \alpha_1\alpha_2 < 0$ then the radicand of Eq. (15) is positive and greater than $(\sigma_1 X_E + \sigma_2 Y_E)^2$. Thus the roots are real with one positive and one negative. The critical point (X_E, Y_E) is an unstable saddle point. Coexistence is not possible. This is the case in the specific example given by Eqs. (2); there $\sigma_1 = 1$, $\alpha_1 = 1$, $\sigma_2 = \frac{1}{4}$, $\alpha_2 = \frac{3}{4}$ and $\sigma_1\sigma_2 - \alpha_1\alpha_2 = -\frac{1}{2}$. On the other hand, if $\sigma_1\sigma_2 - \alpha_1\alpha_2 > 0$, then the radicand of Eq. (15) is less than $(\sigma_1 X_E + \sigma_2 Y_E)^2$. Thus the roots are real, negative, and unequal, or complex with negative real parts. A simple analysis of the radicand of Eq. (15) shows that the roots cannot be complex (see Problem 2). Thus the critical point is an asymptotically stable improper node. Coexistence is possible if $\sigma_1\sigma_2 > \alpha_1\alpha_2$.

Let us relate this result to Figures 9.28c and 9.28d. In Figure 9.28c we have

$$\frac{\epsilon_1}{\sigma_1} > \frac{\epsilon_2}{\alpha_2} \quad \text{or} \quad \epsilon_1\alpha_2 > \epsilon_2\sigma_1 \quad \text{and} \quad \frac{\epsilon_2}{\sigma_2} > \frac{\epsilon_1}{\alpha_1} \quad \text{or} \quad \epsilon_2\alpha_1 > \epsilon_1\sigma_2. \tag{16}$$

These inequalities coupled with the condition that X_E and Y_E given by Eqs. (12) be positive yield the inequality $\sigma_1\sigma_2 < \alpha_1\alpha_2$. Hence in this case the mixed state is an unstable saddle point. Corresponding to Figure 9.28d, we have

$$\frac{\epsilon_2}{\alpha_2} > \frac{\epsilon_1}{\sigma_1} \quad \text{or} \quad \epsilon_2\sigma_1 > \epsilon_1\alpha_2 \quad \text{and} \quad \frac{\epsilon_1}{\alpha_1} > \frac{\epsilon_2}{\sigma_2} \quad \text{or} \quad \epsilon_1\sigma_2 > \epsilon_2\alpha_1. \tag{17}$$

Now, the condition X_E and Y_E positive yields $\sigma_1\sigma_2 > \alpha_1\alpha_2$. Hence this mixed state is asymptotically stable. For this case we can also show that the critical points $(0, 0)$, $(\epsilon_1/\sigma_1, 0)$ and $(0, \epsilon_2/\sigma_2)$ are unstable. Thus no matter what the initial values of $x \neq 0$ and $y \neq 0$ are, the two species approach an equilibrium state of coexistence given by Eqs. (12).

Equations (1) provide the biological interpretation of the result that $\sigma_1\sigma_2 > \alpha_1\alpha_2$ leads to coexistence and $\alpha_1\alpha_2 > \sigma_1\sigma_2$ does not allow coexistence. The σ's are a measure of the inhibitory effect the growth of each species has on its own growth rate, while the α's are a measure of the inhibiting effect the growth of each species has on the other species (interaction). Thus when $\sigma_1\sigma_2 > \alpha_1\alpha_2$ interaction is "small" and the species can coexist; when $\alpha_1\alpha_2 > \sigma_1\sigma_2$ interaction (competition) is "large" and the species cannot coexist—one must die out.

Predator-Prey. As a second example we consider the classical predator-prey problem. We study an ecological situation involving two species, one of which preys on the other (does not compete with it for food but actually preys on it) while the other lives on a different source of food. An example is foxes and rabbits in a closed forest; the foxes prey on the rabbits, the rabbits live on the vegetation in the forest. Other examples are bass in a lake as predators and redear (sunfish) as prey, and lady bugs as predators and aphids (insects that suck the juice of plants) as prey. Let $H(t)$ and $P(t)$ be the populations of prey and predator, respectively, at time t.

We build as simple a model of the interaction as possible. We make the following assumptions:

1. In the absence of the predator the prey grows without bound; thus $dH/dt = aH$, $a > 0$ for $P = 0$.

2. In the absence of the prey the predator dies out, thus $dP/dt = -cP$, $c > 0$, for $H = 0$.

3. The increase in the number of predators is wholly dependent on the food supply (the prey) and the prey are consumed at a rate proportional to the number of encounters between predators and prey. Thus, for example, if the number of prey is doubled the number of encounters is doubled. Encounters decrease the number of prey and increase the number of predators. A fixed proportion of prey is killed in each encounter, and the rate of population growth of the predator is enhanced by a factor proportional to the amount of prey consumed.

As a consequence, we have the equations

$$dH/dt = aH - \alpha HP = H(a - \alpha P),$$

$$dP/dt = -cP + \gamma HP = P(-c + \gamma H). \tag{18}$$

The constants a, c, α, and γ are positive; a and c are the growth rate of the prey and the death rate of the predator, respectively, and α and γ are measures of the effect of the interaction between the two species. Equations (18) are known as the Lotka-Volterra equations. They were developed in papers by Lotka* in 1925 and Volterra† in 1926. Although these equations are simple, they do characterize a wide class of problems. Ways of making them more realistic are discussed at the end of this section and in the problems.

What happens for given initial values of $P > 0$ and $H > 0$? Will the predators eat all of their prey and in turn die out, will the predators die out because of a too low level of prey and then the prey grow without bound, will an equilibrium state be reached, or will a cyclic fluctuation of prey and predator occur?

---

* A. J. Lotka (1880–1949), an American biophysicist born in Austria, was the author of the first book on mathematical biology published in 1924; it is now available as *Elements of Mathematical Biology*, Dover, New York, 1956.
† Vito Volterra (1860–1940) was a very distinguished Italian mathematician. In mathematics he is particularly famous for his work on integral equations. Indeed, one of the major classes of integral equations has been given his name in honor of his extensive investigations. He also made valuable contributions to mathematical biology. His theory of competing species was motivated by data collected by a friend, D'Ancona, covering fish catches in the Adriatic Sea. A translation of his 1926 paper can be found in an appendix of Chapman, R. N., *Animal Ecology with Special Reference to Insects*, McGraw-Hill, New York, 1931.

The critical points of Eqs. (18) are the solutions of

$$H(a - \alpha P) = 0,$$
$$P(-c + \gamma H) = 0. \tag{19}$$

These solutions are

$$H = 0, P = 0 \quad \text{and} \quad H = c/\gamma, P = a/\alpha. \tag{20}$$

We will examine the predator-prey model (18) in the neighborhood of each critical point. It is very easy to show that the critical point $(0, 0)$ is an unstable saddle point. Entrance to the saddle point is along the line $P = 0$; all other trajectories recede from the critical point.

To study the critical point $(c/\gamma, a/\alpha)$ we let

$$H = \frac{c}{\gamma} + u, \qquad P = \frac{a}{\alpha} + v. \tag{21}$$

Substituting for H and P in Eqs. (18), we obtain

$$du/dt = -\alpha \frac{c}{\gamma} v - \alpha uv,$$
$$dv/dt = \gamma \frac{a}{\alpha} u + \gamma uv. \tag{22}$$

This is an almost linear system, and the corresponding linear system is

$$du/dt = -\alpha \frac{c}{\gamma} v,$$
$$dv/dt = \gamma \frac{a}{\alpha} u. \tag{23}$$

The characteristic equation is

$$r^2 + ac = 0 \quad \text{so} \quad r = \pm i\sqrt{ac}. \tag{24}$$

Since the roots of the characteristic equation are pure imaginary, the critical point is a (stable) center of the linear system. The trajectories of the linear system are closed curves corresponding to solutions that are periodic in time. They neither approach nor recede from the critical point. In particular, the trajectories can readily be shown to be ellipses in the following way. From Eqs. (23), we have

$$\frac{dv}{du} = -\frac{(\gamma a/\alpha)u}{(\alpha c/\gamma)v}, \tag{25}$$

$$\frac{\gamma a}{\alpha} u \, du + \frac{\alpha c}{\gamma} v \, dv = 0,$$

$$\frac{\gamma a}{\alpha} u^2 + \frac{\alpha c}{\gamma} v^2 = C, \tag{26}$$

FIGURE 9.29   Trajectories for the linearized predator-prey problem.

where C is an arbitrary nonnegative constant of integration. Several of these trajectories are sketched in Figure 9.29.

While the critical point is a stable center of the linear system, we need to assess its character for the almost linear system. Here, as we know, our theory for almost linear systems fails. The effect of the nonlinear terms may be to change the center into a stable spiral point, or into an unstable spiral point, or it may remain as a stable center. Fortunately, for the predator-prey problem we can actually solve the nonlinear equations (18) and determine what happens. Dividing the second of Eqs. (18) by the first equation, we obtain

$$\frac{dP}{dH} = \frac{P(-c + \gamma H)}{H(a - \alpha P)}.\tag{27}$$

Upon separating the variables in Eq. (27), we have

$$\frac{a - \alpha P}{P} dP = \frac{-c + \gamma H}{H} dH,$$

from which it follows that

$$a \ln P - \alpha P = -c \ln H + \gamma H + \ln C,\tag{28}$$

where C is a constant of integration. We cannot solve Eq. (28) explicitly for P in terms of H or for H in terms of P, but it can be shown that the graph of this equation for a fixed value of C is a closed curve (not an ellipse, of course) enclosing the critical point $(c/\gamma, a/\alpha)$*. Thus the predator and prey have a cyclic variation about the critical point.

We can analyze this cyclic variation in more detail when the deviation from the point $(c/\gamma, a/\alpha)$ is small; that is, when it is permissible to linearize the perturbation equations (22) for u and v. As we have noted, the trajectories are the

---

* Volterra gave a clever elementary geometric proof of this result. Shorter (but mathematically more advanced) proofs have also been discovered.

family of ellipses given by Eq. (26). We can also verify, either by solving Eqs. (23) or by direct substitution, that the solution of Eqs. (23) is

$$u(t) = \frac{c}{\gamma} K \cos(\sqrt{ac}\, t + \phi), \qquad v(t) = \frac{a}{\alpha}\sqrt{\frac{c}{a}} K \sin(\sqrt{ac}\, t + \phi), \qquad (29)$$

where the constants K and ϕ are determined by the initial conditions. Thus

$$H(t) = \frac{c}{\gamma} + \frac{c}{\gamma} K \cos(\sqrt{ac}\, t + \phi),$$

$$P(t) = \frac{a}{\alpha} + \frac{a}{\alpha}\sqrt{\frac{c}{a}} K \sin(\sqrt{ac}\, t + \phi). \tag{30}$$

These equations are valid for the elliptical trajectories close to the critical point $(c/\gamma, a/\alpha)$. We can use them to draw several conclusions about the cyclic variation of the predator and prey on such trajectories.

1. The size of the predator and prey populations varies sinusoidally with period $2\pi/\sqrt{ac}$. This period of oscillation is independent of the initial conditions.

2. The predator and prey populations are out of phase by one quarter of a cycle. The prey leads and the predator lags as one might expect. This is discussed in Problem 5.

3. The amplitude of the oscillations are Kc/γ for the prey and $a\sqrt{c}K/\alpha\sqrt{a}$ for the predator and hence depend on the initial conditions as well as the parameters of the problem.

4. The average numbers of predators and prey over one complete cycle are c/γ and a/α, respectively. These are the same as the equilibrium populations (see Problem 6).

Cyclic variations of predator and prey as predicted by Eqs. (18) are often observed in nature. One striking example is described by Odum (pp. 191–192): based on the records of the Hudson Bay Company of Canada, the abundance of lynx and snowshoe hare as indicated by the number of pelts turned in over the period 1845–1935 shows a distinct periodic variation with a period of nine to ten years. The peaks of abundance are followed by very rapid declines, and the peaks of abundance of the lynx and hare are out of phase, with that of the hare preceding that of the lynx by a year or more.

The Volterra-Lotka model of the predator-prey problem has revealed a cyclic variation that was perhaps intuitively expected. On the other hand, the use of the Volterra-Lotka model in other situations can lead to conclusions that are not intuitively obvious. An example revealing a possible danger in using insecticides is given in Problem 8.

One criticism of the Volterra-Lotka predator-prey model is that in the absence of the predator the prey will grow without bound. This can be corrected by allowing for the natural inhibiting effect that an increasing population has on the growth rate of the population; for example, by modifying the first of Eqs. (18) so that when $P = 0$ it reduces to a logistic equation for H (see Problem 9). The models of two competing species and predator-prey discussed here can be modified to allow for the effect of time delays; probabilistic and statistical effects can also be included. Finally, we mention that there are *discrete* analogs of each of the problems we have discussed corresponding to species that breed only at certain times. The mathematics of the discrete problems are often interesting and some of the results are unexpected. These generalizations are discussed in the references given at the end of this chapter as well as in other books on mathematical biology and ecology.

We conclude with a final warning. Using only elementary phase plane theory for one and two nonlinear ordinary differential equations, we have been able to illustrate several of the fundamental principles of simple biological systems. But one should not be misled—ecology is not this simple.

PROBLEMS

1. Consider the linear system of equations

$$du/dt = -u/2 - v/2, \qquad dv/dt = -3u/8 - v/8$$

associated with the critical point $(\frac{1}{2}, \frac{1}{2})$ of Figure 9.27a.

(a) Determine the general solution of this system of equations.

(b) The trajectories that enter the saddle point correspond to the solution in which only the negative exponential is present. Show for this solution that $dv/du = (dv/dt) \div (du/dt) \to (\sqrt{57} - 3)/8$ as $t \to \infty$.

2. Show that

$$(\sigma_1 X_E + \sigma_2 Y_E)^2 - 4(\sigma_1 \sigma_2 - \alpha_1 \alpha_2) X_E Y_E = (\sigma_1 X_E - \sigma_2 Y_E)^2 + 4\alpha_1 \alpha_2 X_E Y_E.$$

Hence conclude that the roots given by Eq. (15) can never be complex.

3. Two species of fish that compete with each other for food, but do not prey on each other, are bluegill and redear. Suppose that a pond is stocked with bluegill and redear and let x and y be the populations of bluegill and redear, respectively, at time t. Suppose further that the competition is modeled by the equations

$$dx/dt = x(\epsilon_1 - \sigma_1 x - \alpha_1 y),$$
$$dy/dt = y(\epsilon_2 - \sigma_2 y - \alpha_2 x).$$

(a) If $\epsilon_2/\alpha_2 > \epsilon_1/\sigma_1$ and $\epsilon_2/\sigma_2 > \epsilon_1/\alpha_1$ show that the only equilibrium populations in the pond are no fish, no redear, or no bluegill. What will happen?

(b) If $\epsilon_1/\sigma_1 > \epsilon_2/\alpha_2$ and $\epsilon_1/\alpha_1 > \epsilon_2/\sigma_2$, show that the only equilibrium populations in the pond are no fish, no redear, or no bluegill. What will happen?

4. Consider the competition between bluegill and redear mentioned in Problem 3. Suppose that $\epsilon_2/\alpha_2 > \epsilon_1/\sigma_1$ and $\epsilon_1/\alpha_1 > \epsilon_2/\sigma_2$, so, as shown in the text, there is a stable equilibrium point at which both species can coexist. It is convenient to rewrite the equations of Problem 3 in terms of the carrying capacities of the pond for bluegill ($B = \epsilon_1/\sigma_1$) in the absence of redear and for redear ($R = \epsilon_2/\sigma_2$) in the absence of bluegill.

(a) Show that the equations of Problem 3 take the form

$$\frac{dx}{dt} = \epsilon_1 x \left(1 - \frac{1}{B} x - \frac{\gamma_1}{B} y \right),$$

$$\frac{dy}{dt} = \epsilon_2 y \left(1 - \frac{1}{R} y - \frac{\gamma_2}{R} x \right),$$

where $\gamma_1 = \alpha_1/\sigma_1$ and $\gamma_2 = \alpha_2/\sigma_2$. Determine the coexistence equilibrium point (X_E, Y_E) in terms of B, R, γ_1, and γ_2.

(b) Now suppose that a fisherman fishes only for bluegill with the effect that B is reduced. What effect does this have on the equilibrium populations? Is it possible, by fishing, to reduce the population of bluegill to such a level that they will die out?

5. In this problem we examine the phase difference between the cyclic variations of the predator and prey populations as given by Eqs. (30) of the text. Suppose we assume that $K > 0$ and that T is measured from the time that the prey population (H) is a maximum; then $\phi = 0$. Show that the predator population (P) is a maximum at $t = \pi/2\sqrt{ac} = T/4$, where T is the period of the oscillation. When is the prey population increasing most rapidly, decreasing most rapidly, a minimum? Answer the same questions for the predator population. Draw a typical elliptic trajectory enclosing the point $(c/\gamma, a/\alpha)$, and mark these points on it.

6. The average sizes of the prey and predator populations are defined as

$$\bar{H} = \frac{1}{T} \int_A^{A+T} H(t)\, dt, \qquad \bar{P} = \frac{1}{T} \int_A^{A+T} P(t)\, dt,$$

respectively, where T is the period of a full cycle and A is any nonnegative constant. Show for trajectories near the critical point that $\bar{H} = c/\gamma$, $\bar{P} = a/\alpha$.

7. Suppose that the predator-prey equations (18) of the text govern foxes (P) and rabbits (H) in a forest. A trapping company is engaging in trapping foxes and rabbits for their pelts. Explain why it is reasonable for the company to conduct its operation in such a way as to move the population of each species closer to the center $(c/\gamma, a/\alpha)$. When is it best to trap foxes? Rabbits? Rabbits and foxes? Neither?
Hint: See Problem 5. A mathematical argument is not required.

8. Suppose that an insect population (H) is controlled by a natural predator population (P) according to the model (18), so that there are small cyclic variations of the populations about the critical point $(c/\gamma, a/\alpha)$. Show that it is self-defeating to employ an insecticide if the insecticide also kills the predator. Assume that the insecticide kills both prey and predator at rates proportional to each population, respectively. To ban insecticides on the basis of this very simple model would certainly be ill-advised. On the other hand, it is also rash to ignore the possible genuine existence of a phenomenon suggested by a simple model.

9. As was mentioned in the text, one improvement in the predator-prey model is to modify the equation for the prey so that it has the form of a logistic equation in the absence of the predator. Thus in place of Eqs. (18) we consider the model system

$$dH/dt = H(a - \sigma H - \alpha P),$$
$$dP/dt = P(-c + \gamma H),$$

where a, σ, α, c, and γ are positive constants. Determine all critical points and discuss their nature and stability. Assume that $a/\sigma \gg c/\gamma$. What happens for initial data $H \neq 0$, $P \neq 0$?

9.6 LIAPOUNOV'S SECOND METHOD

In Section 9.4 we showed how the stability of a critical point of an almost linear system can usually be determined from a study of the corresponding linear system. However, no conclusion can be drawn when the critical point is a center of the corresponding linear system. Examples of this situation are the undamped pendulum, Eqs. (1) and (2), and the predator-prey problem discussed in Section 9.5. Also, for an asymptotically stable critical point it may be important to investigate the region of asymptotic stability; that is, the domain such that all solutions starting within that domain approach the critical point. Since the theory of almost linear systems is a local theory, it provides no information about this question.

In this section we discuss another approach, known as *Liapounov's second method* or *direct method*. The method is referred to as a direct method because no knowledge of the solution of the system of differential equations is required. Rather, conclusions about the stability or instability of a critical point are obtained by constructing a suitable auxiliary function. The technique is a very powerful one that provides a more global type of information; for example, an estimate of the extent of the region of asymptotic stability of a critical point. In addition, Liapounov's second method can also be used to study systems of equations that are not almost linear; however, we will not discuss such problems.

Basically Liapounov's second method is a generalization of the physical principles that for a conservative system, (i) a rest position is stable if the potential energy is a local minimum, otherwise it is unstable, and (ii) the total energy is a constant during any motion. To illustrate these concepts, again consider the undamped pendulum (a conservative mechanical system), which is governed by the equation

$$\frac{d^2\theta}{dt^2} + \frac{g}{l}\sin\theta = 0. \tag{1}$$

The corresponding system of first order equations is

$$\frac{dx}{dt} = y, \qquad \frac{dy}{dt} = -\frac{g}{l}\sin x, \tag{2}$$

where $x = \theta$ and $y = d\theta/dt$. Omitting an arbitrary constant, the potential energy U is the work done in lifting the pendulum above its lowest position, namely $mgl(1 - \cos\theta)$ (see Figure 9.22 of Section 9.4). Hence

$$U(x, y) = mgl(1 - \cos x). \tag{3}$$

The critical points of the system (2) are $x = \pm n\pi$, $y = 0$, $n = 0, 1, 2, 3, \ldots$, corresponding to $\theta = \pm n\pi$, $d\theta/dt = 0$. Physically, we expect that the points

$x = 0, y = 0$; $x = \pm 2\pi$, $y = 0$; ... corresponding to $\theta = 0$, $\pm 2\pi$, ... for which the pendulum bob is vertical with the weight down will be stable; and that the points $x = \pm \pi$, $y = 0$; $x = \pm 3\pi$, $y = 0$; ... corresponding to $\theta = \pm \pi$, $\pm 3\pi$, ... for which the pendulum bob is vertical with the weight up will be unstable. This agrees with statement (i), for at the former points U is a minimum equal to zero, and at the latter points U is a maximum equal to $2mgl$.

Next consider the total energy V, which is the sum of the potential energy U and the kinetic energy $\frac{1}{2}ml^2(d\theta/dt)^2$. In terms of x and y

$$V(x, y) = mgl(1 - \cos x) + \tfrac{1}{2}ml^2 y^2. \tag{4}$$

On a trajectory corresponding to a solution $x = \phi(t)$, $y = \psi(t)$ of Eqs. (2), V can be considered as a function of t. The derivative of $V[\phi(t), \psi(t)]$ is called the rate of change of V following the trajectory. By the chain rule

$$\frac{dV[\phi(t), \psi(t)]}{dt} = V_x[\phi(t), \psi(t)]\frac{d\phi(t)}{dt} + V_y[\phi(t), \psi(t)]\frac{d\psi(t)}{dt}$$

$$= (mgl \sin x)\frac{dx}{dt} + ml^2 y \frac{dy}{dt}, \tag{5}$$

where it is understood that $x = \phi(t)$, $y = \psi(t)$. But dx/dt and dy/dt can be obtained in terms of x and y from Eqs. (2). Substituting in Eqs. (5) for dx/dt and dy/dt, we find that $dV/dt = 0$. Hence V is a constant along any trajectory of the system (2), which is in agreement with our earlier remark (ii) that the total energy is constant during any motion of a conservative system.

It is important to note that at any point (x, y) the rate of change of V along the trajectory was computed *without actually solving* the system (2). It is precisely this fact that allows us to use Liapounov's second method for systems whose solutions we do not know, and hence makes it such an important technique.

At the stable critical points, $x = \pm 2n\pi$, $y = 0$, $n = 0, 1, 2, \ldots$, the energy V is zero. If the initial state, say (x_1, y_1), of the pendulum is sufficiently near a stable critical point, then the energy $V(x_1, y_1)$ is small, and the motion (trajectory) associated with this energy stays close to the critical point. It can be shown that if $V(x_1, y_1)$ is sufficiently small, then the trajectory is closed and contains the critical point. For example, suppose that (x_1, y_1) is near $(0, 0)$ and that $V(x_1, y_1)$ is very small. The equation of the trajectory with energy $V(x_1, y_1)$ is

$$V(x, y) = mgl(1 - \cos x) + \tfrac{1}{2}ml^2 y^2 = V(x_1, y_1).$$

For x small we have $1 - \cos x = 1 - (1 - x^2/2! + \ldots) \simeq x^2/2$. Thus the equation of the trajectory is approximately

$$\tfrac{1}{2}mgl x^2 + \tfrac{1}{2}ml^2 y^2 = V(x_1, y_1)$$

or

$$\frac{x^2}{2V(x_1, y_1)/mgl} + \frac{y^2}{2V(x_1, y_1)/ml^2} = 1.$$

This is an ellipse enclosing the critical point $(0, 0)$; the smaller $V(x_1, y_1)$ is, the smaller are the major and minor axes of the ellipse. Physically, the closed trajectory corresponds to a solution that is periodic in time—the motion is a small oscillation about the equilibrium point.

If damping is present, however, it is natural to expect that the amplitude of the motion decays in time and that the stable critical point (center) becomes an asymptotically stable critical point (spiral point). See the sketch of the trajectories for the damped pendulum in Figure 9.25 of Section 9.4. This can almost be argued from a consideration of dV/dt. For the damped pendulum, the total energy is still given by Eq. (4), but now from Eqs. (12) of Section 9.4 $dx/dt = y$ and $dy/dt = -(g/l)\sin x - (c/lm)y$. Substituting for dx/dt and dy/dt in Eq. (5) gives $dV/dt = -cly^2 \leq 0$. Thus the energy is nonincreasing along any trajectory and, except for the line $y = 0$, the motion is such that the energy decreases, and hence each trajectory must approach a point of minimum energy—a stable equilibrium point. If $dV/dt < 0$ instead of $dV/dt \leq 0$, it is reasonable to expect that this would be true for all trajectories that start sufficiently close to the origin.

To pursue these ideas further, consider the autonomous system

$$\frac{dx}{dt} = F(x, y), \qquad \frac{dy}{dt} = G(x, y), \tag{6}$$

and suppose that the point $x = 0$, $y = 0$ is an asymptotically stable critical point. Then there exists some domain D containing $(0, 0)$ such that every trajectory that starts in D must approach the origin as $t \to \infty$. Suppose that there exists an "energy" function V such that $V(x, y) \geq 0$ for (x, y) in D with $V = 0$ only at the origin. Since each trajectory in D approaches the origin as $t \to \infty$, then following any particular trajectory, V decreases to zero as t approaches infinity. The type of result we want to prove is essentially the converse: if, on every trajectory, V decreases to zero as t increases, then the trajectories must approach the origin as $t \to \infty$, and hence the origin is asymptotically stable. First, however, it is necessary to make several definitions.

Let V be defined on some domain D containing the origin. Then V is said to be *positive definite* on D if $V(0, 0) = 0$ and $V(x, y) > 0$ for all other points in D. Similarly, V is said to be *negative definite* on D if $V(0, 0) = 0$ and $V(x, y) < 0$ for all other points in D. If the inequalities $>$ and $<$ are replaced by $\geq$ and $\leq$, then V is said to be *positive semidefinite* and *negative semidefinite*, respectively. We emphasize that in speaking of a positive definite (negative definite, ...) function on a domain D containing the origin, the function must be zero at the origin in addition to satisfying the proper inequality at all other points in D.

Example 1. The function
$$V(x, y) = \sin (x^2 + y^2)$$
is positive definite on $x^2 + y^2 < \pi/2$ since $V(0, 0) = 0$ and $V(x, y) > 0$ for $0 < x^2 + y^2 < \pi/2$. However, the function
$$V(x, y) = (x + y)^2$$
is only positive semidefinite since $V(x, y) = 0$ on the line $y = -x$.

We will also want to consider the function

$$\dot{V}(x, y) = V_x(x, y)F(x, y) + V_y(x, y)G(x, y). \tag{7}$$

We choose this notation because $\dot{V}(x, y)$ can be identified as the rate of change of V along the trajectory of the system (6) that passes through the point (x, y). That is, if $x = \phi(t)$, $y = \psi(t)$ is a solution of the system (6), then

$$\frac{dV[\phi(t), \psi(t)]}{dt} = V_x[\phi(t), \psi(t)]\frac{d\phi(t)}{dt} + V_y[\phi(t), \psi(t)]\frac{d\psi(t)}{dt}$$

$$= V_x(x, y)F(x, y) + V_y(x, y)G(x, y)$$

$$= \dot{V}(x, y). \tag{8}$$

The function $\dot{V}$ is sometimes referred to as the derivative of V with respect to the system (6).

We now state two Liapounov theorems, the first dealing with stability, the second with instability.

Theorem 9.3.   *Suppose that the autonomous system* (6) *has an isolated critical point at the origin. If there exists a function V that is continuous and has continuous first partial derivatives, is positive definite, and for which the function $\dot{V}$ given by Eq.* (7) *is negative definite on some domain D in the xy plane containing* $(0, 0)$, *then the origin is an asymptotically stable critical point. If $\dot{V}$ is negative semidefinite, then the origin is a stable critical point.*

Theorem 9.4.   *Let the origin be an isolated critical point of the autonomous system* (6). *Let V be a function that is continuous and has continuous first partial derivatives. Suppose that $V(0, 0) = 0$ and that in every neighborhood of the origin there is at least one point at which V is positive (negative). Then if there exists a domain D containing the origin such that the function $\dot{V}$ as given by Eq.* (7) *is positive definite (negative definite) on D, then the origin is an unstable critical point.*

The function V is called a *Liapounov function*. Before sketching geometric proofs of Theorems 9.3 and 9.4, we note that the difficulty in using these theorems is that they tell us nothing about how to construct a Liapounov function, assuming that one exists. In cases where the autonomous system (6) represents a physical problem, it is natural first to consider the actual total energy function of the system as a possible Liapounov function. However, we emphasize that Theorems 9.3 and 9.4 are applicable in cases where the concept of physical energy is not pertinent. In such cases a judicious trial-and-error approach may be necessary.

Now consider the second part of Theorem 9.3; that is, the case $\dot{V}(x, y) \leq 0$. Let $c \geq 0$ be a constant and consider the curve in the xy plane given by the equation $V(x, y) = c$. For $c = 0$ the curve reduces to the single point $x = 0$, $y = 0$. However, for $c > 0$ and sufficiently small, it can be shown by using the continuity of V that we will obtain a closed curve containing the origin as illustrated

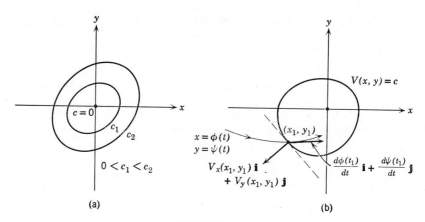

FIGURE 9.30

in Figure 9.30a. There may, of course, be other curves in the xy plane corresponding to the same value of c, but these are not of interest. Further, again by continuity, as c gets smaller and smaller the closed curves $V(x, y) = c$ enclosing the
origin shrink to the origin. We will show that a trajectory starting inside of a
closed curve $V(x, y) = c$ cannot cross to the outside. Thus, given a circle of radius
ϵ about the origin, by taking c sufficiently small we can ensure that every trajectory
starting inside of the closed curve $V(x, y) = c$ stays within the circle of radius ϵ;
indeed, it will stay within the closed curve $V(x, y) = c$ itself. Thus the origin will
be a stable critical point.

It is shown in the calculus that the vector*

$$V_x(x, y)\mathbf{i} + V_y(x, y)\mathbf{j}, \tag{9}$$

where $\mathbf{i}$ and $\mathbf{j}$ are unit vectors in the positive x and y directions, is normal to the
curve $V(x, y) = c$. Furthermore, since $V(x, y)$ increases outward from the origin,
the vector (9) will point away from the origin, as indicated in Figure 9.30b. Next
consider the angle of intersection of a trajectory corresponding to a solution
$x = \phi(t)$, $y = \psi(t)$ of the system (6) and a closed curve $V(x, y) = c$. Suppose
that the point of intersection is $x_1 = \phi(t_1)$, $y_1 = \psi(t_1)$. From Eq. (7) and the fact
that $d\phi(t_1)/dt = F(x_1, y_1)$, $d\psi(t_1)/dt = G(x_1, y_1)$ we have

$$\dot{V}(x_1, y_1) = V_x(x_1, y_1)\frac{d\phi(t_1)}{dt} + V_y(x_1, y_1)\frac{d\psi(t_1)}{dt}$$

$$= [V_x(x_1, y_1)\mathbf{i} + V_y(x_1, y_1)\mathbf{j}] \cdot \left[\frac{d\phi(t_1)}{dt}\mathbf{i} + \frac{d\psi(t_1)}{dt}\mathbf{j}\right],$$

where $[d\phi(t_1)/dt]\mathbf{i} + [d\psi(t_1)/dt]\mathbf{j}$ is the tangent vector to the trajectory $x = \phi(t)$,
$y = \psi(t)$ at the point (x_1, y_1) (see Figure 9.30b). The right-hand side of the above
equation is the scalar product of two vectors and hence can be interpreted as the

* The vector $V_x(x, y)\mathbf{i} + V_y(x, y)\mathbf{j}$ is known as the gradient of V or grad V.

product of their magnitudes times the cosine of the angle between them. Since, by hypothesis, $\dot{V}(x_1, y_1) \leq 0$, it follows that the cosine of the angle is less than or equal to zero so the angle lies between $90°$ and $270°$. Hence the direction of motion on the trajectory must be inward or, at worst, tangent to the curve. Trajectories starting inside a closed curve $V(x, y) = c$ (no matter how small c is) cannot escape, so the origin is a stable point. If $\dot{V}(x, y) < 0$, then the trajectories passing through points on the curve are actually pointed inward. As a consequence, it can be shown that trajectories starting sufficiently close to the origin must approach the origin, and hence the origin is asymptotically stable.

A geometric proof of Theorem 9.4 follows by somewhat similar arguments. Briefly, suppose that $\dot{V}$ is positive definite, and suppose that given any circle about the origin there is an interior point (x_1, y_1) at which $V(x_1, y_1) > 0$. Consider a trajectory that starts at (x_1, y_1). Along this trajectory it follows from Eq. (8) that V must increase, since $\dot{V}(x, y) > 0$; furthermore, since $V(x_1, y_1) > 0$ the trajectory cannot approach the origin because $V(0, 0) = 0$. This shows that the origin cannot be asymptotically stable. By further exploiting the fact that $\dot{V}(x, y) > 0$, it is possible to show that the origin is an unstable point; however, we will not pursue this argument.

To illustrate the use of Theorem 9.3 we consider the question of the stability of the critical point $(0, 0)$ of the undamped pendulum equations (2). While the system (2) is almost linear, the point $(0, 0)$ is a center of the corresponding linear system, so no conclusion can be drawn from Theorem 9.2. Since the mechanical system is conservative, it is natural to suspect that the total energy function V given by Eq. (4) will be a Liapounov function. For example, if we take D to be the domain $-\pi/2 < x < \pi/2$, $-\infty < y < \infty$, then V is positive definite. As we have seen $\dot{V}(x, y) = 0$, so it follows from the second part of Theorem 9.3 that the critical point $(0, 0)$ of Eqs. (2) is a stable critical point. The damped pendulum is discussed in Problem 4.

From a practical point of view one is usually interested more in asymptotic stability. For example, for the automatic control for the wing flap mentioned in Section 9.1, it is not sufficient that the control keep the flap close to a certain setting—rather, it must damp out deviations from the correct setting. Clearly it is desirable to know the allowable deviations from the set position that the automatic control can damp out; that is, to know the region of asymptotic stability. One of the simplest results dealing with this question is given by the following theorem.

Theorem 9.5. *Let the origin be an isolated critical point of the autonomous system* (6). *Let the function V be continuous and have continuous first partial derivatives. If there is a bounded domain D_K containing the origin where $V(x, y) < K$, V is positive definite, and $\dot{V}$ is negative definite, then every solution of Eqs.* (6) *that starts at a point in D_K approaches the origin as t approaches infinity.*

In other words, Theorem 9.5 says that if $x = \phi(t)$, $y = \psi(t)$ is the solution of Eqs. (6) for initial data lying in D_K then (x, y) approaches the critical point

$(0, 0)$ as $t \to \infty$. Thus D_K gives a region of asymptotic stability; of course, it may not be the entire region of asymptotic stability. This theorem is proved by showing that (i) there are no periodic solutions of the system (6) in D_K, and (ii) there are no other critical points in D_K. It then follows that trajectories starting in D_K cannot escape and, hence, must tend to the origin as t tends to infinity.

Theorems 9.3 and 9.4 give sufficient conditions for stability and instability, respectively. However, these conditions are not necessary, nor does our failure to determine a suitable Liapounov function mean that there is not one. Unfortunately, there are no general methods for the construction of Liapounov functions; however, there has been extensive work on the construction of Liapounov functions for special classes of equations. One simple result from elementary algebra, which is often useful in constructing positive definite or negative definite functions, is stated without proof in the following theorem.

Theorem 9.6. *The function*

$$V(x, y) = ax^2 + bxy + cy^2 \tag{10}$$

is positive definite if, and only if,

$$a > 0 \quad and \quad 4ac - b^2 > 0, \tag{11}$$

and is negative definite if, and only if,

$$a < 0 \quad and \quad 4ac - b^2 > 0. \tag{12}$$

The use of Theorem 9.6 is illustrated in the following example.

Example 2. Show that the critical point $(0, 0)$ of the autonomous system

$$\frac{dx}{dt} = -x - xy^2, \qquad \frac{dy}{dt} = -y - yx^2 \tag{13}$$

is asymptotically stable.

We try to construct a Liapounov function of the form (10). Then $V_x(x, y) = 2ax + by$, $V_y(x, y) = bx + 2cy$ so

$$\dot{V}(x, y) = (2ax + by)(-x - xy^2) + (bx + 2cy)(-y - yx^2)$$

$$= -[2a(x^2 + x^2y^2) + b(2xy + xy^3 + yx^3) + 2c(y^2 + x^2y^2)].$$

If we choose $b = 0$, and a and c to be any positive numbers, then $\dot{V}$ is negative definite and V is positive definite by Theorem 9.6. Thus by Theorem 9.3 the origin is an asymptotically stable critical point.

PROBLEMS

1. By constructing suitable Liapounov functions of the form $ax^2 + cy^2$, where a and c are to be determined, show that for each of the following systems the critical point at the origin is of the indicated type.

   (a) $dx/dt = -x^3 + xy^2,$ $dy/dt = -2x^2y - y^3;$ asymptotically stable

(b) $dx/dt = -\frac{1}{2}x^3 + 2xy^2$, $dy/dt = -y^3$;   asymptotically stable

(c) $dx/dt = -x^3 + 2y^3$, $dy/dt = -2xy^2$;   stable (at least)

(d) $dx/dt = x^3 - y^3$, $dy/dt = 2xy^2 + 4x^2y + 2y^3$;   unstable

2. Consider the system of equations

$$\frac{dx}{dt} = y - xf(x, y), \qquad \frac{dy}{dt} = -x - yf(x, y),$$

where f is continuous and has continuous first partial derivatives. Show that if $f(x, y) > 0$ in some neighborhood of the origin, then the origin is an asymptotically stable critical point, and if $f(x, y) < 0$ in some neighborhood of the origin, then the origin is an unstable critical point.

Hint: Construct a Liapounov function of the form $c(x^2 + y^2)$.

3. A generalization of the undamped pendulum equation is

$$\frac{d^2u}{dt^2} + g(u) = 0, \tag{i}$$

where $g(0) = 0, g(u) > 0$ for $0 < u < k$, and $g(u) < 0$ for $-k < u < 0$; that is, $ug(u) > 0$ for $u \neq 0$, $-k < u < k$. Notice that $g(u) = \sin u$ has this property on $(-\pi/2, \pi/2)$.

(a) Letting $x = u$, $y = du/dt$, write Eq. (i) as a system of two equations, and show that $x = 0$, $y = 0$ is a critical point.

(b) Show that

$$V(x, y) = \frac{1}{2}y^2 + \int_0^x g(s)\, ds, \qquad -k < x < k, \tag{ii}$$

is positive definite, and use this result to show that the critical point $(0, 0)$ is stable. Note that the Liapounov function V given by Eq. (ii) corresponds to the energy function $V(x, y) = \frac{1}{2}y^2 + (1 - \cos x)$ for the case $g(u) = \sin u$.

4. By introducing suitable dimensionless variables, the system of nonlinear equations for the damped pendulum [Eqs. (12) of Section 9.4] can be written as

$$dx/dt = y, \qquad dy/dt = -y - \sin x.$$

(a) Show that the origin is a critical point.

*(b) Show that while $V(x, y) = x^2 + y^2$ is positive definite, $\dot{V}(x, y)$ takes on both positive and negative values in any domain containing the origin, so that V is not a Liapounov function.

Hint: $x - \sin x > 0$ for $x > 0$ and $x - \sin x < 0$ for $x < 0$. Consider these cases with y positive but y so small that y^2 can be ignored compared to y.

(c) Using the energy function $V(x, y) = \frac{1}{2}y^2 + (1 - \cos x)$ obtained from Eq. (ii) of Problem 3, show that the origin is a stable critical point. Note, however, that even though there is damping and we can expect that the origin is asymptotically stable, it is not possible to draw this conclusion using the present Liapounov function.

*(d) To show asymptotic stability it is necessary to construct a better Liapounov function than the one used in part (c). Show that $V(x, y) = \frac{1}{2}(x + y)^2 + x^2 + \frac{1}{2}y^2$ is such a Liapounov function, and conclude that the origin is an asymptotically stable critical point.

Hint: From Taylor's formula with a remainder it follows that $\sin x = x - \alpha x^3/3!$, where

α depends on x but $0 < \alpha < 1$ for $-\pi/2 < x < \pi/2$. Then letting $x = r\cos\theta$, $y = r\sin\theta$ show that $\dot{V}(r\cos\theta, r\sin\theta) = -r^2[1 + h(r, \theta)]$ where $|h(r, \theta)| < 1$ if r is sufficiently small.

In Problems 5 and 6 we will prove part of Theorem 9.2: if the critical point $(0, 0)$ of the almost linear system

$$\frac{dx}{dt} = ax + by + F_1(x, y), \qquad \frac{dy}{dt} = cx + dy + G_1(x, y) \tag{i}$$

is an asymptotically stable critical point of the corresponding linear system

$$\frac{dx}{dt} = ax + by, \qquad \frac{dy}{dt} = cx + dy, \tag{ii}$$

then it is an asymptotically stable critical point of the almost linear system (i).

5. Consider the linear system (ii).

(a) Since $(0, 0)$ is an asymptotically stable critical point, show that $a + d < 0$ and $ad - bc > 0$. (See Problem 17 of Section 9.3.)

(b) Construct a Liapounov function $V(x, y) = Ax^2 + Bxy + Cy^2$, such that V is positive definite and $\dot{V}$ is negative definite. One way to insure that $\dot{V}$ is negative definite is to choose A, B, and C so that $\dot{V}(x, y) = -x^2 - y^2$. Show that this leads to the result

$$A = -\frac{c^2 + d^2 + (ad - bc)}{2\Delta}, \qquad B = \frac{bd + ac}{\Delta}, \qquad C = -\frac{a^2 + b^2 + (ad - bc)}{2\Delta},$$

where $\Delta = (a + d)(ad - bc)$.

(c) Using the result of part (a) show that $A > 0$, and then show (several steps of algebra are required) that

$$4AC - B^2 = \frac{(a^2 + b^2 + c^2 + d^2)(ad - bc) + 2(ad - bc)^2}{\Delta^2} > 0.$$

Thus by Theorem 9.6, V is positive definite.

6. In this problem we show that the Liapounov function constructed in the previous problem is also a Liapounov function for the almost linear system (i). We must show that there is some region containing the origin for which $\dot{V}$ is negative definite.

(a) Show that

$$\dot{V}(x, y) = -(x^2 + y^2) + (2Ax + By)F_1(x, y) + (Bx + 2Cy)G_1(x, y).$$

(b) Recall that $F_1(x, y)/r \to 0$ and $G_1(x, y)/r \to 0$ as $r = (x^2 + y^2)^{1/2} \to 0$. This means that given any $\epsilon > 0$ there exists a circle $r = R$ about the origin such that for $0 \le r < R$, $|F_1(x, y)| < \epsilon r$, and $|G_1(x, y)| < \epsilon r$. Letting M be the maximum of $|2A|$, $|B|$, and $|2C|$, show by introducing polar coordinates that R can be chosen so that $\dot{V}(x, y) < 0$ for $r < R$. *Hint:* Choose ϵ sufficiently small in terms of M.

7. In this problem we prove a part of Theorem 9.2 relating to instability.

(a) Show that if $p = (a + d) > 0$ and $q = ad - bc > 0$ then the critical point $(0, 0)$ of the linear system (ii) is unstable.

(b) The same result holds for the almost linear system (i). Following Problems 5 and 6 construct a positive definite function V such that $\dot{V}(x, y) = x^2 + y^2$ and hence is positive definite, and then invoke Theorem 9.4.

*9.7   PERIODIC SOLUTIONS AND LIMIT CYCLES

On several occasions we have mentioned the possible existence of periodic solutions of nonlinear autonomous systems. Such solutions, whose trajectories form *closed curves* in the phase plane, play an important role in many physical phenomena. A periodic solution often represents a "final state" toward which all "neighboring" solutions tend as the transients due to the initial conditions die out. In some cases, for example, in an electronic oscillator the flow of current is periodic even though resistance is present. The existence of these periodic solutions can only be explained by a consideration of the nonlinear terms in the governing equations—no linearized mathematical model with a resistance term can predict a periodic solution.

A special case of a periodic solution is a constant solution $x = x_0$, $y = y_0$, corresponding to a critical point of the autonomous system. Such a solution is clearly periodic with any period. In this section when we speak of a periodic solution we mean a nonconstant periodic solution.

Recall that the solutions of the linear autonomous system

$$\frac{dx}{dt} = ax + by, \qquad \frac{dy}{dt} = cx + dy \tag{1}$$

are periodic if and only if the roots of the characteristic equation

$$r^2 - (a + d)r + (ad - bc) = 0 \tag{2}$$

are pure imaginary. In this case the critical point at the origin will necessarily be a stable critical point, as discussed in Section 9.3. We emphasize that if the roots of Eq. (2) are pure imaginary, then *every* solution of the linear system (1) is periodic, and if the roots are not pure imaginary then Eqs. (1) have *no* periodic solutions.

The situation can be quite different for a nonlinear autonomous system. A standard example is the system

$$\frac{dx}{dt} = y + x - x(x^2 + y^2), \qquad \frac{dy}{dt} = -x + y - y(x^2 + y^2). \tag{3}$$

It is not difficult to show that $x = 0$, $y = 0$ is the only critical point of the system (3), and also that the system is almost linear in the neighborhood of the origin. Further, from a consideration of the corresponding linear system [$a = 1$, $b = 1$, $c = -1$, $d = 1$ in Eq. (1)], it follows that the origin is an unstable spiral point for the system (3). Thus any solution that starts near the origin in the phase plane will spiral away from the origin. Since there are no other critical points, our first thought might be that all solutions of Eqs. (3) have trajectories which spiral out to infinity. However, we will now show that far away from the origin the trajectories are directed inward, so that something else must be happening.

It is convenient to introduce the polar coordinates r and θ where

$$x = r \cos \theta, \qquad y = r \sin \theta. \tag{4}$$

Multiplying the first of Eqs. (3) by x, the second by y, and adding gives

$$x\frac{dx}{dt} + y\frac{dy}{dt} = (x^2 + y^2) - (x^2 + y^2)^2. \tag{5}$$

Since $r^2 = x^2 + y^2$ and $r\,dr/dt = x\,dx/dt + y\,dy/dt$, we obtain

$$r\frac{dr}{dt} = r^2(1 - r^2). \tag{6}$$

Thus if $r > 1$, then $dr/dt < 0$, and the direction of motion on a trajectory is inward. Similarly, if $r < 1$, then the direction of motion is outward. Clearly, the circle $r = 1$ in the phase plane has a special significance for this system.

To obtain an equation for θ, we multiply the first of Eqs. (3) by y, the second by x, and subtract, obtaining

$$y\frac{dx}{dt} - x\frac{dy}{dt} = x^2 + y^2. \tag{7}$$

Upon computing dx/dt and dy/dt from Eqs. (4), we find that the left-hand side of Eq. (7) is $-r^2\,d\theta/dt$, so Eq. (7) reduces to

$$\frac{d\theta}{dt} = -1. \tag{8}$$

The system of equations (6) and (8) for r and θ is equivalent to the original system (3). One solution of the system (6) and (8) is

$$r = 1, \qquad \theta = -t + t_0, \tag{9}$$

where t_0 is an arbitrary constant. As t increases, a point satisfying Eqs. (9) moves clockwise around the unit circle. Thus the autonomous system (3) has a periodic solution. Other solutions can be obtained by solving Eq. (6) by separation of variables; if $r \neq 0$ and $r \neq 1$, then

$$\frac{dr}{r(1 - r^2)} = dt. \tag{10}$$

Equation (10) can be integrated by using partial fractions to rewrite the left-hand side. Omitting these algebraic calculations, we find that the solutions of Eqs. (8) and (10) are

$$r = \frac{1}{\sqrt{1 + c_0 e^{-2t}}}, \qquad \theta = -t + t_0, \tag{11}$$

where c_0 and t_0 are arbitrary constants. In this case the solutions (11) also contain the solution (9), which is obtained by setting $c_0 = 0$ in Eq. (11).

The solution satisfying the initial conditions $r = \rho$, $\theta = \alpha$ at $t = 0$ is given by

$$r = \frac{1}{\sqrt{1 + [(1/\rho^2) - 1]e^{-2t}}}, \qquad \theta = -(t - \alpha). \tag{12}$$

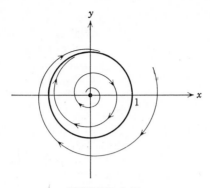

FIGURE 9.31

If $0 < \rho < 1$, $r \rightarrow 1$ from the inside as $t \rightarrow \infty$; if $\rho > 1$, $r \rightarrow 1$ from the outside as $t \rightarrow \infty$. Thus the trajectories spiral toward the circle $r = 1$ as $t \rightarrow \infty$. A schematic sketch of the trajectories is shown in Figure 9.31.

Not only does the circle $r = 1$ correspond to a periodic solution of the autonomous system (3), but also nonclosed trajectories spiral toward it as $t \rightarrow \infty$. A closed curve in the phase plane which has nonclosed curves spiraling toward it, either from the inside or outside, as $t \rightarrow \infty$ is called a *limit cycle*.* Thus the circle $r = 1$ is a limit cycle of the autonomous system (3). If all trajectories that start near a closed trajectory (both inside and outside) spiral toward the closed trajectory as $t \rightarrow \infty$, we say that the limit cycle is *stable*. This type of stability is sometimes referred to as *orbital stability*. If the trajectories on one side spiral toward the closed trajectory, while on the other side they spiral away as $t \rightarrow \infty$, we say that the limit cycle is *semistable*. If the trajectories on both sides of the closed trajectory spiral away as $t \rightarrow \infty$, we say that the closed trajectory is *unstable*. It is also possible to have closed trajectories that are neither approached nor receded from; for example, the family of periodic solutions of a linear autonomous system. In this case we say that the closed trajectory is *neutrally stable*.

For the example just considered the existence of a stable limit cycle was established by direct calculation; in most cases this is not possible. However, there are several general theorems concerning the existence or nonexistence of limit cycles of a nonlinear autonomous system. While proofs of these theorems are too advanced for a book of this level, we can state these theorems and illustrate how they may be used. We consider the autonomous system

$$\frac{dx}{dt} = F(x, y), \qquad \frac{dy}{dt} = G(x, y). \tag{13}$$

Theorem 9.7. *Let the functions F and G have continuous first partial derivatives in a domain D in the xy plane. A closed trajectory of the system* (13) *must necessarily enclose at least one critical point. If it encloses only one critical point, the critical point cannot be a saddle point.*

* Slightly different definitions are also used; the present one is sufficient for our purposes.

This theorem is illustrated by Figure 9.31. The closed trajectory $r = 1$ encloses the critical point $(0, 0)$. Also note that each closed trajectory for the linearized pendulum problem (Figure 9.10 of Section 9.2) encloses the critical point $(0, 0)$. One way in which this theorem is extremely useful is a negative one. If a given region contains no critical points, then there can be no closed trajectory (periodic solution) lying entirely in the region. Also, for the competing species problem whose trajectories are sketched in Figure 9.27b of Section 9.5, the theorem tells us that there is no closed trajectory in the first quadrant since the single critical point $(\frac{1}{2}, \frac{1}{2})$ in this quadrant is a saddle point.

A second result about the nonexistence of periodic solutions (closed trajectories) is given by the following theorem.*

Theorem 9.8.   *Let the functions F and G have continuous first partial derivatives in a simply connected domain D in the xy plane. If $F_x + G_y$ has the same sign throughout D, then there is no periodic solution of Eqs.* (13) *lying entirely in D.*

Simply connected domains are discussed on page 41; briefly, such a domain is one with no holes. Note that if $F_x + G_y$ changes sign in D, no conclusion is possible; there may or may not be periodic solutions in D. For the example problem (3), a simple calculation shows that

$$F_x(x, y) + G_y(x, y) = 2 - 4(x^2 + y^2). \tag{14}$$

Thus we can conclude that there are no periodic solutions of Eqs. (3) in the domain $0 \le r < 1/\sqrt{2}$. In fact, however, we know that there are none in the larger domain $0 \le r < 1$.

The following theorem gives conditions that ensure the existence of a periodic solution.

Theorem 9.9 (*Poincaré-Bendixson*† *Theorem*).   *Let the functions F and G have continuous first partial derivatives in a domain D in the xy plane. Let D_1 be a bounded subdomain in D, and let R be the region that consists of D_1 plus its boundary (all points in R are in D). Suppose that R contains no critical point of the system* (13). *If there exists a constant t_0 such that $x = \phi(t)$, $y = \psi(t)$ is a solution of the system* (13) *that exists and stays in R for all $t \ge t_0$, then either $x = \phi(t)$, $y = \psi(t)$ is a periodic solution (closed trajectory), or $x = \phi(t)$, $y = \psi(t)$ spirals toward a closed trajectory as $t \to \infty$. In either case, the system* (13) *has a periodic solution in R.*

Note that if R does contain a closed trajectory, then necessarily, by Theorem 9.7, this trajectory must enclose a critical point. However, this critical point cannot be in R. Thus R cannot be simply connected; it must have a hole. As an appli-

---

* For students who have had Green's theorem in the plane, which relates line integrals over closed curves to double integrals over the area inside the curve, the proof of Theorem 9.8 is not difficult (see Problem 7).

† Ivar Otto Bendixson (1861–1935) was a Swedish mathematician. This theorem appeared in a paper published by him in *Acta Mathematica* in 1901.

cation of the Poincaré-Bendixson theorem, consider the system (3) again. Since the origin is a critical point, it must be excluded. For example, we might consider the region R defined by $\frac{1}{2} \leq r \leq 2$. Next we must show that there is a solution whose trajectory stays in R for all t greater than or equal to some t_0. But this follows immediately from Eq. (6). For $r = \frac{1}{2}$, $dr/dt > 0$, so r increases and for $r = 2$, $dr/dt < 0$, so r decreases. Thus any solution of Eqs. (3) starting (the t_0 of Theorem 9.9 can be taken as the initial time) in the region $\frac{1}{2} \leq r \leq 2$ must stay in that region. Hence, by Theorem 9.9, there is a periodic solution of the system (3) whose trajectory is a closed curve in the region $\frac{1}{2} \leq r \leq 2$. For this example, we found from Eq. (6) that there are periodic solutions whose trajectories lie on the circle $r = 1$; however, in general we will not be able to solve the autonomous system. The difficulty in using the Poincaré-Bendixson theorem is the determination of a region R not containing critical points and for which a trajectory stays inside.

In Problem 18 of Section 9.4 we introduced the generalized Liénard (1869–1958) equation

$$\frac{d^2u}{dt^2} + f(u)\frac{du}{dt} + g(u) = 0. \qquad (15)$$

This equation can be thought of as describing a spring-mass system with a damping term $f(u)\,du/dt$, which is nonlinear unless f is a constant function, and with a spring force $g(u)$, which is nonlinear unless $g(u)$ is proportional to u. A particularly important special case of Eq. (15) is the van der Pol (1889–1959) equation, which governs the flow of current u in a triode oscillator,

$$\frac{d^2u}{dt^2} - \mu(1 - u^2)\frac{du}{dt} + u = 0, \qquad (16)$$

where μ is a positive constant. Experimentally it is observed that the flow of current is periodic. This periodic flow can only be explained by a consideration of the nonlinear equation. If we consider the linearized form of Eq. (16) obtained by neglecting the term $u^2\,du/dt$ we have

$$\frac{d^2u}{dt^2} - \mu\frac{du}{dt} + u = 0. \qquad (17)$$

Because of the term $-\mu\,du/dt$, Eq. (17) has no periodic solutions. Indeed, since $\mu > 0$, all solutions of Eq. (17) *grow* with increasing time. However, for the nonlinear equation (16) if u becomes greater than one, then the coefficient $-\mu(1 - u^2)$ of du/dt becomes *positive*, and this term damps the current. Thus in the nonlinear equation (16) the "resistance" term $-\mu(1 - u^2)\,du/dt$ amplifies the current if $u < 1$ and damps it if $u > 1$. This alternate action of the resistance term explains how the van der Pol equation can have a periodic solution.

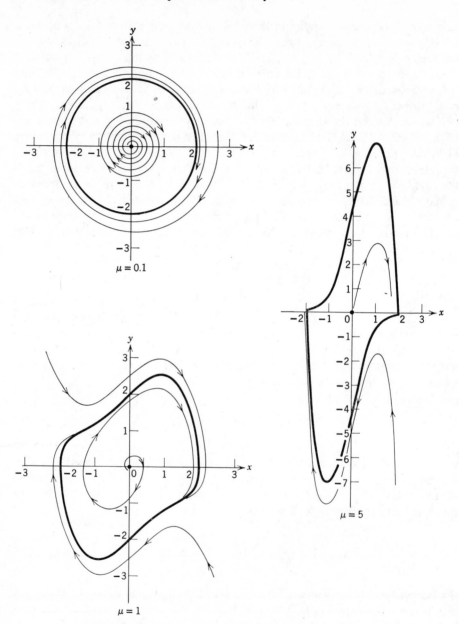

$\mu = 0.1$

$\mu = 1$

$\mu = 5$

FIGURE 9.32   Limit cycles of the van der Pol equation.

Liénard established the existence of periodic solutions of Eq. (15) for certain types of functions f and g, which included the case of the van der Pol equation. A more general result, proved by Levinson* and Smith in 1942, is the following.

Theorem 9.10. *Let f be even $[f(-u) = f(u)]$ and continuous for all u. Let g be odd $[g(-u) = -g(u)]$, with $g(u) > 0$ for all $u > 0$, and have a continuous first derivative for all u. Let*

$$F(u) = \int_0^u f(s)\, ds, \qquad G(u) = \int_0^u g(s)\, ds. \tag{18}$$

If

(a) *$G(u) \to \infty$ as $u \to \infty$;*

(b) *there exists a positive number u_0 such that $F(u) < 0$ for $0 < u < u_0$, $F(u) > 0$ for $u > u_0$, and $F(u)$ is monotonically increasing for $u > u_0$ with $F(u) \to \infty$ as $u \to \infty$;*

then

(a) *the generalized Liénard equation (15) has a unique periodic solution (unique in the sense that there are no other closed trajectories);*

(b) *the corresponding trajectory is a closed curve encircling the origin in the phase plane $x = u$, $y = du/dt$;*

(c) *all other trajectories in the xy phase plane, except that corresponding to the critical point $(0, 0)$, spiral toward the closed trajectory as $t \to \infty$.*

It is a straightforward calculation to show that the conditions of Theorem 9.10 are satisfied for the van der Pol equation (16). For example, the function F satisfies hypothesis (b):

$$F(u) = -\mu \int_0^u (1 - s^2)\, ds = -\mu\left(u - \frac{u^3}{3}\right)$$

is less than zero for $0 < u < \sqrt{3}$, is greater than zero for $u > \sqrt{3}$, and increases monotonically to infinity as u increases to infinity. Thus the van der Pol equation *has* a periodic solution.

In Figure 9.32 a few of the trajectories in the phase plane $x = u$, $y = du/dt$ are sketched for the van der Pol equation for the cases $\mu = 0.1$, 1, and 5, respectively. The limit cycle is shown by a heavy line. For μ small the limit cycle is nearly a circle of radius 2 (the trajectories of the linear equation obtained by setting $\mu = 0$ are circles), but for μ large the limit cycle is considerably different.

---

* Norman Levinson (1912–1975) was a prominent American mathematician who received the Bocher prize from the American Mathematical Society in 1954 for his contributions to the theory of linear, nonlinear, ordinary, and partial differential equations. He received the Chauvenet Prize from The Mathematical Association of America in 1971 for his expository paper, "A Motivated Account for an Elementary Proof of the Prime Number Theorem," *The American Mathematical Monthly*, Vol. 76, 1969, pp. 225–245.

PROBLEMS

1. For each of the following autonomous systems, expressed in polar coordinates, determine all periodic solutions, all limit cycles, and discuss their stability.

(a) $dr/dt = r^2(1 - r^2)$,　　　　$d\theta/dt = 1$

(b) $dr/dt = r(1 - r)^2$,　　　　$d\theta/dt = -1$

(c) $dr/dt = r(r - 1)(r - 3)$,　　$d\theta/dt = 1$

(d) $dr/dt = r(1 - r)(r - 2)$,　　$d\theta/dt = -1$

(e) $dr/dt = \sin \pi r$,　　　　　$d\theta/dt = 1$

(f) $dr/dt = r|r - 2|(r - 3)$,　　$d\theta/dt = -1$

2. Show that if $x = r \cos \theta$, $y = r \sin \theta$, then $y\, dx/dt - x\, dy/dt = -r^2\, d\theta/dt$.

3. (a) Show that the system

$$dx/dt = -y + xf(r)/r, \qquad dy/dt = x + yf(r)/r$$

has periodic solutions corresponding to the zeros of $f(r)$. What is the direction of motion on these closed curves?

(b) Determine all periodic solutions of the above system and discuss their stability if $f(r) = r(r - 2)^2(r^2 - 4r + 3)$.

4. Determine the periodic solutions, if any, of the system

$$\frac{dx}{dt} = y + \frac{x}{\sqrt{x^2 + y^2}}(x^2 + y^2 - 2), \qquad \frac{dy}{dt} = -x + \frac{y}{\sqrt{x^2 + y^2}}(x^2 + y^2 - 2).$$

5. Using Theorem 9.8 show that a sufficient condition for the linear autonomous system

$$dx/dt = ax + by, \qquad dy/dt = cx + dy$$

not to have a periodic solution, other than $x = 0$, $y = 0$, is $a + d \neq 0$.

6. Show that each of the following systems has no periodic solutions other than constant solutions.

(a) $dx/dt = x + y + x^3 - y^2$,　　$dy/dt = -x + 2y + x^2y + \frac{1}{3}y^3$

(b) $dx/dt = -2x - 3y - xy^2$,　　$dy/dt = y + x^3 - x^2y$

7. Prove Theorem 9.8 by completing the following argument. According to Green's theorem in the plane, if C is a closed curve which is sufficiently "smooth" and if F and G are continuous and have continuous first partial derivatives, then

$$\oint_C [F(x, y)\, dy - G(x, y)\, dx] = \iint_R [F_x(x, y) + G_y(x, y)]\, dA,$$

where R is the region enclosed by C. Assume that $x = \phi(t)$, $y = \psi(t)$ is a solution of the system (13) which is periodic with period T. Let C be the closed curve given by $x = \phi(t)$, $y = \psi(t)$ for $0 \leq t \leq T$. Show that for this curve the line integral is zero.

8. Show that each of the following differential equations or systems of equations has a periodic solution. A prime is used to denote d/dt.

(a) $u'' + (u^2 - 1) u' + u^3 = 0$ (b) $u'' + (3u^4 - 4u^2) u' + u = 0$

(c) $u'' - (2 - u^2) u' + u^3 + \sin u = 0$

(d) $u'' + \alpha(u^{2m} - k) u' + \beta u^{2n-1} = 0$

     (α, β, and k positive constants; n, m positive integers)

(e) $x' = y$, $y' = -x + y - x^5 - 2x^2 y$

REFERENCES

Several books on nonlinear differential equations present a number of applications in addition to theory. A few of these are:

LaSalle, J., and Lefschetz, S., *Stability by Liapunov's Direct Method with Applications*, Academic Press, New York, 1961.

Minorsky, N., *Nonlinear Oscillations*, van Nostrand, Princeton, 1962.

Stoker, J. J., *Nonlinear Vibrations*, Interscience, New York, 1950.

A few of the more theoretically oriented books are:

Birkhoff, G., and Rota, G. C., *Ordinary Differential Equations*, Second Edition, Blaisdell, New York, 1969.

Brauer, F. and Nohel, J., *Qualitative Theory of Ordinary Differential Equations*, W. A. Benjamin, New York, 1969.

Cesari, L., *Asymptotic Behavior and Stability Problems in Ordinary Differential Equations*, Springer-Verlag, Berlin, 1959.

Coddington, E. A. and Levinson, N., *Theory of Ordinary Differential Equations*, McGraw-Hill, New York, 1955.

Hirsch, W. H. and Smale, S., *Differential Equations, Dynamical Systems and Linear Algebra*, Academic Press, New York, 1974.

Hurewicz, W., *Lectures on Ordinary Differential Equations*, Wiley, New York, 1958.

Struble, R. A., *Nonlinear Differential Equations*, McGraw-Hill, New York, 1962.

Of these books, the ones by Struble and Brauer and Nohel are the more elementary.

A classic reference on ecology is:

Odum, E. P., *Fundamentals of Ecology*, Third Edition, W. B. Saunders, Philadelphia, 1971.

Two books dealing with ecology and population dynamics on a more mathematical level are:

May, R. M., *Stability and Complexity in Model Ecosystems*, Princeton University Press, Princeton, 1973.

Pielou, E. C., *An Introduction to Mathematical Ecology*, Wiley-Interscience, New York, 1969.

Answers to Problems

CHAPTER ONE

Page 8

1. (a) 2nd order, linear
   (c) 4th order, linear
   (e) 2nd order, nonlinear
   (b) 2nd order, nonlinear
   (d) 1st order, nonlinear
   (f) 3rd order, linear

3. (a) $r = -2$ (b) $r = \pm 1$ (c) $r = 2, -3$ (d) $r = 0, 1, 2$

4. (a) $r = -1, -2$ (b) $r = 1, 4$

CHAPTER TWO

Section 2.1, Page 17

1. $y = ce^{-3x} + (x/3) - (1/9) + e^{-2x}$ 2. $y = ce^{2x} + x^3 e^{2x}/3$

3. $y = ce^{-x} + 1 + \dfrac{x^2}{2} e^{-x}$ 4. $y = \dfrac{c}{x} + \dfrac{3}{4}\dfrac{\cos 2x}{x} + \dfrac{3}{2}\sin 2x$

5. $y = ce^x + 2xe^x$ 6. $y = \dfrac{c}{x^2} - \dfrac{\cos x}{x} + \dfrac{\sin x}{x^2}$

7. $y = 3e^x + 2(x - 1)e^{2x}$ 8. $y = \tfrac{1}{2}(x^2 - 1)e^{-2x}$

9. $y = e^{-x}\displaystyle\int_0^x \dfrac{e^t}{1 + t^2}\, dt$ 10. $y = \dfrac{\sin x}{x^2}$

11. $y = (x + 2)e^{2x}$ 12. $y = x^{-2}[(\pi^2/4) - 1 - x\cos x + \sin x]$

13. $x = ce^{-y} + e^y/2$, an implicit relation for $y = \phi(x)$

18. (a) See Problem 2.   (b) See Problem 4.

Section 2.2, Page 22

1. $y = (c + \sin x)/x - \cos x$ 2. $y = (c - \cos x)/x^3$

3. $y = (c - 2x\cos x + 2\sin x)\cos x$ 4. $y = [c + (x - 1)e^x]/x^2$

5. $y = (3x^2 - 4x + 6 + x^{-2})/12,$ $x > 0$

6. $y = x^{-1}(e^x + 1 - e),$ $x > 0$ 7. $y = (2x + 1 - \pi)/\sin x,$ $0 < x < \pi$

8. $y = x^{-2}(\sin x - x \cos x),$ $x > 0$

9. $y = cx^{-2} + x^{-1};$ $\lim_{x \to 0} |y| = \infty$ for all c

10. $y = cx + x^2;$ $\lim_{x \to 0} y = 0$ for all c

11. $y = cx + 2x^{3/2};$ $\lim_{x \to 0} y = 0$ for all c, but y'', y''', ..., all $\to \pm\infty$ as $x \to 0$

12. $y = \dfrac{c + \sin x}{x};$ $\lim_{x \to 0} y = \begin{cases} \pm\infty, & c \neq 0 \\ 1, & c = 0 \end{cases}$

13. (a) $y_1(x) = \dfrac{1}{\mu(x)};$ $y_2(x) = \dfrac{1}{\mu(x)} \displaystyle\int^x \mu(s)g(s)\, ds$

15. (a) $y = \frac{1}{2}(1 - e^{-2x})$ for $0 \leq x \leq 1;$ $y = \frac{1}{2}(e^2 - 1)e^{-2x}$ for $x > 1$

(b) $y = e^{-2x}$ for $0 \leq x \leq 1;$ $y = e^{-(x+1)}$ for $x > 1$

17. (a) $y = (\frac{2}{5}x^{-1} + cx^4)^{-\frac{1}{2}}$ (b) $y = \left(\dfrac{\sigma}{\epsilon} + ce^{-\epsilon x}\right)^{-1}$

(c) $y = \left(\dfrac{\sigma}{\epsilon} + ce^{-2\epsilon x}\right)^{-\frac{1}{2}}$ (d) $y = \left[2e^{-2\epsilon x}\displaystyle\int^x e^{2\epsilon s}f(s)\, ds + ce^{-2\epsilon x}\right]^{-\frac{1}{2}}$

Section 2.3, Page 30

1. (a) $2x + 5y > 0$ or $2x + 5y < 0$ (b) $x^2 + y^2 < 1$

(c) Everywhere   (d) $x + y > 0$ or $x + y < 0$

(e) $1 - x^2 + y^2 > 0,$ or $1 - x^2 + y^2 < 0,$ $x \neq 0, y \neq 0$

(f) Everywhere

2. $|x| < 1$ 3. $y = [2(x - \frac{7}{8})]^{-\frac{1}{2}};$ $x > \frac{7}{8}$

4. $x > -(1 - e^{-3/2})^{1/3}$

5. For given c, solutions are valid for $|x| < |c|/2$.

(0, 4): $y = (16 - 4x^2)^{1/2};$ (1, -1): $y = -(5 - 4x^2)^{1/2}$

6. (a) $y_1(x)$ is solution for $x \geq 2;$ $y_2(x)$ is solution for all x

(b) f_y is not continuous at $(2, -1)$

8. (a) $\lim\limits_{x\to\infty} \phi(x; y_0) = 0$ for all y_0

   (b) $\lim\limits_{x\to\infty} \phi(x; y_0) = -1$ if $y_0 < 0$; $\phi(x; 0) = 0$ for all x;

   $\lim\limits_{x\to\infty} \phi(x; y_0) = 1$ if $y_0 > 0$

   (c) $\lim\limits_{x\to\infty} \phi(x; y_0) = 1$ if $y_0 < 2$; $\phi(x; 2) = 2$ for all x;

   $\phi(x; y_0) \to \infty$ as x approaches some finite value if $y_0 > 2$

   (d) $\lim\limits_{x\to\infty} \phi(x; y_0) = \epsilon/\sigma$ if $y_0 > 0$; $\phi(x; 0) = 0$ for all x;

   $\phi(x; y_0) \to -\infty$ as x approaches some finite value if $y_0 < 0$

9. (c) $y = 3 - x$ 10. (c) $y = x^2/2 + x + 1$

11. (c) $y = e^{-x^2/2} \int_1^x e^{t^2/2}\, dt$ 12. (c) $y = (1 - x)\widehat{e^x} - 1$

Section 2.4, Page 35

1. $3y^2 - 2x^3 = c$; $y \neq 0$

2. $3y^2 - 2 \ln |1 + x^3| = c$; $1 + x^3 \neq 0$; $y \neq 0$

3. $y^{-1} + \cos x = c$ if $y \neq 0$;   also $y = 0$;   everywhere

4. $\arctan y - x - x^2/2 = c$;   everywhere

5. $2 \tan 2y - 2x - \sin 2x = c$ if $\cos 2y \neq 0$;   also $y = \pm(2n + 1)\pi/4$
   for any integer n;   everywhere

6. $y = \sin [\ln |x| + c]$ if $x \neq 0$ and $|y| < 1$;   also $y = \pm 1$; $x \neq 0$ and $|y| < 1$

7. $y^2 - x^2 + 2(e^y - e^{-x}) = c$; $y + e^y \neq 0$

8. $3y + y^3 - x^3 = c$;   everywhere

9. $y = \frac{1}{3} \arcsin (3 \cos^2 x)$, $0.95 < x < 2.19$ approximately

10. $y = [2(1 - x)e^x - 1]^{1/2}$, $-1.68 < x < 0.77$ approximately

11. $r = 2e^\theta$, $-\infty < \theta < \infty$

12. $y = -[2 \ln (1 + x^2) + 4]^{1/2}$, $-\infty < x < \infty$

13. $y = [3 - 2\sqrt{1 + x^2}]^{-1/2}$, $|x| < \frac{1}{2}\sqrt{5}$

14. $y = -\frac{1}{2} + \frac{1}{2}\sqrt{4x^2 - 15}$, $x > \frac{1}{2}\sqrt{15}$

15. $2y^3 - 3(\arcsin x)^2 = c$

16. $y = \frac{a}{c} x + \frac{bc - ad}{c^2} \ln |cx + d| + k$; $c \neq 0$, $cx + d \neq 0$

17. $x = \frac{c}{a} y + \frac{ad - bc}{a^2} \ln |ay + b| + k$; $a \neq 0$, $ay + b \neq 0$

18. $|y + 2x|^3 |y - 2x| = c$

19. (a) $y = y_0 e^{\epsilon t}$ (b) $y = \dfrac{\epsilon/\sigma}{1 + \left(\dfrac{\epsilon}{\sigma y_0} - 1\right)e^{-\epsilon t}}$; $y_0 \neq 0,\ \epsilon/\sigma$

Section 2.5, Page 40

1. $x^2 + 3x + y^2 - 2y = c$ 2. Not exact

3. $3x^3 + xy - x - 2y^2 = c$ 4. $x^2 y^2 + 2xy = c$

5. $ax^2 + 2bxy + cy^2 = k$ 6. Not exact

7. $e^x \sin y + 2y \cos x = c$; also $y = 0$ 8. Not exact

9. $e^{xy} \cos 2x + x^2 - 3y = c$ 10. $y \ln x + 3x^2 - 2y = c$

11. Not exact   12. $x^2 + y^2 = c$

13. (a) $b = 3$; $x^2 y^2 + 2x^3 y = c$ (b) $b = 1$; $e^{2xy} + x^2 = c$

14. $\displaystyle\int^y N(x, t)\, dt + \int^x \left[M(s, y) - \int^y N_s(s, t)\, dt \right] ds = c$

Section 2.6, Page 44

1. $x^2 + 2 \ln |y| - y^{-2} = c$; also $y = 0$ 2. $e^x \sin y + 2y \cos x = c$

3. $xy^2 - (y^2 - 2y + 2)e^y = c$ 4. $\mu(x) = e^{3x}$; $(3x^2 y + y^3)e^{3x} = c$

5. $\mu(x) = e^{-x}$; $y = ce^x + 1 + e^{2x}$ 6. $\mu(y) = y$; $xy + y \cos y - \sin y = c$

7. $\mu(y) = e^{2y}/y$; $xe^{2y} - \ln |y| = c$;   also $y = 0$

8. $\mu(y) = \sin y$; $e^x \sin y + y^2 = c$ 9. $\mu(x, y) = xy$; $x^3 y + 3x^2 + y^3 = c$

11. $\mu(t) = \exp\displaystyle\int^t R(s)\, ds$, where $t = xy$

Section 2.7, Page 48

1. $y = cx + x \ln |x|$ 2. $y = cx^2$

3. $\arctan (y/x) - \ln |x| = c$ 4. $x^2 + y^2 - cx^3 = 0$

5. $|y - x| = c |y + 3x|^5$ 6. $|y + x| |y + 4x|^2 = c$

7. $-\dfrac{2x}{x + y} = \ln c |x + y|$ 8. $\dfrac{x}{x + y} + \ln |x| = c$

9. (a) $|y - x| = c |y + x|^3$ 10. $|y + x + 4| |y + 4x + 13|^2 = c$

   (b) $|y - x + 3| = c |y + x + 1|^3$

11. $-2 \dfrac{x - 2}{x + y - 3} = \ln c |x + y - 3|$ 12. $x^2 y^2 + 2x^3 y = c$

15. (a) No (b) Yes (c) Yes (d) No

Section 2.8, Page 49

1. $y = \dfrac{c}{x^2} + \dfrac{x^3}{5}$ 2. $\arctan \dfrac{y}{x} - \ln \sqrt{x^2 + y^2} = c$

3. $x^2 + xy - 3y - y^3 = 0$

4. $x = ce^y + ye^y$

5. $x^2y + xy^2 + x = c$

6. $y = x^{-1}(1 - e^{1-x})$

7. $(x^2 + y^2 + 1)e^{-y^2} = c$

8. $y = (4 + \cos 2 - \cos x)/x^2$

9. $x^2y + x + y^2 = c$

10. $\dfrac{y^2}{x^3} + \dfrac{y}{x^2} = c$

11. $x^3/3 + xy + e^y = c$

12. $y = ce^{-x} + e^{-x} \ln(1 + e^x)$

13. $2(y/x)^{\frac{1}{2}} - \ln|x| = c$

14. $x^2 + 2xy + 2y^2 = 34$

15. $y = \dfrac{c}{\cosh^2(x/2)}$

16. $\dfrac{2}{\sqrt{3}} \arctan \dfrac{2y - x}{\sqrt{3}\,x} - \ln|x| = c$

17. $y = ce^{3x} - e^{2x}$

18. $y = cx^{-2} - x$

19. $3y - 2xy^3 - 10x = 0$

20. $e^x + e^{-y} = c$

21. $e^{-y/x} + \ln|x| = c$

22. $y^3 + 3y - x^3 + 3x = 2$

23. $\dfrac{1}{y} = -x\displaystyle\int^x \dfrac{e^{2t}}{t^2}\,dt + cx$; also $y = 0$

24. $\sin^2 x \sin y = c$

25. $x^2/y + \arctan(y/x) = c$

26. $x^2 + 2x^2y - y^2 = c$

27. $\sin x \cos 2y - \frac{1}{2}\sin^2 x = c$

28. $2xy + xy^3 - x^3 = c$

29. $2\arcsin(y/x) - \ln|x| = c$

30. $xy^2 - \ln|y| = 0$

31. $x + \ln|x| + x^{-1} + y - 2\ln|y| = c$;   also $y = 0$

32. $x^3y^2 + xy^3 = -4$

33. (b) $x = \int G'(p)/p\,dp$

34. $x + e^{-y} = c$

36. (a) $y = x + (c - x)^{-1}$

(b) $y = x^{-1} + 2x(c - x^2)^{-1}$

(c) $y = \sin x + (c \cos x - \frac{1}{2}\sin x)^{-1}$

Section 2.9, Page 57

1. (a) $T = (\ln 2)/r$ years   (b) $T = 13.86$ years   (c) $r = 6.93$ percent

2. (a) $S(t) = k(e^{rt} - 1)/r$ (b) $k \cong \$7826$

3. (a) \$51,548   (b) \$61,548   (c) \$71,548   (d) \$81,548

4. (a) $S(t) = \dfrac{k}{r} + \left(S_0 - \dfrac{k}{r}\right)e^{rt}$ (b) $k_0 = rS_0$

(c) $T = \dfrac{1}{r}\ln\left(1 - \dfrac{k_0}{k}\right)^{-1}$ years   (d) $T \cong 21.97$ years

(e) $k = rS_0\,e^{rT}/(e^{rT} - 1)$ (f) $S_0 \cong \$63,212$

5. In six months: four million dollars. After only one year, his wealth will be infinite.

6. (a) 13.20 years   (b) 29.6 milligrams

7. $\dfrac{-11.7 \ln 2}{\ln (2/3)} \cong 20.0$ days

8. $1620 \dfrac{\ln (4/3)}{\ln 2} \cong 672.4$ years

9. (a) $k = -0.000121$ (years)$^{-1}$ (b) $Q(t) = Q_0 e^{-0.000121t}$ (c) 13,300 years

10. $p = (6.0 \times 10^8)e^{0.005135t}$;   A.D. 2376

11. (a) $p = p_0 e^{\epsilon t}$; $\lim\limits_{t \to \infty} p = \infty$

(b) $p = \dfrac{p_0 e^{\epsilon t}}{1 + \dfrac{\sigma}{\epsilon} p_0(e^{\epsilon t} - 1)}$; $\lim\limits_{t \to \infty} p = \dfrac{\epsilon}{\sigma}$

12. (a) $u = u_0 + (u_1 - u_0)e^{kt}$ (b) $t = \dfrac{\ln (13/8)}{\ln (13/12)}$ min $\cong 6.07$ min

13. $r = 3 - t$ mm, $0 < t < 3$ 14. $t = 100 \ln 100$ min $\cong 460.5$ min

15. $Q = 50e^{-0.2}(1 - e^{-0.2})$ lb $\cong 7.42$ lb

16. $Q = 200 + t - \dfrac{100(200)^2}{(200 + t)^2}$ lb, $t < 300$; $c = \frac{121}{125}$ lb/gal; $\lim\limits_{t \to \infty} c = 1$ lb/gal

17. (a) $x = pq[e^{\alpha(q-p)t} - 1]/[qe^{\alpha(q-p)t} - p]$ (b) $x = p^2\alpha t/(p\alpha t + 1)$

18. $x = px_0/[(p-x_0)e^{-p\alpha t} + x_0]$

19. (a) $x = x_0 e^{-\beta t}$ (b) $y = y_0 \exp[-\alpha x_0(1 - e^{-\beta t})/\beta]$ (c) $y_0 \exp(-\alpha x_0/\beta)$

20. (a) $x = 0.04(1 - e^{-t/12000})$ (b) $\tau \cong 36$ minutes

21. (a) $c(t) = k + (P/r) + [c_0 - k - (P/r)]e^{-rt/V}$; $\lim\limits_{t \to \infty} c(t) = k + (P/r)$

(b) $T = (V \ln 2)/r$; $T = (V \ln 10)/r$

(c) Superior, $T = 430$ years;   Michigan, $T = 71.4$ years;
Erie, $T = 6.05$ years;   Ontario, $T = 17.6$ years

22. (a) $dy/dx = 2y/x$ (b) $dy/dx = -x/2y$ (c) $x^2 + 2y^2 = c$

23. (a) $y^2 - x^2 = c$ (b) $x^2 + (y - c)^2 = c^2$

(c) $x - y = c(x + y)^3$ (d) $x^2 - 2cy = c^2$; $c > 0$

24. (a) $y - 3x = c$ (b) $\ln \sqrt{x^2 + y^2} + \arctan (y/x) = c$, or $r = ke^{-\theta}$

25. $y - b = c(x - a)$ 26. $(x - 2)^2 + y^2 = 9$

27. $x = cy^2$

Section 2.10, Page 67

1. (a) $x_m = v_0^2/2g$ (b) $t_m = v_0/g$ (c) $t = 2v_0/g$

2. $v = \dfrac{mg}{k} + \left(v_0 - \dfrac{mg}{k}\right)e^{-kt/m}$; $v_l = \dfrac{mg}{k}$

3. $t = (m \ln 10)/k$

4. (a) $v = 5(1 - e^{-0.2t})$ ft/sec   (b) $v_l = 5$ ft/sec

5. $2\dfrac{m}{k}(\sqrt{v_0} - \sqrt{v}) + 2\dfrac{m^2g}{k^2}\ln\left|\dfrac{mg - k\sqrt{v_0}}{mg - k\sqrt{v}}\right| = t;$ $v_l = \left(\dfrac{mg}{k}\right)^2$

6. $v = \sqrt{\dfrac{mg}{k}\dfrac{e^{2\sqrt{kg/m}\,t} - 1}{e^{2\sqrt{kg/m}\,t} + 1}};$ $v_l = \sqrt{\dfrac{mg}{k}}$ 7. $v_l = \left(\dfrac{mg}{k}\right)^{1/r}$

8. (a) $x_m = -\dfrac{m^2g}{k^2}\ln\left(1 + \dfrac{kv_0}{mg}\right) + \dfrac{mv_0}{k}$ (b) $t_m = \dfrac{m}{k}\ln\left(1 + \dfrac{kv_0}{mg}\right)$

9. (a) $v_l = 2\,a^2g(\rho - \rho')/9\mu$ (b) $e = 4\,\pi a^3g(\rho - \rho')/3E$

10. (a) $w \cong 64$ lb   (b) $w_m \cong 16.9$ lb

11. $\dfrac{x_m}{R} = -1 + \left(1 - \dfrac{v_0^2}{v_e^2}\right)^{-1} = \dfrac{v_0^2}{v_e^2}\left(1 + \dfrac{v_0^2}{v_e^2} + \dfrac{v_0^4}{v_e^4} + \cdots\right)$

12. $v_e = \left(\dfrac{2gR}{1 + \xi}\right)^{1/2};$ altitude $\cong 1536$ miles   14. $v = 5011$ ft/sec

15. $v = \dfrac{s\beta}{k}(1 - w^{k/\beta}) + \dfrac{gm_0w}{\beta - k}(1 - w^{(k-\beta)/\beta}),$ where $w = \dfrac{m_0 - \beta t}{m_0}$

CHAPTER THREE

Section 3.1, Page 86

1. (a) $y = c_1x^{-1} + c_2 + \ln x$ (b) $y = c_1 \ln x + c_2 + x$

   (c) $c' = k^2 > 0,\ y = \dfrac{1}{k}\ln\left|\dfrac{k - x}{k + x}\right| + c_2;\ c_1 = -k^2 < 0,\ y = \dfrac{2}{k}\tan^{-1}\dfrac{x}{k} + c_2;$

   $c_1 = 0,\ y = -2x^{-1} + c_2,\ y = c$

   (d) $y = \pm\tfrac{2}{3}(x - 2c_1)\sqrt{x + c_1} + c_2;$ *Hint:* $\mu(v) = v^{-3}$ is an integrating factor

2. (a) $y^2 = c_1x + c_2$ (b) $y = c_1 \sin(x + c_2) = k_1 \sin x + k_2 \cos x$

   (c) $\tfrac{1}{3}y^3 - 2c_1y + c_2 = 2x$ (d) $x + c_2 = \pm\tfrac{2}{3}(y - 2c_1)(y + c_1)^{1/2}$

3. (a) $y = \tfrac{4}{3}(x + 1)^{3/2} - \tfrac{1}{3}$ (b) $y = 2(1 - x)^{-2}$

   (c) $y = c_1 \arctan x + c_2 + 3\ln x - \tfrac{3}{2}\ln(x^2 + 1)$ (d) $y = \tfrac{1}{2}x^2 + \tfrac{3}{2}$

4. (a) Any interval not containing $x = 0$ (b) Any interval

   (c) Any interval not containing $x = 0$ or $x = 1$

   (d) Any interval not containing $x = 0$ (e) Any interval

   (f) Any interval not containing $x = (2n + 1)\pi/2,\ n = 0, \pm1, \pm2, \ldots$

5. $\phi''(0) = -p(0)a_1 - q(0)a_0,$

   $\phi'''(0) = [p^2(0) - p'(0) - q(0)]a_1 + [p(0)q(0) - q'(0)]a_0,$ yes

6. (a) $y'' - y = 0$ (b) $y'' + y = 0$

   (c) $(1 - x \cot x)y'' - xy' + y = 0$ (d) $y'' - 2y' + y = 0$

   (e) $x^2y'' - 2xy' + 2y = 0$ (f) $y'' - y = 0$

Section 3.2, Page 93

2. $y = \tfrac{2}{3}e^x + \tfrac{1}{3}e^{-2x},$ $y = 0$

8. (a) $b + cx$ (b) $(c - a)\sin x + b \cos x$

   (c) $(ar^2 + br + c)e^{rx}$ (d) $ar(r - 1)x^{r-2} + brx^{r-1} + cx^r$

9. (a) $(2a + 2b + c)x^2$ (b) $(ar^2x^2 + brx + c)e^{rx}$

(c) $[ar^2 + (b - a)r + c]x^r$ (d) $-a + b + c \ln x$

11. (a) $(n - m)e^{(m+n)x}$ (b) -1 (c) x^2e^x (d) $-e^{2x}$ (e) 0

12. (a) $-\infty < x < \infty$ (b) $-\infty < x < \infty$

(c) $-\infty < x < \infty$ (d) $x > 0$ or $x < 0$

14. (a) $y = (e^x - e^{-x})/2$ (b) $y = \sinh x$

(c) $y = 4e^{-2x} - 3e^{-3x}$ (d) $y = 1 - e^{-(x-1)}$

17. If x_1 is an inflection point and $y = \phi(x)$ is a solution, then from the differential equation $p(x_1)\phi'(x_1) + q(x_1)\phi(x_1) = 0$.

18. (a) Yes, $y = c_1 e^{-x^2/2} \int^x e^{t^2/2}\, dt + c_2 e^{-x^2/2}$ (b) No

(c) Yes, $y = \dfrac{1}{\mu(x)}\left[c_1 \int^x \dfrac{\mu(t)}{t}\, dt + c_2 \right]$, $\mu(x) = \exp\left[-\int^x \left(\dfrac{1}{t} + \dfrac{\cos t}{t} \right) dt \right]$

(d) Yes, $y = c_1 x^{-1} + c_2 x$

19. (a) $x^2 \mu'' + 3x\mu' + (1 + x^2 - \nu^2)\mu = 0$

(b) $(1 - x^2)\mu'' - 2x\mu' + \alpha(\alpha + 1)\mu = 0$ (c) $\mu'' - x\mu = 0$

(d) $x^2 \mu'' + 4x\mu' + \left(-\dfrac{x^2}{4} + kx + \dfrac{9}{4} - m^2 \right)\mu = 0$

21. The Legendre equation and the Airy equation are self-adjoint.

22. $v(x) = \exp\left[-\dfrac{1}{2}\int^x p(t)\, dt \right]$, $f(x) = q(x) - \tfrac{1}{4}p^2(x) - \tfrac{1}{2}p'(x)$

Section 3.3, Page 99

5. $W(c_1 y_1, c_2 y_2) = c_1 c_2 W(y_1, y_2) \neq 0$

9. If y_1 and y_2 are linearly independent, then $W(y_1, y_2)$ cannot change sign. Show that $W(y_1, y_2)$ will change sign if y_1 has none or more than one zero between consecutive zeros of y_2.

Section 3.4, Page 102

1. $y_2(x) = e^{-2x}$ 2. $y_2(x) = xe^{-x}$

3. $y_2(x) = x^{-1}$, $x > 0$ or $x < 0$ 4. $y_2(x) = x^{-2}$, $x > 0$ or $x < 0$

5. $y_2(x) = x^{-1} \ln x$ 6. $y_2(x) = xe^x$

7. $y_2(x) = \sin x$ 8. $y_2(x) = x^{-1/2} \cos x$

10. $y_2(x) = \dfrac{3x}{4} + \dfrac{3x^2 - 1}{8} \ln \dfrac{1 - x}{1 + x}$

11. $f(\xi) = c_1 e^{-\delta \xi^2/2} \int_0^\xi e^{\delta s^2/2}\, ds + c_2 e^{-\delta \xi^2/2}$ 13. $y_2(x) = x^{-2}$

Section 3.5, Page 107

1. $y = c_1 e^x + c_2 e^{-3x}$ 2. $y = c_1 e^{-x/2} + c_2 x e^{-x/2}$

3. $y = c_1 e^{x/2} + c_2 e^{-x/3}$ 4. $y = c_1 e^{x/2} + c_2 e^x$

5. $y = c_1 e^x + c_2 e^{-x}$

6. $y = c_1 e^x + c_2 x e^x$

7. $y = c_1 + c_2 e^{-5x}$

8. $y = c_1 e^{(9+3\sqrt{5})x/2} + c_2 e^{(9-3\sqrt{5})x/2}$

9. $y = c_1 e^{(1+\sqrt{3})x} + c_2 e^{(1-\sqrt{3})x}$

10. $y = c_1 e^{-x} + c_2 x e^{-x}$

11. $y = e^x$

12. $y = 2x e^{3x}$

13. $y = \frac{1}{10} e^{-9(x-1)} + \frac{9}{10} e^{x-1}$

Section 3.5.1, Page 110

1. $y = c_1 e^x \cos x + c_2 e^x \sin x$

2. $y = c_1 e^x \cos \sqrt{5}\, x + c_2 e^x \sin \sqrt{5}\, x$

3. $y = c_1 e^{2x} + c_2 e^{-4x}$

4. $y = c_1 e^{-x} \cos x + c_2 e^{-x} \sin x$

5. $y = c_1 e^{x/3} + c_2 x e^{x/3}$

6. $y = e^{-3x}(c_1 \cos 2x + c_2 \sin 2x)$

7. $y = \frac{1}{2} \sin 2x$

8. $y = e^{-2x} \cos x + 2 e^{-2x} \sin x$

14. $\pm 2^{1/4} e^{i\pi/8}$; $\pm 2^{1/4} e^{-i\pi/8}$; $\pm e^{i\pi/4} = \pm (1+i)/\sqrt{2}$

15. (a) $y = c_1 e^{ix} + c_2 e^{-2ix} = (c_1 \cos x + c_2 \cos 2x) + i(c_1 \sin x - c_2 \sin 2x)$, no

   (b) $y = c_1 e^{r_1 x} + c_2 e^{r_2 x}$ where $r_1 = -1 + 2^{1/4} e^{-i\pi/8}$ and $r_2 = -1 - 2^{1/4} e^{-i\pi/8}$, no

17. (a) Yes, $y = c_1 \cos z + c_2 \sin z$, $z = \int^x e^{-t^2/2}\, dt$ (b) No

   (c) Yes, $y = c_1 e^{-x^2/4} \cos (\sqrt{3}\, x^2/4) + c_2 e^{-x^2/4} \sin (\sqrt{3}\, x^2/4)$

18. $y = c_1 \cos \ln x + c_2 \sin \ln x$

Section 3.6.1, Page 120

1. $y = e^x - \frac{1}{2} e^{-2x} - x - \frac{1}{2}$

2. $y = c_1 e^{3x} + c_2 e^{-x} - \frac{1}{2} e^{2x}$

3. $y = \frac{7}{10} \sin 2x - \frac{19}{40} \cos 2x + \frac{1}{4} x^2 - \frac{1}{8} - \frac{3}{5} e^x$

4. $y = c_1 + c_2 e^{-2x} + \frac{3}{2} x - \frac{1}{2} \sin 2x - \frac{1}{2} \cos 2x$

5. $y = c_1 \cos 3x + c_2 \sin 3x + \frac{1}{18}(x^2 - \frac{2}{3} x + \frac{1}{9}) e^{3x} + \frac{2}{3}$

6. $y = 4x e^x - 3 e^x + \frac{1}{6} x^3 e^x + 4$

7. $y = c_1 e^{-x} + c_2 e^{-x/2} + (x^2 - 6x + 14) - \frac{3}{10} \sin x - \frac{9}{10} \cos x$

8. $y = c_1 \cos x + c_2 \sin x - \frac{1}{3} x \cos 2x - \frac{5}{9} \sin 2x$

9. $y = c_1 e^{-x} + c_2 x e^{-x} + \frac{1}{25} e^x (3 \cos x + 4 \sin x)$

10. $u = c_1 \cos \omega_0 t + c_2 \sin \omega_0 t + \dfrac{1}{\omega_0^2 - \omega^2} \cos \omega t$

11. $u = c_1 \cos \omega_0 t + c_2 \sin \omega_0 t + \dfrac{1}{2\omega_0} t \sin \omega_0 t$

12. $u = e^{-\mu t/2}(c_1 \cos \sqrt{4\omega_0^2 - \mu^2}\, t + c_2 \sin \sqrt{4\omega_0^2 - \mu^2}\, t)$

$$+ \dfrac{(\omega_0^2 - \omega^2) \cos \omega t + \mu\omega \sin \omega t}{(\omega_0^2 - \omega^2)^2 + \mu^2 \omega^2}$$

13. $y = c_1 e^{-x/2} \cos (\sqrt{3}\, x/2) + c_2 e^{-x/2} \sin (\sqrt{3}\, x/2) + \frac{1}{2} - \frac{1}{13} \sin 2x + \frac{3}{26} \cos 2x$

14. $y = c_1 e^{-x/2} \cos(\sqrt{15}\, x/2) + c_2 e^{-x/2} \sin(\sqrt{15}\, x/2) + \frac{1}{6} e^x - \frac{1}{4} e^{-x}$

15. $y = c_1 e^{-x} + c_2 e^{2x} + \frac{1}{6} x e^{2x} + \frac{1}{8} e^{-2x}$

16. $y_p(x) = x(A_0 x^4 + A_1 x^3 + A_2 x^2 + A_3 x + A_4) + x(B_0 x^2 + B_1 x + B_2) e^{-3x}$
$+ D \sin 3x + E \cos 3x$

17. $y_p(x) = A_0 x + A_1 + x(B_0 x + B_1) \sin x + x(D_0 x + D_1) \cos x$

18. $y_p(x) = e^x(A \cos 2x + B \sin 2x) + (D_0 x + D_1) e^{2x} \sin x + (E_0 x + E_1) e^{2x} \cos x$

19. $y_p(x) = A e^{-x} + x(B_0 x^2 + B_1 x + B_2) e^{-x} \cos x + x(D_0 x^2 + D_1 x + D_2) e^{-x} \sin x$

20. $y_p(x) = A_0 x^2 + A_1 x + A_2 + x^2(B_0 x + B_1) e^{2x} + (D_0 x + D_1) \sin 2x$
$+ (E_0 x + E_1) \cos 2x$

21. $y_p(x) = x(A_0 x^2 + A_1 x + A_2) \sin 2x + x(B_0 x^2 + B_1 x + B_2) \cos 2x$

22. $y_p(x) = (A_0 x^2 + A_1 x + A_2) e^x \sin 2x + (B_0 x^2 + B_1 x + B_2) e^x \cos 2x$
$+ e^{-x}(D \cos x + E \sin x) + F e^x$

23. $y = c_1 \cos \lambda x + c_2 \sin \lambda x + \sum\limits_{m=1}^{N} \dfrac{a_m}{\lambda^2 - m^2 \pi^2} \sin m \pi x$ 24. No

25. $y = \begin{cases} t, & 0 \le t \le \pi \\ -\left(1 + \dfrac{\pi}{2}\right) \sin t - \dfrac{\pi}{2} \cos t + \dfrac{\pi}{2} e^{(\pi - t)}, & t > \pi \end{cases}$

Section 3.6.2, Page 125

1. $y_p(x) = e^x$
2. $y_p(x) = -\frac{2}{3} x e^{-x}$
3. $y_p(x) = \frac{3}{2} x^2 e^{-x}$
4. $y_p(x) = -(\cos x) \ln(\tan x + \sec x)$
5. $y_p(x) = (\sin 3x) \ln(\tan 3x + \sec 3x) - 1$
6. $y_p(x) = -e^{-2x} \ln x$
7. $y_p(x) = \frac{3}{4}(\sin 2x) \ln \sin 2x - \frac{3}{2} x \cos 2x$
8. $y = c_1 x + c_2 x e^x - 2x^2$
9. $y = c_1(1 + x) + c_2 e^x + \frac{1}{2} e^{2x}(x - 1)$
10. $y = c_1 x^{-\frac{1}{2}} \cos x + c_2 x^{-\frac{1}{2}} \sin x - \frac{3}{2} x^{\frac{1}{2}} \cos x$
11. $y = c_1 e^x + c_2 x - \frac{1}{2} e^{-x}(2x - 1)$

12. $y_p(x) = \displaystyle\int^x [e^{3(x-t)} - e^{2(x-t)}] g(t)\, dt$
13. $y_p(x) = x^{-\frac{1}{2}} \displaystyle\int^x \dfrac{g(t) \sin(x - t)}{t^{3/2}}\, dt$

16. (a) $y = c_1 x + c_2 x^2 + 4x^2 \ln x$
(b) $y = c_1 x^{-1} + c_2 x^{-5} + \frac{1}{12} x$

Section 3.7.1, Page 133

1. $u = \frac{1}{4} \cos 8t$ ft, $\frac{1}{4}$ ft, 8 rad/sec, $\pi/4$ sec; t in sec

2. $u = \frac{5}{7} \sin 14t$ cm; t in sec

3. $u = \dfrac{1}{4\sqrt{2}} \sin 8\sqrt{2}\, t - \dfrac{1}{12} \cos 8\sqrt{2}\, t$ ft, $\dfrac{\sqrt{22}}{24}$ ft, $8\sqrt{2}$ rad/sec, $\dfrac{\pi\sqrt{2}}{8}$ sec; t in sec

5. Approximately 1 percent
6. $c = \sqrt{10}/2$ lb sec/ft

7. $r = \sqrt{A^2 + B^2}$, $r \cos \theta = B$, $r \sin \theta = -A$; $R = r$;

   $\delta = \theta + (4n + 1)\pi/2$, $n = 0, 1, 2, \ldots$

8. $\omega_0 = \sqrt{k/m}$;   (a) $u = u_0 \cos \omega_0 t$ (b) $u = \dfrac{\dot{u}_0}{\omega_0} \cos\left(\omega_0 t - \dfrac{\pi}{2}\right)$

   (c) $u = \sqrt{u_0^2 + \left(\dfrac{\dot{u}_0}{\omega_0}\right)^2} \cos(\omega_0 t - \delta)$, $\delta = \arctan \dfrac{\dot{u}_0/\omega_0}{u_0}$

9. $2\pi\sqrt{l/g}$ 10. $\rho l \ddot{x} + \hat{\rho} g x = 0$, $2\pi\sqrt{\rho l / \hat{\rho} g}$

11. $u = e^{-10t}\left(2 \cos 4\sqrt{6}\, t + \dfrac{5}{\sqrt{6}} \sin 4\sqrt{6}\, t\right)$ cm; t in sec

12. $u = \dfrac{1}{8\sqrt{31}} e^{-2t} \sin 2\sqrt{31}\, t$ ft; t in sec

13. $c > 8$ lb sec/ft, overdamped; $c = 8$ lb sec/ft, critically damped;
    $c < 8$ lb sec/ft, underdamped

14. $\mu = (4km - c^2)^{1/2}/2m$

   (a) $u = u_0 \sqrt{1 + \left(\dfrac{c}{2m\mu}\right)^2}\, e^{-ct/2m} \cos(\mu t - \delta)$, $\delta = \arctan \dfrac{c}{2m\mu}$

   (b) $u = \dfrac{\dot{u}_0}{\mu} e^{-ct/2m} \cos\left(\mu t - \dfrac{\pi}{2}\right)$

   (c) $u = \sqrt{u_0^2 + \left(\dfrac{\dot{u}_0}{\mu} + \dfrac{cu_0}{2m\mu}\right)^2}\, e^{-ct/2m} \cos(\mu t - \delta)$, $\delta = \arctan \dfrac{\dot{u}_0/\mu + cu_0/2\mu m}{u_0}$

16. $\dot{u}_0 < -cu_0/2m$ 17. (a) $2\pi/\sqrt{31}$ (b) $c = 5$ lb sec/ft

Section 3.7.2, Page 138

1. $-2 \sin 8t \sin t$ 2. $2 \sin \frac{1}{2}t \cos \frac{13}{2}t$

3. $u = \frac{151}{1482} \cos 16t + \frac{16}{247} \cos 3t$ ft, t in sec; $\omega = 16$ rad/sec

4. $u = \frac{64}{45}(\cos 7t - \cos 8t) = \frac{128}{45} \sin \frac{1}{2}t \sin \frac{15}{2}t$ ft; t in sec

5. $\omega_0 = \sqrt{k/m}$

   (a) $u = \left[u_0 - \dfrac{F}{m(\omega_0^2 - \omega^2)}\right] \cos \omega_0 t + \dfrac{F}{m(\omega_0^2 - \omega^2)} \cos \omega t$

   (b) $u = \dfrac{\dot{u}_0}{\omega_0} \sin \omega_0 t + \dfrac{F_0}{m(\omega_0^2 - \omega^2)}(\cos \omega t - \cos \omega_0 t)$

   (c) $u = \left[u_0 - \dfrac{F}{m(\omega_0^2 - \omega^2)}\right] \cos \omega_0 t + \dfrac{\dot{u}_0}{\omega_0} \sin \omega_0 t + \dfrac{F}{m(\omega_0^2 - \omega^2)} \cos \omega t$

7. $u = \begin{cases} (F_0/m)(t - \sin t), & 0 \le t \le \pi \\ (F_0/m)[(2\pi - t) - 3\sin t], & \pi < t \le 2\pi \\ -(4F_0/m)\sin t, & 2\pi < t \end{cases}$

8. $u = \frac{8}{901}(30\cos 2t + \sin 2t)$ ft, t in sec; $m = 4$ slugs

9. $u = e^{-ct/2m}(c_1 \cos \mu t + c_2 \sin \mu t) + \frac{F_0}{\Delta}[m(\omega_0^2 - \omega^2)\sin \omega t - c\omega \cos \omega t]$

$\mu = (4km - c^2)^{1/2}/2m, \quad \omega_0^2 = k/m, \quad \Delta = m^2(\omega_0^2 - \omega^2)^2 + c^2\omega^2$

(a) $c_1 = u_0 + \frac{c\omega F_0}{\Delta}, \qquad c_2 = \frac{cu_0}{2m\mu} + \frac{F_0\omega}{\mu\Delta}\left[\frac{c^2}{2m} - m(\omega_0^2 - \omega^2)\right]$

(b) $c_1 = \frac{c\omega F_0}{\Delta}, \qquad c_2 = \frac{\dot{u}_0}{\mu} + \frac{F_0\omega}{\mu\Delta}\left[\frac{c^2}{2m} - m(\omega_0^2 - \omega^2)\right]$

(c) $c_1 = u_0 + \frac{c\omega F_0}{\Delta}, \qquad c_2 = \frac{\dot{u}_0}{\mu} + \frac{cu_0}{2m\mu} + \frac{F_0\omega}{\mu\Delta}\left[\frac{c^2}{2m} - m(\omega_0^2 - \omega^2)\right]$

Section 3.8, Page 142

4. (a) $Q = 10^{-6}(2e^{-500t} - e^{-1000t})$ coulombs; t in sec
   (b) $Q = 10^{-6}(1 - 500t)e^{-500t}$ coulombs; t in sec
   (c) $Q = 10^{-6}e^{-100t}(\cos 200t + \frac{1}{2}\sin 200t)$ coulombs; t in sec
5. $R = 10^3$ ohms
6. $Q = \psi(t) = 10^{-6}(e^{-4000t} - 4e^{-1000t} + 3)$ coulombs; t in sec
   $\psi(0.001) \cong 1.52 \times 10^{-6}$; $\psi(0.01) \cong 3 \times 10^{-6}$; $Q \to 3 \times 10^{-6}$ as $t \to \infty$
7. $I \cong 0.0273 \cos(120\pi t + \delta)$ amperes;
   $\delta = \arctan(-0.9)$, δ in the fourth quadrant; t in sec
8. $\omega = 1/\sqrt{LC}$

9. (d) $I = \dfrac{E_0}{\sqrt{R^2 + (\omega L - 1/\omega C)^2}}\sin(\omega t - \delta), \qquad \delta = \arctan\dfrac{\omega L - 1/\omega C}{R}$

CHAPTER FOUR

Section 4.1, Page 151

1. (a) $\rho = 1$ (b) $\rho = 2$ (c) $\rho = \infty$
   (d) $\rho = \frac{1}{2}$ (e) $\rho = \frac{1}{2}$ (f) $\rho = 1$

2. (a) $\displaystyle\sum_{n=0}^{\infty}\frac{(-1)^n x^{2n+1}}{(2n+1)!}$, $\rho = \infty$ (b) $\displaystyle\sum_{n=0}^{\infty}\frac{x^n}{n!}$, $\rho = \infty$

   (c) $1 + (x - 1)$, $\rho = \infty$ (d) $1 - 2(x + 1) + (x + 1)^2$, $\rho = \infty$

   (e) $\displaystyle\sum_{n=1}^{\infty}(-1)^{n+1}\frac{(x-1)^n}{n}$, $\rho = 1$ (f) $\displaystyle\sum_{n=0}^{\infty}(-1)^n x^n$, $\rho = 1$

   (g) $\displaystyle\sum_{n=0}^{\infty}x^n$, $\rho = 1$ (h) $\displaystyle\sum_{n=0}^{\infty}(-1)^{n+1}(x - 2)^n$, $\rho = 1$

3. $y' = 1 + 2^2x + 3^2x^2 + 4^2x^3 + \cdots + (n + 1)^2x^n + \cdots$
   $y'' = 2^2 + 3^2 \cdot 2x + 4^2 \cdot 3x^2 + 5^2 \cdot 4x^3 + \cdots + (n + 2)^2(n + 1)x^n + \cdots$

4. $y' = a_1 + 2a_2x + 3a_3x^2 + 4a_4x^3 + \cdots + (n + 1)a_{n+1}x^n + \cdots$

$$= \sum_{n=1}^{\infty} na_nx^{n-1} = \sum_{n=0}^{\infty}(n + 1)a_{n+1}x^n$$

$y'' = 2a_2 + 6a_3x + 12a_4x^2 + 20a_5x^3 + \cdots + (n + 2)(n + 1)a_{n+2}x^n + \cdots$

$$= \sum_{n=2}^{\infty} n(n - 1)a_nx^{n-2} = \sum_{n=0}^{\infty}(n + 2)(n + 1)a_{n+2}x^n$$

6. $a_n = (-2)^na_0/n!, \qquad n = 1, 2, \ldots; \qquad a_0e^{-2x}$

Section 4.2, Page 160

1. (a) $a_{n+2} = \dfrac{a_n}{(n + 2)(n + 1)}$

$$y_1(x) = 1 + \frac{x^2}{2!} + \frac{x^4}{4!} + \frac{x^6}{6!} + \cdots = \sum_{n=0}^{\infty} \frac{x^{2n}}{(2n)!} = \cosh x$$

$$y_2(x) = x + \frac{x^3}{3!} + \frac{x^5}{5!} + \frac{x^7}{7!} + \cdots = \sum_{n=0}^{\infty} \frac{x^{2n+1}}{(2n + 1)!} = \sinh x$$

(b) $a_{n+2} = \dfrac{a_n}{n + 2}$

$$y_1(x) = 1 + \frac{x^2}{2} + \frac{x^4}{2 \cdot 4} + \frac{x^6}{2 \cdot 4 \cdot 6} + \cdots = \sum_{n=0}^{\infty} \frac{x^{2n}}{2^n n!}$$

$$y_2(x) = x + \frac{x^3}{3} + \frac{x^5}{3 \cdot 5} + \frac{x^7}{3 \cdot 5 \cdot 7} + \cdots = \sum_{n=0}^{\infty} \frac{2^n n! \, x^{2n+1}}{(2n + 1)!}$$

(c) $(n + 2)a_{n+2} - a_{n+1} - a_n = 0$
   $y_1(x) = 1 + \frac{1}{2}(x - 1)^2 + \frac{1}{6}(x - 1)^3 + \frac{1}{6}(x - 1)^4 + \cdots$
   $y_2(x) = (x - 1) + \frac{1}{2}(x - 1)^2 + \frac{1}{2}(x - 1)^3 + \frac{1}{4}(x - 1)^4 + \cdots$

(d) $a_{n+4} = -\dfrac{k^2a_n}{(n + 4)(n + 3)}; \qquad a_2 = a_3 = 0$

$$y_1(x) = 1 - \frac{k^2x^4}{3 \cdot 4} + \frac{k^4x^8}{3 \cdot 4 \cdot 7 \cdot 8} - \frac{k^6x^{12}}{3 \cdot 4 \cdot 7 \cdot 8 \cdot 11 \cdot 12} + \cdots$$

$$= 1 + \sum_{m=0}^{\infty} \frac{(-1)^{m+1}(k^2x^4)^{m+1}}{3 \cdot 4 \cdot 7 \cdot 8 \cdots (4m + 3)(4m + 4)}$$

$$y_2(x) = x - \frac{k^2x^5}{4 \cdot 5} + \frac{k^4x^9}{4 \cdot 5 \cdot 8 \cdot 9} - \frac{k^6x^{13}}{4 \cdot 5 \cdot 8 \cdot 9 \cdot 12 \cdot 13} + \cdots$$

$$= x\left[1 + \sum_{m=0}^{\infty} \frac{(-1)^{m+1}(k^2x^4)^{m+1}}{4 \cdot 5 \cdot 8 \cdot 9 \cdots (4m + 4)(4m + 5)}\right]$$

Hint: Let $n = 4m$ in the recurrence relation, $m = 1, 2, 3, \ldots$.

(e) $(n + 2)(n + 1)a_{n+2} - n(n + 1)a_{n+1} + a_n = 0,$ $n \geq 1;$ $a_2 = -\frac{1}{2}a_0$
$y_1(x) = 1 - \frac{1}{2}x^2 - \frac{1}{6}x^3 - \frac{1}{24}x^4 + \cdots,$ $y_2(x) = x - \frac{1}{6}x^3 - \frac{1}{12}x^4 + \cdots$

(f) $a_{n+2} = -\dfrac{(n^2 - 2n + 4)a_n}{2(n + 1)(n + 2)},$ $n \geq 2;$ $a_2 = -a_0,$ $a_3 = -\frac{1}{4}a_1$

$y_1(x) = 1 - x^2 + \frac{1}{6}x^4 - \frac{1}{30}x^6 + \cdots$

$y_2(x) = x - \frac{1}{4}x^3 + \frac{7}{160}x^5 - \frac{19}{1920}x^7 + \cdots$

2. $y = \displaystyle\sum_{n=0}^{\infty} \frac{2^n n! \, x^{2n+1}}{(2n + 1)!},$ $y = 2\displaystyle\sum_{n=0}^{\infty} \frac{x^{2n}}{2^n n!} + \displaystyle\sum_{n=0}^{\infty} \frac{2^n n! \, x^{2n+1}}{(2n + 1)!}$

3. $y_1(x) = 1 - \frac{1}{3}(x - 1)^3 - \frac{1}{12}(x - 1)^4 + \frac{1}{18}(x - 1)^6 + \cdots$

$y_2(x) = (x - 1) - \frac{1}{4}(x - 1)^4 - \frac{1}{20}(x - 1)^5 + \frac{1}{28}(x - 1)^7 + \cdots$

Section 4.2.1, Page 166

1. (a) $\phi''(0) = -1,$ $\phi'''(0) = 0,$ $\phi^{iv}(0) = 3$
   (b) $\phi''(0) = 0,$ $\phi'''(0) = -2,$ $\phi^{iv}(0) = 0$
   (c) $\phi''(1) = 0,$ $\phi'''(1) = -6,$ $\phi^{iv}(1) = 42$
   (d) $\phi''(0) = 0,$ $\phi'''(0) = -a_0,$ $\phi^{iv}(0) = -4a_1$

2. (a) $\rho = \infty, \rho = \infty$ (b) $\rho = 1, \rho = 3, \rho = 1$
   (c) $\rho = 1, \rho = \sqrt{3}$ (d) $\rho = 1$

3. (a) $\rho = \infty$ (b) $\rho = \infty$ (c) $\rho = \infty$
   (d) $\rho = \infty$ (e) $\rho = 1$ (f) $\rho = \sqrt{2}$

4. (a) $y_1(x) = 1 - \dfrac{\alpha^2}{2!}x^2 - \dfrac{(2^2 - \alpha^2)\alpha^2}{4!}x^4 - \dfrac{(4^2 - \alpha^2)(2^2 - \alpha^2)\alpha^2}{6!}x^6 - \cdots$

$$- \dfrac{[(2m - 2)^2 - \alpha^2] \cdots (2^2 - \alpha^2)\alpha^2}{(2m)!}x^{2m} - \cdots$$

$$y_2(x) = x + \dfrac{1 - \alpha^2}{3!}x^3 + \dfrac{(3^2 - \alpha^2)(1 - \alpha^2)}{5!}x^5 + \cdots$$

$$+ \dfrac{[(2m - 1)^2 - \alpha^2] \cdots (1 - \alpha^2)}{(2m + 1)!}x^{2m+1} + \cdots$$

   (b) $y_1(x)$ or $y_2(x)$ terminates with x^n as $\alpha = n$ is even or odd.
   (c) $n = 0, y = 1;$ $n = 1, y = x;$ $n = 2, y = 1 - 2x^2;$
   $n = 3, y = x - \frac{4}{3}x^3$

5. $y_1(x) = 1 - \dfrac{x^3}{3!} + \dfrac{x^5}{5!} + \dfrac{x^6}{2 \cdot 3 \cdot 5 \cdot 6} + \cdots$

$y_2(x) = x - \dfrac{x^4}{3 \cdot 4} + \dfrac{x^6}{2 \cdot 3 \cdot 5 \cdot 6} + \cdots$

6. $y_1(x) = 1 - \frac{1}{6}x^3 + \frac{1}{12}x^4 - \frac{1}{40}x^5 + \cdots,$ $y_2(x) = x - \frac{1}{12}x^4 + \frac{1}{20}x^5 + \cdots$
   $\rho = \infty$

7. Cannot specify arbitrary initial conditions at $x = 0$; hence $x = 0$ is a singular point.

8. $f'(0) = \lim\limits_{h \to 0} \dfrac{e^{-1/h^2} - 0}{h} = 0,$ $f''(0) = \lim\limits_{h \to 0} \dfrac{(2/h^3)e^{-1/h^2} - 0}{h} = 0,$ etc.

9. (a) $y = 1 + x + \dfrac{x^2}{2!} + \cdots + \dfrac{x^n}{n!} + \cdots = e^x$

(b) $y = 1 + \dfrac{x^2}{2} + \dfrac{x^4}{2 \cdot 4} + \dfrac{x^6}{2 \cdot 4 \cdot 6} + \cdots + \dfrac{x^{2n}}{2^n \cdot n!} + \cdots$

(c) $y = 1 + x + \frac{1}{2}x^2 + \frac{1}{2}x^3 + \cdots$

(d) $y = 1 + x + x^2 + \cdots + x^n + \cdots = \dfrac{1}{1 - x}$

(e) $y = a_0 \left(1 + x + \dfrac{x^2}{2!} + \cdots + \dfrac{x^n}{n!} + \cdots \right) + 2! \left(\dfrac{x^3}{3!} + \dfrac{x^4}{4!} + \cdots + \dfrac{x^n}{n!} + \cdots \right)$

$= a_0 e^x + 2 \left(e^x - 1 - x - \dfrac{x^2}{2} \right) = ce^x - 2 - 2x - x^2$

(f) $y = a_0 \left(1 - \dfrac{x^2}{2} + \dfrac{x^4}{2^2 2!} - \dfrac{x^6}{2^3 3!} + \cdots + \dfrac{(-1)^n x^{2n}}{2^n n!} + \cdots \right)$

$+ \left(x + \dfrac{x^2}{2} - \dfrac{x^3}{3} + \dfrac{x^4}{2 \cdot 4} + \dfrac{x^5}{3 \cdot 5} + \cdots \right)$

$= a_0 e^{-x^2/2} + \left(x + \dfrac{x^2}{2} - \dfrac{x^3}{3} + \dfrac{x^4}{2 \cdot 4} + \dfrac{x^5}{3 \cdot 5} + \cdots \right)$

Section 4.3, Page 172

	$x = -1$	$x = 0$	$x = 1$
1.	Ordinary	Regular singular	Ordinary
2.	Regular singular	Regular singular	Regular singular
3.	Regular singular	Irregular singular	Regular singular
4.	Regular singular	Irregular singular	Regular singular
5.	Irregular singular	Ordinary	Regular singular
6.	Ordinary	Ordinary	Irregular singular
7.	Ordinary	Ordinary	Ordinary
8.	Regular singular	Regular singular	Irregular singular

9. Regular singular point   10. Regular singular point
11. (a) Regular singular point   (b) Regular singular point
14. (a) Regular singular point   (b) Irregular singular point
    (c) Irregular singular point   (d) Irregular singular point

Section 4.4, Page 178

1. (a) $y = c_1 x^{-1} + c_2 x^{-2}$ (b) $y = c_1 |x + 1|^{-\frac{1}{2}} + c_2 |x + 1|^{-\frac{3}{4}}$
   (c) $y = c_1 x^2 + c_2 x^2 \ln |x|$
   (d) $y = c_1 x^{-1} \cos (2 \ln |x|) + c_2 x^{-1} \sin (2 \ln |x|)$
   (e) $y = c_1 x + c_2 x \ln |x|$ (f) $y = c_1 (x - 1)^{-3} + c_2 (x - 1)^{-4}$

(g) $y = c_1 |x|^{(-5+\sqrt{29})/2} + c_2 |x|^{(-5-\sqrt{29})/2}$

(h) $y = c_1 |x|^{3/2} \cos(\tfrac{1}{2}\sqrt{3} \ln |x|) + c_2 |x|^{3/2} \sin (\tfrac{1}{2}\sqrt{3} \ln |x|)$

(i) $y = c_1 x^3 + c_2 x^3 \ln |x|$

(j) $y = c_1(x - 2)^{-2} \cos (2 \ln |x - 2|) + c_2(x - 2)^{-2} \sin (2 \ln |x - 2|)$

(k) $y = c_1 |x|^{-1/2} \cos (\tfrac{1}{2}\sqrt{15} \ln |x|) + c_2 |x|^{-1/2} \sin (\tfrac{1}{2}\sqrt{15} \ln |x|)$

(l) $y = c_1 x + c_2 x^4$

3. $\alpha < 1$

5. (a) $y = c_1 x^{-1} + c_2 x^2$ (b) $y = c_1 x^2 + c_2 x^2 \ln x + \tfrac{1}{4} \ln x + \tfrac{1}{4}$

(c) $y = c_1 x^{-1} + c_2 x^{-5} + \tfrac{1}{12} x$ (d) $y = c_1 x + c_2 x^2 + 3 x^2 \ln x + \ln x + \tfrac{3}{2}$

(e) $y = c_1 \cos (2 \ln x) + c_2 \sin (2 \ln x) + \tfrac{1}{3} \sin (\ln x)$

(f) $y = c_1 x^{-3/2} \cos (\tfrac{3}{2} \ln x) + c_2 x^{-3/2} \sin (\tfrac{3}{2} \ln x)$

7. $x > 0$: $c_1 = k_1$, $c_2 = k_2$; $x < 0$: $c_1(-1)^{r_1} = k_1$, $c_2 = k_2$

8. (a) $y = c_1 x^{\frac{1}{2}-i(1+\sqrt{3}/2)} + c_2 x^{\frac{1}{2}-i(1-\sqrt{3}/2)}$

$$= c_1 x^{1/2} \left\{ \cos \left[\left(1 + \tfrac{1}{2}\sqrt{3} \right) \ln x \right] - i \sin \left[\left(1 + \tfrac{1}{2}\sqrt{3} \right) \ln x \right] \right\}$$

$$+ c_2 x^{1/2} \left\{ \cos \left[\left(1 - \tfrac{1}{2}\sqrt{3} \right) \ln x \right] - i \sin \left[\left(1 - \tfrac{1}{2}\sqrt{3} \right) \ln x \right] \right\}$$

(b) $y = c_1 x^{2i} + c_2 x^{-i}$

$= c_1[\cos (2 \ln x) + i \sin (2 \ln x)] + c_2[\cos (\ln x) - i \sin (\ln x)]$

(c) $y = c_1 x^{1+i} + c_2 x^{-1-i}$

$= c_1 x[\cos (\ln x) + i \sin (\ln x)] + c_2 x^{-1}[\cos (\ln x) - i \sin (\ln x)]$

Section 4.5, Page 184

1. (a) $r(2r - 1) = 0$; $a_n = - \dfrac{a_{n-2}}{(n + r)[2(n + r) - 1]}$; $r_1 = \tfrac{1}{2}, r_2 = 0$

$$y_1(x) = x^{1/2}\left[1 - \frac{x^2}{2 \cdot 5} + \frac{x^4}{2 \cdot 4 \cdot 5 \cdot 9} - \frac{x^6}{2 \cdot 4 \cdot 6 \cdot 5 \cdot 9 \cdot 13} + \cdots \right.$$

$$\left. + \frac{(-1)^n x^{2n}}{2^n n! \, 5 \cdot 9 \cdot 13 \cdots (4n + 1)} + \cdots \right]$$

$$y_2(x) = 1 - \frac{x^2}{2 \cdot 3} + \frac{x^4}{2 \cdot 4 \cdot 3 \cdot 7} - \frac{x^6}{2 \cdot 4 \cdot 6 \cdot 3 \cdot 7 \cdot 11} + \cdots$$

$$+ \frac{(-1)^n x^{2n}}{2^n n! \, 3 \cdot 7 \cdot 11 \cdots (4n - 1)} + \cdots$$

(b) $r^2 - \frac{1}{9} = 0;$ $a_n = -\dfrac{a_{n-2}}{(n+r)^2 - \frac{1}{9}}$; $r_1 = \frac{1}{3}, r_2 = -\frac{1}{3}$

$$y_1(x) = x^{1/3}\left[1 - \frac{1}{1!\,(1+\frac{1}{3})}\left(\frac{x}{2}\right)^2 + \frac{1}{2!\,(1+\frac{1}{3})(2+\frac{1}{3})}\left(\frac{x}{2}\right)^4 + \cdots\right.$$

$$\left. + \frac{(-1)^m}{m!\,(1+\frac{1}{3})(2+\frac{1}{3})\cdots(m+\frac{1}{3})}\left(\frac{x}{2}\right)^{2m} + \cdots\right]$$

$$y_2(x) = x^{-1/3}\left[1 - \frac{1}{1!\,(1-\frac{1}{3})}\left(\frac{x}{2}\right)^2 + \frac{1}{2!\,(1-\frac{1}{3})(2-\frac{1}{3})}\left(\frac{x}{2}\right)^4 + \cdots\right.$$

$$\left. + \frac{(-1)^m}{m!\,(1-\frac{1}{3})(2-\frac{1}{3})\cdots(m-\frac{1}{3})}\left(\frac{x}{2}\right)^{2m} + \cdots\right]$$

Hint: Let $n = 2m$ in the recurrence relation, $m = 1, 2, 3, \ldots$.

(c) $r(r-1) = 0;$ $a_n = -\dfrac{a_{n-1}}{(n+r)(n+r-1)}$; $r_1 = 1, r_2 = 0$

$$y_1(x) = x\left[1 - \frac{x}{1!\,2!} + \frac{x^2}{2!\,3!} + \cdots + \frac{(-1)^n}{n!\,(n+1)!}x^n + \cdots\right]$$

(d) $r^2 = 0;$ $a_n = \dfrac{a_{n-1}}{(n+r)^2}$; $r_1 = r_2 = 0$

$$y_1(x) = 1 + \frac{x}{(1!)^2} + \frac{x^2}{(2!)^2} + \cdots + \frac{x^n}{(n!)^2} + \cdots$$

(e) $r(3r-1) = 0;$ $a_n = -\dfrac{a_{n-2}}{(n+r)[3(n+r)-1]}$; $r_1 = \frac{1}{3}, r_2 = 0$

$$y_1(x) = x^{1/3}\left[1 - \frac{1}{1!\,7}\left(\frac{x^2}{2}\right) + \frac{1}{2!\,7\cdot 13}\left(\frac{x^2}{2}\right)^2 + \cdots\right.$$

$$\left. + \frac{(-1)^m}{m!\,7\cdot 13\cdots(6m+1)}\left(\frac{x^2}{2}\right)^m + \cdots\right]$$

$$y_2(x) = 1 - \frac{1}{1!\,5}\left(\frac{x^2}{2}\right) + \frac{1}{2!\,5\cdot 11}\left(\frac{x^2}{2}\right)^2 + \cdots$$

$$+ \frac{(-1)^m}{m!\,5\cdot 11\cdots(6m-1)}\left(\frac{x^2}{2}\right)^m + \cdots$$

Hint: Let $n = 2m$ in the recurrence relation, $m = 1, 2, 3, \ldots$.

(f) $r^2 - 2 = 0;$ $a_n = -\dfrac{a_{n-1}}{(n+r)^2 - 2}$; $r_1 = \sqrt{2},\ r_2 = -\sqrt{2}$

$$y_1(x) = x^{\sqrt{2}}\left[1 - \frac{x}{1(1 + 2\sqrt{2})} + \frac{x^2}{2!\,(1 + 2\sqrt{2})(2 + 2\sqrt{2})} + \cdots\right.$$

$$\left. + \frac{(-1)^n}{n!\,(1 + 2\sqrt{2})(2 + 2\sqrt{2})\cdots(n + 2\sqrt{2})}\,x^n + \cdots\right]$$

$$y_2(x) = x^{-\sqrt{2}}\left[1 - \frac{x}{1(1 - 2\sqrt{2})} + \frac{x^2}{2!\,(1 - 2\sqrt{2})(2 - 2\sqrt{2})} + \cdots\right.$$

$$\left. + \frac{(-1)^n}{n!\,(1 - 2\sqrt{2})(2 - 2\sqrt{2})\cdots(n - 2\sqrt{2})}\,x^n + \cdots\right]$$

2. $r^2 = 0;$ $r_1 = 0,\ r_2 = 0$

$$y_1(x) = 1 + \frac{\alpha(\alpha + 1)}{2 \cdot 1^2}(x - 1) - \frac{\alpha(\alpha + 1)[1 \cdot 2 - \alpha(\alpha + 1)]}{(2 \cdot 1^2)(2 \cdot 2^2)}(x - 1)^2 + \cdots$$

$$+ (-1)^{n+1}\frac{\alpha(\alpha + 1)[1 \cdot 2 - \alpha(\alpha + 1)]\cdots[n(n-1) - \alpha(\alpha + 1)]}{2^n(n!)^2}(x - 1)^n + \cdots$$

3. $r^2 = 0;$ $r_1 = 0,\ r_2 = 0;$ $a_n = \dfrac{(n - 1 - \lambda)a_{n-1}}{n^2}$

$$y_1(x) = 1 + \frac{-\lambda}{(1!)^2}x + \frac{(-\lambda)(1 - \lambda)}{2!}x^2 + \cdots$$

$$+ \frac{(-\lambda)(1 - \lambda)\cdots(n - 1 - \lambda)}{(n!)^2}x^n + \cdots$$

For $\lambda = n$, the coefficients of all terms past x^n are zero.

6. (b) $[(n - 1)^2 - 1]b_n = -b_{n-2}$, and it is impossible to determine b_2.

Section 4.5.1, Page 190

1. (a) $x = 0;$ $r(r - 1) = 0;$ $r_1 = 1,\ r_2 = 0$
   (b) $x = 0;$ $r^2 - 3r + 2 = 0;$ $r_1 = 2,\ r_2 = 1$
   (c) $x = 0;$ $r(r - 1) = 0;$ $r_1 = 1,\ r_2 = 0$
       $x = 1;$ $r(r + 5) = 0;$ $r_1 = 0,\ r_2 = -5$
   (d) none
   (e) $x = 0;$ $r^2 + 2r - 2 = 0;$ $r_1 = -1 + \sqrt{3},\ r_2 = -1 - \sqrt{3}$
   (f) $x = 0;$ $r(r - \tfrac{3}{4}) = 0;$ $r_1 = \tfrac{3}{4},\ r_2 = 0$
       $x = -2;$ $r(r - \tfrac{5}{4}) = 0;$ $r_1 = \tfrac{5}{4},\ r_2 = 0$
   (g) $x = 0;$ $r^2 + 1 = 0;$ $r_1 = i,\ r_2 = -i$
   (h) $x = -1;$ $r^2 - 7r + 3 = 0;$ $r_1 = (7 + \sqrt{37})/2,\ r_2 = (7 - \sqrt{37})/2$

2. $y_1(x) = x - \dfrac{x^3}{4!} + \dfrac{x^5}{6!} + \cdots$

3. $r_1 = \frac{1}{2}, r_2 = 0$

$y_1(x) = (x - 1)^{\frac{1}{2}}[1 - \frac{3}{4}(x - 1) + \frac{53}{480}(x - 1)^2 + \cdots], \qquad \rho = 1$

4. (c) *Hint*: $(n - 1)(n - 2) + (1 + \alpha + \beta)(n - 1) + \alpha\beta$

$$= (n - 1 + \alpha)(n - 1 + \beta)$$

(d) *Hint*: $(n - \gamma)(n - 1 - \gamma) + (1 + \alpha + \beta)(n - \gamma) + \alpha\beta$

$$= (n - \gamma + \alpha)(n - \gamma + \beta)$$

Section 4.7, Page 204

1. (a) $y_1(x) = \sum_{n=0}^{\infty} \frac{(-1)^n x^n}{n! \, (n + 1)!}$, $y_2(x) = -y_1(x) \ln x + \frac{1}{x}\left[1 - \sum_{n=1}^{\infty} \frac{H_n + H_{n-1}}{n! \, (n - 1)!}(-1)^n x^n\right]$

(b) $y_1(x) = \sum_{n=0}^{\infty} \frac{(-1)^n x^n}{(n!)^2}$, $\qquad y_2(x) = y_1(x) \ln x - \frac{2}{x} \sum_{n=1}^{\infty} \frac{(-1)^n H_n}{(n!)^2} x^n$

(c) $y_1(x) = \sum_{n=0}^{\infty} \frac{(-1)^n 2^n}{(n!)^2} x^n$, $\qquad y_2(x) = y_1(x) \ln x - 2 \sum_{n=1}^{\infty} \frac{(-1)^n 2^n H_n}{(n!)^2} x^n$

(d) $y_1(x) = \frac{1}{x} \sum_{n=0}^{\infty} \frac{(-1)^n}{n! \, (n + 1)!} x^n$,

$y_2(x) = -y_1(x) \ln x + \frac{1}{x^2}\left[1 - \sum_{n=1}^{\infty} \frac{H_n + H_{n-1}}{n! \, (n - 1)!}(-1)^n x^n\right]$

2. $y_1(x) = x^{\frac{3}{2}}\left[1 + \sum_{m=1}^{\infty} \frac{(-1)^m}{m! \, (1 + \frac{3}{2})(2 + \frac{3}{2}) \cdots (m + \frac{3}{2})}\left(\frac{x}{2}\right)^{2m}\right]$

$y_2(x) = x^{-\frac{3}{2}}\left[1 + \sum_{m=1}^{\infty} \frac{(-1)^m}{m! \, (1 - \frac{3}{2})(2 - \frac{3}{2}) \cdots (m - \frac{3}{2})}\left(\frac{x}{2}\right)^{2m}\right]$

Hint: Let $n = 2m$ in the recurrence relation, $m = 1, 2, 3, \ldots$. For $r = -\frac{3}{2}$, $a_1 = 0$ and a_3 is arbitrary.

CHAPTER FIVE

Section 5.1, Page 208

1. $\phi^{(n)}(0) = -p_1(0)y_0^{(n-1)} - \cdots - p_n(0)y_0$

$\phi^{(n+1)}(0) = p_1(0)[p_1(0)y_0^{(n-1)} + \cdots + p_n(0)y_0] - p_1'(0)y_0^{(n-1)} - \cdots$

$$- p_n(0)y_0' - p_n'(0)y_0$$

2. (a) $y''' = -\cos x$ (b) $y''' + y' = 0$

(c) $y''' - 2y'' - y' + 2y = 0$ (d) $x^3 y''' - 3x^2 y'' + 6xy' - 6y = 0$

(e) $y''' + y' = 1$ (f) $y^{iv} - y'' = 0$

3. (a) $-\infty < x < \infty$ (b) $x > 0$ or $x < 0$

(c) $x > 1$, or $0 < x < 1$, or $x < 0$ (d) $x > 0$

Section 5.2, Page 211

4. (a) 1   (b) 1   (c) $-6e^{-2x}$ (d) e^{-2x} (e) $6x$

5. $\sin^2 x = \frac{1}{10}(5) - \frac{1}{2}\cos 2x$

6. (a) $a_0[n(n-1)(n-2)\cdots 1] + a_1[n(n-1)\cdots 2]x + \cdots + a_n x^n$

(b) $(a_0 r^n + a_1 r^{n-1} + \cdots + a_n)e^{rx}$

(c) $e^x, e^{-x}, e^{2x}, e^{-2x}$;   yes, $W(e^x, e^{-x}, e^{2x}, e^{-2x}) \neq 0, -\infty < x < \infty$

8. (a) $y = c_1 x + c_2 x^2 + c_3 x^3$ (b) $y = c_1 x^2 + c_2 x^3 + c_3(x+1)$

Section 5.3, Page 216

1. (a) $\sqrt{2}e^{i[(\pi/4)+2m\pi]}$ (b) $2e^{i[(2\pi/3)+2m\pi]}$ (c) $e^{i(\pi+2m\pi)}$

(d) $e^{i[(3\pi/2)+2m\pi]}$ (e) $2e^{i[(11\pi/6)+2m\pi]}$ (f) $\sqrt{2}e^{i[(5\pi/4)+2m\pi]}$

2. (a) $1, \frac{1}{2}(-1+i\sqrt{3}), \frac{1}{2}(-1-i\sqrt{3})$ (b) $2^{1/4}e^{-\pi i/8}, 2^{1/4}e^{7\pi i/8}$

(c) $1, i, -1, -i$ (d) $(\sqrt{3}+i)/\sqrt{2}, -(\sqrt{3}+i)/\sqrt{2}$

3. $y = c_1 e^x + c_2 x e^x + c_3 e^{-x}$ 4. $y = c_1 e^x + c_2 x e^x + c_3 x^2 e^x$

5. $y = c_1 e^x + c_2 e^{2x} + c_3 e^{-x}$ 6. $y = c_1 + c_2 x + c_3 e^{2x} + c_4 x e^{2x}$

7. $y = c_1 \cos x + c_2 \sin x + e^{\sqrt{3}x/2}(c_3 \cos \frac{1}{2}x + c_4 \sin \frac{1}{2}x)$

$$+ e^{-\sqrt{3}x/2}(c_5 \cos \tfrac{1}{2}x + c_6 \sin \tfrac{1}{2}x)$$

8. $y = c_1 e^x + c_2 e^{-x} + c_3 e^{2x} + c_4 e^{-2x}$

9. $y = c_1 e^x + c_2 x e^x + c_3 x^2 e^x + c_4 e^{-x} + c_5 x e^{-x} + c_6 x^2 e^{-x}$

10. $y = c_1 + c_2 x + c_3 e^x + c_4 e^{-x} + c_5 \cos x + c_6 \sin x$

11. $y = c_1 + c_2 e^x + c_3 e^{2x} + c_4 \cos x + c_5 \sin x$

12. $y = c_1 + c_2 e^{2x} + e^{-x}(c_3 \cos \sqrt{3}\,x + c_4 \sin \sqrt{3}\,x)$

13. $y = e^x[(c_1 + c_2 x)\cos x + (c_3 + c_4 x)\sin x] + e^{-x}[(c_5 + c_6 x)\cos x + (c_7 + c_8 x)\sin x]$

14. $y = 2 - 2\cos x + \sin x$

15. $y = \frac{1}{2}(\cosh x - \cos x) + \frac{1}{2}(\sinh x - \sin x)$

17. (a) $y = c_1 x + c_2 x^2 + c_3 x^{-1}$ (b) $y = c_1 x + c_2 x \ln x + c_3 x(\ln x)^2$

(c) $y = c_1 x + c_2 \cos(\ln x) + c_3 \sin(\ln x)$

(d) $y = c_1 + c_2 x^{-1} + x[c_3 \cos(2\ln x) + c_4 \sin(2\ln x)]$

(e) $y = c_1 + c_2 x + c_3 x \ln x$ (f) $y = c_1 x^2 + c_2 x^{-1} + c_3 x^{-2}$

20. (a) Stable   (b) Not stable   (c) Not stable

(d) Not stable   (e) Not stable   (f) Stable

Section 5.4, Page 222

1. $y = \frac{3}{16}(1 - \cos 2x) + \frac{1}{8}x^2$

2. $y = c_1 e^x + c_2 x e^x + c_3 e^{-x} + \frac{1}{2}x e^{-x} + 3$

3. $y = c_1 e^{-x} + c_2 \cos x + c_3 \sin x + \frac{1}{2}x e^{-x} + 4(x-1)$

4. $y = c_1 + c_2 e^x + c_3 e^{-x} + \cos x$

5. $y = (x-4)\cos x - (\frac{3}{2}x + 4)\sin x + 3x + 4$

6. $y = c_1 + c_2 x + c_3 e^{-2x} + c_4 e^{2x} - \frac{1}{3}e^x - \frac{1}{48}x^4 - \frac{1}{16}x^2$

7. $y = c_1 \cos x + c_2 \sin x + c_3 x \cos x + c_4 x \sin x + 3 + \frac{1}{9}\cos 2x$

8. $y = 1 + \frac{1}{4}(x^2 + 3x) - x e^x$

9. $y = c_1 + c_2 x + c_3 x^2 + c_4 e^{-x} + e^{x/2}\left(c_5 \cos \dfrac{\sqrt{3}}{2}x + c_6 \sin \dfrac{\sqrt{3}}{2}x\right) + \dfrac{x^4}{24}$

10. $y = c_1 + c_2 x + c_3 x^2 + c_4 e^{-x} + \frac{1}{20} \sin 2x + \frac{1}{40} \cos 2x$

11. $y = c_1 e^x + c_2 e^{-x} + c_3 \cos x + c_4 \sin x - 3x - \frac{1}{4} x \sin x$

12. $y_p(x) = x(A_0 x^3 + A_1 x^2 + A_2 x + A_3) + B x^2 e^x$

13. $y_p(x) = x(A_0 x + A_1)e^{-x} + B \cos x + C \sin x$

14. $y_p(x) = A x^2 e^x + B \cos x + C \sin x$

15. $y_p(x) = x(A_0 x^2 + A_1 x + A_2) + (B_0 x + B_1) \cos x + (C_0 x + C_1) \sin x$

16. $y_p(x) = A x^2 + (B_0 x + B_1)e^x + x(C \cos 2x + D \sin 2x)$

17. $y_p(x) = A e^x + (B_0 x + B_1)e^{-x} + x e^{-x}(C \cos x + D \sin x)$

18. $t_0 = a_0, \qquad t_n = a_0 \alpha^n + a_1 \alpha^{n-1} + \cdots + a_{n-1} \alpha + a_n$

Section 5.5, Page 224

1. $y_p(x) = -\ln \cos x - (\sin x) \ln (\sec x + \tan x)$

2. $y_p(x) = -x^2/2$ 3. $y_p(x) = e^{4x}/30$ 4. $y_p(x) = x^4/15$

5. $y_p(x) = \frac{1}{2} \int^x [e^{x-t} - \sin(x - t) - \cos(x - t)]g(t)\, dt$

6. $y_p(x) = \frac{1}{2} \int^x [\sinh(x - t) - \sin(x - t)]g(t)\, dt$

7. $y_p(x) = \frac{1}{2} \int^x \left(\frac{x}{t^2} - 2\frac{x^2}{t^3} + \frac{x^3}{t^4} \right) g(t)\, dt$

CHAPTER SIX

Section 6.1, Page 231

1. (a) Piecewise continuous   (b) Neither   (c) Continuous
   (d) Piecewise continuous

2. (a) $1/s^2$, $s > 0$ (b) $2/s^3$, $s > 0$ (c) $n!/s^{n+1}$, $s > 0$

3. $s/(s^2 + a^2)$, $s > 0$

4. (a) $\dfrac{s}{s^2 - b^2}$, $s > |b|$ (b) $\dfrac{b}{s^2 - b^2}$, $s > |b|$

   (c) $\dfrac{s - a}{(s - a)^2 - b^2}$, $s - a > |b|$ (d) $\dfrac{b}{(s - a)^2 - b^2}$, $s - a > |b|$

5. (a) $\dfrac{b}{s^2 + b^2}$, $s > 0$ (b) $\dfrac{s}{s^2 + b^2}$, $s > 0$

   (c) $\dfrac{b}{(s - a)^2 + b^2}$, $s > a$ (d) $\dfrac{s - a}{(s - a)^2 + b^2}$, $s > a$

6. (a) $\dfrac{1}{(s - a)^2}$, $s > a$ (b) $\dfrac{2as}{(s^2 + a^2)^2}$, $s > 0$

(c) $\dfrac{s^2 + a^2}{(s - a)^2(s + a)^2}$, $s > |a|$ (d) $\dfrac{n!}{(s - a)^{n+1}}$, $s > a$

(e) $\dfrac{a\,n!\,(2s)^n}{(s^2 + a^2)^{n+1}}$, $s > 0$ (f) $\dfrac{a\,n!\,(2s)^n}{(s^2 - a^2)^{n+1}}$, $s > |a|$

7. (a) Converges   (b) Converges   (c) Diverges

9. (c) $\Gamma(3/2) = \sqrt{\pi}/2$; $\Gamma(11/2) = 945\sqrt{\pi}/32$

Section 6.2, Page 239

1. $y = \frac{1}{5}(e^{3t} + 4e^{-2t})$ 2. $y = 2e^{-t} - e^{-2t}$

3. $y = e^t \sin t$ 4. $y = e^{2t} - te^{2t}$

5. $y = te^t - t^2e^t + \frac{2}{3}t^3e^t$ 6. $y = \cosh t$

7. $y = 2e^t \cosh \sqrt{3}t - (2/\sqrt{3})\, e^t \sinh \sqrt{3}t$

8. $y = \dfrac{1}{\omega^2 - 4}\, [(\omega^2 - 5) \cos \omega t + \cos 2t]$

9. $y = \frac{1}{5}(\cos t - 2 \sin t + 4e^t \cos t - 2e^t \sin t)$

10. $y = \frac{1}{5}(e^{-t} - e^t \cos t + 7e^t \sin t)$

11. $Y(s) = \dfrac{s}{s^2 + 4} + \dfrac{1 - e^{-\pi s}}{s(s^2 + 4)}$ 12. $Y(s) = \dfrac{1}{s^2(s^2 + 1)} - \dfrac{e^{-s}(s + 1)}{s^2(s^2 + 1)}$

15. (a) $F(s) = \dfrac{n!}{(s - a)^{n+1}}$ (b) $F(s) = \dfrac{6s^2 - 2}{(s^2 + 1)^3}$ (c) $F(s) = \dfrac{n!}{s^{n+1}}$

17. (a) $Y' - s^2Y = -s$ (b) $s^2Y'' + 2sY' - [s^2 + \alpha(\alpha + 1)]Y = -1$

Section 6.3, Page 246

2. (a) $F(s) = 2e^{-2s}/s^3$ (b) $F(s) = e^{-s}(s^2 + 2)/s^3$

(c) $F(s) = \dfrac{e^{-\pi s}}{s^2} - \dfrac{e^{-2\pi s}}{s^2}(1 + \pi s)$ (d) $F(s) = \dfrac{1}{s}(e^{-s} + 2e^{-3s} - 6e^{-4s})$

(e) $F(s) = s^{-2}\,[(1 - s)e^{-2s} - (1 + s)e^{-3s}]$

(f) $F(s) = (1 - e^{-s})/s^2$

3. (a) $f(t) = t^3e^{2t}$ (b) $f(t) = \frac{1}{3}u_2(t)[e^{t-2} - e^{-2(t-2)}]$

(c) $f(t) = 2u_2(t)e^{t-2} \cos (t - 2)$ (d) $f(t) = u_2(t) \sinh 2(t - 2)$

(e) $f(t) = u_1(t)e^{2(t-1)} \cosh (t - 1)$ (f) $f(t) = u_1(t) + u_2(t) - u_3(t) - u_4(t)$

5. (a) $f(t) = 2(2t)^n$ (b) $f(t) = \frac{1}{2}e^{-t/2} \cos t$

(c) $f(t) = \frac{1}{6}e^{t/3}(e^{2t/3} - 1)$ (d) $f(t) = \frac{1}{2}e^2\, u_4(t)e^{(t-4)/2}$

6. (a) $F(s) = s^{-1}(1 - e^{-s}),\quad s > 0$

(b) $F(s) = s^{-1}(1 - e^{-s} + e^{-2s} - e^{-3s}),\quad s > 0$

(c) $F(s) = \dfrac{1}{s}[1 - e^{-s} + \cdots + e^{-2ns} - e^{-(2n+1)s}] = \dfrac{1 - e^{-(2n+2)s}}{s(1 + e^{-s})},\quad s > 0$

(d) $F(s) = \dfrac{1}{s}\displaystyle\sum_{n=0}^{\infty}(-1)^n e^{-ns} = \dfrac{1/s}{1 + e^{-s}},\quad s > 0$

8. (a) $\mathscr{L}\{f(t)\} = \dfrac{1/s}{1 + e^{-s}},\quad s > 0$ (b) $\mathscr{L}\{f(t)\} = \dfrac{1 - e^{-s}}{s(1 + e^{-s})},\quad s > 0$

(c) $\mathscr{L}\{f(t)\} = \dfrac{1 - (1 + s)e^{-s}}{s^2(1 - e^{-s})},\quad s > 0$

(d) $\mathscr{L}\{f(t)\} = \dfrac{1 + e^{-s\pi}}{(1 + s^2)(1 - e^{-s\pi})},\quad s > 0$

9. (a) $\mathscr{L}\{f(t)\} = s^{-1}(1 - e^{-s}),\quad s > 0$ (b) $\mathscr{L}\{g(t)\} = s^{-2}(1 - e^{-s}),\quad s > 0$

(c) $\mathscr{L}\{h(t)\} = s^{-2}(1 - e^{-s})^2,\quad s > 0$

10. (b) $\mathscr{L}\{p(t)\} = \dfrac{1 - e^{-s}}{s^2(1 + e^{-s})},\quad s > 0$

Section 6.3.1, Page 252

1. $y = 1 - \cos t + \sin t - u_{\pi/2}(t)[1 - \sin t]$

2. $y = e^{-t}\sin t + \tfrac{1}{2}u_\pi(t)[1 + e^{-(t-\pi)}\cos t + e^{-(t-\pi)}\sin t]$

   $- \tfrac{1}{2}u_{2\pi}(t)[1 - e^{-(t-2\pi)}\cos t - e^{-(t-2\pi)}\sin t]$

3. $y = \tfrac{1}{6}[1 - u_{2\pi}(t)](2\sin t - \sin 2t)$ 4. $y = \tfrac{1}{6}(2\sin t - \sin 2t) - \tfrac{1}{6}u_\pi(t)(2\sin t + \sin 2t)$

5. $y = 1 - u_1(t)[1 - e^{-(t-1)} - (t - 1)e^{-(t-1)}]$

6. $y = e^{-t} - e^{-2t} + u_2(t)[\tfrac{1}{2} - e^{-(t-2)} + \tfrac{1}{2}e^{-2(t-2)}]$

7. $y = \cos t + u_\pi(t)(1 + \cos t)$

8. $y = 1 + \displaystyle\sum_{n=1}^{\infty}(-1)^n u_{n\pi}(t)[1 - \cos(t - n\pi)]$

9. (a) $F(s) = s^{-2}(1 - e^{-s})$ (b) $y = t - u_1(t)[t - 1 - \sin(t - 1)]$

Section 6.4, Page 257

1. $y = e^{-t}\cos t + e^{-t}\sin t - u_\pi(t)e^{-(t-\pi)}\sin t$
2. $y = \tfrac{1}{2}u_\pi(t)\sin 2t - \tfrac{1}{2}u_{2\pi}(t)\sin 2t$
3. $y = 2te^{-t} + u_{2\pi}(t)[1 - e^{-(t-2\pi)} - (t - 2\pi)e^{-(t-2\pi)}]$
4. $y = \cosh t + 2u_1(t)\sinh(t - 1)$

5. $y = \frac{1}{2}\sqrt{2}\, e^{-t} \sin \sqrt{2}\, t + \frac{1}{4}e^{-t} \cos \sqrt{2}\, t + \frac{1}{4}(\sin t - \cos t)$

$$+ \frac{1}{2}\sqrt{2}\, u_\pi(t)e^{-(t-\pi)} \sin \sqrt{2}(t - \pi)$$

6. $y = \cos \omega t - \omega^{-1} u_{\pi/\omega}(t) \sin \omega t$ 7. $y = [1 + u_\pi(t)] \sin t$

Section 6.5, Page 260

3. $\sin t * \sin t = \frac{1}{2}(\sin t - t \cos t)$ is negative when $t = 2\pi$, for example.

4. (a) $F(s) = 2/s^2(s^2 + 4)$ (b) $F(s) = 1/(s + 1)(s^2 + 1)$

(c) $F(s) = 1/s^2(s - 1)$ (d) $F(s) = s/(s^2 + 1)^2$

5. (a) $f(t) = \dfrac{1}{6} \displaystyle\int_0^t (t - \tau)^3 \sin \tau \, d\tau$ (b) $f(t) = \displaystyle\int_0^t e^{-(t-\tau)} \cos 2\tau \, d\tau$

(c) $f(t) = \dfrac{1}{2} \displaystyle\int_0^t (t - \tau)e^{-(t-\tau)} \sin 2\tau \, d\tau$ (d) $f(t) = \displaystyle\int_0^t \sin (t - \tau)g(\tau) \, d\tau$

6. $y = \frac{1}{2}(1 - e^{-t} \sin t - e^{-t} \cos t) - \frac{1}{2}u_{2\pi}(t)[1 - e^{-(t-2\pi)} \sin t - e^{-(t-2\pi)} \cos t]$

7. $y = \displaystyle\int_0^t e^{-(t-\tau)} \sin (t - \tau) \sin \alpha\tau \, d\tau$ 8. $\Phi(s) = \dfrac{F(s)}{1 + K(s)}$

9. (c) $\phi(t) = \frac{1}{3}(4 \sin 2t - 2 \sin t)$ (d) $u(t) = \frac{1}{3}(2 \sin t - \sin 2t)$

CHAPTER SEVEN

Section 7.1, Page 269

2. $x_1' - x_2 = 0, \qquad x_1(0) = u_0$
$x_2' + p(t)x_2 + q(t)x_1 = g(t), \qquad x_2(0) = u_0'$

Section 7.2, Page 275

1. $x_1 = c_1 e^{-3t} + c_2 e^{2t}$ 2. $x_1 = c_1 e^{3t} + c_2 e^{-t} + \frac{1}{4}e^t$
   $x_2 = -4c_1 e^{-3t} + c_2 e^{2t}$ $x_2 = 2c_1 e^{3t} - 2c_2 e^{-t} - 2e^t$

3. $x_1 = -\frac{2}{3} \cos t - \frac{4}{3} \sin t + \frac{2}{3} \sin 2t + \frac{2}{3} \cos 2t - 5t$
   $x_2 = \qquad\quad - \frac{2}{3} \sin t + \frac{1}{3} \sin 2t \qquad\quad - 2t + 1$

4. $x_1 = c_1 e^{2t} + c_2(t - 1)e^{2t} + \frac{1}{8} + \frac{1}{4}t + \frac{3}{4}t^2$
   $x_2 = -c_1 e^{2t} - c_2 t e^{2t} \qquad - \frac{3}{8} - t - \frac{1}{4}t^2$

5. $x_1 = e^t + 7te^t + 3t^2 e^t$ 6. $x_1 = c_1 + c_2 t \qquad\qquad\qquad + \frac{4}{3}t^3$
   $x_2 = -e^t + 2te^t + \frac{3}{2}t^2 e^t$ $x_2 = 2c_1 + c_2(2t - \frac{1}{2}) + t - 2t^2 + \frac{8}{3}t^3$

7. $x_1 = c_1 e^t \cos 2t + c_2 e^t \sin 2t + \frac{4}{13}e^{-t} \sin t - \frac{7}{13}e^{-t} \cos t$
   $x_2 = c_1 e^t(\cos 2t + \sin 2t) + c_2 e^t(-\cos 2t + \sin 2t) - \frac{2}{13}e^{-t} \sin t - \frac{16}{13}e^{-t} \cos t$

8. $x_1 = e^{-2t} \cos t + 3e^{-2t} \sin t$
   $x_2 = \qquad\qquad\quad 2e^{-2t} \sin t$

9. $x_1 = \qquad c_2 e^t$
$x_2 = -\sqrt{2}c_1 e^{-2t} + \qquad c_3 e^t$
$x_3 = \qquad c_1 e^{-2t} + \sqrt{2}c_3 e^t$

10. $x_1 = \qquad c_1 e^t + c_2 e^{-2t} + \quad c_3 e^{3t}$
$x_2 = -4c_1 e^t - c_2 e^{-2t} + 2c_3 e^{3t}$
$x_3 = \quad -c_1 e^t - c_2 e^{-2t} + \quad c_3 e^{3t}$

11. $x_1 = \quad 4c_1 e^{-2t} + 3c_2 e^{-t}$
$x_2 = -5c_1 e^{-2t} - 4c_2 e^{-t} + c_3 e^{2t}$
$x_3 = -7c_1 e^{-2t} - 2c_2 e^{-t} - c_3 e^{2t}$

13. $x_1 = \quad c_1 + 2c_2 e^t + c_3 e^{2t}$
$x_2 = 2c_1 + \quad c_2 e^t$

14. $x_1 = \qquad c_1 e^{-2t} \qquad + \frac{1}{2}t - \frac{5}{4}$
$x_2 = -\frac{3}{2}c_1 e^{-2t} + c_2 + \frac{1}{4}t^2 + \frac{1}{4}t$

15. $x_1 = \quad c_1 e^{-t} + c_2 e^{2t} + 3c_3 e^{4t} + \frac{1}{4}$
$x_2 = 3c_1 e^{-t} \qquad - c_3 e^{4t} + \frac{1}{4}$

16. $x_1 = -\cos t - \sin t$
$x_2 = \qquad - \sin t$

17. $x_1 = c_1 e^{2t}$
$x_2 = \qquad c_2 e^{-2t}$

18. No solutions

19. Infinitely many solutions: x_1 arbitrary; $x_2 = \frac{1}{4}(e^{2t} - e^{-2t}) - x_1$

20. No solutions

21. Infinitely many solutions: any solution of $(D + 1)x_1 + Dx_2 = 0$

23. (a) Five   (b) Three

Section 7.3, Page 285

1. (a) $\begin{pmatrix} 6 & -6 & 3 \\ 5 & 9 & -2 \\ 2 & 3 & 8 \end{pmatrix}$

(b) $\begin{pmatrix} -15 & 6 & -12 \\ 7 & -18 & -1 \\ -26 & -3 & -5 \end{pmatrix}$

(c) $\begin{pmatrix} 6 & -12 & 3 \\ 4 & 3 & 7 \\ 9 & 12 & 0 \end{pmatrix}$

(d) $\begin{pmatrix} -8 & -9 & 11 \\ 14 & 12 & -5 \\ 5 & -8 & 5 \end{pmatrix}$

2. (a) $\begin{pmatrix} 1 - i & -7 + 2i \\ -1 + 2i & 2 + 3i \end{pmatrix}$

(b) $\begin{pmatrix} 3 + 4i & 6i \\ 11 + 6i & 6 - 5i \end{pmatrix}$

(c) $\begin{pmatrix} -3 + 5i & 7 + 5i \\ 2 + i & 7 + 2i \end{pmatrix}$

(d) $\begin{pmatrix} 8 + 7i & 4 - 4i \\ 6 - 4i & -4 \end{pmatrix}$

3. (a) $\begin{pmatrix} -2 & 1 & 2 \\ 1 & 0 & -1 \\ 2 & -3 & 1 \end{pmatrix}$ (b) $\begin{pmatrix} 1 & 3 & -2 \\ 2 & -1 & 1 \\ 3 & -1 & 0 \end{pmatrix}$ (c), (d) $\begin{pmatrix} -1 & 4 & 0 \\ 3 & -1 & 0 \\ 5 & -4 & 1 \end{pmatrix}$

4. (a) $\begin{pmatrix} 3 - 2i & 2 - i \\ 1 + i & -2 + 3i \end{pmatrix}$

(b) $\begin{pmatrix} 3 + 2i & 1 - i \\ 2 + i & -2 - 3i \end{pmatrix}$

(c) $\begin{pmatrix} 3 + 2i & 2 + i \\ 1 - i & -2 - 3i \end{pmatrix}$

5. $\begin{pmatrix} 10 & 6 & -4 \\ 0 & 4 & 10 \\ 4 & 4 & 6 \end{pmatrix}$

6. (a) $\begin{pmatrix} 7 & -11 & -3 \\ 11 & 20 & 17 \\ -4 & 3 & -12 \end{pmatrix}$ (b) $\begin{pmatrix} 5 & 0 & -1 \\ 2 & 7 & 4 \\ -1 & 1 & 4 \end{pmatrix}$ (c) $\begin{pmatrix} 6 & -8 & -11 \\ 9 & 15 & 6 \\ -5 & -1 & 5 \end{pmatrix}$

8. (a) $4i$ (b) $12 - 8i$ (c) $2 + 2i$ (d) 16

10. $\begin{pmatrix} \frac{3}{11} & -\frac{4}{11} \\ \frac{2}{11} & \frac{1}{11} \end{pmatrix}$ 11. $\begin{pmatrix} \frac{1}{6} & \frac{1}{12} \\ -\frac{1}{2} & \frac{1}{4} \end{pmatrix}$

12. $\begin{pmatrix} 1 & -3 & 2 \\ -3 & 3 & -1 \\ 2 & -1 & 0 \end{pmatrix}$ 13. $\begin{pmatrix} \frac{1}{3} & \frac{1}{3} & 0 \\ \frac{1}{3} & -\frac{1}{3} & \frac{1}{3} \\ -\frac{1}{3} & 0 & \frac{1}{3} \end{pmatrix}$

14. Singular

15. $\begin{pmatrix} \frac{1}{2} & -\frac{1}{4} & \frac{1}{8} \\ 0 & \frac{1}{2} & -\frac{1}{4} \\ 0 & 0 & \frac{1}{2} \end{pmatrix}$

16. $\begin{pmatrix} \frac{1}{10} & \frac{3}{10} & \frac{1}{10} \\ -\frac{2}{10} & \frac{4}{10} & -\frac{2}{10} \\ -\frac{7}{10} & -\frac{1}{10} & \frac{3}{10} \end{pmatrix}$ 17. Singular

18. $\begin{pmatrix} 1 & 1 & 0 & 1 \\ 1 & 0 & 1 & 1 \\ 1 & 1 & 1 & 1 \\ 0 & 1 & 0 & 1 \end{pmatrix}$ 19. $\begin{pmatrix} 6 & \frac{13}{5} & -\frac{8}{5} & \frac{2}{5} \\ 5 & \frac{11}{5} & -\frac{6}{5} & \frac{4}{5} \\ 0 & -\frac{1}{5} & \frac{1}{5} & \frac{1}{5} \\ -2 & -\frac{4}{5} & \frac{4}{5} & -\frac{1}{5} \end{pmatrix}$

22. (a) $\begin{pmatrix} 7e^t & 5e^{-t} & 10e^{2t} \\ -e^t & 7e^{-t} & 2e^{2t} \\ 8e^t & 0 & -e^{2t} \end{pmatrix}$

(b) $\begin{pmatrix} 2e^{2t} - 2 + 3e^{3t} & 1 + 4e^{-2t} - e^t & 3e^{3t} + 2e - e^{4t} \\ 4e^{2t} - 1 - 3e^{3t} & 2 + 2e^{-2t} + e^t & 6e^{3t} + e^t + e^{4t} \\ -2e^{2t} - 3 + 6e^{3t} & -1 + 6e^{-2t} - 2e^t & -3e^{3t} + 3e^t - 2e^{4t} \end{pmatrix}$

(c) $\begin{pmatrix} e^t & -2e^{-t} & 2e^{2t} \\ 2e^t & -e^{-t} & -2e^{2t} \\ -e^t & -3e^{-t} & 4e^{2t} \end{pmatrix}$ (d) $(e-1)\begin{pmatrix} 1 & 2e^{-1} & \frac{1}{2}(e+1) \\ 2 & e^{-1} & -\frac{1}{2}(e+1) \\ -1 & 3e^{-1} & e+1 \end{pmatrix}$

Section 7.4, Page 297

1. $x_1 = -\frac{1}{3}, x_2 = \frac{7}{3}, x_3 = -\frac{1}{3}$ 2. No solution
3. $x_1 = -c, x_2 = c + 1, x_3 = c$, where c is arbitrary
4. $x_1 = c, x_2 = -c, x_3 = -c$, where c is arbitrary
5. $x_1 = 0, x_2 = 0, x_3 = 0$
6. (a) Linearly independent   (b) $\mathbf{x}^{(1)} - 5\mathbf{x}^{(2)} + 2\mathbf{x}^{(3)} = \mathbf{0}$
   (c) $2\mathbf{x}^{(1)} - 3\mathbf{x}^{(2)} + 4\mathbf{x}^{(3)} - \mathbf{x}^{(4)} = \mathbf{0}$
   (d) Linearly independent   (e) $\mathbf{x}^{(1)} + \mathbf{x}^{(2)} - \mathbf{x}^{(4)} = \mathbf{0}$
8. (a) $3\mathbf{x}^{(1)}(t) - 6\mathbf{x}^{(2)}(t) + \mathbf{x}^{(3)}(t) = \mathbf{0}$ (b) Linearly independent

10. $\lambda_1 = 2, \mathbf{x}^{(1)} = \begin{pmatrix} 1 \\ 3 \end{pmatrix}; \qquad \lambda_2 = 4, \mathbf{x}^{(2)} = \begin{pmatrix} 1 \\ 1 \end{pmatrix}$

11. $\lambda_1 = 1 + 2i, \mathbf{x}^{(1)} = \begin{pmatrix} 1 \\ 1 - i \end{pmatrix}; \qquad \lambda_2 = 1 - 2i, \mathbf{x}^{(2)} = \begin{pmatrix} 1 \\ 1 + i \end{pmatrix}$

12. $\lambda_1 = -3, \mathbf{x}^{(1)} = \begin{pmatrix} 1 \\ -1 \end{pmatrix}; \qquad \lambda_2 = -1, \mathbf{x}^{(2)} = \begin{pmatrix} 1 \\ 1 \end{pmatrix}$

13. $\lambda_1 = 0, \mathbf{x}^{(1)} = \begin{pmatrix} 1 \\ i \end{pmatrix}; \qquad \lambda_2 = 2, \mathbf{x}^{(2)} = \begin{pmatrix} 1 \\ -i \end{pmatrix}$

14. $\lambda_1 = 2, \mathbf{x}^{(1)} = \begin{pmatrix} \sqrt{3} \\ 1 \end{pmatrix}; \qquad \lambda_2 = -2, \mathbf{x}^{(2)} = \begin{pmatrix} 1 \\ -\sqrt{3} \end{pmatrix}$

15. $\lambda_1 = \lambda_2 = -3, \mathbf{x}^{(1)} = \begin{pmatrix} 1 \\ 1 \end{pmatrix}$

16. $\lambda_1 = 1, \mathbf{x}^{(1)} = \begin{pmatrix} 2 \\ -3 \\ 2 \end{pmatrix}; \qquad \lambda_2 = 1 + 2i, \mathbf{x}^{(2)} = \begin{pmatrix} 0 \\ 1 \\ -i \end{pmatrix};$

   $\lambda_3 = 1 - 2i, \mathbf{x}^{(3)} = \begin{pmatrix} 0 \\ 1 \\ i \end{pmatrix}$

17. $\lambda_1 = 1, \mathbf{x}^{(1)} = \begin{pmatrix} 1 \\ 0 \\ -1 \end{pmatrix}; \qquad \lambda_2 = 2, \mathbf{x}^{(2)} = \begin{pmatrix} -2 \\ 1 \\ 0 \end{pmatrix}; \qquad \lambda_3 = 3, \mathbf{x}^{(3)} = \begin{pmatrix} 0 \\ 1 \\ -1 \end{pmatrix}$

18. $\lambda_1 = \lambda_2 = \lambda_3 = 2, \mathbf{x}^{(1)} = \begin{pmatrix} 0 \\ 1 \\ -1 \end{pmatrix}$

19. $\lambda_1 = -1$, $\mathbf{x}^{(1)} = \begin{pmatrix} 1 \\ 0 \\ -1 \end{pmatrix}$; $\lambda_2 = -1$, $\mathbf{x}^{(2)} = \begin{pmatrix} 0 \\ 2 \\ -1 \end{pmatrix}$; $\lambda_3 = 8$, $\mathbf{x}^{(3)} = \begin{pmatrix} 2 \\ 1 \\ 2 \end{pmatrix}$

20. (a) $\mathbf{T} = \begin{pmatrix} 1 & 1 \\ 3 & 1 \end{pmatrix}$ (b) $\mathbf{T} = \begin{pmatrix} 1 & 1 \\ 1-i & 1+i \end{pmatrix}$ (c) $\mathbf{T} = \begin{pmatrix} 1 & 1 \\ i & -i \end{pmatrix}$

Section 7.5, Page 303

2. (c) $W(t) = c \exp \int [p_{11}(t) + p_{22}(t)]\, dt$

6. (a) $W(t) = t^2$
   (b) $\mathbf{x}^{(1)}$ and $\mathbf{x}^{(2)}$ are linearly independent at each point except $t = 0$; they are linearly independent on every interval.
   (c) At least one coefficient must be discontinuous at $t = 0$.

   (d) $\mathbf{x}' = \begin{pmatrix} 0 & 1 \\ -2t^{-2} & 2t^{-1} \end{pmatrix} \mathbf{x}$

7. (a) $W(t) = t(t-2)e^t$
   (b) $\mathbf{x}^{(1)}$ and $\mathbf{x}^{(2)}$ are linearly independent at each point except $t = 0$ and $t = 2$; they are linearly independent on every interval.
   (c) There must be at least one discontinuous coefficient at $t = 0$ and $t = 2$.

   (d) $\mathbf{x}' = \begin{pmatrix} 0 & 1 \\ \dfrac{2-2t}{t^2-2t} & \dfrac{t^2-2}{t^2-2t} \end{pmatrix} \mathbf{x}$

Section 7.6, Page 309

1. $\mathbf{x} = c_1 \begin{pmatrix} 1 \\ 2 \end{pmatrix} e^{-t} + c_2 \begin{pmatrix} 2 \\ 1 \end{pmatrix} e^{2t}$

2. $\mathbf{x} = c_1 \begin{pmatrix} 3 \\ 4 \end{pmatrix} + c_2 \begin{pmatrix} 1 \\ 2 \end{pmatrix} e^{-2t}$

3. $\mathbf{x} = c_1 \begin{pmatrix} 1 \\ 1 \end{pmatrix} e^{t} + c_2 \begin{pmatrix} 1 \\ 3 \end{pmatrix} e^{-t}$

4. $\mathbf{x} = c_1 \begin{pmatrix} 1 \\ -4 \end{pmatrix} e^{-3t} + c_2 \begin{pmatrix} 1 \\ 1 \end{pmatrix} e^{2t}$

5. $\mathbf{x} = c_1 \begin{pmatrix} 1 \\ -1 \end{pmatrix} e^{-3t} + c_2 \begin{pmatrix} 1 \\ 1 \end{pmatrix} e^{-t}$

6. $\mathbf{x} = c_1 \begin{pmatrix} 1 \\ i \end{pmatrix} + c_2 \begin{pmatrix} 1 \\ -i \end{pmatrix} e^{2t}$

7. $\mathbf{x} = c_1 \begin{pmatrix} 1 \\ 1 \\ 1 \end{pmatrix} e^{4t} + c_2 \begin{pmatrix} 1 \\ -2 \\ 1 \end{pmatrix} e^{t} + c_3 \begin{pmatrix} 1 \\ 0 \\ -1 \end{pmatrix} e^{-t}$

8. $\mathbf{x} = c_1 \begin{pmatrix} 1 \\ -4 \\ 1 \end{pmatrix} e^{-t} + c_2 \begin{pmatrix} 1 \\ 0 \\ -1 \end{pmatrix} e^{-t} + c_2 \begin{pmatrix} 2 \\ 1 \\ 2 \end{pmatrix} e^{8t}$

9. $\mathbf{x} = c_1 \begin{pmatrix} 4 \\ -5 \\ -7 \end{pmatrix} e^{-2t} + c_2 \begin{pmatrix} 3 \\ -4 \\ -2 \end{pmatrix} e^{-t} + c_3 \begin{pmatrix} 0 \\ 1 \\ -1 \end{pmatrix} e^{2t}$

10. $\mathbf{x} = c_1 \begin{pmatrix} 1 \\ -4 \\ -1 \end{pmatrix} e^{t} + c_2 \begin{pmatrix} 1 \\ -1 \\ -1 \end{pmatrix} e^{-2t} + c_3 \begin{pmatrix} 1 \\ 2 \\ 1 \end{pmatrix} e^{3t}$

11. $\mathbf{x} = -\dfrac{3}{2}\begin{pmatrix} 1 \\ 3 \end{pmatrix} e^{2t} + \dfrac{7}{2}\begin{pmatrix} 1 \\ 1 \end{pmatrix} e^{4t}$ 12. $\mathbf{x} = \dfrac{1}{2}\begin{pmatrix} 1 \\ 1 \end{pmatrix} e^{-t} + \dfrac{1}{2}\begin{pmatrix} 1 \\ 5 \end{pmatrix} e^{3t}$

13. $\mathbf{x} = \begin{pmatrix} 0 \\ -2 \\ 1 \end{pmatrix} e^{t} + 2\begin{pmatrix} 1 \\ 1 \\ 0 \end{pmatrix} e^{2t}$

14. $\mathbf{x} = \dfrac{62}{17}\begin{pmatrix} 1 \\ 2 \\ -1 \end{pmatrix} e^{t} + \dfrac{27}{17}\begin{pmatrix} 1 \\ -2 \\ 1 \end{pmatrix} e^{-t} + \dfrac{15}{17}\begin{pmatrix} 2 \\ 1 \\ -8 \end{pmatrix} e^{4t}$

16. $\mathbf{x} = c_1 \begin{pmatrix} 1 \\ 1 \end{pmatrix} t + c_2 \begin{pmatrix} 1 \\ 3 \end{pmatrix} t^{-1}$ 17. $\mathbf{x} = c_1 \begin{pmatrix} 1 \\ 3 \end{pmatrix} t^2 + c_2 \begin{pmatrix} 1 \\ 1 \end{pmatrix} t^4$

18. $\mathbf{x} = c_1 \begin{pmatrix} 3 \\ 4 \end{pmatrix} + c_2 \begin{pmatrix} 1 \\ 2 \end{pmatrix} t^{-2}$ 19. $\mathbf{x} = c_1 \begin{pmatrix} 1 \\ 2 \end{pmatrix} t^{-1} + c_2 \begin{pmatrix} 2 \\ 1 \end{pmatrix} t^2$

23. $\mathbf{x} = c_1 \begin{pmatrix} 1 \\ 1 \end{pmatrix} e^{-3t} + c_2 \left[\begin{pmatrix} 1 \\ 1 \end{pmatrix} te^{-3t} - \begin{pmatrix} 0 \\ \frac{1}{4} \end{pmatrix} e^{-3t} \right]$

24. $\mathbf{x} = c_1 \begin{pmatrix} 2 \\ 1 \end{pmatrix} e^{t} + c_2 \left[\begin{pmatrix} 2 \\ 1 \end{pmatrix} te^{t} + \begin{pmatrix} 1 \\ 0 \end{pmatrix} e^{t} \right]$

25. (a) $\begin{pmatrix} I \\ V \end{pmatrix} = c_1 \begin{pmatrix} 1 \\ 3 \end{pmatrix} e^{-2t} + c_2 \begin{pmatrix} 1 \\ 1 \end{pmatrix} e^{-t}$ 26. (a) $\left(\dfrac{1}{CR_2} - \dfrac{R_1}{L} \right)^2 - \dfrac{4}{CL} > 0$

Section 7.7, Page 316

1. $\mathbf{x} = c_1 e^{t}\begin{pmatrix} \cos 2t \\ \cos 2t + \sin 2t \end{pmatrix} + c_2 e^{t}\begin{pmatrix} \sin 2t \\ -\cos 2t + \sin 2t \end{pmatrix}$

2. $\mathbf{x} = c_1 e^{-t}\begin{pmatrix} 2\cos 2t \\ \sin 2t \end{pmatrix} + c_2 e^{-t}\begin{pmatrix} -2\sin 2t \\ \cos 2t \end{pmatrix}$

3. $\mathbf{x} = c_1 \begin{pmatrix} 5\cos t \\ 2\cos t + \sin t \end{pmatrix} + c_2 \begin{pmatrix} 5\sin t \\ -\cos t + 2\sin t \end{pmatrix}$

4. $\mathbf{x} = c_1 e^{t/2} \begin{pmatrix} 5 \cos \frac{3}{2} t \\ 3(\cos \frac{3}{2} t + \sin \frac{3}{2} t) \end{pmatrix} + c_2 e^{t/2} \begin{pmatrix} 5 \sin \frac{3}{2} t \\ 3(-\cos \frac{3}{2} t + \sin \frac{3}{2} t) \end{pmatrix}$

5. $\mathbf{x} = c_1 \begin{pmatrix} 2 \\ -3 \\ 2 \end{pmatrix} e^t + c_2 e^t \begin{pmatrix} 0 \\ \cos 2t \\ \sin 2t \end{pmatrix} + c_3 e^t \begin{pmatrix} 0 \\ \sin 2t \\ -\cos 2t \end{pmatrix}$

6. $\mathbf{x} = c_1 \begin{pmatrix} 2 \\ -2 \\ 1 \end{pmatrix} e^{-2t} + c_2 e^{-t} \begin{pmatrix} -\sqrt{2} \sin \sqrt{2}\, t \\ \cos \sqrt{2}\, t \\ -\cos \sqrt{2}\, t - \sqrt{2} \sin \sqrt{2}\, t \end{pmatrix}$
$\qquad + c_3 e^{-t} \begin{pmatrix} \sqrt{2} \cos \sqrt{2}\, t \\ \sin \sqrt{2}\, t \\ \sqrt{2} \cos \sqrt{2}\, t - \sin \sqrt{2}\, t \end{pmatrix}$

7. $\mathbf{x} = e^{-t} \begin{pmatrix} \cos t - 3 \sin t \\ \cos t - \sin t \end{pmatrix}$
8. $\mathbf{x} = e^{-2t} \begin{pmatrix} \cos t - 5 \sin t \\ -2 \cos t - 3 \sin t \end{pmatrix}$

9. $\mathbf{x} = c_1 t^{-1} \begin{pmatrix} \cos (\sqrt{2} \ln t) \\ \sqrt{2} \sin (\sqrt{2} \ln t) \end{pmatrix} + c_2 t^{-1} \begin{pmatrix} \sin (\sqrt{2} \ln t) \\ -\sqrt{2} \cos (\sqrt{2} \ln t) \end{pmatrix}$

10. $\mathbf{x} = c_1 \begin{pmatrix} 5 \cos (\ln t) \\ 2 \cos (\ln t) + \sin (\ln t) \end{pmatrix} + c_2 \begin{pmatrix} 5 \sin (\ln t) \\ -\cos (\ln t) + 2 \sin (\ln t) \end{pmatrix}$

11. (b) $\begin{pmatrix} I \\ V \end{pmatrix} = c_1 e^{-t/2} \begin{pmatrix} \cos t/2 \\ 4 \sin t/2 \end{pmatrix} + c_2 e^{-t/2} \begin{pmatrix} \sin t/2 \\ -4 \cos t/2 \end{pmatrix}$

(c) $c_1 = 2, c_2 = -\frac{3}{4}$ in answer to part (b).

(d) $\lim_{t \to \infty} I(t) = \lim_{t \to \infty} V(t) = 0;$ No

12. (b) $\begin{pmatrix} I \\ V \end{pmatrix} = c_1 e^{-t} \begin{pmatrix} \cos t \\ -\cos t - \sin t \end{pmatrix} + c_2 e^{-t} \begin{pmatrix} \sin t \\ -\sin t + \cos t \end{pmatrix}$

(c) Use $c_1 = 2$ and $c_2 = 3$ in answer to part (b).

(d) $\lim_{t \to \infty} I(t) = \lim_{t \to \infty} V(t) = 0;$ No

Section 7.8, Page 321

1. $\mathbf{x} = c_1 \begin{pmatrix} 2 \\ 1 \end{pmatrix} e^t + c_2 \left[\begin{pmatrix} 2 \\ 1 \end{pmatrix} t e^t + \begin{pmatrix} 1 \\ 0 \end{pmatrix} e^t \right]$
2. $\mathbf{x} = c_1 \begin{pmatrix} 1 \\ 2 \end{pmatrix} + c_2 \left[\begin{pmatrix} 1 \\ 2 \end{pmatrix} t - \begin{pmatrix} 0 \\ \frac{1}{2} \end{pmatrix} \right]$

3. $\mathbf{x} = c_1 \begin{pmatrix} -3 \\ 4 \\ 2 \end{pmatrix} e^{-t} + c_2 \begin{pmatrix} 0 \\ 1 \\ -1 \end{pmatrix} e^{2t} + c_3 \left[\begin{pmatrix} 0 \\ 1 \\ -1 \end{pmatrix} t e^{2t} + \begin{pmatrix} 1 \\ 0 \\ 1 \end{pmatrix} e^{2t} \right]$

4. $\mathbf{x} = c_1 \begin{pmatrix} 1 \\ 1 \\ 1 \end{pmatrix} e^{2t} + c_2 \begin{pmatrix} 1 \\ 0 \\ -1 \end{pmatrix} e^{-t} + c_3 \begin{pmatrix} 0 \\ 1 \\ -1 \end{pmatrix} e^{-t}$

5. $\mathbf{x} = \begin{pmatrix} 3 + 4t \\ 2 + 4t \end{pmatrix} e^{-3t}$ 6. $\mathbf{x} = \begin{pmatrix} -1 \\ 2 \\ -33 \end{pmatrix} e^{t} + 4 \begin{pmatrix} 0 \\ 1 \\ -6 \end{pmatrix} t e^{t} + 3 \begin{pmatrix} 0 \\ 0 \\ 1 \end{pmatrix} e^{2t}$

7. $\mathbf{x} = c_1 \begin{pmatrix} 0 \\ 1 \\ -1 \end{pmatrix} e^{2t} + c_2 \left[\begin{pmatrix} 0 \\ 1 \\ -1 \end{pmatrix} t e^{2t} + \begin{pmatrix} 1 \\ 0 \\ 1 \end{pmatrix} e^{2t} \right]$

$+ c_3 \left[\begin{pmatrix} 0 \\ 1 \\ -1 \end{pmatrix} t^2 e^{2t} + 2 \begin{pmatrix} 1 \\ 0 \\ 1 \end{pmatrix} t e^{2t} + 2 \begin{pmatrix} 1 \\ 0 \\ 2 \end{pmatrix} e^{2t} \right]$

8. $\mathbf{x} = c_1 \begin{pmatrix} 2 \\ 1 \end{pmatrix} t + c_2 \left[\begin{pmatrix} 2 \\ 1 \end{pmatrix} t \ln t + \begin{pmatrix} 1 \\ 0 \end{pmatrix} t \right]$

9. $\mathbf{x} = c_1 \begin{pmatrix} 1 \\ 1 \end{pmatrix} t^{-3} + c_2 \left[\begin{pmatrix} 1 \\ 1 \end{pmatrix} t^{-3} \ln t - \begin{pmatrix} 0 \\ \frac{1}{4} \end{pmatrix} t^{-3} \right]$

11. (a) $\mathbf{x}^{(1)}(t) = \begin{pmatrix} 1 \\ 0 \\ 2 \end{pmatrix} e^{t}$, $\mathbf{x}^{(2)}(t) = \begin{pmatrix} 0 \\ 2 \\ -3 \end{pmatrix} e^{t}$ (e) $\mathbf{x}^{(3)}(t) = \begin{pmatrix} 2 \\ 4 \\ -2 \end{pmatrix} t e^{t} + \begin{pmatrix} 0 \\ 0 \\ -1 \end{pmatrix} e^{t}$

13. (b) $\begin{pmatrix} I \\ V \end{pmatrix} = -\begin{pmatrix} 1 \\ -2 \end{pmatrix} e^{-t/2} + \left[\begin{pmatrix} 1 \\ -2 \end{pmatrix} t e^{-t/2} + \begin{pmatrix} 2 \\ 0 \end{pmatrix} e^{-t/2} \right]$

Section 7.9, Page 328

1. $\Phi(t) = \begin{pmatrix} -\frac{1}{3}e^{-t} + \frac{4}{3}e^{2t} & \frac{2}{3}e^{-t} - \frac{2}{3}e^{2t} \\ -\frac{2}{3}e^{-t} + \frac{2}{3}e^{2t} & \frac{4}{3}e^{-t} - \frac{1}{3}e^{2t} \end{pmatrix}$ 2. $\Phi(t) = \begin{pmatrix} 3 - 2e^{-2t} & -\frac{3}{2} + \frac{3}{2}e^{-2t} \\ 4 - 4e^{-2t} & -2 + 3e^{-2t} \end{pmatrix}$

3. $\Phi(t) = \begin{pmatrix} \frac{3}{2}e^{t} - \frac{1}{2}e^{-t} & -\frac{1}{2}e^{t} + \frac{1}{2}e^{-t} \\ \frac{3}{2}e^{t} - \frac{3}{2}e^{-t} & -\frac{1}{2}e^{t} + \frac{3}{2}e^{-t} \end{pmatrix}$

4. $\Phi(t) = \begin{pmatrix} \frac{1}{5}e^{-3t} + \frac{4}{5}e^{2t} & -\frac{1}{5}e^{-3t} + \frac{1}{5}e^{2t} \\ -\frac{4}{5}e^{-3t} + \frac{4}{5}e^{2t} & \frac{4}{5}e^{-3t} + \frac{1}{5}e^{2t} \end{pmatrix}$

5. $\Phi(t) = \begin{pmatrix} \cos t + 2 \sin t & -5 \sin t \\ \sin t & \cos t - 2 \sin t \end{pmatrix}$

6. $\Phi(t) = \begin{pmatrix} e^{-t} \cos 2t & -2e^{-t} \sin 2t \\ \frac{1}{2}e^{-t} \sin 2t & e^{-t} \cos 2t \end{pmatrix}$

7. $\Phi(t) = \begin{pmatrix} -\frac{1}{2}e^{2t} + \frac{3}{2}e^{4t} & \frac{1}{2}e^{2t} - \frac{1}{2}e^{4t} \\ -\frac{3}{2}e^{2t} + \frac{3}{2}e^{4t} & \frac{3}{2}e^{2t} - \frac{1}{2}e^{4t} \end{pmatrix}$ 8. $\Phi(t) = \begin{pmatrix} e^t + 2te^t & -4te^t \\ te^t & e^t - 2te^t \end{pmatrix}$

9. $\Phi(t) = \begin{pmatrix} -2e^{-2t} + 3e^{-t} & -e^{-2t} + e^{-t} & -e^{-2t} + e^{-t} \\ \frac{5}{2}e^{-2t} - 4e^{-t} + \frac{3}{2}e^{2t} & \frac{5}{4}e^{-2t} - \frac{4}{3}e^{-t} + \frac{13}{12}e^{2t} & \frac{5}{4}e^{-2} - \frac{4}{3}e^{-t} + \frac{1}{12}e^{2t} \\ \frac{7}{2}e^{-2t} - 2e^{-t} - \frac{3}{2}e^{2t} & \frac{7}{4}e^{-2t} - \frac{2}{3}e^{-t} - \frac{13}{12}e^{2t} & \frac{7}{4}e^{-2t} - \frac{2}{3}e^{-t} - \frac{1}{12}e^{2t} \end{pmatrix}$

10. $\Phi(t) = \begin{pmatrix} \frac{1}{6}e^t + \frac{1}{3}e^{-2t} + \frac{1}{2}e^{3t} & -\frac{1}{3}e^t + \frac{1}{3}e^{-2t} & \frac{1}{2}e^t - e^{-2t} + \frac{1}{2}e^{3t} \\ -\frac{2}{3}e^t - \frac{1}{3}e^{-2t} + e^{3t} & \frac{4}{3}e^t - \frac{1}{3}e^{-2t} & -2e^t + e^{-2t} + e^{3t} \\ -\frac{1}{6}e^t - \frac{1}{3}e^{-2t} + \frac{1}{2}e^{3t} & \frac{1}{3}e^t - \frac{1}{3}e^{-2t} & -\frac{1}{2}e^t + e^{-2t} + \frac{1}{2}e^{3t} \end{pmatrix}$

Section 7.10, Page 333

1. $\mathbf{x} = c_1 \begin{pmatrix} 1 \\ 1 \end{pmatrix} e^t + c_2 \begin{pmatrix} 1 \\ 3 \end{pmatrix} e^{-t} + \frac{3}{2} \begin{pmatrix} 1 \\ 1 \end{pmatrix} te^t - \frac{1}{4} \begin{pmatrix} 1 \\ 3 \end{pmatrix} e^t + \begin{pmatrix} 1 \\ 2 \end{pmatrix} t - \begin{pmatrix} 0 \\ 1 \end{pmatrix}$

2. $\mathbf{x} = c_1 \begin{pmatrix} \sqrt{3} \\ 1 \end{pmatrix} e^{2t} + c_2 \begin{pmatrix} 1 \\ -\sqrt{3} \end{pmatrix} e^{-2t} - \begin{pmatrix} 2/3 \\ 1/\sqrt{3} \end{pmatrix} e^t + \begin{pmatrix} -1 \\ 2/\sqrt{3} \end{pmatrix} e^{-t}$

3. $\mathbf{x} = \frac{1}{5}(2t - \frac{3}{2} \sin 2t - \frac{1}{2} \cos 2t + c_1) \begin{pmatrix} 5 \cos t \\ 2 \cos t + \sin t \end{pmatrix}$

$\quad + \frac{1}{5}(-t - \frac{1}{2} \sin 2t + \frac{3}{2} \cos 2t + c_2) \begin{pmatrix} 5 \sin t \\ -\cos t + 2 \sin t \end{pmatrix}$

4. $\mathbf{x} = c_1 \begin{pmatrix} 1 \\ -4 \end{pmatrix} e^{-3t} + c_2 \begin{pmatrix} 1 \\ 1 \end{pmatrix} e^{2t} - \begin{pmatrix} 0 \\ 1 \end{pmatrix} e^{-2t} + \frac{1}{2} \begin{pmatrix} 1 \\ 0 \end{pmatrix} e^t$

5. $\mathbf{x} = c_1 \begin{pmatrix} 1 \\ 2 \end{pmatrix} + c_2 \left[\begin{pmatrix} 1 \\ 2 \end{pmatrix} t - \frac{1}{2} \begin{pmatrix} 0 \\ 1 \end{pmatrix} \right] - 2 \begin{pmatrix} 1 \\ 2 \end{pmatrix} \ln t + \begin{pmatrix} 2 \\ 5 \end{pmatrix} t^{-1} - \begin{pmatrix} \frac{1}{2} \\ 0 \end{pmatrix} t^{-2}$

6. $\mathbf{x} = (\frac{1}{2} \sin^2 t + c_1) \begin{pmatrix} 5 \cos t \\ 2 \cos t + \sin t \end{pmatrix}$

$\quad + (-\frac{1}{2}t - \frac{1}{2} \sin t \cos t + c_2) \begin{pmatrix} 5 \sin t \\ -\cos t + 2 \sin t \end{pmatrix}$

7. $\mathbf{x} = [\frac{1}{5} \ln (\sin t) - \ln (-\cos t) - \frac{2}{5}t + c_1] \begin{pmatrix} 5 \cos t \\ 2 \cos t + \sin t \end{pmatrix}$

$\quad + [\frac{2}{5} \ln (\sin t) - \frac{4}{5}t + c_2] \begin{pmatrix} 5 \sin t \\ -\cos t + 2 \sin t \end{pmatrix}$

8. (a) $\Psi(t) = \begin{pmatrix} e^{-t/2}\cos\frac{1}{2}t & e^{-t/2}\sin\frac{1}{2}t \\ 4e^{-t/2}\sin\frac{1}{2}t & -4e^{-t/2}\cos\frac{1}{2}t \end{pmatrix}$ (b) $\mathbf{x} = e^{-t/2}\begin{pmatrix} \sin\frac{1}{2}t \\ 4 - 4\cos\frac{1}{2}t \end{pmatrix}$

9. (a) $\Psi(t) = \begin{pmatrix} e^{-2t} & 2e^{-t} \\ -e^{-2t} & -e^{-t} \end{pmatrix}$

(b) $\mathbf{x} = \Psi(t)\mathbf{u}(t)$, where

$$\mathbf{u}(t) = \begin{pmatrix} -\frac{7}{2}e^{2t} + \frac{7}{2} \\ 4e^{t} - 4 \end{pmatrix}, \quad 0 \le t \le 1; \quad \mathbf{u}(t) = \begin{pmatrix} -\frac{7}{2}e^{2} + \frac{7}{2} \\ 4e - 4 \end{pmatrix}, \quad t \ge 1$$

10. $\mathbf{x} = c_1\begin{pmatrix}1\\1\end{pmatrix}t + c_2\begin{pmatrix}1\\3\end{pmatrix}t^{-1} - \begin{pmatrix}2\\3\end{pmatrix} + \frac{1}{2}\begin{pmatrix}1\\3\end{pmatrix}t - \begin{pmatrix}1\\1\end{pmatrix}t\ln t - \frac{1}{3}\begin{pmatrix}4\\3\end{pmatrix}t^2$

11. $\mathbf{x} = c_1\begin{pmatrix}2\\1\end{pmatrix}t^2 + c_2\begin{pmatrix}1\\2\end{pmatrix}t^{-1} + \begin{pmatrix}3\\2\end{pmatrix}t + \frac{1}{10}\begin{pmatrix}-2\\1\end{pmatrix}t^4 - \frac{1}{2}\begin{pmatrix}2\\3\end{pmatrix}$

14. (a) $\mathbf{x} = c_1\begin{pmatrix}1\\2\end{pmatrix}e^{3t} + c_2\begin{pmatrix}1\\2\end{pmatrix}e^{-t} + \frac{1}{4}\begin{pmatrix}1\\-8\end{pmatrix}e^{t}$

(b) $\mathbf{x} = c_1\begin{pmatrix}1\\1\end{pmatrix}e^{t} + c_2\begin{pmatrix}1\\3\end{pmatrix}e^{-t} + \begin{pmatrix}1\\0\end{pmatrix}e^{t} + 2\begin{pmatrix}1\\1\end{pmatrix}te^{t}$

CHAPTER EIGHT

Section 8.2, Page 342

1. (a) $y = \phi(x) = (1 + e^{2x})/2$;   1.111, 1.246, 1.411, 1.613
   (b) 1.1, 1.22, 1.364, 1.537   (c) 1.05, 1.105, 1.166, 1.233
2. (a) $y = \phi(x) = e^{2x} + \frac{1}{2}x$;   1.271, 1.592, 1.972, 2.426
   (b) 1.25, 1.54, 1.878, 2.274   (c) 1.125, 1.260, 1.406, 1.564
3. (a) 1.1, 1.222, 1.375, 1.573   (b) 1.05, 1.105, 1.167, 1.236
4. $y_6 = 8.903824$, $y_7 = 12.505354$ 5. 1.595, 2.464
8. (a) $\phi_3(x) = 1 + x + x^2 + \frac{2}{3}x^3$; $\phi_2(0.4) = 1.56$, $\phi_3(0.4) = 1.603$
   (b) $\phi_3(x) = 1 + \frac{5}{2}x + 2x^2 + \frac{4}{3}x^3 - \frac{1}{6}x^4$; $\phi_2(0.4) = 2.299$, $\phi_3(0.4) = 2.401$
   (c) $\phi_2(x) = 1 + x + x^2 + \frac{2}{3}x^3 + \frac{1}{6}x^4 + \frac{2}{15}x^5 + \frac{1}{63}x^7$; $\phi_2(0.4) = 1.608$

Section 8.3, Page 349

1. $e_{n+1} = [2\phi(\bar{x}_n) - 1]h^2$, $|e_{n+1}| \le \left[1 + 2\max_{0\le x\le 1}|\phi(x)|\right]h^2$,

$e_{n+1} = e^{2\bar{x}_n}h^2$, $|e_1| \le 0.012$, $|e_4| \le 0.022$

2. $e_{n+1} = [2\phi(\bar{x}_n) - \bar{x}_n]h^2$, $|e_{n+1}| \le \left[1 + 2\max_{0\le x\le 1}|\phi(x)|\right]h^2$,

$e_{n+1} = 2e^{2\bar{x}_n}h^2$, $|e_1| \le 0.024$, $|e_4| \le 0.044$

3. $e_{n+1} = [\bar{x}_n + \bar{x}_n{}^2 \phi(\bar{x}_n) + \phi^3(\bar{x}_n)]h^2$

4. (a) $\phi(x) = 1 + (1/5\pi) \sin 5\pi x$ (b) 1.2, 1.0, 1.2
   (c) 1.1, 1.1, 1.0, 1.0   (d) $h < 1\sqrt{50\pi} \cong 0.08$

7. (a) $y_1(h = 0.2) = 1.4$, $y_2(h = 0.1) = 1.44$; $\phi(0.2) \cong 1.48$, exact 1.4918
   (b) $y_1(h = 0.2) = 1.2$, $y_2(h = 0.1) = 1.22$; $\phi(0.2) \cong 1.24$, exact 1.246
   (c) $y_1(h = 0.2) = 1.5$, $y_2(h = 0.1) = 1.54$; $\phi(0.2) \cong 1.58$, exact 1.592
   (d) $y_1(h = 0.2) = 1.2$, $y_2(h = 0.1) = 1.222$; $\phi(0.2) \cong 1.244$

9. (a) 56.75310   (b) 62.027322

10. (a) 1.05, 1.11, 1.17, 1.24   (b) 1.13, 1.27, 1.42, 1.58
    (c) 1.05, 1.11, 1.17, 1.24

11. (a) 0   (b) 60   (c) −92.16

12. $0.224 \neq 0.225$ 13. (b) $\phi_2(x) - \phi_1(x) = 0.001e^x \to \infty$ as $x \to \infty$

Section 8.4, Page 355

1. 1.110, 1.244, 1.408   2. 1.270, 1.589, 1.967
3. 1.111, 1.252, 1.437
5. $e_{n+1} = 38e^{4\bar{x}_n}(h^3/3)$, $|e_{n+1}| \leq 691.6h^3$ on $0 \leq x \leq 1$, $|e_1| \leq 0.0189$
6. (a) $e_{n+1} = 2e^{2\bar{x}_n}(h^3/3)$, $|e_{n+1}| \leq 4.926h^3$ on $0 \leq x \leq 1$, $|e_1| \leq 0.00081$
   (b) $e_{n+1} = 4e^{2\bar{x}_n}(h^3/3)$, $|e_{n+1}| \leq 9.852h^3$ on $0 \leq x \leq 1$, $|e_1| \leq 0.0016$
8. (a) 1.110, 1.244, 1.408   (b) 1.270, 1.588, 1.965
   (c) 1.110, 1.250, 1.433

Section 8.5, Page 358

1. 1.110, 1.244, 1.408   2. 1.270, 1.588, 1.966
3. 1.110, 1.249, 1.431   4. $y_2 = 2.464$, $y_3 = 3.738$
7. $\phi(1) \cong 64.586856$, exact 64.897803

Section 8.6, Page 361

1. 1.249, 1.598   2. 1.590, 2.423
3. 1.251, 1.698   7. $\phi(1) \cong 64.885876$, exact 64.897803

Section 8.8, Page 373

1. $y_{4p} = 1.61267$, $y_{4c} = 1.61276$ 2. $y_{4p} = 2.42536$, $y_{4c} = 2.42553$
3. $y_{4p} = 1.69396$, $y_{4c} = 1.69613$

4. Hint: $\phi'(x) \cong y_n' + \dfrac{y_n' - y_{n-1}'}{h}(x - x_n)$.

6. (a) $y_{4p} = 1.61269$, $y_{4c} = 1.61276$
   (b) $y_{4p} = 2.42539$, $y_{4c} = 2.42554$
   (c) $y_{4p} = 1.69412$, $y_{4c} = 1.69604$

Section 8.9, Page 376

1. (a) $x_1 = 1.1$, $x_2 = 1.26$; $y_1 = 0.4$, $y_2 = 0.76$
   (b) $x_1 = 1.2$, $x_2 = 1.451$; $y_1 = 1.1$, $y_2 = 1.32$
   (c) $x_1 = 0.8$, $x_2 = 0.582$; $y_1 = 1.1$, $y_2 = 1.18$

2. 1.30, -0.24

3. (a) $x_1 = 1.30$, $y_1 = 0.76$ (b) $x_1 = 1.50$, $y_1 = 1.26$
   (c) $x_1 = 0.56$, $y_1 = 1.16$

4. $x_1 = 1.32007$, $y_1 = -0.250667$

5. $x_{4p} = 1.99423$, $y_{4p} = -0.662181$; $x_{4c} = 1.99521$, $y_{4c} = -0.662442$

6. $x' = y$, $y' = -t^2 y - 3x + t$; $x(0) = 1$, $y(0) = 2$
   $x_1 = 1.2$, $x_2 = 1.37$; $y_1 = 1.7$, $y_2 = 1.348$

CHAPTER NINE

Section 9.1, Page 384

1. $A_0 < 0$, $A \to -\infty$; $A_0 = 0$, $A = 0$; $0 < A_0 < \epsilon/\sigma$, $A \to \epsilon/\sigma$;
   $A_0 = \epsilon/\sigma$, $A = \epsilon/\sigma$; $A_0 > \epsilon/\sigma$, $A \to \epsilon/\sigma$

2. $A_0 < 0$, $A \to 0$; $A_0 = 0$, $A = 0$; $0 < A_0 < \epsilon/\sigma$, $A \to 0$;
   $A_0 = \epsilon/\sigma$, $A = \epsilon/\sigma$; $A_0 > \epsilon/\sigma$, $A \to \infty$

3. $A_0 = 0$, $A = 0$; $A_0 > 0$, $A \to \infty$

4. $A_0 < -\epsilon/\sigma$, $A \to -\epsilon/\sigma$; $A_0 = -\epsilon/\sigma$, $A = -\epsilon/\sigma$;
   $-\epsilon/\sigma < A_0 < 0$, $A \to -\epsilon/\sigma$; $A_0 = 0$, $A = 0$; $A_0 > 0$, $A \to \infty$

5. $A_0 = 0$, $A = 0$; $0 < A_0 < 1$, $A \to 1$; $A_0 = 1$, $A = 1$;
   $1 < A_0 < 2$, $A \to 1$; $A_0 = 2$, $A = 2$; $A_0 > 2$, $A \to \infty$

6. $A_0 = 0$, $A = 0$; $0 < A_0 < 1$, $A \to 0$; $A_0 = 1$, $A = 1$;
   $1 < A_0 < 2$, $A \to 2$; $A_0 = 2$, $A = 2$; $A_0 > 2$, $A \to 2$

7. $A_0 > 1$, $A \to \infty$; $A_0 = 1$, $A = 1$; $0 < A_0 < 1$, $A \to 0$;
   $A_0 = 0$, $A = 0$; $-1 < A_0 < 0$, $A \to -1$; $A_0 = -1$, $A = -1$;
   $A_0 < -1$, $A \to -1$

8. $A_0 = 0$, $A = 0$; $0 < A_0 < K$, $A \to K$; $A_0 = K$, $A = K$;
   $A_0 > K$, $A \to K$

9. (b) $dN/dt = \alpha N(N_E - N) - \beta N$; $\beta/\alpha < N_E$;
   critical point is $(N_E - \beta/\alpha) < N_E$
   (c) $dN/dt = \alpha N(N_E - N) - \beta$; $4\beta/\alpha < N_E^2$;
   $N \to 0$ as $t \to \infty$ if $N_0 < \frac{1}{2}(N_E - B)$,
   $N \to \frac{1}{2}(N_E + B)$ as $t \to \infty$ if $N_0 > \frac{1}{2}(N_E - B)$; $B = (N_E^2 - 4\beta/\alpha)^{1/2}$

13. $A_0 = 0$, $A = 0$; $A_0 = \epsilon/\sigma$, $A = \epsilon/\sigma$;

$$0 < A_0 < \frac{\epsilon}{\sigma}, \quad A = \frac{(\epsilon/\sigma)A_0 e^{\epsilon t}}{A_0 e^{\epsilon t} + (\epsilon/\sigma - A_0)} \to 0 \text{ as } t \to \infty;$$

$$A_0 > \frac{\epsilon}{\sigma}, \quad A = \frac{(\epsilon/\sigma)A_0 e^{\epsilon t}}{A_0 e^{\epsilon t} - (A_0 - \epsilon/\sigma)} \to \infty \text{ as } t \to \frac{1}{|\epsilon|} \ln \frac{A_0}{A_0 - \epsilon/\sigma}$$

14. $d^2A/dt^2 = A(2\sigma A - \epsilon)(\sigma A - \epsilon)$

15. (a) $du/dt = A_E(\epsilon - \sigma A_E) + (\epsilon - 2\sigma A_E)u - \sigma u^2$

   (b) $du/dt = A_E(\epsilon - \sigma A_E) + (\epsilon - 2\sigma A_E)u$

16. $A_E = 0$ is linearly stable, $A_E = \epsilon/\sigma$ is linearly unstable

Section 9.2, Page 395

1. (a) $x = 4e^{-t}$, $y = 2e^{-2t}$; $y = x^2/8$

   (b) $x = 4e^{-t}$, $y = 2e^{2t}$, $y = 32x^{-2}$; $x = 4e^{-t}$, $y = 0$,

   (c) $x = 4\cos t$, $y = 4\sin t$; $x = -4\sin t$, $y = 4\cos t$

2. (a) $(0, 0)$ (b) $(0, 0)$, $(-\frac{1}{2}, 1)$

   (c) $(0, 0)$, $(1, 0)$, $(0, 2)$, $(\frac{1}{2}, \frac{1}{2})$ (d) $(0, 0)$

   (e) $(\pm n\pi, 0)$, $n = 0, 1, 2, \ldots$ (f) $(0, 0)$

4. (a) $x = Ae^{-t}$, $y = Be^{-2t}$; $y = Bx^2/A^2$, $A \neq 0$

   (b) $x = Ae^{-t}$, $y = Be^{2t}$; $y = BA^2/x^2$, $A \neq 0$

   (c) $x = Ae^{t}$, $y = Be^{-2t}$; $y = BA^2/x^2$, $A \neq 0$

   (d) $x = Ae^{-t}$, $y = Be^{2t} - Ae^{-t}$; $y = BA^2/x^2 - x$, $A \neq 0$

11. $x = (\epsilon_1\sigma_2 - \epsilon_2\alpha_1)/(\sigma_1\sigma_2 - \alpha_1\alpha_2)$, $y = (\epsilon_2\sigma_1 - \epsilon_1\alpha_2)/(\sigma_1\sigma_2 - \alpha_1\alpha_2)$

12. $(x, y) \to (\epsilon_1/\sigma_1, 0)$ as $t \to \infty$

13. Except for a special set of initial points on two curves, $(x, y) \to (\epsilon_1/\sigma_1, 0)$ or $(0, \epsilon_2/\sigma_2)$ depending on the initial point.

Section 9.3, Page 407

1. Saddle point, unstable   2. Improper node, unstable

3. Saddle point, unstable   4. Improper node, asymptotically stable

5. Spiral point, asymptotically stable   6. Center, stable

7. Spiral point, unstable   8. Improper node, asymptotically stable

9. $x_0 = 1$, $y_0 = 1$; saddle point, unstable

10. $x_0 = -1$, $y_0 = 0$; improper node, asymptotically stable

11. $x_0 = -2$, $y_0 = 1$; spiral point, asymptotically stable

12. $x_0 = \gamma/\delta$, $y_0 = \alpha/\beta$; center, stable

13. $c^2 > 4\,km$, improper node, asymptotically stable; $c^2 = 4\,km$, improper node, asymptotically stable; $c^2 < 4\,km$, spiral point, asymptotically stable

Section 9.4, Page 418

1. Spiral point, asymptotically stable   2. PN, IN, or SpP, unstable

3. Spiral point, asymptotically stable   4. Spiral point, unstable

5. Saddle point, unstable   6. Saddle point, unstable

7. $\mu < 2$, spiral point, unstable; $\mu = 2$, PN, IN, or SpP, unstable;
   $\mu > 2$, improper node, unstable

8. Spiral point, unstable   9. C or SpP, indeterminate

10. Improper node, asymptotically stable

11. (a) $(0, 0)$, PN, IN, or SpP, unstable; $(-1, 1)$, saddle point, unstable

   (b) $(1, 1)$, PN, IN, or SpP, asymptotically stable; $(-1, -1)$, saddle point, unstable

   (c) $(0, 0)$, improper node, unstable; $(0, \frac{3}{2})$, improper node, asymptotically stable; $(1, 0)$, saddle point, unstable; $(-1, 2)$, saddle point, unstable

   (d) $(1, 1)$, spiral point, asymptotically stable; $(-1, 1)$, saddle point, unstable

15. (b) Nothing.

18. Linear system is $dx/dt = y$, $dy/dt = -c(0)y - g'(0)x$; $r^2 + c(0)r + g'(0) = 0$.

19. (a) $x_E = (\epsilon_1\sigma_2 - \epsilon_2\alpha_1)/(\sigma_1\sigma_2 - \alpha_1\alpha_2)$, $y_E = (\epsilon_2\sigma_1 - \epsilon_1\alpha_2)/(\sigma_1\sigma_2 - \alpha_1\alpha_2)$

   (c) Asymptotically stable improper node

20. (a) $x_E = (\epsilon_1\sigma_2 - \epsilon_2\alpha_1)/(\sigma_1\sigma_2 - \alpha_1\alpha_2)$, $y_E = (\epsilon_2\sigma_1 - \epsilon_1\alpha_2)/(\sigma_1\sigma_2 - \alpha_1\alpha_2)$

   (c) Unstable saddle point.

Section 9.5, Page 432

1. (a) $\begin{pmatrix} u \\ v \end{pmatrix} = c_1 \begin{pmatrix} 1 \\ (\sqrt{57} - 3)/8 \end{pmatrix} e^{-(5 + \sqrt{57})t/16} + c_2 \begin{pmatrix} -1 \\ (\sqrt{57} + 3)/8 \end{pmatrix} e^{(\sqrt{57} - 5)t/16}$

3. (a) Critical points are $x = 0$, $y = 0$; $x = \epsilon_1/\sigma_1$, $y = 0$; $x = 0$, $y = \epsilon_2/\sigma_2$. Mixed state (X_E, Y_E) is not possible since one or the other of X_E and Y_E is negative. $x \to 0$, $y \to \epsilon_2/\sigma_2$ as $t \to \infty$, the redear survive.

   (b) Same as part (a) except $x \to \epsilon_1/\sigma_1$, $y \to 0$ as $t \to \infty$, the bluegill survive.

4. (a) $X_E = (B - \gamma_1 R)/(1 - \gamma_1\gamma_2)$, $Y_E = (R - \gamma_2 B)/(1 - \gamma_1\gamma_2)$.

   (b) X_E is reduced, Y_E is increased; yes, if B becomes less than $\gamma_1 R$, then $x \to 0$ and $y \to R$ as $t \to \infty$.

5. $t = 0, T, 2T, \ldots$: H is a max., dP/dt is a max.

   $t = T/4, 5T/4, \ldots$: dH/dt is a min., P is a max.

   $t = T/2, 3T/2, \ldots$: H is a min., dP/dt is a min.

   $t = 3T/4, 7T/4, \ldots$: dH/dt is a max., P is a min.

7. Trap foxes in half cycle when $dP/dt > 0$, trap rabbits in half cycle when $dH/dt > 0$, trap rabbits and foxes in quarter cycle when $dP/dt > 0$ and $dH/dt > 0$, trap neither in quarter cycle when $dP/dt < 0$, $dH/dt < 0$.

8. $dH/dt = aH - \alpha HP - \beta H$, $dP/dt = -cP + \gamma HP - \delta P$, where β and δ are constants of proportionality. New center is $H = (c + \delta)/\gamma > c/\gamma$ and $P = (a - \beta)/\alpha < a/\alpha$ so equilibrium value of prey is increased and equilibrium value of predator is decreased!

9. Let $A = a/\sigma - c/\gamma$. The critical points are $(0, 0)$, $(a/\sigma, 0)$, and $(c/\gamma, \sigma A/\alpha)$. If $A > 0$ then $(0, 0)$ is a saddle point, $(a/\sigma, 0)$ is a saddle point, and $(c/\gamma, \sigma A/\alpha)$ is an asymptotically stable node if $(c\sigma/\gamma)^2 - 4c\sigma A \geq 0$ or an asymptotically stable spiral point if $(c\sigma/\gamma)^2 - 4c\sigma A < 0$. $(H, P) \to (c/\gamma, \sigma A/\alpha)$ as $t \to \infty$. If $A < 0$ then $(0, 0)$ is a saddle point, $(a/\sigma, 0)$ is an asymptotically stable node, and $(c/\gamma, \sigma A/\alpha)$ is a saddle point. $(H, P) \to (a/\sigma, 0)$ as $t \to \infty$.

Section 9.7, Page 450

1. (a) $r = 1$, $\theta = t + t_0$, stable limit cycle

   (b) $r = 1$, $\theta = -t + t_0$, semistable limit cycle

(c) $r = 1$, $\theta = t + t_0$, stable limit cycle; $r = 3$, $\theta = t + t_0$, unstable periodic solution

(d) $r = 1$, $\theta = -t + t_0$, unstable periodic solution;
$r = 2$, $\theta = -t + t_0$, stable limit cycle

(e) $r = (2n - 1)$, $\theta = t + t_0$, $n = 1, 2, 3, \ldots$, stable limit cycle;
$r = 2n$, $\theta = t + t_0$, $n = 1, 2, 3, \ldots$, unstable periodic solution

(f) $r = 2$, $\theta = -t + t_0$, semistable limit cycle;
$r = 3$, $\theta = -t + t_0$, unstable periodic solution

3. (a) Counterclockwise

   (b) $r = 1$, $\theta = t + t_0$, stable limit cycle;
   $r = 2$, $\theta = t + t_0$, semistable limit cycle;
   $r = 3$, $\theta = t + t_0$, unstable periodic solution

4. $r = \sqrt{2}$, $\theta = -t + t_0$, unstable periodic solution

Index